Green Technologies for Wastewater Treatment and Bioenergy Production

Currently, most of the physical and chemical technologies generally employed for the treatment of wastewater are costly and inefficient. Researchers have been working to find green technologies for the sustainable and economically viable treatment methods for wastewater, and *Green Technologies for Wastewater Treatment and Bioenergy Production* focuses on these latest advancements. These include bioremediation, bioaugmentation, wetlands, and algal treatment technologies, which are considered to contribute toward a circular bioeconomy and to be crucial in attaining various sustainable development goals.

- Examines the production of biofuels and other forms of bioenergy derived from wastewater.
- Explains how these emerging green technologies contribute toward a circular bioeconomy and aid in achieving various sustainable development goals.

Green Technologies for Wastewater Treatment and Bioenergy Production

Edited by
Imran Ahmad

CRC Press
Taylor & Francis Group
Boca Raton London New York

CRC Press is an imprint of the
Taylor & Francis Group, an **informa** business

Designed cover image: Shutterstock

First edition published 2025
by CRC Press
2385 NW Executive Center Drive, Suite 320, Boca Raton FL 33431

and by CRC Press
4 Park Square, Milton Park, Abingdon, Oxon, OX14 4RN

CRC Press is an imprint of Taylor & Francis Group, LLC

ISBN: 978-1-032-60467-1 (hbk)
ISBN: 978-1-032-60468-8 (pbk)
ISBN: 978-1-003-45928-6 (ebk)

DOI: 10.1201/9781003459286

Typeset in Times
by codeMantra

To the brave people who endured oppression…

To the children whose innocence was taken too soon,

To the women whose strength shone amidst unimaginable suffering,

To the helpless and disabled who faced unthinkable challenges with dignity,

This book is dedicated to your courage, sacrifice, and unyielding spirit.

Though the world may not always see your pain, your legacy endures in every heart that beats for justice and every soul that seeks peace. You are not forgotten, and your struggle will continue, inspiring future generations to relentlessly pursue and uphold the values of freedom, dignity, and hope.

May your sacrifice be a beacon for a just and peaceful world.

With profound respect and remembrance.

Contents

Foreword

As global environmental challenges intensify, sustainable solutions are becoming increasingly urgent. Wastewater treatment is a critical area where innovative technologies are essential for environmental protection and resource recovery. Traditional systems, while effective at removing pollutants, often consume significant energy and release greenhouse gases. However, the emergence of green technologies is transforming wastewater into a valuable resource, contributing to cleaner environments and sustainable energy production.

Green Technologies for Wastewater Treatment and Bioenergy Production explores these emerging solutions, blending fundamental research with real-world applications. The book covers advanced techniques that not only treat wastewater efficiently but also convert it into bioenergy. It introduces both traditional and cutting-edge processes, such as aerobic and anaerobic systems, microbial nanomaterials, and metagenomics, emphasizing resource recovery and minimizing environmental impacts.

Nature-based solutions like constructed wetlands are highlighted, illustrating how these systems purify wastewater and support ecosystem health. The innovative use of microalgae in photobioreactors, for nutrient removal, carbon sequestration, and biofuel production is a rapidly growing field aligned with sustainable water management. The symbiotic relationships between microalgae and bacteria offer enhanced efficiency in treatment systems, while strategies like wastewater-fertigated short-rotation coppice systems link waste management with biomass and bioenergy production.

This book also covers biofuel production, focusing on algae's role in producing sustainable alternatives to fossil fuels. Bioethanol, biodiesel, glycerol, and biomethane are discussed, along with the potential of biohydrogen to drive energy independence. The economic aspects of these technologies are explored, ensuring their viability in the broader context of environmental sustainability and economic growth.

This book offers a comprehensive framework for the future, making it a vital resource for environmental engineers, scientists, policymakers, and industry leaders. It bridges research and practice, providing a path toward a future where wastewater is not only treated but transformed into a resource that benefits both the environment and society.

Abdelfatah Abomohra
Hamburg, Germany
27 September 2024

Preface

This book was initiated last year during my work on a green technology project. As I delved deeper into the possibilities of environmentally sustainable and economically viable solutions for wastewater treatment and bioenergy production, the idea to compile and share cutting-edge developments in this field began to take shape. Green technologies are not just promising alternatives but essential tools for reshaping how we approach sustainability and resource recovery in the 21st century.

This book is a testament to the power of global collaboration, with contributions from leading experts worldwide. Their unique insights and regional perspectives bring a comprehensive view to the forefront of this dynamic field. Together, we cover advanced and forward-thinking topics, such as using microalgae-bacteria consortia for wastewater treatment—a breakthrough method that cleans water while generating renewable energy. Nanomaterials, metagenomics, and circular bioeconomy frameworks push the boundaries of current knowledge, offering new avenues for transforming waste into valuable resources.

This book serves as a profound tribute to the transformative power of interdisciplinary collaboration. It bridges the gap between environmental sustainability and economic practicality, illustrating how the production of biomethane, biodiesel, and biohydrogen from wastewater can address waste management challenges and fuel the energy needs of our future. These approaches embody the principles of the circular bioeconomy, where waste is not discarded but repurposed to power communities, industries, and ecosystems.

The diverse perspectives and innovative solutions presented in this book have the power to inspire and transform. They can inspire researchers, engineers, and policymakers to reimagine the potential of green technologies. I hope that this volume serves as a valuable resource, equipping you with the knowledge and inspiration to advance sustainable practices in wastewater treatment and bioenergy production. Together, we can foster a future where technology and nature work in harmony for the benefit of all.

Imran Ahmad

Acknowledgments

First, I express my deepest gratitude to *Allah, the Almighty*, whose boundless mercy and grace have given me the strength, patience, and determination to complete this book. Without His guidance and blessings, none of this would have been possible. I am forever grateful for the strength and clarity of mind granted to me throughout this journey.

This book, Green Technologies for Wastewater Treatment and Bioenergy Production, is the culmination of dedicated research, collaboration, and an unwavering commitment to advancing environmentally sustainable solutions. It is not merely a personal milestone but a shared achievement made possible by the invaluable guidance, support, and encouragement of many individuals.

To my esteemed colleagues, mentors, friends, and loved ones, I extend my deepest gratitude for their invaluable contributions. Their expertise and insights into wastewater treatment and bioenergy production have been instrumental in shaping the chapters of this book, ensuring its technical rigor and relevance in promoting sustainability and the circular bioeconomy.

I also acknowledge the countless blessings of my beloved mother, whose prayers and love have been a constant source of strength and guidance. I want to express my deepest gratitude to Dr. Rizwan Ahmad for his unwavering support and belief in me throughout this journey. After our father's passing, he stepped in with strength and wisdom, offering me the love and guidance I needed to continue this path. His encouragement, especially in challenging times, gave me the strength to pursue and see this project through.

A special note of appreciation goes to the esteemed authors from various countries who contributed their expertise to this edited volume. Your contributions, which span a wide range of environmentally friendly technologies and processes, bring diverse perspectives and valuable knowledge to this work. Each chapter is a testament to the global collaboration required to tackle pressing environmental challenges, and you are an integral part of this global community. I deeply appreciate your commitment to sharing your research and advancing solutions that promote sustainability.

Special thanks also go to the research teams and collaborators whose innovation and dedication to sustainable practices have been an inspiration. I reserve my most profound gratitude for my beloved wife, whose unwavering support, motivation, and strength have been my greatest source of inspiration. Through every challenge and every late night, she stood by my side with love and encouragement, reminding me that no goal is out of reach. Her belief in me has been my guiding force, and I dedicate much of this work to her love and patience.

To all those contributing to a cleaner, more sustainable future, this book is a testament to our shared dedication and a source of inspiration for the journey ahead. Your efforts and commitment drive this work, and I am honored to dedicate it to you.

With sincere gratitude.

About the Editor

Dr. Imran Ahmad is a dedicated researcher, academic, and scientific consultant deeply committed to environmental sustainability and algae biotechnology. With an MPhil and PhD from Universiti Teknologi Malaysia, his expertise lies in biological wastewater treatment, microalgal bioremediation, and resource recovery. His work bridges research and application, focusing on innovative solutions that address environmental challenges while contributing to the principles of a circular bioeconomy. He has been actively involved in designing and optimizing photobioreactors, anaerobic reactors, and microalgae-based treatment systems that offer sustainable alternatives for wastewater management and carbon sequestration.

He works as a Scientific Consultant and Bioprocess Engineer with ProAlgaeTech, a startup focused on microalgal carbon sequestration and wastewater bioremediation. His contributions extend beyond academic research, engaging in industrial applications that translate scientific advancements into real-world impact. He has been integral to international projects, including the Indo-European Horizon 2020 PAVITR project and Liquid Trees, where his insights have helped shape strategies for efficient wastewater treatment and bioenergy production. He also obtained a utility innovation for a modified grease trap design for restaurant wastewater, further demonstrating his ability to develop practical engineering solutions. His research has resulted in various high-impact publications, reflecting his commitment to advancing knowledge in environmental engineering.

His work has been recognized at both national and international levels. He has received the My Membrane Award from the Malaysia Membrane Society, UTM, and has been acknowledged by Japan's Algae Industry Incubation Consortium. He was also nominated as the Secretary of the International Conference on Environmental Science and Green Technology in 2022. Alongside his research contributions, he has spent more than eight years in academia, mentoring students, designing laboratory infrastructure, and fostering interdisciplinary collaborations. His experience in teaching and research has allowed him to contribute to shaping the future of environmental engineering education.

Beyond professional achievements, he remains committed to developing sustainable technologies that create meaningful change. His work is driven by the belief that scientific research should lead to practical, lasting solutions that benefit the environment and society. Whether through academia, consultancy, or industry partnerships, he continues exploring ways to integrate innovation with sustainability, ensuring that his contributions have a lasting impact on green technology.

Contributors

R. Ben Abdallah
University of Carthage
Tunis, Tunisia

Gholamareza Abdi
Department of Biotechnology
Persian Gulf Research Institute
Persian Gulf University
Bushehr, Iran

Mehboob Ahmed
Institute of Microbiology
and Molecular Genetics
University of the Punjab
Lahore, Pakistan

Sameh S. Ali
Botany Department, Faculty
of Science
Tanta University
Tanta, Egypt

Rania Al-Tohamy
Biofuels Institute
School of the Environment
and Safety Engineering
Jiangsu University
Zhenjiang, China

Majid Al-Zahrani
Biological Sciences Department
College of Science and Art at Rabigh
King Abdulaziz University
Rabigh, Saudi Arabia

Rasit Bakir
Department of Chemistry
Selcuk University Konya
Konya, Turkey

Mukul Machhindra Barwant
Yashwantrao Chavan Institute of Science
Satara, India

Ghadir A. El-Chaghaby
BioAnalysis Lab., Regional Center
for Food and Feed
Agricultural Research Center
Giza, Egypt

Tamer Elsamahy
Biofuels Institute
School of the Environment and Safety
Engineering
Jiangsu University
Zhenjiang, China

A. Evangelina
Department of Microbiology
The American College
Madurai, India

S. Fattouch
University of Carthage
Tunis, Tunisia

Ebru Guller
Department of Chemistry
Selcuk University
Konya, Turkey

Imran Hashmi
Institute of Environmental Science
& Engineering (IESE)
School of Civil & Environmental
Engineering (SCEE)
National University of Sciences
& Technology (NUST)
Islamabad, Pakistan

L. H. S. Hassan
Department of Botany and Microbiology
Faculty of Science
Menoufia University
Shebin El-kom, Egypt

Saadia Ijaz
Department of Microbiology and Molecular Genetics
The Women University Multan
Multan, Pakistan

Nirmala Kumari Jangid
Department of Chemistry, DESM
National Institute of Education
NCERT
New Delhi, India

Faria Javed
Institute of Environmental Science & Engineering (IESE)
School of Civil & Environmental Engineering (SCEE)
National University of Sciences & Technology (NUST)
Islamabad, Pakistan

Avinash Shivaji Jondhale
MJM Arts Commerce and Science College
Karanjali Nashik, India

Hatice Kübra Kara
Department of Chemistry
Selcuk University
Konya, Turkey

Hüseyin Kara
Department of Chemistry
Selcuk University
Konya, Turkey

Vaneeta Chandrashekhar Karande
Yashwantrao Chavan Institute of Science
Satara, India

Srinivas Kasulla
Arka BRENStech Private Limited
Gurugram, India

Navjeet Kaur
Department of Chemistry & Division of Research and Development
Lovely Professional University
Phagwara, India

Mehwish Khalid
Institute of Environmental Science & Engineering (IESE)
School of Civil & Environmental Engineering (SCEE)
National University of Sciences & Technology (NUST)
Islamabad, Pakistan

Michael Kornaros
Laboratory of Biochemical Engineering & Environmental Technology (LBEET)
Department of Chemical Engineering
University of Patras
Patras, Greece

Sudesh Kumar
Department of Chemistry, DESM
National Institute of Education
NCERT
New Delhi, India

Chetna Kumari
Department of Chemistry
Banasthali Vidyapith
Banasthali, India

C. S. Liew
Universiti Teknologi Petronas
Perak, Malaysia

J. W. Lim
Universiti Teknologi Petronas
Perak, Malaysia

Yehia A.-G. Mahmoud
Botany Department, Faculty of Science
Tanta University
Tanta, Egypt

S. J. Malik
Arka BRENStech Private Limited
Gurugram, India

N. Belalkhel
University of Carthage
Tunis, Tunisia

Raziye Nur Özçiçek
Department of Chemistry
Selcuk University
Konya, Turkey

Prahalad Prasad Paroha
Associate Professor Department of Basic Science and Humanities
Pranveer Singh Institute of Technology
Kanpur, India

Varsha Sadashiv Patil
Art's Science and Commerce College
Rahata, India

Ilangovan Pugazhenthi
Department of Chemistry
Annamacharya Institute of Technology and Sciences (Autonomous)
Kadapa, India

R. Raksasat
Universiti Teknologi Petronas
Perak, Malaysia

S. Ramanathan
Universiti Teknologi Malaysia
Johor Bahru, Malaysia

Sayed Rashad
BioAnalysis lab.
Regional Center for Food and Feed
Agricultural Research Center
Giza, Egypt

H. Rawindran
Universiti Teknologi Malaysia
Johor Bahru, Malaysia

O. Rebai
University of Carthage
Tunis, Tunisia

Michael Schagerl
Department of Functional and Evolutionary Ecology
University of Vienna
Vienna, Austria

Shalagha Sharma
Department of Biological Sciences
Shobhit deemed to be University
Meerut, India

Sanam Iram Soomro
Department of Chemistry
Selcuk University Konya
Konya, Turkey
and
Department of Community Health Sciences
Aga Khan University
Karachi, Pakistan

U. Suparmaniam
Universiti Teknologi Petronas
Perak, Malaysia

J. Immanuel Suresh
Department of Microbiology
The American College
Madurai, India

Anurag Tewari
Department of Basic Science and Humanities
Pranveer Singh Institute of Technology
Kanpur, India

1 An Introduction to Green Technologies for Wastewater Treatment

Faria Javed, Mehwish Khalid, and Imran Hashmi

1.1 INTRODUCTION AND BACKGROUND

Rapid urbanization and industrialization generate massive volumes of wastewater, which is increasingly being used for irrigation in urban and peri-urban agriculture. It produces a tremendous number of economic activities; greatly modifies the quality of the water in natural water bodies; and supports countless livelihoods, particularly those of poor farmers (Marshall et al., 2007). Urbanization and industry are the main causes of the increased risk of consuming tainted water and coping with sanitation problems in most developing countries. The growing problem of water shortage has a negative influence on human livelihoods, economic progress, and global environmental quality. Therefore, it is now essential for the modern environment to identify inexpensive remediation techniques for the protection of water or to stop it from becoming contaminated (Rajasulochana & Preethy, 2016).

Many traditional methods have been used to clean wastewater since ancient times, but they are quite costly and not very effective (Narmadha & Kavitha, 2012; Shivajirao, 2012; Turovskiy, 1998). Conventional wastewater treatment systems are being replaced by newer, greener technical methods. Modern, environmentally friendly techniques are being implemented to replace antiquated wastewater treatment systems (Rajasulochana & Preethy, 2016).

Phosphorus and nitrogen are the two primary chemical pollutants that are discovered in wastewater. Although there are additional chemical contaminants such as heavy metals, detergents, and pesticides, nitrogen and phosphorus are the most often occurring limiting components in eutrophication. Some techniques concentrate on lowering heavy metals, while others target lowering phosphorus and nitrogen. It has been found that toxic chemicals, heavy metals, nitrogen, phosphorus, etc. cannot be reduced using the conventional approaches. Most chemicals cannot be treated with a single step using a special technique (Rajasulochana & Preethy, 2016).

Metals from wastewater are often removed using filtration, flocculation, activated charcoal, and ion exchange resins (Karthikeyan et al., 2005; Vijayaraghavan et al., 2007; Wang & Chen, 2009). The three major objectives of treatment between 1900 and the early 1970s were to (i) treat biodegradable organics [biochemical oxygen demand (BOD) removal]; (ii) remove pathogenic bacteria that cause disease;

DOI: 10.1201/9781003459286-1

and (iii) remove floatable and suspended debris from wastewater. Wastewater treatment was dominated from the early 1970s to the early 1990s by aesthetic and environmental concerns.

The previous tasks, which involved reducing and eliminating BOD, suspended particles, and hazardous microbes, were completed in larger amounts. Moreover, initiatives to eliminate contaminants such as nitrogen and phosphorus were initiated, particularly in a few of the lakes and streams. Significant global efforts have been made to improve surface waterway standards through more thorough and efficient wastewater treatment. This initiative was spurred by two factors: (i) a better understanding of how wastewater discharges affect the environment and (ii) knowledge of the harmful long-term impacts of wastewater discharges of specific elements.

Around 1990, wastewater treatment began to focus on the health risks associated with toxic and potentially dangerous synthetic substances released into the atmosphere because of increased knowledge and information. The objectives for enhancing water quality from the 1970s remain the same, but the emphasis now is on identifying and eliminating hazardous and trace chemicals that may have long-term detrimental effects on human health and the ecosystem.

As a result, even if the initial treatment goals are still valid today, the required level of care has virtually increased and new goals and objectives have been introduced. A typical Dewats framework includes purified water, solids removal (or usage), and optional and necessary medications. In regions with poor soil and high groundwater levels, a septic tank can be utilized as the primary treatment to remove settleable materials and provide limited anaerobic treatment. Modifications to the aforementioned framework enable the profluent to be treated with oxygen while preventing drifting particulates from getting into the optional treatment.

Even though they are small and don't need much assistance, they are prone to failure and, even when they do function, they could nevertheless leave behind a waste stream that is rich in microbes. In areas with deeply porous soils, auxiliary treatment options with sand channels provide effective microbial expulsion; nevertheless, they are ineffective in other areas with particularly permeable soil types. The use of natural frameworks to remove radioactive elements and heavy metals from setups has received a great deal of attention (Rajasulochana & Preethy, 2016).

All organic treatment methods take advantage of the ability of microorganisms to use various wastewater elements as building blocks for cell amalgamation and as a source of energy for microbial digesting. This metabolic process can remove impurities from a variety of basic materials and byproducts. The type of hazardous metals that are still present in wastewater treatment facilities (WWTP) influences the choice of expulsion technique. Various methods have been used to achieve specific results, such as electrochemical treatment, switch assimilation, post precipitation, particle exchange, adsorption using activated carbon or other suitable sorbents, and evaporation (Matsumoto et al., 2007).

Since there are now fewer water resources available due to rapid industrialization and population growth, the issue of rising wastewater and water treatment expenses has surfaced (Lee et al., 2021; Shin et al., 2020). Furthermore, the large wastewater discharges from industry are problematic for humans and a serious threat to future generations (Qu et al., 2019; Lee et al., 2019). In response to a rise in serious environmental pollution in recent years, businesses and governments are reassessing their traditional methods of purifying water and wastewater and are shifting their focus to the development of green technology for a more sustainable future (Yadav et al., 2019).

Numerous tools can be used in the wastewater sustainability assessment. Possible foundation for assessment methods includes energy flow, economic analysis, and life cycle. The use of a balanced set of indicators in response to the complexity of the pollutants in wastewater provides an all-encompassing evaluation that is necessary, even if most assessments only look at one tool at a time (Metcalf & Eddy, 2003). Muga and Mihelcik (2007) state that for full sustainability in wastewater management, choosing a particular wastewater treatment method should not be based just on technical knowledge but also on the surrounding human and environmental activities. Concentrating on a particular contaminant is not the best course of action because there is no "one-size-fits-all" method for treating wastewater. Achieving sufficient disinfection to protect public health, minimize the production of disinfection byproducts (DBPs), lower greenhouse gas emissions, and lessen the neighborhood's impact requires striking a compromise between opposing variables (Paul & Swaim, 2016).

One of the most important sustainable development objectives for the foreseeable future is to improve the quality of the environment (Ridzuan et al., 2020).

The circular economy, in which wastewater has traditionally been seen as a beneficial resource rather than a burden, is greatly influenced by wastewater treatment (Baawain et al., 2020; Meena et al., 2019; Machineni, 2019). Reusing and recycling materials and goods is given priority in the circular economy in order to reduce the demand for natural resources and promote environmental sustainability. Wastewater treatment has grown in importance as a source of nutrients, energy, and clean water (López-Morales & Rodríguez-Tapia, 2019; Baawain et al., 2020; Mahtab et al., 2021; Ashraf et al., 2021). It is a vital source of biogas that may be used for both domestic and commercial purposes. It reduces the usage of fossil fuels and the burden on natural resources by producing energy (Kamali et al., 2019; Jodar-Abellan et al., 2019; Chaúque & Rott, 2021; Suzenet et al., 2002).

Most governments throughout the world have identified ways to turn waste into a valuable resource rather than throwing it into nearby lakes, rivers, or oceans (Jin et al., 2014). Because of its many advantages, politicians searching for alternative sources of economic growth have also shown a significant interest in it.

Fito and Van Hulle (2021) assert that in many countries, wastewater management helps meet the sustainable development goals (SDGs). SDG 6 is primarily concerned with ensuring that everyone has access to sanitary facilities and clean drinking water, with a focus on ecosystem preservation, wastewater treatment, and sustainable water resource management (Silva, 2023).

The main tenets of green technologies for water and wastewater treatment include resource recovery, sustainability, zero or low emissions, biological process, cradle-to-cradle design, lowering environmental stress, energy reduction, and preservation of natural resources (Ngo et al., 2020) (Figure 1.1).

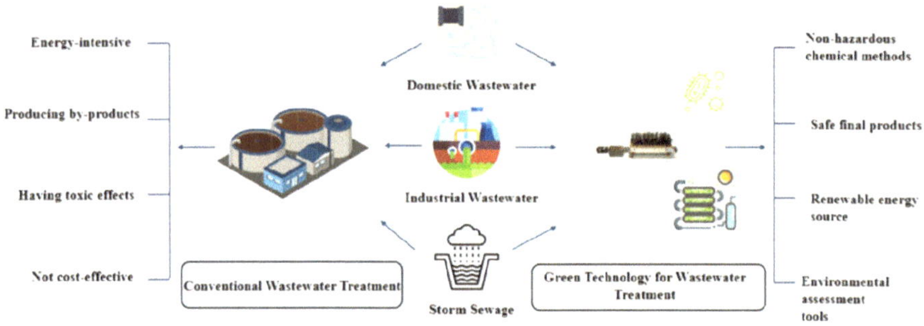

FIGURE 1.1 Conventional vs green technologies for wastewater treatment.

1.2 WASTEWATER COMPOSITION, SOURCES, CHARACTERISTICS, AND IMPACTS

Wastewater is characterized as contaminated water resulting from human activity and rainfall runoff that has been impacted by residential, commercial, and industrial use. Because the makeup of all wastewaters is so diverse and ever changing, it is quite challenging to come up with a single definition for the word. 99.9% of wastewater is made up of water, with the remaining 0.1% being eliminated. This 0.1% is made up of inorganic chemicals, microbes, and organic stuff. Numerous habitats, including lakes, ponds, streams, rivers, estuaries, and seas are exposed to the wastewater. Since hazardous materials wash off of parking lots, rooftops, and roadways, wastewater also includes storm runoff.

Wastewater comes from a variety of sources. Sanitary sewage, often known as domestic sewage, is the used water from homes and flats. Used water from manufacturing or chemical processes is known as industrial sewage. Runoff from precipitation that gathers in a network of pipelines or open channels is known as storm sewage or stormwater.

The weight of domestic wastewater is somewhat higher than 99.9% water. Less than 0.1% of the mixture is made up of various suspended and dissolved contaminants. The nature of these contaminants and the huge quantities of sewage in which they are carried make disposal of residential wastewater a substantial technological difficulty, even though they make up a relatively small fraction of the sewage by weight.

Putrescible organic debris and plant nutrients are the main contaminants, but it's also highly likely that household sewage contains bacteria that cause illness. Depending on the type of industrial operation, industrial wastewater typically comprises distinct and easily recognized chemical components. As storm sewage moves across the ground, it picks up organic components, suspended and dissolved sediments, grit, nutrients, and different compounds.

Effluent characteristics vary from industry to industry; hence, they require different treatment methods. Wastewater pollutants are generally divided into two categories: physicochemical and microbiological. Particularly concerning physicochemical features of wastewater are pH, dissolved oxygen (DO), chemical oxygen demand (COD) and BOD, suspended and dissolved particles, nitrogen (nitrite, nitrate, and ammonia), phosphate, and metals (Larsdotter, 2006). In wastewater influents, viruses, bacteria, fungus, protozoa, and helminths are the most common microorganisms.

One of the main causes of many different types of water contamination is wastewater. They are constantly produced in the absence of proper care. Pollution arises from inadequate treatment of wastewater by poorly designed wastewater treatment plants, which then release the wastewater into receiving water sources. The receiving water body is deteriorating because of the low quality of wastewater released.

While wastewater-associated microorganisms serve a variety of useful functions in the systems, they are also thought to have a major role in the development of several waterborne illnesses. Furthermore, a wide range of anthropogenic substances, many of which have the potential to disrupt hormones, have been found to be present in wastewater. In order to prevent pathogenic risks to both the aquatic ecology and users of these water resources, wastewater should be effectively treated before being released.

The ecosystem and human physiology are affected in the short and long terms when raw or inadequately treated wastewater is released onto watercourses. Enforcing water and environmental laws to safeguard rural and urban areas and adequately treating wastewater before release are the only ways to lessen the impact of wastewater on the environment, human physiology, and public health. In order to guarantee the discharge of uncontaminated wastewater into receiving bodies of water, meticulous planning, sufficient treatment, frequent monitoring, and suitable laws are needed (Owhonka et al., 2021).

1.3 CONVENTIONAL TREATMENT TECHNOLOGIES

Municipal and industrial wastewater treatment systems are based primarily on conventional wastewater treatment technologies (Bera et al., 2022). Before wastewater is released into the environment, these procedures are intended to remove a sizable amount of toxins and pollutants.

Primary Treatment: According to Awaleh and Soubaneh (2014), primary treatment is the first stage of wastewater treatment, with the main goal being the removal of big particles and floating debris. Screening and grit removal are the two primary therapy procedures. To get rid of big items like sticks, plastics, and rags, wastewater is passed through screens (Markin et al., 2020). Sand and gravel are examples of heavy inorganic particles that need to be settled in grit chambers or sedimentation tanks to prevent damage to downstream equipment (Jasim, 2020).

Secondary Treatment: The elimination of suspended particles and organic matter from wastewater is the main goal of secondary treatment procedures (Sonune & Ghate, 2004). Microorganisms play a crucial role in these processes by breaking down organic pollutants. Trickling filters and the activated sludge process are common secondary treatment techniques. In aeration tanks, wastewater is combined with a microbial culture (activated sludge) to promote the biological breakdown of organic contaminants. Organic matter is biologically oxidized by spreading wastewater over a bed of pebbles or plastic medium covered in microbial biofilm (Malakar et al., 2017). Through a series of meticulously designed stages, activated sludge, a byproduct of secondary wastewater treatment, is transformed into a useful resource for agriculture in the form of biofertilizer (Engida et al., 2020). Dewatering techniques are used to reduce the activated sludge's moisture content after the secondary treatment. Following that, pathogens are eliminated, and the amount of organic matter is decreased using stabilizing techniques such as aerobic or anaerobic digestion (AD) (Seruga et al., 2020). The sludge can be composted to enhance its nutrient profile and ensure that any potential pollutants are eliminated once it has stabilized. Increased additions of organic or mineral additives can improve the biofertilizer's composition (Chojnacka et al., 2020). Strict quality control measures are implemented to meet regulatory standards and ensure the efficacy and safety of the final product. Following biofertilizer manufacture, these fertilizers are tested on various plant species to determine their rate of growth and yield.

Tertiary Treatment: A next step called tertiary treatment is used to further improve the quality of wastewater that has been treated, especially for certain uses like water reuse or discharge into sensitive areas (Pintilie et al., 2016). Filtration, disinfection, and the removal of excess nutrients like phosphate and nitrogen are examples of tertiary treatment techniques. To get rid of any leftover suspended solids and tiny particles, methods like sand filters, multimedia filters, or membrane filtration (microfiltration, ultrafiltration) are used (Bairagi & Ali, 2020). Pathogens and undesired nutrients are removed, and treated wastewater is subjected to either physical disinfection (UV irradiation) or chemical disinfection (chlorination, ozonation). This ensures the wastewater's microbiological safety. Tertiary treatment is mostly used to purify water so that it is safe to drink using a variety of techniques.

1.4 LIMITATIONS OF CONVENTIONAL TREATMENT TECHNOLOGIES

There are now a number of problems with conventional wastewater treatment technologies, such as increasing chemical use, sludge disposal, and increased energy and space requirements. Major operational problems in these systems also include the inability to treat more wastewater than the system's restricted design capacity, the effective removal of resistant organic components, and a labor shortage. To overcome the operational and technological constraints of conventional wastewater treatment technologies, researchers are attempting to create new categories of advanced wastewater treatment procedures (Anekwe et al., 2022).

1.5 GREEN WASTEWATER TREATMENT TECHNOLOGIES

Many scientists and engineering groups have concentrated their research efforts on sustainable water and wastewater treatment using greener materials and cost-effective and highly efficient procedures in the area of sustainable water and wastewater purification technology. They have shown how to use green technologies, such as physical (like eco-friendly adsorbents), biological (like activated sludge), and chemical processes (like oxidant and catalyst), to remove organic and inorganic contaminants from water and WWTP (Lee & Chon, 2022).

Green technology offers scientific and technological solutions that lessen harmful impacts on the environment and people in a variety of sectors, such as manufacturing, energy, and agriculture (Fu et al., 2023; Saleh et al., 2021).

These days, technology is widely accepted because of its benefits, which include reduced or zero greenhouse gas emissions, energy and resource conservation, less environmental degradation, and improved utilization of renewable resources. Bioremediation, biomineralization, phytoremediation, bioaugmentation, bioflocculation, biosurfactant, biocatalyst, membrane bioreactor (MBR), wetlands, nanogreen technology (NGT), etc., are some of the popular green technologies that are used to protect and save the environment from harmful pollutants. Various biological organisms like microbes (*Acinetobacter, Pseudomonas*, etc.), algae (*Chlorella, Desmodesmus*, etc.), plants (*Amaranthus*, *Camellia sinensis*, etc.), or their byproducts (exopolysaccharides, enzymes, phenolic compounds, nanomaterials, etc.) are involved in various mechanisms of green technology. This section of the chapter will briefly discuss the different green technologies that will be explained in detail in the subsequent chapters.

1.6 AEROBIC AND ANAEROBIC TREATMENT TECHNOLOGIES

Since biological water treatment methods are used to create safe drinking water, they are essential parts of a wastewater treatment strategy. The methods used for this are bioremediation, anaerobic digestion (AD), and aerobic digestion. Here is a summary of these procedures.

1.7 AEROBIC PROCESSES

When oxygen or air is widely accessible in wastewater in the dissolved form, aerobic and facultative bacteria cause biodegradable organic matter to break down aerobically (Goli et al., 2019; Samer, 2015). The temperature, duration of retention, availability of oxygen, and the bacterial biological activity—all restrict the process's scope. Furthermore, the rate at which organic pollutants are biologically oxidized may be accelerated by the addition of particular chemicals that are necessary for bacterial growth.

Phosphates, nitrates, volatile organic compounds, dissolved and suspended organic compounds, COD, BOD, and other pollutants can all be eliminated with this method. Up to 90% of the biodegradable organics in the environment can be eliminated. The method's drawback is that it generates a large volume of bio-solids, which calls for more expensive management and treatment. The aerobic process is carried out via activated sludge technologies, aeration lagoons, and oxidation ponds (Samer, 2015).

1.8 ANAEROBIC PROCESSES OR AD

In the biological process known as anaerobic waste treatment, organic contaminants are broken down by microorganisms in the absence of oxygen. Anaerobic breakdown, or putrefaction, occurs when there is no free DO present in the wastewater. This process involves facultative and anaerobic bacteria breaking down complex organic materials into sulfur-based organic compounds, carbon, and nitrogen. Biogas, which includes nitrogen, ammonia, hydrogen sulfide, and methane, is created by this series of biochemical reactions. By using this method, the number of bacteria in wastewater is reduced (Jin et al., 2022; Khan et al., 2018; Chan et al., 2009). When treating streams with high levels of organic matter [expressed as high BOD, COD, or total suspended solids (TSS)], anaerobic technologies are typically employed prior to aerobic treatment. An established low-energy method for treating industrial wastewater is anaerobic treatment. The AD method is attractive because it is an environmentally friendly process that processes wastewater, produces renewable energy, and yields byproducts that can be used as agricultural fertilizers (Ruiz & Flotats, 2014). The following benefits of the AD procedure over the aerobic wastewater treatment method are less fertilizers needed and less biological sludge produced, which only needs to be dried as a last step in treatment (Buitrón et al., 2014). Additionally, it requires no oxygen and a modest reactor capacity, which lowers the power required to deliver oxygen in the aerobic approach and frees the system's organic loading from oxygen supply constraints. Therefore, in AD, a higher loading rate can be employed, enabling a quicker response to the administration of substrate following extended intervals without feeding and several months of semi-feeding techniques.

Because of these advantages to the system, AD is a practical solution for treating seasonal industrial effluent and removing off-gas that pollutes the air. Anaerobic filter reactors, expanded granular sludge bed (EGSB), anaerobic baffled reactor (ABR), anaerobic sludge bed (UASB) reactors, and anaerobic lagoons are a few examples of anaerobic treatment systems (Anekwe et al., 2022).

Furthermore, anaerobic–aerobic systems have been remarkably used for many years in the treatment of municipal and industrial wastes. While conventional anaerobic–aerobic treatment plants were historically used for the majority of wastewater treatment, high-rate anaerobic–aerobic bioreactors have been used more and more recently for wastewaters with high COD. The great potential for resource recovery, low energy and chemical consumption, low sludge generation, cheap equipment requirements, and high operational simplicity are only a few benefits of anaerobic–aerobic treatments.

However, it has been discovered that traditional anaerobic–aerobic systems have operational limits with regard to long hydraulic retention time (HRT), space requirements, and biogas capture facilities. These constraints are addressed by the use of recently designed high-rate bioreactors, which offer improved organic matter removal at shorter HRT and higher methane yields for the production of biogas. Integrated anaerobic–aerobic bioreactors, which combine the aerobic and anaerobic process in a single bioreactor, have received a lot of interest as a way to meet severe limits regarding space, smells, and minimal sludge generation.

Anaerobic–aerobic sequencing batch reactors (SBR), integrated bioreactors with physical separation of the anaerobic–aerobic zone, integrated bioreactors without physical separation of the anaerobic–aerobic zone, and combined anaerobic–aerobic culture systems are the four types of integrated bioreactors.

The total degradation efficiency can be increased by combining anaerobic and aerobic degradation mechanisms in a single bioreactor. A wide range of high-organic-strength industrial and municipal wastewater is anticipated to be treated by the compact integrated bioreactors because of their straightforward yet economical technology, ability to produce renewable energy, and exceptional treatment efficiency. To assess the effectiveness of these intriguing reactors on a broader scale, more work is needed as the majority of integrated bioreactors documented in the literature lack widespread industrial application. In addition, further enhancements like installing a biogas collecting system and using packing medium or suspended carriers are seen to be crucial.

Because of this, integrated bioreactors are able to obtain carbon credits through the Kyoto Protocol's Clean Development Mechanism (CDM), changing the perspective of wastewater management from "treatment and disposal" to "beneficial utilization" and "profitable endeavor."

1.9 CONSTRUCTED WETLANDS FOR WASTEWATER TREATMENT

Built wetlands use soil, plants, and microbiological activities to treat wastewater in a manner similar to that of natural wetland ecosystems (Gopal, 1999). These systems support biodiversity and ecological advantages while offering efficient treatment. Subsurface flow (SSF) wetlands, hybrid systems, and free water surface (FWS) wetlands are the different types of artificial wetlands and natural treatment systems. Shallow surface water in free-flowing surface wetlands allows wastewater to flow through, promoting the establishment of wetland plants and microbial activity that removes pollutants (Zhang et al., 2010). SSF wetlands allow wastewater to be treated by biological, physical, and chemical processes by allowing it to percolate through a bed of gravel or soil covered with wetland vegetation (Vymazal, 2011). To improve pollution removal, hybrid systems combine artificial wetlands with conventional treatment technologies like filtration or activated sludge (Liu et al., 2015).

Furthermore, artificial wetlands can absorb carbon just as well as natural wetlands. Using artificial wetlands sustainably as a carbon sink may be one of the most successful ways to slow down global warming (Rosli et al., 2017).

1.10 PHOTOBIOREACTORS FOR WASTEWATER TREATMENT

Typically, microalgae are grown to treat wastewater, produce a variety of food products, produce biofuels, remove hazardous waste, produce biofertilizers, and reduce atmospheric flue gases like carbon dioxide (Pérez-López et al., 2014). Numerous viable species have already been thoroughly studied technologically, with their physical, chemical, and biological characteristics known, and many more are being studied daily. The primary cause of this is the inefficient design of photobioreactor (PBR), which can offer the following benefits: low capital and energy input; cleaning; easy maintenance; temperature control owing to seasonal and diurnal variations; uniform light distribution; uniform mixing; degasification; and harvesting at the lowest possible cost (Singh & Sharma, 2012). Numerous PBR designs are available for the production of microalgae. Each reactor is distinct, with pros and cons of its own (Singh & Sharma, 2012). Despite these benefits and the availability of PBR designs, the total economics and operational complexity of these microbes mostly limit their use in industrial settings.

Numerous useful end products are hidden in the wastewater and must be retrieved. Many PBR designs have been tried in the past, but sadly, not many of them have been successful in being produced on an industrial basis. The degree of performance exhibited by each reactor varies, contingent upon the adept selection of various criteria such as species, light, environment, mixing, and reactor type. Before building a reactor, the following important factors should be taken into account: the necessary end products; ease of construction; simplicity; low power consumption; small footprint; cost effectiveness; and ease of cleaning and maintenance.

Few key points that should be taken care of:

- A shortage of land for other uses, such as crop cultivation, may result from the increased usage of land for PBR design. Reactors with vertical rather than horizontal designs hence have a lower land footprint.
- High pollution, uneven lighting, and restrictions on the size of the land footprint can all be faced by open systems. As a result, it can be applied to the treatment of wastewater in places with cheap land prices and discharge.
- Because of its circular design, cylindrical or columnar reactors are a great option for treating wastewater on expensive land areas while ensuring even distribution of light and mixing.
- Using closed PBRs, including tubular and flat panels, for wastewater treatment can be costly, labor intensive, and requires a lot of land. For the production of contamination-free biomass and the extraction of high-value end products, such systems ought to be chosen (Ashok et al., 2019).

1.11 MICROALGAE—BACTERIA CONSORTIUM FOR WASTEWATER TREATMENT

The function of wastewater treatment has gradually changed from being primarily focused on sanitation to also involving nutrient recovery and removal. To be sure, conventional activated sludge treatment works quite well in eliminating pathogens as well as dissolved organic carbon, which contributes to pathogen renewal. Conventional activated sludge treatment requires the integration of supplementary steps or processes, such as nitrification/denitrification, anammox, and chemical precipitation, to remove nutrients. Despite being extensively utilized, these treatment systems are intricate, costly, and energy intensive (Nhut & Le, 2020; Wollmann et al., 2019). For instance, the two essential nutrients, phosphorus and nitrogen, can only be extracted individually using processes like chemical precipitation and anammox, respectively. Eutrophication and other serious ecological harm are caused by inadequate nitrogen and phosphorus removal from wastewater.

Many academics are interested in the work that cooperation of microalgae and bacteria is doing to produce energy-efficient and effective nutrient removal technologies (Mohsenpour et al., 2021; Udaiyappan et al., 2020). There are many different ways that bacteria and microalgae interact, ranging from mutualism to competition (antagonism) (Fallahi et al., 2021). Microalgae and bacteria engage in symbiotic relationships that involve nutrition exchange, cell-to-cell communication, and chemical compound stimulation. Additionally, microalgae can identify vitamins, siderophores made by bacteria, and quorum-sensing signal molecules. This leads to symbiotic interactions that enhance bioactivity, tolerance, and elimination. The effectiveness of treating wastewater is increased when microalgae and bacteria have a symbiotic connection.

A microalgae–bacteria consortium is therefore a practical choice for the treatment of wastewater and the production of bioenergy. Conversely, the antagonistic effect is frequently insignificant in light of the overall positive association. When bacteria and decaying microalgae fight for oxygen, hostile interactions result. Furthermore, a variety of inhibitory chemicals that can damage the growth of the partner can be produced by bacteria and microalgae. It is interesting to note that, to prevent cross-contamination, the majority of research on the interactions between microalgae and bacteria has been carried out in well-controlled laboratory settings (Aditya et al., 2022).

1.12 MICROALGAE FOR BIOHYDROGEN PRODUCTION

As a photosynthetic organism, microalgae are referred to as third-generation feedstock for the manufacturing of biofuels. Because of their ability to sequester CO_2, microalgae have recently gained a lot of attention for cultivation. These algae turn into biomass that is mostly made up of lipids, proteins, and carbs, which can be used to produce biofuel. Microalgae have the ability to grow more quickly than terrestrial plants and can be used in non-arable regions. 1.5–2.2 kg of CO_2 can be fixed to produce 1 kg of algal biomass, and algal biomass has a 50% carbon content on average.

Algal culture has several benefits, but for the best results, it frequently requires freshwater, a variety of macro- and micronutrients, and CO_2. These elements make production more expensive overall and less appealing as a substitute. The use of wastewater as an alternative to freshwater and nutrient medium in microalgae culture could help to mitigate these issues. The removal of nutrients from wastewater and the production of biomass rich in lipids, proteins, and carbs that can be utilized for a variety of purposes are two benefits of this process. All nutrients—both organic and inorganic forms of carbon and nitrogen—necessary for the growth of microalgae were present in the effluent (Nagarajan et al., 2020) for biohydrogen production. Thus, the culture of microalgae in wastewater could have two benefits: It could cleanse the wastewater and produce algal biomass at the same time (Satheesh et al., 2023).

1.13 WASTEWATER TREATMENT WITH SUSPENDED AND NON-SUSPENDED ALGAE

The majority of studies have concentrated on suspended microalgae that are growing in shallow manmade sewage ponds. Nonetheless, the use of non-suspended algae has received more attention in recent years, either as attached algal communities (biofilm/periphyton) growing in shallow artificial streams or on the surfaces of rotating biological contactors (RBC/biodiscs) or as unialgal cultures immobilized in a polymeric matrix.

1.14 TREATMENT OF WASTEWATERS VIA SEAWEEDS

Algae are a broad class of photosynthetic organisms that live in freshwater and marine settings. They can be unicellular (microalgae) or multicellular (macroalgae) (Bharathiraja et al., 2015).

Seaweeds, also referred to as macroalgae, are quickly expanding creatures that resemble plants and can reach lengths of up to 60 m in certain species (McHugh, 2003). Seaweed has garnered attention recently as the possible agent for adsorption-based wastewater treatment. The primary cause of macroalgae's high capacity to bind contaminants like trace metals is the presence of sulfated polysaccharides in its cellular wall, specifically in its fibril matrix and intercellular gaps. Actually, the polysaccharide's carboxyl, sulfate, and hydroxyl groups are potent ion exchangers and are crucial locations for the complexation of metal cation (Vasconcelos & Leal, 2001). The aforementioned chemical compositions of seaweed cell walls vary significantly across different species, and the most advantageous kind is determined by its intended usage.

It has been demonstrated that biosorption with naturally occurring seaweeds may effectively remove contaminants from a variety of sources, including heavy metals; nitrogen; and phosphorous and phenolic compounds; as well as dyes from the printing, textile, and paper industries. The effective removal of contaminants from wastewater through biosorption using seaweed is becoming a viable option for existing technologies due to its environmentally friendly character and the low cost and availability of raw materials (Arumugam et al., 2018).

1.15 WASTEWATER-FERTIGATED SHORT-ROTATION COPPICE SYSTEMS (WFSRCS)

Short-rotation coppice (SRC)-based vegetated filters have the potential for pollution management and the bio-based circular economy. They can be utilized to treat a variety of industrial and municipal wastes while economically and sustainably providing important biomass. According to studies, wfSRCs have a great chance of replacing traditional treatment plants as practical and affordable wastewater treatment options.

Studies conducted by scientists show high potential for the elimination of BOD, COD, total nitrogen (TN), total phosphorus (TP), and increased biomass production. Further research trials are necessary to confirm whether the method can also be used to treat micropollutants. Using various irrigation techniques, raw or processed wastewater is directly applied to the surface or higher soil layer in wfSRC systems. Both the root zone and the upper soil layer undergo treatment. While most nutrients are absorbed by plants or stored and changed in the soil, organic matter is broken down by soil microorganisms and biofilms on roots. Similar to other nature-based treatment options, wfSRCs have sizable environmental footprints. A maximum of $15\,m^2$ of wfSRCs are needed to treat wastewater from one person, with a daily discharge of 100 L/day (Mosiej et al., 2012). In most situations, wastewater application to wfSRCs in the Northern Hemisphere is restricted to the tree-growing season because cold weather reduces plant nutrient uptake and treatment effectiveness. Given the efficiency of the soil–root filter system and its ability to buffer nutrients, wfSRCs can be operated year round in climates with lengthy growing seasons (10–11 months) and the absence of prolonged cold (Hänel et al., 2022).

1.16 MICROBIAL NANOMATERIALS FOR THE TREATMENT OF WASTEWATERS

Traditional methods for fabricating nanomaterials have been thoroughly studied during the last few decades. However, there are problems with these conventional approaches, namely their high energy consumption and usage of hazardous materials. As is common knowledge, microbes have the ability to cleanse their surroundings and bind and concentrate dissolved metals. The production of nanomaterials using bioinspired techniques is inexpensive, has low energy consumption, and is environmentally benign.

Nanotechnology presents enormous opportunities for the treatment of wastewater and the removal of contaminants and microorganisms. Dendrimers, carbonaceous nanomaterials, zeolites, and metals incorporating nanoparticles (NPs) are among the four categories of nanoscale materials that are thought to be effective in treating wastewater. Recently, there has been a great deal of interest in studies on microorganism-based biogenic NPs because of their high catalytic reactivity.

Compared to polycrystalline materials, microbe-based nanomaterials have a bigger surface area and more active sites, which results in superior catalytic activity. Therefore, the utilization of microbes in nanomaterials can offer a novel method for decontaminating and treating wastewater (Thakur & Mukherjee, 2021).

1.17 METAGENOMICS: HIDDEN DETAILS OF MICROBES INVOLVED IN WASTEWATER TREATMENT

Earth is a large, diversified planet with an area of 510.1 million km^2 There are many different biological species present here, with bacteria being the most prevalent. To comprehend these bacteria's place in the global ecology, research is being done on them. Research in the scientific discipline of metagenomics is still in its infancy. Some methods, such as the use of next-generation sequencing techniques for one or more genome collections, provide a platform for the discovery of novel information on microbial genomics and the connection between environmental occupation and genome evolution.

All of the wastewater is collected in a reservoir at the wastewater treatment plant. The majority of the microorganisms in this effluent are unknown. Exposing the hidden traits of unidentified microorganisms could result in the resolution of current medical issues, which would be beneficial to humanity (Gaglani et al., 2021).

1.18 CHALLENGES IN WASTEWATER TREATMENT AND FUTURE RESEARCH DIRECTIONS

To effectively solve the many issues that the wastewater treatment field faces, research, advancement, and innovation must continue (Sharma et al., 2020).

The difficulties and potential avenues for future research point to important areas that require more study in order to advance wastewater treatment technology and procedures and get past current constraints. Wastewater treatment faces substantial problems from emerging contaminants, including medicines, personal care items, micro-plastics, heavy metals, pesticides, herbicides, polychlorinated biphenyls (PCBs), dioxins, NPs, radioactive chemicals, and endocrine-disrupting substances (Necibi et al., 2021). Subsequent investigations ought to concentrate on creating novel treatment methods and procedures that can efficiently eliminate these pollutants. Enhancement of wastewater treatment processes in terms of cost effectiveness and efficiency should be part of the advanced treatments. In order to retain high treatment performance while maximizing treatment efficiency, minimizing the formation of undesirable byproducts, reducing energy consumption, and minimizing the usage of chemicals, future research should investigate novel methodologies and technologies. This entails figuring out how to intensify processes, creating creative reactor designs, and incorporating cutting-edge technology to raise system efficiency as a whole (Rubio et al., 2010). Subsequent investigations ought to focus on bridging the gap between theory and practice by overseeing experiments at the pilot stage, evaluating the viability of novel technologies in practical settings, proving their scalability, and formulating protocols for their effective application in wastewater treatment plants. There is growing interest in the recovery of resources from wastewater, including energy, nutrients, and water (Van Lier, 2008). Subsequent investigations ought to explore methods for enhancing the financial feasibility of resource extraction technologies, refining the procedures for recovering resources, and creating novel strategies for employing wastewater treatment byproducts in diverse areas. Investigating the possibilities of cutting-edge technologies is part of this (Chin et al., 2022).

1.19 CONCLUSION

By using less energy, wastewater treatment contributes significantly to environmental protection, public health protection, and the sustainable use of water resources. Several topics related to wastewater treatment have been covered in this chapter, such as the significance of treatment procedures and the dangers of using untreated wastewater for the environment and public health. Moreover, the application of both traditional and cutting-edge treatment methods, resource recovery, energy efficiency, and obstacles, as well as future research objectives and problems in wastewater treatment have been explored. Wastewater treatment is essential for eliminating toxic substances and xenobiotic pollutants from wastewater; saving ecosystems from harm; and stopping the spread of waterborne illnesses like cholera, typhoid, diarrhea, hepatitis, gastroenteritis, amebiasis, giardiasis, scabies, and worm infections. Advanced green technologies, such as micro- and macroalgae, microbial nanomaterials, wastewater-fertigated short-rotation coppice, and constructed wetlands, offer higher levels of treatment and opportunities for resource recovery than conventional technologies, such as primary, secondary, and tertiary treatments for wastewater treatment plants. A byproduct of secondary treatment called activated sludge can be used to make biofertilizers.

Resource recovery and energy productivity are crucial components of sustainable wastewater treatment. Recovering nutrients for reuse, using wastewater treatment plants, and producing biogas through AD can all reduce energy use, produce renewable energy, and support the circular economy. Decentralized wastewater treatment systems are feasible alternatives in places with low population density or inadequate infrastructure because of their flexibility, cost effectiveness, and benefits related to resource recovery. Reducing environmental impacts and improving wastewater management are possible when decentralized technologies are integrated with traditional treatment plants. Policies, procedures, and legal frameworks are essential for directing wastewater treatment methods, encouraging sustainability, establishing benchmarks, and offering financial incentives for adherence to rules and innovations. Sustainable wastewater management methods are driven by national rules, international standards, and economic instruments. These factors assure environmental protection. The need for improved treatment efficiency, the existence of new contaminants, and closing the gap between research and implementation are some of the issues that wastewater treatment still faces. Future studies should concentrate on creating technology to deal with newly discovered pollutants, encouraging resource recovery, streamlining treatment procedures, and improving climate change adaptability. The treatment of wastewater using green technology is a developing sector that calls for constant innovation, teamwork, collaboration between academic institutions and industry, and research in order to produce effective and sustainable solutions. Wastewater treatment can maintain public health, resource productivity, environmental preservation, and sustainable water resource management by putting forward-thinking technologies, addressing new issues, and enacting sensible regulations.

REFERENCES

Aditya, L., Mahlia, T. I., Nguyen, L. N., Vu, H. P., & Nghiem, L. D. (2022). Microalgae-bacteria consortium for wastewater treatment and biomass production. *Science of the Total Environment, 838*, 155871.

Anekwe, I. M. S., Adedeji, J., Akpasi, S. O., & Kiambi, S. L. (2022). Available technologies for wastewater treatment. *Wastewater Treatment*, 33–54.

Arumugam, N., Chelliapan, S., Kamyab, H., Thirugnana, S., Othman, N., & Nasri, N. S. (2018). Treatment of wastewater using seaweed: a review. *International Journal of Environmental Research and Public Health, 15*(12), 2851.

Ashok, V., Gupta, S. K., & Shriwastav, A. (2019). Photobioreactors for wastewater treatment. In S. K. Gupta, & F. Bux (eds.), *Application of Microalgae in Wastewater Treatment: Volume 1: Domestic and Industrial Wastewater Treatment* (pp. 383–409). Springer Nature.

Ashraf, A., Ramamurthy, R., & Rene, E. R. (2021). Wastewater treatment and resource recovery technologies in the brewery industry: current trends and emerging practices. *Sustainable Energy Technologies and Assessments*, *47*, 101432.

Awaleh, M. O., & Soubaneh, Y. D. (2014). Waste water treatment in chemical industries: the concept and current technologies. *Hydrology: Current Research*, *5*(1), 1–12.

Baawain, M. S., Al-Mamun, A., Omidvarborna, H., Al-Sabti, A., & Choudri, B. S. (2020). Public perceptions of reusing treated wastewater for urban and industrial applications: challenges and opportunities. *Environment, Development and Sustainability*, *22*, 1859–1871.

Bairagi, S., & Ali, S. W. (2020). Conventional and advanced technologies for wastewater treatment. In S. Ul-Islam (ed.), *Environmental Nanotechnology for Water Purification* (pp. 33–56). John Wiley & Sons, Incorporated.

Bera, S. P., Godhaniya, M., & Kothari, C. (2022). Emerging and advanced membrane technology for wastewater treatment: a review. *Journal of Basic Microbiology*, *62*(3–4), 245–259.

Bharathiraja, B., Chakravarthy, M., Kumar, R. R., Yogendran, D., Yuvaraj, D., Jayamuthunagai, J., ... & Palani, S. (2015). Aquatic biomass (algae) as a future feed stock for bio-refineries: A review on cultivation, processing and products. *Renewable and Sustainable Energy Reviews, 47*, 634–653.

Buitrón, G., Kumar, G., Martinez-Arce, A., & Moreno, G. (2014). Hydrogen and methane production via a two-stage processes (H2-SBR+ CH4-UASB) using tequila vinasses. *International Journal of Hydrogen Energy, 39*(33), 19249–19255.

Chan, Y. J., Chong, M. F., Law, C. L., & Hassell, D. G. (2009). A review on anaerobic–aerobic treatment of industrial and municipal wastewater. *Chemical Engineering Journal, 155*(1-2), 1–18.

Chaúque, B. J. M., & Rott, M. B. (2021). Solar disinfection (SODIS) technologies as alternative for large-scale public drinking water supply: advances and challenges. *Chemosphere*, *281*, 130754.

Chin, M. Y., Phuang, Z. X., Woon, K. S., Hanafiah, M. M., Zhang, Z., & Liu, X. (2022). Life cycle assessment of bioelectrochemical and integrated microbial fuel cell systems for sustainable wastewater treatment and resource recovery. *Journal of Environmental Management*, *320*, 115778.

Chojnacka, K., Witek-Krowiak, A., Moustakas, K., Skrzypczak, D., Mikula, K., & Loizidou, M. (2020). A transition from conventional irrigation to fertigation with reclaimed wastewater: prospects and challenges. *Renewable and Sustainable Energy Reviews*, *130*, 109959.

Engida, T., Alemu, T., Wu, J., Xu, D., Zhou, Q., & Wu, Z. (2020). Analysis of constructed wetlands technology performance efficiency for the treatment of floriculture industry wastewater, in Ethiopia. *Journal of Water Process Engineering*, *38*, 101586.

Fallahi, A., Rezvani, F., Asgharnejad, H., Nazloo, E. K., Hajinajaf, N., & Higgins, B. (2021). Interactions of microalgae-bacteria consortia for nutrient removal from wastewater: a review. *Chemosphere*, *272*, 129878.

Fito, J., & Van Hulle, S. W. (2021). Wastewater reclamation and reuse potentials in agriculture: towards environmental sustainability. *Environment, Development and Sustainability*, *23*(3), 2949–2972.

Fu, K., Li, Y., Mao, H., & Miao, Z. (2023). Firms' production and green technology strategies: The role of emission asymmetry and carbon taxes. *European Journal of Operational Research*, *305*(3), 1100–1112.

Gaglani, P., Srivastava, A., & Bhattacharya, S. (2021). Metagenomics: a field revealing the secrets of the microbial world of wastewater treatment plants. In M. P. Shah, & S. Rodriguez-Couto (eds.), *Wastewater Treatment Reactors* (pp. 159–184). Elsevier.

Goli, A., Shamiri, A., Khosroyar, S., Talaiekhozani, A., Sanaye, R., Azizi, K. (2019). A review on different aerobic and anaerobic treatment methods in dairy industry wastewater. *Journal of Environmental Treatment Techniques*, *7*(1), 113–141.

Gopal, B. (1999). Natural and constructed wetlands for wastewater treatment: potentials and problems. *Water Science and Technology*, *40*(3), 27–35.

Hänel, M., Istenič, D., Brix, H., & Arias, C. A. (2022). Wastewater-fertigated short-rotation coppice, a combined scheme of wastewater treatment and biomass production: A state-of-the-art review. *Forests*, *13*(5), 810.

Jasim, N. A. (2020). The design for wastewater treatment plant (WWTP) with GPS X modelling. *Cogent Engineering*, *7*(1), 1723782.

Jin, L., Zhang, G., & Tian, H. (2014). Current state of sewage treatment in China. *Water research*, *66*, 85–98.

Jin, Z., Zhao, Z., Liang, L., & Zhang, Y. (2022). Effects of ferroferric oxide on azo dye degradation in a sulfate-containing anaerobic reactor: From electron transfer capacity and microbial community. *Chemosphere*, *286*, 131779.

Jodar-Abellan, A., López-Ortiz, M. I., & Melgarejo-Moreno, J. (2019). Wastewater treatment and water reuse in Spain. Current situation and perspectives. *Water*, *11*(8), 1551.

Kamali, M., Persson, K. M., Costa, M. E., & Capela, I. (2019). Sustainability criteria for assessing nanotechnology applicability in industrial wastewater treatment: current status and future outlook. *Environment International*, *125*, 261–276.

Karthikeyan, T., Rajgopal, S., & Miranda, L. R. (2005). Chromium (VI) adsorption from aqueous solution by Hevea Brasilinesis sawdust activated carbon. *Journal of Hazardous Materials*, *124*(1–3), 192–199.

Khan, M. A., Ngo, H. H., Guo, W., Liu, Y., Zhang, X., Guo, J., ... & Wang, J. (2018). Biohydrogen production from anaerobic digestion and its potential as renewable energy. *Renewable Energy, 129*, 754–768.

Larsdotter, K. (2006). Wastewater treatment with microalgae-a literature review. *Vatten*, *62*(1), 31.

Lee, Y. G., & Chon, K. (2022). Green technologies for sustainable water and wastewater treatment: removal of organic and inorganic contaminants. *Separations*, *9*(11), 335.

Lee, Y. G., Park, Y., Lee, G., Kim, Y., & Chon, K. (2019). Enhanced degradation of pharmaceutical compounds by a microbubble ozonation process: effects of temperature, pH, and humic acids. *Energies*, *12*(22), 4373.

Lee, Y. G., Shin, J., Kwak, J., Kim, S., Son, C., Cho, K. H., & Chon, K. (2021). Effects of NaOH activation on adsorptive removal of herbicides by biochars prepared from ground coffee residues. *Energies*, *14*(5), 1297.

Liu, R., Zhao, Y., Doherty, L., Hu, Y., & Hao, X. (2015). A review of incorporation of constructed wetland with other treatment processes. *Chemical Engineering Journal, 279*, 220-230.

López-Morales, C. A., & Rodríguez-Tapia, L. (2019). On the economic analysis of wastewater treatment and reuse for designing strategies for water sustainability: lessons from the Mexico Valley Basin. *Resources, Conservation and Recycling*, *140*, 1–12.

Machineni, L. (2019). Review on biological wastewater treatment and resources recovery: attached and suspended growth systems. *Water Science and Technology*, *80*(11), 2013–2026.

Mahtab, M. S., Farooqi, I. H., & Khursheed, A. (2021). Zero Fenton sludge discharge: a review on reuse approach during wastewater treatment by the advanced oxidation process. *International Journal of Environmental Science and Technology*, *19*, 1–14.

Malakar, S., Saha, P. D., Baskaran, D., & Rajamanickam, R. (2017). Comparative study of biofiltration process for treatment of VOCs emission from petroleum refinery wastewater: a review. *Environmental Technology & Innovation*, *8*, 441–461.

Markin, A. A., Lepikhin, A. V., & Ulianov, S. V. (2020). Sewage treatment. In A. N. Soldatov & S. L. Minkova (eds.), *INNOVATICA-2020: Proceedings of the XVI International School-Conference for Students, Graduate Students, and Young Scientists* (pp. 504–507). Tomsk State University.

Marshall, F. M., Holden, J., Ghose, C., Chisala, B., Kapungwe, E., Volk, J., ... & Singh, R. P. (2007). Contaminated irrigation water and food safety for the urban and peri-urban poor: appropriate measures for monitoring and control from field research in India and Zambia. *Inception Report DFID Enkar*, *8160*(3), 1–12.

Matsumoto, N., Uemoto, H., & Saiki, H. (2007). Case study of electrochemical metal removal from actual sediment, sludge, sewage and scallop organs and subsequent pH adjustment of sediment for agricultural use. *Water Research*, *41*(12), 2541–2550.

McHugh, D. J. (2003). A guide to the seaweed industry. FAO Fisheries Technical Paper 441.

Meena, R. A. A., Kannah, R. Y., Sindhu, J., Ragavi, J., Kumar, G., Gunasekaran, M., & Banu, J. R. (2019). Trends and resource recovery in biological wastewater treatment system. *Bioresource Technology Reports*, *7*, 100235.

Metcalf and Eddy. (2003). *Wastewater Engineering: Treatment and Reuse*. McGraw-Hill.

Mohsenpour, S. F., Hennige, S., Willoughby, N., Adeloye, A., & Gutierrez, T. (2021). Integrating micro-algae into wastewater treatment: a review. *Science of the Total Environment*, *752*, 142168.

Mosiej, J., Karczmarczyk, A., Wyporska, K., & Rodzkin, A. (2012). Biomass production in energy forests: short rotation plantations. In L. Rydén & I. Karlsson (eds.), *Rural Development and Land Use Ecosystem Health and Sustainable Agriculture* 3 (pp. 234–242). The Baltic University Programme, Uppsala University.

Muga, E. H., & Mihelcik, R. J. (2007). Sustainability of wastewater treatment technologies. *Journal of Environmental Management*, *88*(3), 437–447.

Nagarajan, D., Lee, D. J., Chen, C. Y., & Chang, J. S. (2020). Resource recovery from wastewaters using microalgae-based approaches: a circular bioeconomy perspective. *Bioresource Technology*, *302*, 122817.

Narmadha, D., & Kavitha, V. M. S. (2012). Treatment of domestic wastewater using natural flocculants. *International Journal of Life Sciences Biotechnology and Pharma Research*, *1*, 206.

Necibi, M. C., Dhiba, D., & El Hajjaji, S. (2021). Contaminants of emerging concern in African wastewater effluents: occurrence, impact and removal technologies. *Sustainability*, *13*(3), 1125.

Ngo, H. H., Bui, X. T., Nghiem, L. D., & Guo, W. (2020). Green technologies for sustainable water. *Bioresource Technology*, *317*, 123978.

Nhut, H. H., & Le, L. T. (2020). Removal of H_2S in biogas using biotrickling filter: recent development. *Process Safety and Environmental Protection*, *144*, 297–309.

Owhonka, A., Fubara, E. F., & Justice, O. B. (2021). Wastewater quality-it's impact on the environment and human physiology: a review. *International Journal of Advance Research and Innovation*, *9*(4), 43–58.

Paul, D., & Swaim, P. E. (2016). Sustainability principles for drinking water and reuse treatment. *Symposium for Sustainable Infrastructure*, Colorado.

Pérez-López, P., González-García, S., Jeffryes, C., Agathos, S. N., McHugh, E., Walsh, D., ... & Moreira, M. T. (2014). Life cycle assessment of the production of the red antioxidant carotenoid astaxanthin by microalgae: from lab to pilot scale. *Journal of Cleaner Production, 64,* 332–344.

Pintilie, L., Torres, C. M., Teodosiu, C., & Castells, F. (2016). Urban wastewater reclamation for industrial reuse: an LCA case study. *Journal of Cleaner Production*, *139*, 1–14.

Qu, J., Wang, H., Wang, K., Yu, G., Ke, B., Yu, H. Q., … & Gong, H. (2019). Municipal wastewater treatment in China: development history and future perspectives. *Frontiers of Environmental Science & Engineering*, *13*, 1–7.

Rajasulochana, P., & Preethy, V. (2016). Comparison on efficiency of various techniques in treatment of waste and sewage water–a comprehensive review. *Resource-Efficient Technologies*, *2*(4), 175#184.

Ridzuan, N. H. A. M., Marwan, N. F., Khalid, N., Ali, M. H., & Tseng, M. L. (2020). Effects of agriculture, renewable energy, and economic growth on carbon dioxide emissions: evidence of the environmental Kuznets curve. *Resources, Conservation and Recycling*, *160*, 104879.

Rosli, F. A., Lee, K. E., Goh, C. T., Mokhtar, M., Latif, M. T., Goh, T. L., & Simon, N. (2017). The use of constructed wetlands in sequestrating carbon: an overview. *Nature Environment & Pollution Technology*, *16*(3), 813.

Rubio, D. M., Schoenbaum, E. E., Lee, L. S., Schteingart, D. E., Marantz, P. R., Anderson, K. E., … & Esposito, K. (2010). Defining translational research: implications for training. *Academic Medicine*, *85*(3), 470–475.

Ruiz, B., & Flotats, X. J. W. M. (2014). Citrus essential oils and their influence on the anaerobic digestion process: An overview. *Waste Management, 34*(11), 2063–2079.

Saleh, T. A. (2021). Protocols for synthesis of nanomaterials, polymers, and green materials as adsorbents for water treatment technologies. *Environmental Technology & Innovation, 24,* 101821.Samer, M. (2015). Biological and chemical wastewater treatment processes. *Wastewater Treatment Engineering, 150*(10.5772), 61250.

Satheesh, S., Pugazhendi, A., Al-Mur, B. A., & Balasubramani, R. (2023). Biohydrogen production coupled with wastewater treatment using selected microalgae. *Chemosphere*, *334*, 138932.

Seruga, P., Krzywonos, M., & Wilk, M. (2020). Treatment of by-products generated from anaerobic digestion of municipal solid waste. *Waste and Biomass Valorization*, *11*, 4933–4940.

Sharma, H. B., Vanapalli, K. R., Cheela, V. S., Ranjan, V. P., Jaglan, A. K., Dubey, B., … & Bhattacharya, J. (2020). Challenges, opportunities, and innovations for effective solid waste management during and post COVID-19 pandemic. *Resources, Conservation and Recycling*, *162*, 105052.

Shin, J., Lee, Y. G., Lee, S. H., Kim, S., Ochir, D., Park, Y., … & Chon, K. (2020). Single and competitive adsorptions of micropollutants using pristine and alkali-modified biochars from spent coffee grounds. *Journal of Hazardous Materials*, *400*, 123102.

Shivajirao, P. A. (2012). Treatment of distillery wastewater using membrane technologies. *International Journal of Advanced Engineering Research and Studies*, *1*(3), 275–283.

Silva, J. A. (2023). Wastewater treatment and reuse for sustainable water resources management: a systematic literature review. *Sustainability*, *15*(14), 10940.

Singh, R. N., & Sharma, S. (2012). Development of suitable photobioreactor for algae production: a review. *Renewable and Sustainable Energy Reviews*, *16*(4), 2347–2353.

Sonune, A., & Ghate, R. (2004). Developments in wastewater treatment methods. *Desalination*, *167*, 55–63.

Suzenet, G., Tal, A., & Boymanns, D. (2002). Sustainable water management for the city: technologies for improving domestic water supply. *Built Environment*, *28*(2), 138–151.

Thakur, P., & Mukherjee, G. (2021). Microbes-based nanomaterials for the wastewater treatment and decontamination of water. In M. P. Shah, A. Sarkar, & S. Mandal (eds.), *Wastewater Treatment* (pp. 311–325). Elsevier.

Turovskiy, I. (1998). New techniques for wastewater and sludge treatment in…. *Water Engineering & Management*, *145*(5), 40–44.

Udaiyappan, A. F. M., Hasan, H. A., Takriff, M. S., Abdullah, S. R. S., Maeda, T., Mustapha, N. A., … & Hakimi, N. I. N. M. (2020). Microalgae-bacteria interaction in palm oil mill effluent treatment. *Journal of Water Process Engineering*, *35*, 101203.

Van Lier, J. B. (2008). High-rate anaerobic wastewater treatment: diversifying from end-of-the-pipe treatment to resource-oriented conversion techniques. *Water Science and Technology*, *57*(8), 1137–1148.

Vasconcelos, M. T. S., & Leal, M. F. C. (2001). Seasonal variability in the kinetics of Cu, Pb, Cd and Hg accumulation by macroalgae. *Marine Chemistry*, *74*(1), 65–85.

Vijayaraghavan, K., Ahmad, D., & Aziz, M. E. B. A. (2007). Aerobic treatment of palm oil mill effluent. *Journal of Environmental Management*, *82*(1), 24–31.

Vymazal, J. (2011). Constructed wetlands for wastewater treatment: five decades of experience. *Environmental Science & Technology*, *45*(1), 61–69.

Wang, J., & Chen, C. (2009). Biosorbents for heavy metals removal and their future. *Biotechnology Advances*, *27*(2), 195–226.

Wollmann, F., Dietze, S., Ackermann, J. U., Bley, T., Walther, T., Steingroewer, J., & Krujatz, F. (2019). Microalgae wastewater treatment: biological and technological approaches. *Engineering in Life Sciences*, *19*(12), 860–871.

Yadav, M., Gupta, R., & Sharma, R. K. (2019). Green and sustainable pathways for wastewater purification. In S. Ahuja (ed.), *Advances in Water Purification Techniques* (pp. 355–383). Elsevier.

Zhang, B. Y., Zheng, J. S., & Sharp, R. G. (2010). Phytoremediation in engineered wetlands: mechanisms and applications. *Procedia Environmental Sciences*, *2*, 1315–1325.

2 Bioremediation
An Innovative Approach for Wastewater Treatment

Anurag Tewari and Prahalad Prasad Paroha

2.1 INTRODUCTION

Bioremediation refers to the utilization of biological agents to eliminate hazardous substances from the environment (Benjamin et al. 2019; Dixit et al. 2015), making it the most effective approach for managing contaminated surroundings. With different degrees of success, bioremediation has been used at numerous sites across the world. Both in- and out-of-place biological remediation have both experienced significant scientific advancements (Jiang et al. 2012; Chauhan, González-González, and Iqbal 2022). Biological carbon sequestration, endocrine disruptors, landfill stabilization, mixed waste bio-treatment, and other growing contamination issues are all thought to be solved via bioremediation and natural attenuation. Microbes are particularly useful for cleaning up contaminated environments. In the process of bioremediation, various microbes, including fungi, anaerobes, and aerobes, are active (Das and Chandran 2011).

Inadequate management of industries, food, health care, vehicles, etc., could impede the sustainability of the standard of living, resulting in a detrimental impact on the environment (Dubey and Tewari 2018; Stark et al. 2016). In nature, a variety of fungi, bacteria, and microbes work continuously to break down organic molecules, but when pollution emerges, the question of who will handle the cleanup task arises. Since the general health of the environment and the quality of life are closely intertwined, efforts worldwide have been made to sustain and protect the environment (Lemos et al. 2008).

Researchers in the field of environmental biotechnology possess extensive expertise in dealing with a diverse range of pollutants (Zeeman and Lettinga 1999; Adebayo, Adedoyin, and Kirikkaleli 2021). These encompass chlorinated solvents, petroleum hydrocarbons, polynuclear aromatic hydrocarbons, ketones, tri nitro toluene (TNT), inorganic nitrogen compounds (NO_3, NH_4), as well as elements like Tt, Pu, Np, Cr, U, and other heavy metals (Kılkış et al. 2020). The term "bioremediation" pertains to the application of biological methods for the restoration of a compromised environment, employing various biological components. The method makes use of the catabolic capacity of microorganisms that consume oil in the case of oil spills. With encouraging results, a number of researchers have detailed several applications of microorganisms in the bioremediation of wastewater (Seddighi et al. 2023; Buonomano, Barone, and Forzano 2023).

DOI: 10.1201/9781003459286-2

Successful bioremediation in the field requires the integration of appropriate engineering designs to offer adequate growth settings with proper usage of natural or engineered microbial capabilities. But there is a gap between field applications and advances in laboratory research (Batstone and Virdis 2014). The existing gap can be attributed to two primary factors: the challenge in accurately predicting degradation rates and outcomes of pollutants, and the effectiveness of microorganisms for environmental cleanup is restricted by the absence of specialized field research centers focused on studying and demonstrating bioremediation technology, as laboratory experiments have shown (Batstone et al. 2015). Encouragingly, advancements have been achieved in overcoming barriers that previously hindered the effective real-world implementation of bioremediation approaches (Meloni, Martino, and Palma 2020).

It is necessary to clean the environment and put effective decontamination procedures into place since the rise in polluted sites poses a serious threat to both human and environmental health. Decontamination's primary objectives should be the restoration of soil fertility and health, the purification of groundwater, the reduction of negative impacts on the well-being of humans and animals, the reuse of sewerage (especially in locations with major water shortages), and the production of clean air. As a result, there is growing interest in developing effective remediation methods that can partially or totally restore damaged locations (McDonald and Schrattenholzer 2001; Du et al. 2023).

Various primary biological methods can be employed for the restoration of soil and groundwater, encompassing processes such as composting, land farming, bioreactors, bioremediation, and phytoremediation. Irrespective of the chosen approach, the purification of polluted sites can be executed through in-situ methods, involving the treatment of soils and water at their original location, or through ex-situ techniques, which entail excavation, relocation, and subsequent treatment at a different site (Malka et al. 2023; Jain et al. 2022).

In-situ treatments frequently cost less money and need less physical intervention than ex-situ treatments, which are more expensive and affect the environment more. Nowadays, bioremediation and phytoremediation seem to be attractive technologies since they rely on the usage of living things, including microbes, plants, and their enzyme system (Singh, Carliell-Marquet, and Kansal 2012).

Enzymatic mechanisms are the most efficient for enhancing bioremediation when the surrounding environment is conducive to optimization. For ex-situ bioremediation, creating these conditions is simple, but in-situ bioremediation may be more challenging. By accelerating natural biodegradation processes, bioremediation offers a way to remove pollution from the environment (Gikas 2017). Consequently, through the examination of bacterial populations and their reactions to their environment and pollutants, the advancement of our comprehension of microbial genetics to enhance their ability to degrade pollutants, the application of cost-effective bioremediation techniques in real-life situations, and the identification of locations for future research—these opportunities have the potential to significantly enhance water treatment in everyday life (Wang, Cui, and Peng 2019).

Recent and rapid progress in uncovering various pathways employed by microorganisms for the degradation of toxins has illuminated their underlying mechanisms.

These pathways not only contribute to a deeper comprehension but also pave the way for developing sustainable bioremediation strategies to address these contaminants (Hess et al. 1997; Werkneh and Gebru 2023). However, a foundational grasp of molecular biology, regulatory principles, and enzyme structures remains essential. Moreover, substantial efforts are still required to firmly establish a comprehensive insight into the molecular foundations of catabolic sequences. This endeavor involves curtailing enzyme inactivation under conditions of high concentrations of xenobiotics, augmenting the bioavailability of compounds in natural ecosystems, and refining the functional efficacy of chosen enzymes within specific physicochemical parameters (Rajeshwari et al. 2000).

As the proposed study focuses on understanding potential inherent difficulties, the role of enzymes, conventional treatment, innovative methods for bioremediation, as well as new pathways to reduce time and cost both for the treatment of wastewater and develop a strategy or plan in such a way that they will not only help in the treatment of wastewater but also produce valuable energy-driven products (methane, organic manure, etc.), which will be helpful to meet our future energy needs.

2.2 IMPORTANT ENZYMES FOR BIOREMEDIATION

Petroleum hydrocarbons and industrial chemicals are a prevalent problem in today's world, contaminating both soil and water. Bioremediation technologies offer a secure and cost-effective solution by utilizing bacterial activity and hydrolytic enzymes to degrade these toxins (San Martín et al. 2023).

Oil spills and pesticides including carbamates and organophosphates can both be effectively biodegraded using this method. DDT and heptachlor are two examples of organochlorine pesticides that are stable in well aerated soil but quickly degrade in anaerobic environments. Additionally, condensations and alcoholysis are two related reactions that are catalyzed by hydrolases (Ansari et al. 2017). The key benefits of this enzyme class include its rapid accessibility, lack of substrate stereoselectivity, and capacity to withstand the presence of water-miscible solvents. Groups 3 and 4 of the enzyme class may be sub-classified further into hydrolases based on the type of bond being broken down. (Nowak, Enderle, and Varbanov 2015). According to Table 2.1, there are several potential applications for exogenous hydrolyzing enzymes in the chemical, biomedical, food, and feed sectors. These enzymes include amylases, proteases, lipases, DNases, pullunases, and xylanases (Nabernegg et al. 2017).

2.2.1 Microbial Lipases

Lipids produced by a wide range of microbes, animals, and plants are broken down by lipase. Recent studies have demonstrated a close relationship between lipase and the organic contaminants in soil. Total hydrocarbon from contaminated soil was drastically reduced as a result of lipase activity. Research in this field is probably going to advance our understanding of the bioremediation of oil spills (Albuquerque, Torres, and Reis 2010; Astals et al. 2014). Bacteria, plants, actinomycetes, and mammalian cells have all been used to extract lipases. Due to their extensive use in various industries, microbial lipases are more adaptable than the others. Numerous processes,

TABLE 2.1
Industrially Important Enzymes and Their Applications

S. N.	Enzyme	Substrate	Reaction	Applications	References
1.	Oxidoreductase				
i.	Oxygenase				
ii.	Monooxygenase	Lipids, alkaline substances, drugs, and volatile compounds	Use of the substrate as a reducing agent and integration of an oxygen atom	Bioremediation, industrial chemical science, peptide technology, and other related fields	Puchongkawarin et al. (2015)
iii.	Dioxygenase	Cyclic molecules	Production of an unsaturated derivative follows the addition of a pair of oxygen atoms to the base	Medicinal industry, biological remediation, chemical manufacturing, and so on	Catherine, Guwy, and Massanet-Nicolau (2022) and Longis et al. (2023)
iv.	Laccase	Aminophenols, polyamines, andaryldiamines	Oxidation, decarboxylation, and demethylation of a material	Agriculture, textiles, newspaper, nanotechnologies, biological remediation, beauty products	Ostle and Holt (1982)
2.	Peroxidase				
i.	Lignin peroxidase	Organic and hydrogenated compounds	Degradation of substances in the presence of H_2O_2 and methanol	Agriculture, newspaper, industrial chemistry, biological remediation, etc.	Do et al. (2022)
ii.	Manganese peroxidase	Aromatic and lignin compounds	Manganese and hydrogen-peroxide-induced catalysis resulted in to produce manganese chelate molecules	Agriculture, newspaper, nanotechnologies, industrial chemistry, etc.	Kuntke et al. (2012)
iii.	Versatile peroxidase	Aromatic oxybenzene	Catalyzed flow of electrons from base and formation of organic reaction species	Biological catalyst and biological remediation	Hülsen, Batstone, and Keller (2014)
3.	Hydrolase				
i.	Lipase	Aromatic contaminants and spillage	Conversion of fatty acids into 1,2,3-trioxypropane	Spillage reduction, soap, tissues, cardboards, etc.	Lettinga, Rebac, and Zeeman (2001)
ii.	Cellulase	Cellular material	Conversion of material into simple sugars	Agriculture, nanotechnologies, industrial chemistry, etc.	Astals et al. (2013)
iii.	Protease	Amino acids	Hydrolysis of tubulin in the presence of water	Fabrics, laundry, soap, biological remediation, etc.	Chitapornpan et al. (2012)

including hydrolysis, interesterification, esterification, alcoholysis, and aminolysis, can be catalyzed by these enzymes (Verstraete, Van de Caveye, and Diamantis 2009).

The hydrolysis of triacylglycerols into glycerol and free fatty acids is catalyzed by the universally present enzymes known as lipases (Bagheri and Wetterlund 2023). Lipolytic mechanisms transpire at the interface between lipids and water, where equilibration between monomeric, micellar, and emulsion forms of lipolytic substrates is frequently achieved. Based on benchmarks such as (i) heightened enzyme activity upon the formation of triglyceride emulsions and (ii) the presence of a protein loop (lid) encasing the active site, lipases have been categorized into two distinct groupings (Hülsen, Hsieh, and Batstone 2019; Madigan 1999).

2.2.2 Microbial Cellulases

Cellulases have been the focus of extensive research because they have the potential to transform waste cellulosic material into foods to meet the needs of a growing population. Cellulases that are extracellular, connected with the cell membrane, and attached to cells are all produced by some species. There is evidence that some bacteria and fungi express extracellular cellulases, hemicellulases, and pectinases at a constitutively low level. Cellulases are typically a blend of several enzymes (Simona et al. 2022; Sepúlveda-Muñoz, de Godos, and Muñoz 2023).

The hydrolysis process involves three primary categories of cellulases. First, there's cellulase (EG or endo-1,4-β-D-glucanohydrolase), which targets areas with lower crystallinity within the cellulose fiber, thus generating unbound chain ends. Second, exoglucanase, also known as cellobiohydrolase (CBH or 1,4-β-D-glucan cellobiohydrolase), plays a role in further cellulose degradation by removing the cellulose molecules from these released sequence ends (Molina Grima et al. 2003; Alloul et al. 2021). Lastly, there's β-glucosidase, responsible for hydrolyzing cellobiose into glucose units. There are some auxiliary enzymes present in addition to the primary enzymes (Wear et al. 2021).

Cellulases break down cellulose during enzymatic hydrolysis, converting it into reducing sugars that yeasts or bacteria can ferment to produce ethanol. Crystalline cellulose can be broken down by cellulase enzymes to glucose. Since the early 1990s, cellulases have been employed in detergent production. When cleaning and employing cotton-based materials, cellulose microfibrils are created and are removed by cellulases (Lettinga et al. 1999). This is also seen in the textile sector as the material and color brightening processes. *Trichoderma* and *Humicola* fungi produce neutral and acidic cellulases, while *Bacillus* strains create alkaline cellulases. Cellulases have been used in the paper and pulp industries to remove ink during paper recycling (Khatipov et al. 1998).

2.2.3 Microbiological Proteases

Proteases play a crucial role in hydrolyzing protein-based substances, which emerge in the environment due to factors such as animal mortality, molting, and industrial residues from sectors like poultry, fishing, and leather production (Zeeman et al. 2008). These enzymes belong to a class that cleaves peptide bonds within aqueous contexts and also functions within non-aqueous environments. Proteases are utilized in a wide range of businesses, including those in the alimentary, leather, soap, and pharmacological fields. Based on how they break down peptide chains, proteases are divided into enzymes such as exopeptidases. Endopeptidases, including serine, cysteine, aspartic, and metallopeptidases, are further subdivided according to their active site locations (Malik et al. 2020).

Exopeptidases specifically target the amino or carboxyl terminals of chains. Aminopeptidases and carboxypeptidases are examples of exopeptidases responsible for modifying free amino and carboxyl ends, respectively. In contrast, endopeptidases act on internal regions of peptide chains (Wong 2016). The presence of free amino and carboxyl terminals can impede enzyme activity. Proteases have a long history of application, being utilized in processes like cheese production and detergent formulation, where their well-understood reaction mechanisms find utility (Liu et al. 2021).

Many of wastewater's detrimental effects on the environment and public health can be reduced through enzymatic treatment. *Salmonella*, *Vibrio cholerae*, *Salmonella typhi*, and *Giradia* are just a few of the fecal-borne pathogens found in wastewater that can lead to severe gastrointestinal (GI) diseases like cholera, typhoid, and other water-borne diseases (Anand et al. 2021). It has been shown that enzymes can eliminate these infections by weakening their cell membranes. The peptidoglycan layer, which is mostly composed of amino acids, various sugars, and teichoic acid, helps to make the cell wall stiff. Gram-positive cells have a thick, rigid peptidoglycan layer that is encircled by the lipopolysaccharide membrane (Mofijur et al. 2023).

2.3 BIOREMEDIATION MECHANISM

The usage of enzyme-producing proteins could be a viable approach for solving the majority of environmental issues. A number of advantageous traits exist in enzymes. They are the primary agents of all the biota's alterations (Galdames et al. 2021). They are catalysts that have either a restricted (chemo regularity and stereoselectivity) or an extensive preciseness, and thus, they can be applied to a wide range of different compounds in combinations. They may generate significant structural and toxicological changes in pollutants, moreover, the entire transformation into secure end items that are organic. They might carry out procedures for which effective chemical transformations have not yet been developed. Additionally, enzymes may also be more effective than traditional and biological treatments (Guerra et al. 2018).

Enzymes may act intracellular, which means inside or outside the cells from which they originated; extracellularly; or free, meaning the catalyst will be available in liquids and the reaction product will be uniform or stabilized. This indicates that the catalysis will be heterogeneous and coupled to a solid base through multiple linkages. The most efficient way to remove environmental toxins is by using enzymatic activities to improve bioremediation (Baxter-Plant, Mikheenko, and Macaskie 2003; Planamente and Frank 2019).

A number of phenols, biphenols, anilines, benzidines, and other related aromatic compounds can be oxidized by enzymes such as peroxidases and laccases in the presence of hydrogen peroxide and oxygen (Thuku et al. 2009), respectively as shown in Table 2.2. As a result of the isolated enzyme's high substrate specificity

TABLE 2.2
Role of Major Enzymes and Their Actions

S. N.	Enzymes	Sources	Substrate	Applications	References
1	Peroxidase (horseradish peroxidase - HRP)	Plant source, Fungi, soyabean, radish beetroot, and peanut	Hydroxylated polycholoro biphenyls, lignin, and other compounds	Food, paper, packing, textile, and bioremediation processes	Kumar and Chandra (2020)
2	Manganese peroxidase				
3	Lignin peroxidase				
4	Laccase	Lignin-degrading fungi species (*Trameters versicolor*, *T villosa*)	Phenolic compounds, lignin and amino group compounds	Food, paper, packing, pharmaceutical industries, and bioremediation processes	Zhou, Zhang, and Cai (2021) and Janusz et al. (2020)
5	Bacterial oxidoreductases	Sludge methogenesis bacteria	Acid orange 6, acid orange 7, methyl orange, and methyl red	Packing, textile, pharmaceutical industries, and bioremediation processes	Carvalho et al. (2014)
6	Azoreductase	Green algae	Tartrazine and ponceau	Decolorization of azo dyes	Singh, Singh, and Singh (2015)
7	Lipase	Vegetation sources (avocado, walnut, pinenut, coconut, lupin, and lentils)	Aromatic pollutants such as spillage	Bioremediation, industrial chemical science, peptide technology, and other related fields	Nimkande and Bafana (2022)
8	Cellulase	Fungi such as *Trichoderma viride*, lígnomm, *C. pruinosum*, and *Penicillium*	Cellulosic substance	Paper and pulp, textile, cement industries, bio-remediation, etc.	Kuhad, Gupta, and Singh (2011)
9	Protease	*Pseudomonas aeruginosa*, *Vibrio cholerae*, & *Xanthomonas maltophilia*	Proteins	Laundry, biocatalyst, bioremediation, etc.	Nolasco-Soria, Nolasco-Alzaga, and Gisbert (2020)
10	Versatile peroxidase	Fungus *Pleurotus eryngii*	Methoxybenzenes and phenolic aromatics	Industrial biocatalyst, bioremediation, etc.	Bhandari et al. (2021)

and ability to have its potency more precisely controlled, it is chosen for this application over the intact organism that contains the enzyme (Shraddha et al. 2011). Since it is not dependent on microbial growth, the concentration of the isolated enzyme is simpler to manage. The oxidation of substrates to diffusible free radical intermediates, known to take part in a range of non-enzymatic processes such disproportionation, polymerization, and electron transfer, is a distinctive characteristic of peroxidase-catalyzed reactions (Krainer et al. 2016). The one-electron oxidation of aromatic substrates (AH_2) for peroxidases (Trujillo, Alvarez, and Radi 2016) is as follows:

$$E + H_2O_2 \Rightarrow Ei + H_2O$$

$$E + AH_2 \Rightarrow Eii + AH^{\bullet}$$

$$Eii + AH_2 \Rightarrow E + AH^{\bullet} + H_2O$$

When natural enzyme (E) combines with peroxide, "Compound I" (Ei) is created. This compound is then oxidized by two electrons above native enzyme and contains peroxidic oxygen. In order to complete its one-electron oxidation, compound (I) accepts an aromatic molecule in its active site (Blankenhorn 1976; Perez-Benito and Arias 1998). While the enzyme is reduced to "compound II" (Eii), the oxidized aromatic is liberated from the catalytic site as a free radical (AH). Completing the cycle is compound II's oxidation of a second aromatic molecule, which results in the release of a new free radical and the enzyme's own return to its initial state (Hyster 2020). The following describes the overall enzymatic reaction:

$$AH_2 + H_2O \Rightarrow 2AH^{\bullet} + 2H_2O$$

There is a need to find and create sustainable bioremediation solutions for polluting substances as a result of quick advancements in the numerous pathways used by bacteria to degrade contaminants. The molecular biology, regulatory mechanisms, and enzyme architectures, however, still need to be well understood (Held and Sadowski 2016; Taura, Tanaya, and Sirikantaramas 2019). In order to prevent the deactivation of enzymes at high threshold concentrations, to fully comprehend the molecular basis for catabolic sequences, there is still ample opportunity for study to be done (Bedendi et al. 2022).

Currently, researchers are focusing at extracellular enzymes' potential to bioremediate a wide range of substances and contaminants. Because bioremediation is still a relatively young area, the majority of the work done thus far has only involved bench-scale studies. Few field investigations have been conducted yet in an effort to comprehend the mechanism of the enzymatic process. However, utilizing the results of earlier research on enzymatic processes can significantly improve bioremediation (Robinson 2015; Bharagava, Saxena, and Mulla 2020).

2.4 DELIVERY SYSTEM FOR BIOREMEDIATION

The cells or tissues that produce the enzyme can be directly added to the effluent as one way to distribute the enzyme to the target effluent. This method of enzyme distribution is used when the target pollutants are co-metabolized by appropriately adapted strains of microbes or when the tissue that produces the enzyme is added directly to the effluent (Atelge et al. 2020). Utilizing plant tissues or microbial cells for wastewater bioremediation are other options. Microbial cells can be used to grow a particular bacterial species and then insert them directly into the wastewater (Narayanan and Narayan 2019). The following principles serve as the foundation for the delivery system's process:

2.4.1 Bio-Augmentation

Bio-augmentation involves the remediation of contaminated soil or water by introducing a naturally occurring collection of microbial strains or a genetically modified version. Restarting activated sludge bioreactors is frequently done in municipal wastewater treatment facilities (Lim et al. 2022). The majority of commercial cultures use a consortium of microbial cultures based on research that contains all necessary microorganisms (Choerudin et al. 2021). The use of bio-augmentation is used at locations where soil and groundwater are contaminated with chlorinated ethenes, such as tetrachloroethylene and trichloroethylene, in order to ensure that the in-situ microorganisms can entirely breakdown these substances into non-toxic compounds ethylene and chloride. It is difficult to keep track of this system (Ambaye et al. 2022).

2.4.2 Bioreactor

To treat polluted soil and water extracted from a contaminated plume in an off-site manner, slurry reactors or aqueous reactors are employed. The process of remediating contaminated water or solid materials (like soil, sediment, or sludge) within a controlled system constitutes reactor-based bioremediation. The term "slurry bioreactor" pertains to a containment vessel and associated equipment used to establish a three-phase mixing environment (solid, liquid, and gas) that accelerates the bioremediation process for both soil-bound and water-soluble pollutants (Habe et al. 2020). This involves creating water slurry of the contaminated soil and biomass. The controlled conditions provided by the reactor environment lead to enhanced biodegradation rates and outcomes compared to in-situ or solid-phase systems, owing to improved manageability, control, and predictability. Despite their advantages, reactor systems also have limitations. Prior to entering a bioreactor, contaminated soil typically requires pretreatment, which could involve processes such as excavation. Alternatively, the contaminant can be physically extracted (e.g., vacuum extraction) or leached from the soil (Lei and Ferdous 2016).

2.4.3 Anaerobic Digestion

An alternative molecule to oxygen serves as the ultimate electron acceptor during anaerobic digestion, which converts organic materials into biogas (CH_4, CO_2). In contrast, some bacteria (nitrate reducers) transfer electrons to nitrate (NO_3) and reduce it to nitrite (NO_2), nitrous oxide (NO), or nitrogen gas (N_2). Sulfate-reducing bacteria, for instance, transport electrons to sulfate (SO_{42-}), reducing it to H_2S. Mn (IV) and Fe (III) are other potential electron acceptors (Uddin and Wright 2022). Since it maximizes cost effectiveness, is environmentally friendly, reduces the amount of final sludge disposal, and has the ability to produce a net energy gain in the form of methane gas, anaerobic treatment process has emerged as unique and the most advantageous stabilization technique among the many treatment technologies currently available. Larger deterioration levels, however, need more prolonged retention durations and/or additional treatment methods, which are not yet possible with anaerobic treatment technology (Sillero et al. 2022). This means that anaerobic treatment technology can only partially treat waste in a conventional wastewater treatment system. Figure 2.1 illustrates the four distinct phases that are generally acknowledged

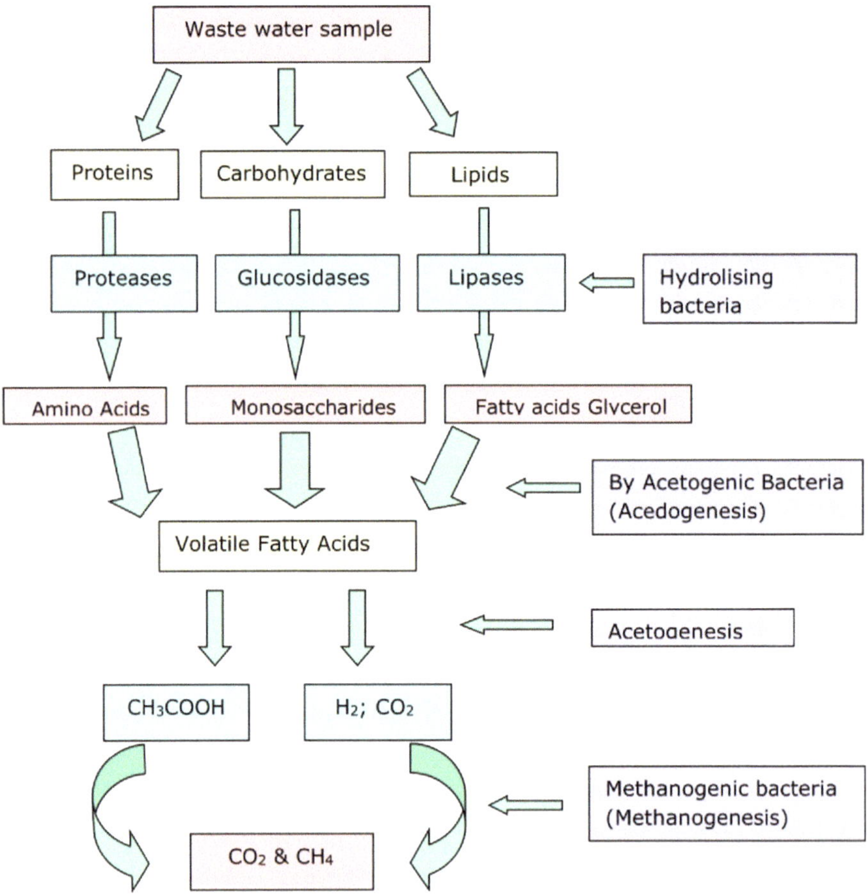

FIGURE 2.1 Anaerobic digestion of complex organic matter present in wastewater.

to exist: By using water (hydrolysis) and hydrolase enzymes (glucosidases, lipases, proteases, sulfatases, phosphatases), complex organic matter is broken down into simple soluble organic molecules; acidogenesis is the chemical breakdown of these single monomeric unit molecules (monosaccharides, amino acids, fatty acids, and glycerol) into volatile fatty acids; (c) acetogenic which on fermentation converts into acetate (El-Zawawy et al. 2011).

2.5 INNOVATIVE BIOREMEDIATION APPROACH FOR WASTEWATER TREATMENT

Biological agents such as microbes, fungi, protists, or their enzymes are used in bioremediation to transform environmental contaminants into less dangerous forms. This method has several benefits, including low cost, high efficiency, reduced production of chemical and biological sludge, selectivity for particular metals, minimum extra nutritional requirements, and the possibility of metal recovery (Bilal, Nguyen, and Iqbal 2021; Rizwan et al. 2014).

Nanomaterials, due to their unique properties, enhance the effectiveness of bioremediation. Their nanoscale size provides a larger surface area for interactions with pollutants, increasing reactivity. Quantum effects and surface plasmon resonance lower the energy requirements for chemical reactions and aid in the detection of toxic substances. Nanoparticles of various shapes and sizes, including metallic and non-metallic types offer versatility in environmental cleanup (Chaudhary et al. 2023). For instance, oxide-coated iron nanoparticles can form complexes with contaminants, facilitating their degradation. Additionally, nanomaterials such as nanoparticles, polymeric nanomaterials, nanocomposites, nanocrystals, and carbon nanotubes, as given in Table 2.3, find their applications in environmental solutions, including sorbents, high-flux membranes, depth filters, antimicrobial agents, photoelectrocatalytic reactions, and environmental sensors (Mauter and Elimelech 2008). These materials exhibit remarkable sorption capabilities, making them valuable for removing sediment-based organic contaminants like polycyclic aromatic hydrocarbons (PAHs) from soil and groundwater. Engineered polymeric nanoparticles enhance the bioavailability of contaminants, allowing for their more efficient removal and degradation (Bhattu and Singh 2023; Bolade, Williams, and Benson 2020).

Furthermore, dendrimers, highly branched macromolecules, show promise in water treatment and pollutant monitoring applications due to their ability to form composite materials with nanomaterials. These composites increase catalytic activity and surface area, making them suitable for cleaning up environmental pollution (Sharma et al. 2022).

The integration of nanomaterials in bioremediation and environmental science offers innovative solutions to address pollution and contamination challenges. These materials, with their unique properties and versatility, hold the potential to revolutionize the field, enabling more efficient, cost-effective, and sustainable environmental cleanup strategies (Nastasi et al. 2023).

TABLE 2.3
Nanomaterials and Their Methods for Synthesis

S. N.	Types of Nanomaterials	Methods for Synthesis	Type of Methods	Examples	References
1	Nanoparticles (NPs)	Pulse laser ablation, pulsed wire discharge method, chemical vapor deposition, laser pyrolysis, photochemical, electrochemical, bio-chemical, polyol method, thermochemical, & ionized cluster beam deposition	Physical & chemical	ZnO, ZnS, CdS, CdSe, CdTe, FeNi, CoNi, Ag, etc.	Tungittiplakorn, Cohen, and Lion (2005)
2	Carbon nanotubes (CNTs)	Arc discharge, laser ablation, pulse laser deposition (PLD), chemical vapor deposition (CVD)	Physical & chemical	CNTs [single-walled carbon nanotube (SWCNT) & multi-walled carbon nanotube (MWCNT)], fullerenes, etc.	Kandah and Meunier (2007)
3	Metal oxide nanoparticles (MoNPs)	Hydrothermal, solvothermal, thermal decomposition, electrochemical deposition, sonochemical synthesis, microemulsion & sol gel	Physical & chemical	$BaSO_4$, Fe_3O_4, MgO, ZnO, etc.	Schrick et al. (2004)
4	Polymeric nanomaterials (PNMs)	High-energy ball milling, electrochemical polymerization, nanoprecipitation technique, solvent evaporation technique & layered filled	Physical & chemical	Micelles; polymersomes; nanofibers; nanogels; nanowires of polypyrrole, polyaniline, poly(ethylene oxide), polythiophene, etc.	Borah et al. (2022)
5	Nanocomposites (NCs)	Ultrasound sonication, sol gel & colloid synthesis	Physical & chemical	Polymers/metal oxide; matrix/rein for cement, ceramic/matrix, & polymer nanocomposites	Gong et al. (2018)
6	Biological nanomaterials (BNMs)	*In vivo* or *in vitro* & plants extract	Biological	Lipoproteins, DNA or RNA, peptides, micro-organism, bionanomaterials, polyglycolic acid (PGA), silk-based biomaterials	Ghosh et al. (1999)

2.6 INTEGRATION OF NANOMATERIALS IN BIOREMEDIATION PROCESS

Nanomaterials have the capacity to break down harmful compounds, dyes, heavy metal ions, and pesticides that are present in contaminated wastewater. Wastewater treatment employs enzymes and microbial cultures consisting of bacteria and yeast, including *Lactobacillus, Leuconostoc, Lactococcus*, *Bacillus*, azoreductase, laccase, bacterial oxidoreductase, lipase, cellulose protease, and lignin peroxidase. These are employed to eliminate hazardous compounds found in wastewater (Sepúlveda-Muñoz et al. 2022). These bacteria have shown exceptional tolerance to various stresses while being immobilized in proteins and a heteropolysaccharide matrix. Numerous research have also reported on the use of various nanomaterials to remove organic matter from wastewater, including chemical oxygen demand (COD), nitrates, and phosphates (Hairom et al. 2021). The recyclability of the applied nanoparticles, microbial consortia, and immobilized microbial strains allows for their subsequent usage. Some studies claim that, in contrast to an enzymatic approach, the natural consortium/biological nanoparticles quickly destroyed the harmful compounds contained in wastewater. Because of their larger surface area, the immobilized microbial cells also performed better when it came to the elimination of harmful substances (He et al. 2014). Reusing natural consortia, immobilized cells, and nanoparticles can help with resource recovery and sustainable wastewater treatment as shown in Figure 2.2.

Nanomaterials have been employed to speed up the process of removing hazardous contaminants from wastewater in the recharge unit, which is how the novel bioremediation process for wastewater has been envisioned with their integration.

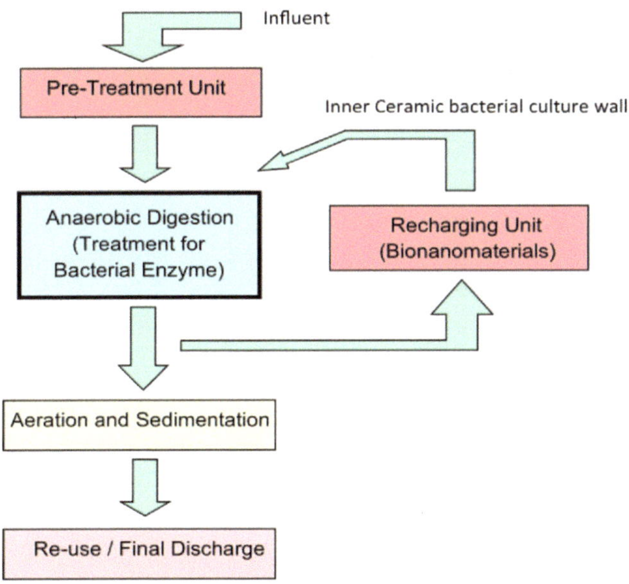

FIGURE 2.2 Innovative setup for bioremediation process.

TABLE 2.4
Bioremediation Process and Integration of Nanomaterials

S. N.	Steps Involved in the Process	Major Work Requirement
1	Pretreatment unit	Analysis and pH adjustment of wastewater, neutralization of wastewater, measurement of COD, total suspended solids (TSS), heavy metals, and identification of the type of culture needed for anaerobic treatment
2	Anaerobic treatment unit	Bacterial culture is used, and the bacteria grow on ceramic membranes to purify wastewater. The recharging unit controls and maintains bacterial growth. Identification and elimination of intermediates in the process of bacterial metabolism. Ceramic membrane development on the chamber's interior walls
3	Recharging unit (bioreactor)	Recharging unit storage of bacterial culture to provide adequate bacterial culture concentration for anaerobic treatment unit
4	Aeration and sedimentation	Storing effluent in open ponds for 6–8 hours to allow for the settling of heavy solids and sludge
5	Reuse/final discharge	Before discharge into a river or open space or for reuse, settling effluent is analyzed. Identifying a possible application for sludge in lucrative energy products

In the innovative bioremediation process that has been proposed, nanomaterials and enzymes have been integrated, and in the recharging unit, the use of nanomaterials has been used to speed up the process of removing toxic pollutants in the delivery process of bioremediation of wastewater (Bhatt et al. 2022), as described in Table 2.4.

There are numerous physical and chemical techniques for cleaning up pollutants, including oxidizing agents, electrochemical processes, pollutant adsorption, ion exchange, and membrane filtration. Traditional procedures could handle the high levels of contaminants as given in Table 2.5, but they were insufficient to bring the contamination down to acceptable levels. The high expense, non-specificity, and potential development of secondary contamination are only a few drawbacks of traditional methods for cleaning up pollutants; as a result, interest has grown in bioremediation, an environmentally benign and biological technique (Ampese et al. 2022).

In order to reduce pollution at the source and the threat to the environment and public health, bioremediation is described as procedures and products that are practicable and cost effective (Muter 2023). It mostly uses intracellular accumulation or enzymatic transformation to break down and detoxify pollutants. Pollutant features including chemical structure; hydrophobicity; polarity; temperature; pH; and redox state; as well as soil factors like aggregate, thickness, dissolved organic matter, and pollutant aging affect biological degradation and the availability of pollutants (Banerjee, Mukherjee, and Mukherjee 2021).

TABLE 2.5
Comparison between Conventional and Innovative Bioremediation Treatment Methods

S. N.	Conventional Bioremediation Method	Innovative Bioremediation Method
1	Basic operation cost is very high	Cost of the process will be lower than the previously developed methods/techniques
2	May require more extensive monitoring	No use of extensive monitoring, effective over wide range of contaminants
3	Lack of (hydraulic) control	Effective at low hydraulic retention time
4	May be difficult to control	Concentration of bacterial culture maintained by bioreactor
5	Requires more retention time	Retention of bacterial culture in anaerobic unit controlled through ceramic membrane
6	Commonly specific enzyme is used for specific pollutants	Bacterial crude culture is used; operation over wide temperature, pH, and salinity range
7	Evolving process, tough to forecast performance	Reduction of solid waste production, effective & simple process as compared to other methods/techniques
8	Unable to operate under mild conditions	Operate under mild, less corrosive conditions than chemical oxidation method
9	Costly and high maintenance cost	Cost effectiveness in the production of enzymes & treatment
10	Prepared culture is used	Direct use of crude enzyme in the process

2.7 CONCLUSION

Bioremediation serves as a robust and environmentally responsible approach for mitigating environmental contamination. It harnesses the potential of biological agents and enzymes, providing a cost-effective means to convert hazardous substances into safer forms, effectively targeting a wide range of pollutants. Various techniques, such as introducing enzyme-producing cells or tissues directly into the effluent and utilizing plant tissues or microbial cells, operate based on principles like bio-augmentation, bioreactor-based treatment, and anaerobic digestion. Nevertheless, these methods require careful planning and implementation in wastewater treatment processes. Microorganisms, functioning in both natural and controlled environments, play a pivotal role in this process. Furthermore, the emergent field of nanomaterials has enhanced the effectiveness and adaptability of bioremediation. These nanomaterials, with their distinct properties, improve pollutant reactivity and detection, providing innovative, sustainable, and highly efficient solutions to address pollution challenges. Integrating nanomaterials into bioremediation not only accelerates pollutant removal but also facilitates resource recovery, promising to protect the environment and public health while minimizing overall environmental impact. Ongoing research into enzyme mechanisms, molecular biology, nanomaterials, fungi, and algae holds the potential to deepen our understanding of underlying mechanisms, particularly in terms of cumulative effects. Additionally, exploring both indigenous and mixed microbial cultures could offer opportunities for large-scale remediation of toxic pollutants. Without a doubt, these advancements in bioremediation will propel the field to new heights, paving the way for a cleaner and more sustainable future.

CONFLICT OF INTEREST

Authors declare no conflict of interest.

DATA AVAILABILITY STATEMENT

The authors declare that the data supporting the findings of this study are not applicable in the paper nor its supplementary information files. Should any raw data files be needed in another format, they are available from the corresponding author upon reasonable request. Source data are provided with this paper.

ACKNOWLEDGMENT

Authors are thankful to the institute for providing support in the study.

REFERENCES

Adebayo, Tomiwa Sunday, Festus Fatai Adedoyin, and Dervis Kirikkaleli. 2021. "Toward a Sustainable Environment: Nexus between Consumption-Based Carbon Emissions, Economic Growth, Renewable Energy and Technological Innovation in Brazil." *Environmental Science and Pollution Research* 28 (37). Springer Science and Business Media Deutschland GmbH: 52272–82. doi:10.1007/S11356-021-14425-0

Albuquerque, M. G. E., C. A. V. Torres, and M. A. M. Reis. 2010. "Polyhydroxyalkanoate (PHA) Production by a Mixed Microbial Culture Using Sugar Molasses: Effect of the Influent Substrate Concentration on Culture Selection." *Water Research* 44 (11). Elsevier Ltd: 3419–33. doi:10.1016/j.watres.2010.03.021

Alloul, Abbas, Maarten Muys, Nick Hertoghs, Frederiek Maarten Kerckhof, and Siegfried E. Vlaeminck. 2021. "Cocultivating Aerobic Heterotrophs and Purple Bacteria for Microbial Protein in Sequential Photo- and Chemotrophic Reactors." *Bioresource Technology* 319 (January). Elsevier Ltd. doi:10.1016/j.biortech.2020.124192

Ambaye, Teklit Gebregiorgis, Alif Chebbi, Francesca Formicola, Shiv Prasad, Franco Hernan Gomez, Andrea Franzetti, and Mentore Vaccari. 2022. "Remediation of Soil Polluted with Petroleum Hydrocarbons and Its Reuse for Agriculture: Recent Progress, Challenges, and Perspectives." *Chemosphere*. doi:10.1016/j.chemosphere.2022.133572

Ampese, Larissa Castro, William Gustavo Sganzerla, Henrique Di Domenico Ziero, Ackmez Mudhoo, Gilberto Martins, and Tânia Forster-Carneiro. 2022. "Research Progress, Trends, and Updates on Anaerobic Digestion Technology: A Bibliometric Analysis." *Journal of Cleaner Production*. doi:10.1016/j.jclepro.2021.130004

Anand, Uttpal, Bhaskar Reddy, Vipin Kumar Singh, Amit Kishore Singh, Kavindra Kumar Kesari, Pooja Tripathi, Pradeep Kumar, Vijay Tripathi, and Jesus Simal-Gandara. 2021. "Potential Environmental and Human Health Risks Caused by Antibiotic-Resistant

Bacteria (ARB), Antibiotic Resistance Genes (ARGs) and Emerging Contaminants (ECs) from Municipal Solid Waste (MSW) Landfill." *Antibiotics* 2021 10 (4). Multidisciplinary Digital Publishing Institute: 374. doi:10.3390/ANTIBIOTICS10040374

Ansari, Ashley J., Faisal I. Hai, William E. Price, Jörg E. Drewes, and Long D. Nghiem. 2017. "Forward Osmosis as a Platform for Resource Recovery from Municipal Wastewater - A Critical Assessment of the Literature." *Journal of Membrane Science* 529. Elsevier B.V.: 195–206. doi:10.1016/j.memsci.2017.01.054

Astals, S., D. J. Batstone, J. Mata-Alvarez, and P. D. Jensen. 2014. "Identification of Synergistic Impacts during Anaerobic Co-Digestion of Organic Wastes." *Bioresource Technology* 169. Elsevier Ltd: 421–27. doi:10.1016/j.biortech.2014.07.024

Astals, S., M. Esteban-Gutiérrez, T. Fernández-Arévalo, E. Aymerich, J. L. García-Heras, and J. Mata-Alvarez. 2013. "Anaerobic Digestion of Seven Different Sewage Sludges: A Biodegradability and Modelling Study." *Water Research* 47 (16). Elsevier Ltd: 6033–43. doi:10.1016/j.watres.2013.07.019

Atelge, M. R., A. E. Atabani, J. Rajesh Banu, David Krisa, M. Kaya, Cigdem Eskicioglu, Gopalakrishnan Kumar, et al. 2020. "A Critical Review of Pretreatment Technologies to Enhance Anaerobic Digestion and Energy Recovery." *Fuel* 270. doi:10.1016/j.fuel.2020.117494

Bagheri, Marzieh, and Elisabeth Wetterlund. 2023. "Introducing Hydrothermal Carbonization to Sewage Sludge Treatment Systems—A Way of Improving Energy Recovery and Economic Performance?" *Waste Management* 170 (October): 131–43. doi:10.1016/J.WASMAN.2023.08.006

Banerjee, Monojit, Sadhin Mukherjee, and Suprabhat Mukherjee. 2021. "Facets of Nanoparticle-Microbe Interactions and Their Roles in Nanobioremediation of Environmental Pollutants: Biochemical, Molecular, and Technological Perspectives." *Nano-Bioremediation: Fundamentals and Applications*, November. Elsevier, 111–45. doi:10.1016/B978-0-12-823962-9.00038-6

Batstone, Damien John, and Bernardino Virdis. 2014. "The Role of Anaerobic Digestion in the Emerging Energy Economy." *Current Opinion in Biotechnology* 27. Elsevier Ltd: 142–49. doi:10.1016/j.copbio.2014.01.013

Batstone, Damien John, Tim Hülsen, . Mehta, C. M. and Jürg Keller. 2015. "Platforms for Energy and Nutrient Recovery from Domestic Wastewater: A Review." *Chemosphere* 140 (December). Pergamon: 2–11. doi:10.1016/J.CHEMOSPHERE.2014.10.021

Baxter-Plant, Victoria S., Iryna P. Mikheenko, and Lynne E. Macaskie. 2003. "Sulphate-Reducing Bacteria, Palladium and the Reductive Dehalogenation of Chlorinated Aromatic Compounds." *Biodegradation* 14 (2). doi:10.1023/A:1024084611555

Bedendi, Giada, Lilian D. De Moura Torquato, Sophie Webb, Cécile Cadoux, Amogh Kulkarni, Selmihan Sahin, Plinio Maroni, Ross D. Milton, and Matteo Grattieri. 2022. "Enzymatic and Microbial Electrochemistry: Approaches and Methods." *ACS Measurement Science Au*. doi:10.1021/acsmeasuresciau.2c00042

Benjamin, Stephen Rathinaraj, Fabio de Lima, Eridan Orlando Pereira Tramontina Florean, and Maria Izabel Florindo Guedes. 2019. "Current Trends in Nanotechnology for Bioremediation." *International Journal of Environment and Pollution* 66 (1–3). Inderscience Publishers: 19–40. doi:10.1504/IJEP.2019.104526

Bhandari, Sobika, Darbin Kumar Poudel, Rishab Marahatha, Sonika Dawadi, Karan Khadayat, Sitaram Phuyal, Shreesti Shrestha, et al. 2021. "Microbial Enzymes Used in Bioremediation." *Journal of Chemistry*. doi:10.1155/2021/8849512

Bharagava, Ram Naresh, Gaurav Saxena, and Sikandar I. Mulla. 2020. "Introduction to Industrial Wastes Containing Organic and Inorganic Pollutants and Bioremediation Approaches for Environmental Management." In Gaurav Saxena, and Ram Naresh Bharagava (eds.), *Bioremediation of Industrial Waste for Environmental Safety*. Springer Singapore, pp. 1–18. doi:10.1007/978-981-13-1891-7_1

Bhatt, Pankaj, Satish Chandra Pandey, Samiksha Joshi, Parul Chaudhary, Vinay Mohan Pathak, Yaohua Huang, Xiaozhen Wu, Zhe Zhou, and Shaohua Chen. 2022. "Nanobioremediation: A Sustainable Approach for the Removal of Toxic Pollutants from the Environment." *Journal of Hazardous Materials* 427 (April). Elsevier: 128033. doi:10.1016/J.JHAZMAT.2021.128033

Bhattu, Monika, and Jagpreet Singh. 2023. "Recent Advances in Nanomaterials Based Sustainable Approaches for Mitigation of Emerging Organic Pollutants." *Chemosphere* 321 (April). Elsevier Ltd. doi:10.1016/j.chemosphere.2023.138072

Bilal, Muhammad, Tuan Anh Nguyen, and Hafiz M. N. Iqbal. 2021. "Nanobioremediation: Status Quo and View Ahead." In Hafiz M. N. Iqbal, Muhammad Bilal, and Tuan Anh Nguyen (eds.), *Nano-Bioremediation: Fundamentals and Applications*. Elsevier, pp. 573–77. doi:10.1016/B978-0-12-823962-9.00027-1

Blankenhorn, Gunter. 1976. "Nicotinamide-Dependent One-Electron and Two-Electron (Flavin) Oxidoreduction: Thermodynamics, Kinetics, and Mechanism." *European Journal of Biochemistry* 67 (1). doi:10.1111/j.1432-1033.1976.tb10634.x

Bolade, Oladotun P., Akan B. Williams, and Nsikak U. Benson. 2020. "Green Synthesis of Iron-Based Nanomaterials for Environmental Remediation: A Review." *Environmental Nanotechnology, Monitoring and Management* 13 (May). Elsevier B.V. doi:10.1016/j.enmm.2019.100279

Borah, Siddhartha Narayan, Niharika Koch, Suparna Sen, Ram Prasad, and Hemen Sarma. 2022. "Novel Nanomaterials for Nanobioremediation of Polyaromatic Hydrocarbons." In Hemen Sarma, Delfina C. Domínguez, and Wen-Yee Lee (eds.), *Emerging Contaminants in the Environment: Challenges and Sustainable Practices*, January. Elsevier, pp. 643–67. doi:10.1016/B978-0-323-85160-2.00024-X

Buonomano, Annamaria, Giovanni Barone, and Cesare Forzano. 2023. "Latest Advancements and Challenges of Technologies and Methods for Accelerating the Sustainable Energy Transition." *Energy Reports* 9 (December). Elsevier Ltd: 3343–55. doi:10.1016/j.egyr.2023.02.015

Carvalho, Gilda, Adrian Oehmen, Maria G. E. Albuquerque, and Maria A. M. Reis. 2014. "The Relationship between Mixed Microbial Culture Composition and PHA Production Performance from Fermented Molasses." *New Biotechnology* 31 (4). Elsevier: 257–63. doi:10.1016/J.NBT.2013.08.010

Catherine, Marie Claire, Alan Guwy, and Jaime Massanet-Nicolau. 2022. "Effect of Acetate Concentration, Temperature, PH and Nutrient Concentration on Polyhydroxyalkanoates (PHA) Production by Glycogen Accumulating Organisms." *Bioresource Technology Reports* 20 (December). Elsevier Ltd. doi:10.1016/j.biteb.2022.101226

Chaudhary, Parul, Lukman Ahamad, Anuj Chaudhary, Govind Kumar, Wen Juan Chen, and Shaohua Chen. 2023. "Nanoparticle-Mediated Bioremediation as a Powerful Weapon in the Removal of Environmental Pollutants." *Journal of Environmental Chemical Engineering* 11 (2). Elsevier Ltd. doi:10.1016/j.jece.2023.109591

Chauhan, Gaurav, Reyna Berenice González-González, and Hafiz M. N. Iqbal. 2022. "Bioremediation and Decontamination Potentials of Metallic Nanoparticles Loaded Nanohybrid Matrices: A Review." *Environmental Research* 204 (March). Academic Press Inc. doi:10.1016/J.ENVRES.2021.112407

Chitapornpan, S., C. Chiemchaisri, W. Chiemchaisri, R. Honda, and K. Yamamoto. 2012. "Photosynthetic Bacteria Production from Food Processing Wastewater in Sequencing Batch and Membrane Photo-Bioreactors." *Water Science and Technology* 65 (3): 504–12. doi:10.2166/WST.2012.740

Choerudin, Choerudin, Fauziyah Istiqomah Arrahmah, Jonatan Kevin Daniel, Takahiro Watari, Takashi Yamaguchi, and Tjandra Setiadi. 2021. "Evaluation of Combined Anaerobic Membrane Bioreactor and Downflow Hanging Sponge Reactor for Treatment of Synthetic Textile Wastewater." *Journal of Environmental Chemical Engineering* 9 (4). doi:10.1016/j.jece.2021.105276

Das, Nilanjana, and Preethy Chandran. 2011. "Microbial Degradation of Petroleum Hydrocarbon Contaminants: An Overview." *Biotechnology Research International* 2011 (September). Hindawi Limited: 1–13. doi:10.4061/2011/941810

Dixit, Ruchita, Wasiullah, Deepti Malaviya, Kuppusamy Pandiyan, Udai B. Singh, Asha Sahu, Renu Shukla, et al. 2015. "Bioremediation of Heavy Metals from Soil and Aquatic Environment: An Overview of Principles and Criteria of Fundamental Processes." *Sustainability (Switzerland)* 7 (2). MDPI: 2189–212. doi:10.3390/SU7022189

Do, Cam Van T., Mai Huong T. Pham, Thanh Yen T. Pham, Cuc T. Dinh, Thu Uyen T. Bui, Thuan Dang Tran, and Van Tuyen Nguyen. 2022. "Microalgae and Bioremediation of Domestic Wastewater." *Current Opinion in Green and Sustainable Chemistry* 34 (April). Elsevier B.V. doi:10.1016/J.COGSC.2022.100595

Dubey, Ashutosh, and Anurag Tewari. 2018. "Performance of Aluminium Electrode in Defluoridation of Water during Electrocoagulation." *Materials Focus* 7 (5). American Scientific Publishers: 657–61. doi:10.1166/MAT.2018.1560

Du, Juntao, Zhiyang Shen, Malin Song, and Michael Vardanyan. 2023. "The Role of Green Financing in Facilitating Renewable Energy Transition in China: Perspectives from Energy Governance, Environmental Regulation, and Market Reforms." *Energy Economics* 120 (April). Elsevier B.V. doi:10.1016/j.eneco.2023.106595

El-Zawawy, Waleed K., Maha M. Ibrahim, Yasser R. Abdel-Fattah, Nadia A. Soliman, and Morsi M. Mahmoud. 2011. "Acid and Enzyme Hydrolysis to Convert Pretreated Lignocellulosic Materials into Glucose for Ethanol Production." *Carbohydrate Polymers* 84 (3). doi:10.1016/j.carbpol.2010.12.022

Galdames, Alazne, Leire Ruiz-Rubio, Pilar Brettes, Javier Etxebarria, and José Luis Vilas Vilela. 2021. "Nanobioremediation for Soil Remediation: An Introduction." In Hafiz M. N. Iqbal, Muhammad Bilal, and Tuan Anh Nguyen (eds.), *Nano-Bioremediation: Fundamentals and Applications*, November. Elsevier, pp. 479–500. doi:10.1016/B978-0-12-823962-9.00017-9

Ghosh, Upal, A. Scott Weber, James N. Jensen, and John R. Smith. 1999. "Granular Activated Carbon and Biological Activated Carbon Treatment of Dissolved and Sorbed Polychlorinated Biphenyls on JSTOR." *Water Environment Research* 7 (2): 232–40. https://www.jstor.org/stable/25045203

Gikas, Petros. 2017. "Towards Energy Positive Wastewater Treatment Plants." *Journal of Environmental Management* 203 (December). Academic Press: 621–29. doi:10.1016/j.jenvman.2016.05.061

Gong, Xiaomin, Danlian Huang, Yunguo Liu, Zhiwei Peng, Guangming Zeng, Piao Xu, Min Cheng, Rongzhong Wang, and Jia Wan. 2018. "Remediation of Contaminated Soils by Biotechnology with Nanomaterials: Bio-Behavior, Applications, and Perspectives." *Critical Reviews in Biotechnology* 38 (3). Taylor and Francis Ltd: 455–68. doi:10.1080/07388551.2017.1368446

Guerra, Fernanda D., Mohamed F. Attia, Daniel C. Whitehead, and Frank Alexis. 2018. "Nanotechnology for Environmental Remediation: Materials and Applications." *Molecules* 23 (7). MDPI AG. doi:10.3390/MOLECULES23071760

Habe, Hiroshi, Yuya Sato, Tomo Aoyagi, Tomohiro Inaba, Tomoyuki Hori, Takaya Hamai, Kentaro Hayashi, Mikio Kobayashi, Takeshi Sakata, and Naoki Sato. 2020. "Design, Application, and Microbiome of Sulfate-Reducing Bioreactors for Treatment of Mining-Influenced Water." *Applied Microbiology and Biotechnology* 104: 6893–903. doi:10.1007/s00253-020-10737-2

Hairom, Nur Hanis Hayati, Chin Fhong Soon, Radin Maya Saphira Radin Mohamed, Marlia Morsin, Nurfarina Zainal, Nafarizal Nayan, Che Zalina Zulkifli, and Nor Hazlyna Harun. 2021. "A Review of Nanotechnological Applications to Detect and Control Surface Water Pollution." *Environmental Technology and Innovation* 24: 102032. doi:10.1016/j.eti.2021.102032

Held, Christoph, and Gabriele Sadowski. 2016. "Thermodynamics of Bioreactions." *Annual Review of Chemical and Biomolecular Engineering* 7: 395–414. doi:10.1146/annurev-chembioeng-080615-034704

Hess, Annatina, Boris Zarda, Dittmar Hahn, Andreas Häner, Dietmar Stax, Patrick Höhener, and Josef Zeyer. 1997. "In Situ Analysis of Denitrifying Toluene- and m-Xylene-Degrading Bacteria in a Diesel Fuel-Contaminated Laboratory Aquifer Column." *Applied and Environmental Microbiology* 63 (6). American Society for Microbiology: 2136–41. doi:10.1128/AEM.63.6.2136-2141.1997

He, Weiwei, Wayne Wamer, Qingsu Xia, Jun Jie Yin, and Peter P. Fu. 2014. "Enzyme-like Activity of Nanomaterials." *Journal of Environmental Science and Health - Part C Environmental Carcinogenesis and Ecotoxicology Reviews* 32 (2). doi:10.1080/10590501.2014.907462

Hülsen, Tim, Damien J. Batstone, and Jürg Keller. 2014. "Phototrophic Bacteria for Nutrient Recovery from Domestic Wastewater." *Water Research* 50 (March). Elsevier Ltd: 18–26. doi:10.1016/j.watres.2013.10.051

Hülsen, Tim, Kent Hsieh, and Damien J. Batstone. 2019. "Saline Wastewater Treatment with Purple Phototrophic Bacteria." *Water Research* 160 (September). Elsevier Ltd: 259–67. doi:10.1016/j.watres.2019.05.060

Hyster, Todd K. 2020. "Radical Biocatalysis: Using Non-Natural Single Electron Transfer Mechanisms to Access New Enzymatic Functions." *Synlett* 31 (3). doi:10.1055/s-0037-1611818

Jain, Marut, Sadaf Aiman Khan, Komal Sharma, Prashant Ram Jadhao, Kamal Kishore Pant, Zyta Maria Ziora, and Mark A. T. Blaskovich. 2022. "Current Perspective of Innovative Strategies for Bioremediation of Organic Pollutants from Wastewater." *Bioresource Technology* 344 (January). Elsevier: 126305. doi:10.1016/J.BIORTECH.2021.126305

Janusz, Grzegorz, Anna Pawlik, Urszula Świderska-Burek, Jolanta Polak, Justyna Sulej, Anna Jarosz-Wilkołazka, and Andrzej Paszczyński. 2020. "Laccase Properties, Physiological Functions, and Evolution." *International Journal of Molecular Sciences* 31: 996. doi:10.3390/ijms21030966

Jiang, Yang, Leonie Marang, Jelmer Tamis, Mark C. M. van Loosdrecht, Henk Dijkman, and Robbert Kleerebezem. 2012. "Waste to Resource: Converting Paper Mill Wastewater to Bioplastic." *Water Research* 46 (17). Elsevier Ltd: 5517–30. doi:10.1016/j.watres.2012.07.028

Kandah, Munther Issa, and Jean Luc Meunier. 2007. "Removal of Nickel Ions from Water by Multi-Walled Carbon Nanotubes." *Journal of Hazardous Materials* 146 (1–2): 283–88. doi:10.1016/J.JHAZMAT.2006.12.019

Khatipov, Emir, Masato Miyake, Jun Miyake, and Yasuo Asada. 1998. "Accumulation of Poly-β-Hydroxybutyrate by Rhodobacter Sphaeroides on Various Carbon and Nitrogen Substrates." *FEMS Microbiology Letters* 162 (1): 39–45. doi:10.1016/S0378-1097(98)00099-8

Kılkış, Şiir, Goran Krajačić, Neven Duić, Marc A. Rosen, and Moh'd Ahmad Al-Nimr. 2020. "Advances in Integration of Energy, Water and Environment Systems towards Climate Neutrality for Sustainable Development." *Energy Conversion and Management* 225 (December). Pergamon: 113410. doi:10.1016/J.ENCONMAN.2020.113410

Krainer, Florian Wolfgang, Barbara Darnhofer, Ruth Birner-Gruenberger, and Anton Glieder. 2016. "Recombinant Production of a Peroxidase-Protein G Fusion Protein in Pichia Pastoris." *Journal of Biotechnology* 219. doi:10.1016/j.jbiotec.2015.12.020

Kuhad, Ramesh Chander, Rishi Gupta, and Ajay Singh. 2011. "Microbial Cellulases and Their Industrial Applications." *Enzyme Research* 2011: 280696. doi:10.4061/2011/280696

Kumar, Adarsh, and Ram Chandra. 2020. "Ligninolytic Enzymes and Its Mechanisms for Degradation of Lignocellulosic Waste in Environment." *Heliyon* 6 (2): e03170. doi:10.1016/j.heliyon.2020.e03170

Kuntke, P., K. M. Śmiech, H. Bruning, G. Zeeman, M. Saakes, T. H. J. A. Sleutels, H. V. M. Hamelers, and C. J. N. Buisman. 2012. "Ammonium Recovery and Energy Production from Urine by a Microbial Fuel Cell." *Water Research* 46 (8). Elsevier Ltd: 2627–36. doi:10.1016/j.watres.2012.02.025

Lei, Ying, and Zannatul Ferdous. 2016. "Design Considerations and Challenges for Mechanical Stretch Bioreactors in Tissue Engineering." *Biotechnology Progress* 32 (3). doi:10.1002/btpr.2256

Lemos, Paulo C., Caterina Levantesi, Luisa S. Serafim, Simona Rossetti, Maria A. M. Reis, and Valter Tandoi. 2008. "Microbial Characterisation of Polyhydroxyalkanoates Storing Populations Selected under Different Operating Conditions Using a Cell-Sorting RT-PCR Approach." *Applied Microbiology and Biotechnology* 78 (2): 351–60. doi:10.1007/S00253-007-1301-5

Lettinga, Gatze, Salih Rebac, Sofia Parshina, Alla Nozhevnikova, Jules B. Van Lier, and Alfons J. M. Stams. 1999. "High-Rate Anaerobic Treatment of Wastewater at Low Temperatures." *Applied and Environmental Microbiology* 65 (4). American Society for Microbiology: 1696–1702. doi:10.1128/AEM.65.4.1696-1702.1999

Lettinga, Gatze, Salih Rebac, and Grietje Zeeman. 2001. "Challenge of Psychrophilic Anaerobic Wastewater Treatment." *Trends in Biotechnology* 19 (9): 363–70. doi:10.1016/S0167-7799(01)01701-2

Lim, Diana, Eric S. Renteria, Drake S. Sime, Young Min Ju, Ji Hyun Kim, Tracy Criswell, Thomas D. Shupe, et al. 2022. "Bioreactor Design and Validation for Manufacturing Strategies in Tissue Engineering." *Bio-Design and Manufacturing* 5 (1): 43–63. doi:10.1007/s42242-021-00154-3.

Liu, Weixing, Cristina Maria Iordan, Francesco Cherubini, Xiangping Hu, and Dafang Fu. 2021. "Environmental Impacts Assessment of Wastewater Treatment and Sludge Disposal Systems under Two Sewage Discharge Standards: A Case Study in Kunshan, China." *Journal of Cleaner Production* 287 (March). Elsevier Ltd. doi:10.1016/j.jclepro.2020.125046

Longis, Marion, Joana Carvalho Pereira, Thomas H. Högl, Peter Neubauer, and Stefan Junne. 2023. "Plug-Flow Anaerobic Digestion with Multi-Position Sensors: The Value of Gradient Measurement for Process Monitoring." *Biomass and Bioenergy* 173 (June). Elsevier Ltd. doi:10.1016/j.biombioe.2023.106803

Madigan, Michael T. 1999. "Isolation and Characterization of Psychrophilic Purple Bacteria From Antarctica." In *The Phototrophic Prokaryotes*. Springer US, pp. 699–706. doi:10.1007/978-1-4615-4827-0_81

Malik, Kamla, Nisha Kumari, Sushil Ahlawat, Upendra Kumar, and Meena Sindhu. 2020. "Extremophile Microorganisms and Their Industrial Applications BT: Microbial Diversity, Interventions and Scope." In Shiwani Guleria Sharma, Neeta Raj Sharma, and Mohit Sharma (eds.), *Microbial Diversity, Interventions and Scope*. Springer Singapore, pp. 137–56. doi:10.1007/978-981-15-4099-8_10

Malka, Lorenc, Flamur Bidaj, Alban Kuriqi, Aldona Jaku, Rexhina Roçi, and Alemayehu Gebremedhin. 2023. "Energy System Analysis with a Focus on Future Energy Demand Projections: The Case of Norway." *Energy* 272 (June). Elsevier Ltd. doi:10.1016/j.energy.2023.127107

Mauter, Meagan S., and Menachem Elimelech. 2008. "Environmental Applications of Carbon-Based Nanomaterials." *Environmental Science and Technology* 42 (16): 5843–59. doi:10.1021/ES8006904

McDonald, Alan, and Leo Schrattenholzer. 2001. "Learning Rates for Energy Technologies." *Energy Policy* 29 (4): 255–61. doi:10.1016/S0301-4215(00)00122-1

Meloni, Eugenio, Marco Martino, and Vincenzo Palma. 2020. "A Short Review on Ni Based Catalysts and Related Engineering Issues for Methane Steam Reforming." *Catalysts* 10 (3). MDPI. doi:10.3390/CATAL10030352

Mofijur, M., M. M. Hasan, Sabrina Sultana, Zobaidul Kabir, F. Djavanroodi, Shams Forruque Ahmed, M. I. Jahirul, Irfan Anjum Badruddin, and T. M. Yunus Khan. 2023. "Advancements in Algal Membrane Bioreactors: Overcoming Obstacles and Harnessing Potential for Eliminating Hazardous Pollutants from Wastewater." *Chemosphere* 336 (September). Elsevier Ltd. doi:10.1016/j.chemosphere.2023.139291

Molina Grima, E., E. H. Belarbi, F. G. Acién Fernández, A. Robles Medina, and Yusuf Chisti. 2003. "Recovery of Microalgal Biomass and Metabolites: Process Options and Economics." *Biotechnology Advances* 20 (7–8). Elsevier Inc.: 491–515. doi:10.1016/S0734-9750(02)00050-2

Muter, Olga. 2023. "Current Trends in Bioaugmentation Tools for Bioremediation: A Critical Review of Advances and Knowledge Gaps." *Microorganisms* 11: 710. doi:10.3390/microorganisms11030710

Nabernegg, Stefan, Birgit Bednar-Friedl, Fabian Wagner, Thomas Schinko, Janusz Cofala, and Yadira Mori Clement. 2017. "The Deployment of Low Carbon Technologies in Energy Intensive Industries: A Macroeconomic Analysis for Europe, China and India." *Energies* 10 (3). MDPI AG. doi:10.3390/EN10030360

Narayanan, C. M., and Vikas Narayan. 2019. "Biological Wastewater Treatment and Bioreactor Design: A Review." *Sustainable Environment Research* 29: 1. doi:10.1186/s42834-019-0036-1

Nastasi, Benedetto, Natasa Markovska, Tomislav Puksec, Neven Duić, and Aoife Foley. 2023. "Techniques and Technologies to Board on the Feasible Renewable and Sustainable Energy Systems." *Renewable and Sustainable Energy Reviews* 182 (August). Elsevier BV: 113428. doi:10.1016/j.rser.2023.113428

Nimkande, Vijay D., and Amit Bafana. 2022. "A Review on the Utility of Microbial Lipases in Wastewater Treatment." *Journal of Water Process Engineering*. doi:10.1016/j.jwpe.2022.102591

Nolasco-Soria, Héctor, Héctor Rodrigo Nolasco-Alzaga, and Enric Gisbert. 2020. "The Importance of Pepsin-like Acid Protease Quantification in Aquaculture Studies: A Revision of Available Procedures and Presentation of a New Protocol for Its Assessment." *Reviews in Aquaculture* 12: 1928–43. doi:10.1111/raq.12417

Nowak, Otto, Peter Enderle, and Petar Varbanov. 2015. "Ways to Optimize the Energy Balance of Municipal Wastewater Systems: Lessons Learned from Austrian Applications." *Journal of Cleaner Production* 88 (February). Elsevier Ltd: 125–31. doi:10.1016/j.jclepro.2014.08.068

Ostle, A. G., and J. G. Holt. 1982. "Nile Blue A as a Fluorescent Stain for Poly-β-Hydroxybutyrate." *Applied and Environmental Microbiology* 44 (1): 238–41. doi:10.1128/AEM.44.1.238-241.1982

Perez-Benito, Joaquin F., and Conchita Arias. 1998. "Kinetics and Mechanism of the Reactions of Superoxochromium(III) Ion with Biological Thiols." *Journal of Physical Chemistry A* 102 (29). doi:10.1021/jp9809336

Planamente, Sara, and Stefanie Frank. 2019. "Bio-Engineering of Bacterial Microcompartments: A Mini Review." *Biochemical Society Transactions*. doi:10.1042/BST20170564

Puchongkawarin, C., C. Gomez-Mont, D. C. Stuckey, and B. Chachuat. 2015. "Optimization-Based Methodology for the Development of Wastewater Facilities for Energy and Nutrient Recovery." *Chemosphere* 140 (December). Elsevier Ltd: 150–8. doi:10.1016/j.chemosphere.2014.08.061

Rajeshwari, K. V., M. Balakrishnan, A. Kansal, Kusum Lata, and V. V. N. Kishore. 2000. "State-of-the-Art of Anaerobic Digestion Technology for Industrial Wastewater Treatment." *Renewable & Sustainable Energy Reviews* 4 (2). Elsevier Science Ltd: 135–56. doi:10.1016/S1364-0321(99)00014-3

Rizwan, Md., Man Singh, Chanchal K. Mitra, and Roshan K. Morve. 2014. "Ecofriendly Application of Nanomaterials: Nanobioremediation." *Journal of Nanoparticles* 2014 (June). Hindawi Limited: 1–7. doi:10.1155/2014/431787

Robinson, Peter K. 2015. "Correction: Enzymes: Principles and Biotechnological Applications." *Essays in Biochemistry* 59. doi:10.1042/bse0590075

San Martín, Javier, Daniel Puyol, Yolanda Segura, Juan A. Melero, and Fernando Martínez. 2023. "A Novel Photoanaerobic Process as a Feasible Alternative to the Traditional Aerobic Treatment of Refinery Wastewater." *Journal of Water Process Engineering* 51 (February). Elsevier Ltd. doi:10.1016/j.jwpe.2022.103352

Schrick, Bettina, Bianca W. Hydutsky, Jennifer L. Blough, and Thomas E. Mallouk. 2004. "Delivery Vehicles for Zerovalent Metal Nanoparticles in Soil and Groundwater." *Chemistry of Materials* 16 (11): 2187–93. doi:10.1021/CM0218108

Seddighi, Sadegh, Edward J. Anthony, Hamed Seddighi, and Filip Johnsson. 2023. "The Interplay between Energy Technologies and Human Health: Implications for Energy Transition." *Energy Reports* 9 (December). Elsevier Ltd: 5592–611. doi:10.1016/j.egyr.2023.04.351

Sepúlveda-Muñoz, Cristian A., Ignacio de Godos, and Raúl Muñoz. 2023. "Wastewater Treatment Using Photosynthetic Microorganisms." *Symmetry*. doi:10.3390/sym15020525

Sepúlveda-Muñoz, Cristian A., Andrés F. Torres-Franco, Ignacio de Godos, and Raúl Muñoz. 2022. "Exploring the Metabolic Capabilities of Purple Phototrophic Bacteria during Piggery Wastewater Treatment." *Journal of Water Process Engineering* 50 (December). Elsevier Ltd. doi:10.1016/j.jwpe.2022.103317

Sharma, Pooja, Surendra Pratap Singh, Hafiz M. N. Iqbal, and Yen Wah Tong. 2022. "Omics Approaches in Bioremediation of Environmental Contaminants: An Integrated Approach for Environmental Safety and Sustainability." *Environmental Research* 211 (August). Academic Press Inc. doi:10.1016/J.ENVRES.2022.113102

Shraddha, Ravi Shekher, Simran Sehgal, Mohit Kamthania, and Ajay Kumar. 2011. "Laccase: Microbial Sources, Production, Purification, and Potential Biotechnological Applications." *Enzyme Research*. doi:10.4061/2011/217861

Sillero, Leonor, William Gustavo Sganzerla, Tânia Forster-Carneiro, Rosario Solera, and Montserrat Perez. 2022. "A Bibliometric Analysis of the Hydrogen Production from Dark Fermentation." *International Journal of Hydrogen Energy*. doi:10.1016/j.ijhydene.2022.06.083

Simona, Crognale, Lorini Laura, Valentino Francesco, Villano Marianna, Marzo Gago Cristina, Tonanzi Barbara, Majone Mauro, and Rossetti Simona. 2022. "Effect of the Organic Loading Rate on the PHA-Storing Microbiome in Sequencing Batch Reactors Operated with Uncoupled Carbon and Nitrogen Feeding." *The Science of the Total Environment* 825 (June). NLM (Medline): 153995. doi:10.1016/j.scitotenv.2022.153995

Singh, Pratima, Cynthia Carliell-Marquet, and Arun Kansal. 2012. "Energy Pattern Analysis of a Wastewater Treatment Plant." *Applied Water Science* 2 (3). Springer Verlag: 221–6. doi:10.1007/S13201-012-0040-7

Singh, Ram Lakhan, Pradeep Kumar Singh, and Rajat Pratap Singh. 2015. "Enzymatic Decolorization and Degradation of Azo Dyes: A Review." *International Biodeterioration and Biodegradation*. doi:10.1016/j.ibiod.2015.04.027

Stark, Jonathan S., Patricia A. Corbett, Glenn Dunshea, Glenn Johnstone, Catherine King, Julie A. Mondon, Michelle L. Power, Angelingifta Samuel, Ian Snape, and Martin Riddle. 2016. "The Environmental Impact of Sewage and Wastewater Outfalls in Antarctica: An Example from Davis Station, East Antarctica." *Water Research* 105 (November). Elsevier Ltd: 602–14. doi:10.1016/j.watres.2016.09.026

Taura, Futoshi, Ryosuke Tanaya, and Supaart Sirikantaramas. 2019. "Recent Advances in Cannabinoid Biochemistry and Biotechnology." *ScienceAsia* 45 (5). doi:10.2306/scienceasia1513-1874.2019.45.399

Thuku, R. N., D. Brady, M. J. Benedik, and B. T. Sewell. 2009. "Microbial Nitrilases: Versatile, Spiral Forming, Industrial Enzymes." *Journal of Applied Microbiology*. doi:10.1111/j.1365-2672.2008.03941.x

Trujillo, Madia, Beatriz Alvarez, and Rafael Radi. 2016. "One- and Two-Electron Oxidation of Thiols: Mechanisms, Kinetics and Biological Fates." *Free Radical Research* 50 (2). doi:10.3109/10715762.2015.1089988

Tungittiplakorn, Warapong, Claude Cohen, and Leonard W. Lion. 2005. "Engineered Polymeric Nanoparticles for Bioremediation of Hydrophobic Contaminants." *Environmental Science and Technology* 39 (5): 1354–8. doi:10.1021/ES049031A

Uddin, Md Mosleh, and Mark Mba Wright. 2022. "Anaerobic Digestion Fundamentals, Challenges, and Technological Advances." *Physical Sciences Reviews*. doi:10.1515/psr-2021-0068

Verstraete, Willy, Pieter Van de Caveye, and Vasileios Diamantis. 2009. "Maximum Use of Resources Present in Domestic 'Used Water.'" *Bioresource Technology* 100 (23): 5537–45. doi:10.1016/j.biortech.2009.05.047

Wang, Zhen, Can Cui, and Sha Peng. 2019. "How Do Urbanization and Consumption Patterns Affect Carbon Emissions in China? A Decomposition Analysis." *Journal of Cleaner Production* 211 (February). Elsevier Ltd: 1201–8. doi:10.1016/j.jclepro.2018.11.272

Wear, Stephanie L., Vicenç Acuña, Rob McDonald, and Carme Font. 2021. "Sewage Pollution, Declining Ecosystem Health, and Cross-Sector Collaboration." *Biological Conservation* 255 (March). Elsevier: 109010. doi:10.1016/J.BIOCON.2021.109010

Werkneh, Adhena Ayaliew, and Shifare Berhe Gebru. 2023. "Development of Ecological Sanitation Approaches for Integrated Recovery of Biogas, Nutrients and Clean Water from Domestic Wastewater." *Resources, Environment and Sustainability* 11 (March). Elsevier B.V. doi:10.1016/j.resenv.2022.100095

Wong, Cecilia. 2016. "Life's Engines: How Microbes Made Earth Habitable." *Leonardo* 49 (5). doi:10.1162/leon_r_01307

Zeeman, G., and G. Lettinga. 1999. "The Role of Anaerobic Digestion of Domestic Sewage in Closing the Water and Nutrient Cycle at Community Level." *Water Science and Technology* 39 (5). Elsevier Science Ltd: 187–94. doi:10.1016/S0273-1223(99)00101-8

Zeeman, Grietje, Katarzyna Kujawa, Titia de Mes, Lucia Hernandez, Marthe de Graaff, Lina Abu-Ghunmi, Adriaan Mels, et al. 2008. "Anaerobic Treatment as a Core Technology for Energy, Nutrients and Water Recovery from Source-Separated Domestic Waste(Water)." *Water Science and Technology* 57 (8): 1207–12. doi:10.2166/WST.2008.101

Zhou, Wenting, Wenxiang Zhang, and Yanpeng Cai. 2021. "Laccase Immobilization for Water Purification: A Comprehensive Review." *Chemical Engineering Journal* 403: 126272. doi:10.1016/j.cej.2020.126272

3 Investigating Ecological Friendly Advancements in Wastewater Treatment for Sustainable Solutions

R. Raksasat, S. Ramanathan, H. Rawindran, and C.S. Liew

LIST OF ABBREVIATIONS

BOD: Biological Oxygen Demand
COD: Chemical Oxygen Demand
CWs: Constructed Wetlands
DAF: Dissolved Air Flotation
MBfRs: Membrane Biofilm Reactors
MBRs: Membrane Bioreactors
MFCs: Microbial Fuel Cells
PBRs: Photobioreactors
RSFs: Rapid Sand Filters
SSFs: Slow Sand Filters
THMs: Trihalomethanes
TKN: Total Kjeldahl Nitrogen
TOC: Total Organic Carbon
Total-N: Total Nitrogen
Total-P: Total Phosphorus
TSS: Total Suspended Solids

3.1 INTRODUCTION: BACKGROUND AND DRIVING FORCES

Urbanization and industrialization in various fields due to the rapid growth of the world's population have resulted in a high volume of wastewater, also known as sewage, being released from several activities that have been undertaken to satisfy human demands. Approximately 359.4 billion cubic meters of global wastewater that is generally discharged from households, agriculture, and industries is produced annually, with a significant amount of effluent being reported to be launched mainly from power plants owing to the cooling water employed [1,2].

DOI: 10.1201/9781003459286-3

Wastewater originating from different sources leads to dissimilar constituents. However, it typically contains organic matter, nutritional compounds, suspended solids, and pathogens [3]. The high contamination of hazardous substances has been found in effluent produced from manufacturing and mining industries [1]. Exposure to wastewater resulting from improper management poses a potential threat to ecosystems and human health. The prevalent practice of discharging the wastewater into natural water bodies necessitates attention, as the microorganisms, including viruses, bacteria, protozoa, fungi, and parasitic organisms found in wastewater, have been reported to induce health issues such as gastroenteritis, giardiasis, cryptosporidiosis, or infections [4]. Consequently, it is crucial to implement appropriate wastewater treatment measures to reduce pollutants and adhere to discharge limits that ensure safety. The permissible concentration of effluent characteristics is typically defined as follows [5]; (i) biological oxygen demand (BOD) value of approximately 250–300 mg/L upon disposal into sewer systems with direct disposal permitted at 10 mg/L [6]; (ii) chemical oxygen demand (COD) of less than 200 ppm, as stipulated by the World Health Organization (WHO); (iii) total suspended solids (TSS) typically limited to about 20 mg/L for a secondary wastewater treatment plant [7]; (iv) total nitrogen (total-N) recommended to be restricted below 15 mg/L [8]; (v) total phosphorus (total-P) pollution should be less than 1 mg/L, as mandated by the Wisconsin Pollutant Discharge Elimination System (WPDES) [9]; and (vi) heavy metals, especially mercury (Hg), lead (Pb), cadmium (Cd), chromium (Cr), and nickel (Ni), are controlled at 0.001, 0.01, 0.003, 0.05, and 0.02 ppm, respectively, as advised by WHO [10].

Wastewater undergoes three main treatment procedures: primary treatment (physical), which removes solid contaminants; secondary treatment (biological), which decomposes microorganisms; and tertiary treatment (chemical), which involves chemical processes [11]. Furthermore, sludge treatment is the concluding process that utilizes solid-liquid separation principles to achieve additional filtration and reduces the concentration of contaminants [12]. Traditional methods for treating wastewater, such as boiling, filtration, sedimentation, chlorination, ion exchange, thermal treatment, adsorption, coagulation, electrochemical degradation, and chemical precipitation, have drawbacks such as the production of by-products, decreased efficiency, increased operational expenses, and higher energy usage which plays a substantial role in causing environmental pollution and climate change [13,14]. This chapter examines environment-friendly technologies, such as microbial fuel cells (MFCs), bioreactors, and natural processes-associated wastewater treatment like constructed wetlands (CWs) that have been created to tackle the environmental issues linked to traditional approaches.

3.2 SOURCES OF WASTEWATER

Municipal wastewater consists of polluted water that arises from various human activities or natural occurrences. For the aim of this chapter, the sources of wastewater creation are classified into four categories: households, farming, industry, and stormwater.

3.2.1 Households

Domestic wastewater is the term used to describe wastewater that comes from residences, as well as public and commercial establishments like hotels, restaurants, and department shops. Blackwater is a category that mainly comes from sanitary activities like using the bathroom. It has high levels of pollution with compounds such as BOD, ammonium, and pathogenic microorganisms. The second type, known as gray water, is produced by routine activities other than toilet usage, such as showering, washing, and using hand basins [15–17]. The average composition of domestic wastewater has been documented to contain 204–440 mg/L of total COD, 50.3–139 mg/L of total organic carbon, 50–70 mg/L of total-N, and 4.88 mg/L of total-P [18,19]. The wastewater produced by individual homes, resulting from human activities, is transported through pipes and sewer lines, and then treated by a decentralized wastewater system, which is well suited for small-scale areas [20,21].

3.2.2 Farming

Key components present in agricultural wastewater include phosphorus and nitrogen-related chemicals, particularly phosphate and ammonium. The latter can be transformed into nitrate by a process called nitrification [22,23]. In wastewater collected from a livestock farm, Feng et al. [22] documented ammonium and phosphate concentrations of 1,257.14 and 25.01 mg/L, respectively. Phosphorus is acknowledged as a non-renewable resource, while nitrate is a vital mineral for plant growth when used as a fertilizer. However, excessive contamination of both compounds in wastewater has resulted in harm to ecosystems and negative impacts on human health. Eutrophication is a phenomenon caused by excessive nutrient enrichment, which leads to the growth of plants and algae, commonly referred to as algal blooms [22,24]. The subsequent decrease in oxygen levels, resulting in the formation of a dead zone, is caused by the excessive growth of plants, making the afflicted area unsuitable for aquatic animals [25]. In addition, the breakdown of both algae and plants results in the emission of carbon dioxide [24].

3.2.3 Industries

The proliferation of industries has greatly expanded in modern times to cater to a wide range of human requirements by manufacturing a variety of items. Nevertheless, wastewater is produced simultaneously during the manufacturing process, with its contents differing depending on the specific industry. Şentürk et al. [26] conducted an analysis of wastewater from a food factory and found the following concentrations: 5,250–5,750 mg/L of COD, 4,000–5,000 mg/L of BOD, 2,000–2,500 mg $CaCO_3$/L of alkalinity, 200–250 mg/L of total Kjeldahl nitrogen (TKN), and 90–100 mg/L of total-P. On the other hand, Haq and Kalamdhad [27] discovered certain qualities of wastewater originating from the paper sector, such as a concentration of 19,600 mg/L of COD, 672 mg/L of BOD, 266 mg/L of total-N, and roughly 51.9 mg/L of heavy metals. Power plants, which transform primary energy sources like coal and nuclear power into electricity, are the primary sources of discharged leachate compared to other types of plants. This can be related to the significant amount of cooling water needed to regulate the system during the entire procedure [1]. Furthermore, brine wastewater, which is released as a result of mining operations during the extraction and refining of raw materials such as coal for power generation, is known to include a large number of suspended particles and display high levels of acidity and salt [1,28,29]. The elevated salinity concentration, which is detrimental to the growth of plants, suggests the presence of significant amounts of salt components, such as sodium ions (Na^+), chloride ions (Cl^-), and sulfate ions (SO_4^{2-}), in the wastewater [28]. Industrial wastewater is commonly processed via centralized systems, such as public sewer systems [20].

3.2.4 Storm Water

Stormwater, which comes from rainfall, melted snow or ice, and the condensation of water vapor from clouds, gets polluted with dangerous particles, like minerals or harmful gasses, that are produced by human activities such as fuel combustion and then released into the atmosphere [30]. Environmental contamination, particularly eutrophication and acidification, occurs when stormwater is absorbed into the ground or flows into water bodies [31,32]. Gulliver et al. [33] found that urban stormwater in the USA typically has the following concentrations: 65 mg/L of COD, 9 mg/L of BOD, 1.5 mg/L of TKN, 0.33 mg/L of total-P, 100 mg/L of total suspended particles, and 0.338 mg/L of heavy metals.

3.3 TRADITIONAL METHODS FOR WASTEWATER TREATMENT

3.3.1 Chlorination and Boiling

Chlorination and boiling represent common methods applied for disinfection treatment in the purification of wastewater. Beyond their cost effectiveness, these processes inhibit bacteria and pathogenic microorganisms [34]. Chlorination treatment can be categorized into four types based on the amount of chlorine used: batch disinfection (fluctuating chlorine consumption), simple chlorination (0.3–0.5 mg/L of free chlorine residual), super-chlorination (3–5 mg/L of free chlorine residual) followed by de-chlorination (chlorine removal), and shock chlorination [35]. Given chlorine's reactivity and facile binding with other elements, the hydrogen atoms of bacterial enzymes and other cells present in wastewater undergo substitution by chlorine atoms. This amendment of enzyme molecules results in the deactivation of bacteria, viruses, parasites, or microorganisms [36]. Furthermore, ammonia in wastewater can be removed through its reaction with added chlorine at a sufficient ratio [37].

However, the highly reactive nature of chlorine leads to the formation of by-products, primarily trihalomethanes (THMs) and haloforms, resulting from the reaction between chlorine and bromide, iodide, as well as natural organic matter in wastewater [34,38,39]. The presence of these by-products is linked to carcinogenicity, teratogenicity, and mutagenicity [34]. Additionally, the application of chlorine requires careful operation due to its strong oxidizing agent characteristics, leading to the separation of hydrogen from water molecules and the subsequent emission of corrosive nascent oxygen and hydrogen chloride, which may cause tissue damage [36,40].

3.3.2 Boiling Treatment

After the chlorination process, boiling water treatment is used to reduce the production of by-products by causing them to evaporate or react with water [34]. The hydrolysis reaction rate increases when the water temperature is raised, as the higher kinetic energy leads to the breakdown and evaporation of the by-product molecules after the halogen atom is substituted with an OH^- group [41]. Ma et al. [38] showed that the presence of water turbulence and bubbling during the heating of water in a boiler can result in the removal of roughly 92% of halomethanes. Boiling treatment, according to research conducted by Batterman et al. [42], can effectively reduce THMs by up to 69%. Nevertheless, it is crucial to acknowledge that the combination of chlorination and boiling treatment requires a prolonged period of time and may not adequately eliminate all pollutants in wastewater.

3.3.3 Coagulation, Flocculation, and Flotation

To remove suspended solid particles, such as COD and BOD, from wastewater, a sequence of procedures can be implemented. These steps include coagulation, flocculation, optional dissolved air flotation (DAF), sedimentation, and either filtration or an activated sludge system. The purpose of these steps is to enhance water purity and quality [43–45]. Aluminum sulfate, often known as alum, is commonly used as a coagulant to destabilize and aggregate microscopic particles. In the DAF process, air bubbles are used instead of chemicals to create heavier solids called floc, which then float to the surface of the water [45]. The coagulant chemicals are evenly distributed throughout the wastewater during the quick mixing in the coagulation process. This is done to balance out the charges on the particles and promote the hydrolysis reactions that cause the development of flocs [46]. During the flocculation operation, as the mixing slows down, the coagulated particles settle due to increased agitation time, which promotes collisions between particles and results in the creation of larger flocs [45,46]. This, in turn, improves the efficiency of separation. Bubakova et al. [47] found that a decrease in shear rate, which refers to the rate of change in velocity of fluid flow between two surfaces in wastewater, enhances the pace at which floc is formed. Guvenc et al. [48] proved the efficacy of coagulation in water treatment. They discovered that the use of poly-aluminum chloride as a coagulant can remove 57.5% of COD, 96.8% of total suspended particles, and 90.5% of color index in chemical industry effluent.

The efficacy of DAF separation is affected by the electrical charges on particles and bubbles applied [49]. Reducing the size of bubbles helps them adhere to smaller particles, leading to an enhanced yield in removing impurities [50]. Abo-El Ela and Nawar [43] showed that the application of DAF following coagulation-flocculation processes can improve the purity of effluent. Coagulation with ferric chloride and alum as coagulant agents, followed by DAF, can achieve a reduction of nearly 94.5% in the pollution of oil droplets, COD, BOD, and turbidity in wastewater [43]. After being formed, the floc is allowed to settle, going through four distinct phases: lag, zone settling, transition, and compression phases [51]. This settling procedure is necessary before the floc is transferred for filtration to effectively remove any leftover contaminants that do not settle. A support medium consisting of porous and granular material, such as sand, is employed to capture suspended particles, including dangerous protozoa, by utilizing the gaps between pores [45]. There are three operational methods for sand filters: slow, fast, and multimedia filters. Slow sand filters (SSFs) are primarily effective for the biological elimination of microorganisms, such as bacteria or viruses. They are commonly employed in centralized or semi-centralized wastewater treatment facilities [52]. Nevertheless, the efficiency of SSFs is constrained in some circumstances, such as when the effluent's turbidity surpasses 10–20 NTU, there is a significant presence of algae, the temperatures are chilly, and the filter skimming and ripening times are prolonged [52]. In addition, the elimination of natural organic matter is not efficient using SSFs, resulting in the generation of by-products [52].

Conversely, rapid sand filters (RSFs) are a commonly used form of filter that require less surface area compared to SSFs. RSFs can be operated as either rapid gravity or rapid pressure filters and typically use coarse gravel as the filter medium [45,53]. Multimedia filters are composed of various filter media materials and commonly employ either two (also known as in-depth filtration) or three filter layers. The process of double-layer filtering involves the use of anthracite coal in the upper layer to catch larger floc particles, along with the inclusion of fine sand in the lower layer to eliminate smaller particles. Additionally, a third layer consisting of garnet mineral is added at the bottom [45]. According to Xi et al. [54], the content of BOD and COD in cheese production effluent can be reduced by over 99% and 85%, respectively, with the use of gravel/sand filtering. Impurity removal can be improved by using multilayer sand filtration [55]. According to a study, a single layer of filtrate was able to eliminate 76% of BOD, whereas a double layer of filtrate was shown to lower it by up to 85% [56]. Kang et al. [55] found that implementing a dual-layer filtering system consisting of coarse sand and fine sand can effectively reduce over 94% of total organic carbon (TOC) and BOD levels in wastewater generated from turkey processing. Although the flotation during floc creation and filtration processes consumes a significant amount of energy, resulting in expensive operation and maintenance expenses, particularly when the system is expanded for larger plants, the effectiveness of wastewater purification is insufficient [52,57,58].

3.3.4 Aeration

The surplus sludge, consisting of microbial aggregates, is then passed to the activated sludge process, which is a biological treatment that employs a sequence of reactors with aeration to eliminate biodegradable contaminants [59,60]. Around 65%–70% of wastewater released by pulp and paper manufacturers is treated using the activated sludge technique [61]. The addition of oxygen to the treatment tank has the purpose of suspending the sludge and stimulating the proliferation of microorganisms. These microorganisms break down organic substances and assimilate them as nutrients [59,62]. The biological contaminants undergo a process of conversion into gaseous and biomass products. These products can be removed from the wastewater using gravity settling in a secondary clarifier. As a result, the concentration of BOD in the treated effluent reduces primarily [59,62]. The performance of the activated sludge treatment is influenced by several factors, including the origin and properties of the wastewater, the design of the plant, and the conditions under which it operates.

According to Bajpai [61], the method typically removes approximately 60%–85% of COD, 85%–98% of BOD, 40%–65% of adsorbable organically bound halogens, 40%–85% of phosphorus, and 20%–50% of nitrogen concentration.

3.4 GREEN TECHNOLOGIES FOR WASTEWATER TREATMENT

Currently, sustainable solutions, also known as green technologies, have gained attention for wastewater treatment to protect ecosystems. Table 3.1 presents a comprehensive overview of several conventional techniques and green technologies used for treating wastewater. Electrochemical procedures and biological treatments are notable approaches used to improve the quality of wastewater.

3.4.1 Electrochemical Treatment Process

Electrochemical treatment technologies, such as electrocoagulation, electro-flotation by H_2 bubbles, electrodialysis, and solar energy-driven electrochemical oxidation-reduction, can be used to purify both organic and inorganic pollutants. These technologies rely on electricity as a reactant instead of chemicals for the treatment process [63,64]. Ebba et al. [63] showed that the effectiveness of electrocoagulation in treating wastewater can be influenced by various factors, including the initial pH of the effluent, the amount of current supplied, the distance between the electrodes, the concentration of electrolytes, and the duration of electrolysis. The experiment achieved a maximum removal of 95.28% for COD, 92.30% for color, and 83.33% for turbidity. This was accomplished by conducting the experiment at pH 7.5, with a current of 0.09 A, an electrolytic concentration of 3 g/L, an electrode distance of 2 cm, and a retention time of 60 minutes [63]. Electrocoagulation is environment-friendly because it requires less energy and has a shorter processing time, while also having the potential to eliminate a substantial number of contaminants. However, it does produce sludge as a by-product and necessitates the replacement of electrodes [63].

3.4.2 Membrane Filtration

Membrane filtration is a pressure-driven treatment method used to separate solid particles, including large molecules like proteins and carbohydrates. The classification of the membrane can be categorized into four groups based on the size of the pores. (i) Hyperfiltration (HF), also known as reverse osmosis (RO), uses a membrane with a pore size of 0.001 μm and operates at pressures up to 100 bar. It is commonly used for removing monovalent salts from seawater or brackish water. (ii) Nanofiltration (NF) is used to separate larger molecular sizes like sugars or divalent and multivalent salts. It operates at pressures of 10–40 bar. (iii) Ultrafiltration (UF) has a pore size range of 0.001–0.1 μm and is used for separating proteins or colloids under pressure conditions of 1–10 bar. (iv) Microfiltration (MF) is used to remove insoluble particles of 0.1–10 μm size, such as microbes. It typically operates at pressures of 0.5–2 bar [65–67]. The membrane allows wastewater to pass through a porous mechanism, except for HF, which acts as a dense membrane where molecules are transported based on their diffusivity and solubility [66].

TABLE 3.1
Summary of Wastewater Treatment

Type	Process	Advantages	Disadvantages	Treatment Phases	Process Type
Conventional treatment	Chlorination	Inexpensive Harmful microorganisms will be inhibited	Formation of by-product Corrosive	Primary and secondary	Chemical
	Boiling	Simple process	Incapable to remove all pollutants	Primary	Physical
	Coagulation, flocculation	Effective to remove pollutants	Formation of by-product		Chemical
	Flotation, sedimentation, filtration	High separation capacity	High cost Frequent maintenance		Physical
	Activated sludge process	Effective to remove pollutants	High operating cost Large amount of sludge is generated	Secondary	Biological
Green technologies	Electrocoagulation Electro-flotation	Simple process Effective to treat wastewater	Electricity is essential		Electrochemical
	Membrane filtration	Short operating time Small space required	Membrane fouling High membrane cost	Tertiary	Physical
	MBfRs	High effluent quality		Secondary	Biological
	MBRs	Small space required High effluent quality			Biological
	MFCs	Energy efficient Environmentally friendly	Low production yield	Secondary or tertiary	Bio-electrochemical
	PBRs	Low risk of contamination occurred	Difficult to clean System conditions need to be controlled		
	Constructed wetlands	Inexpensive	Large space required	Tertiary	Physical, chemical, and biological

Membrane filtration can be conducted using two modes: dead-end filtration and crossflow filtration. In dead-end filtration, the feed stream moves at a right angle in the direction of the membrane, causing a cake layer (biofilm) to form on the membrane surface. This layer slows down the rate at which substances may pass through the membrane, leading to membrane fouling caused by the buildup of particles that are not able to pass through. In contrast, crossflow filtering demonstrates a more consistent permeating flow due to the parallel flow of the feed stream with the membrane surface. This flow helps to carry residual chemicals and reduce the thickness of the cake layer [66]. While the utilization of membranes for water treatment results in lower energy consumption, the presence of membrane fouling requires additional expenditures for maintenance.

Demulsification employing membrane filtration is an effective method for treating oily wastewater, particularly the type that is discharged from petrochemical plants. Bhattacharyya et al. [68] found that MF could effectively eliminate more than 94% of total petroleum hydrocarbon in oily seawater. However, when the concentration of surplus oil exceeded 30 ppm, it caused significant and irreversible fouling of the Sumitomo Electric hollow fiber-membrane modules. Membrane filtration was shown to be less effective than membrane bioreactors (MBRs) in treating oily wastewater. MBRs are suitable to manage oil concentrations up to 93.1 ppm, which is influenced by the biological activity present in MBRs [68].

In traditional activated sludge processes, air is necessary to provide microorganisms with the dissolved oxygen they need to break down COD and ammonium. This results in a significant energy requirement for generating gas bubbles and the release of volatile organic compounds, such as carbon footprint, into the water. To tackle these difficulties, scientists have created membrane biofilm reactors (MBfRs) [69]. The gas is compressed to pass through the pores of the membrane and spread into a biofilm that has developed on the surface of the membrane [69]. The MBfRs method of biological wastewater treatment can be used in both aerobic and anaerobic environments and has been found to effectively remove BOD, COD, and nitrogen [70].

3.4.3 Bioreactors

The utilization of MBRs, which combine membranes with bioreactors, presents several benefits in treating wastewater, especially those with high saline and oily concentrations. These advantages include a reduced sludge production rate and enhanced efficiency in removing contaminants, surpassing the capabilities of conventional methods [68]. Di Bella et al. [71] conducted a study to examine the practicality of MBRs. The findings indicated that MBRs achieved a COD removal efficiency of above 70% in saline effluent. Nevertheless, MBRs face a significant obstacle in terms of their time-intensive nature. The process of cultivating bacterial populations capable of metabolizing hydrocarbons requires a considerable amount of time, which is not ideal for on-site treatment, particularly in the context of oil spill reactions within the marine industry. Efficient effluent quality typically requires a hydraulic retention time of at least 10 hours [68,72].

MFCs are a sustainable technology that uses electroactive bacteria in a bioreactor to transfer electrons outside of the cells through an electrochemical process.

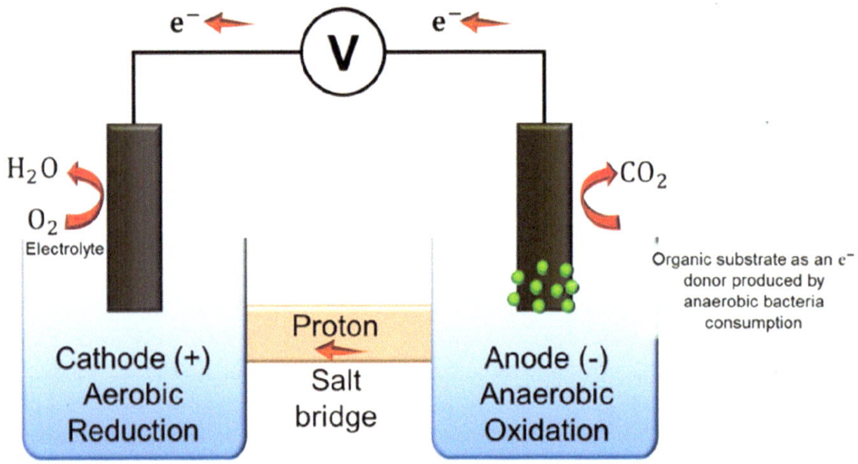

FIGURE 3.1 Simplified mechanism of MFCs electrolysis.

This process not only generates energy but also improves the quality of the wastewater being treated [73,74]. MFCs generally have cathode and anode electrodes that are linked by a salt bridge or proton exchange membrane. In the anodic chamber of MFCs, organic compounds found in wastewater undergo oxidation in the absence of oxygen. Microbes metabolize these chemicals, which promotes their growth and leads to the production of electrons (serving as redox mediators) and protons. Following oxidation, the electrons are sent to the electrode, while the protons disperse into the electrolyte solution within the aerobic cathode chamber [73]. The anodic chamber's potential diminishes as organic molecules undergo oxidation, resulting in the production of electricity due to the disparity in potential between the two electrodes [73]. Figure 3.1 displays the simplified schematic of the MFCs electrolysis transport mechanism. Several studies have documented the capacity of MFCs to reduce COD levels by an average of 83% in several types of effluents, including those from agriculture, the paper industry, and domestic wastewater [75–77].

Bioreactors are often categorized into two operational modes: open and closed systems, which are determined by their level of exposure to the surrounding environment. As a result, an open system has a higher possibility of contamination and can potentially affect the functioning of cells [78]. Photobioreactors (PBRs) are highly suitable enclosed systems for cultivating phototrophic organisms, particularly microalgae. These organisms utilize either artificial light or solar energy to carry out photosynthesis. The light is utilized as an energy source for their growth, requiring the identification of the most favorable light conditions for the PBRs system [79]. In the open PBRs system, one of them referred to as "raceway ponds," cultivation takes place within closed circulation channels formed by shallow artificial ponds arranged in a U-shape [80]. Raceway ponds have a more affordable initial investment, but their effectiveness relies on the variations in temperature and humidity that occur on a daily and seasonal basis [81]. In contrast, closed PBRs have reduced CO_2 leakage, allow for precise control of environmental conditions, and demand less area. Nevertheless, these organisms pose difficulties in terms of purification, necessitating cooling, and functioning in oxygen-deprived environments, as high amounts of oxygen might impede the growth of phototrophs [82]. The properties of wastewater can be improved, particularly in the elimination of total-N and total-P, to comply with discharge regulations by utilizing the bioenergy conversion of microalgae in PBRs [83]. The performance of batch PBRs in reducing nitrogen and phosphorus was assessed by Yan et al. [84]. The findings revealed that the PBRs system can effectively remove 96% of nitrogen pollutants and 98% of phosphorus pollutants through photosynthetic metabolism [84].

3.4.4 Constructed Wetlands

CWs are artificial engineering systems specifically built for the treatment of wastewater, particularly rainwater. These systems consist of three operational zones, as shown in Figure 3.2. They function by extracting solid particles from effluent using natural mechanisms [85,86]. Large particles are first filtered in the inflow area before entering the macrophyte zone, where aquatic plants are densely packed in shallow places to remove small and dissolved contaminants through rhizome networks and bed substrate media. The surplus water is directed toward the final functioning area, known as the high-flow bypass channel, which covers the entirety of the built wetlands system [85]. CWs can be classified based on their dominant vegetative life form, which includes free-floating, floating-leaved, rooted emergent, and submerged plants. The systems can be classified into two categories based on the water level in reference to the surface: free water surface and subsurface systems [86,87]. The subsurface flow direction can be classified as either vertical (either upward or downhill) or horizontal flow [86]. Sheng-Bing et al. [88] credited a three-stage CWs system, which used a combination of gravel, zeolite, and fly ash as a filtration medium,

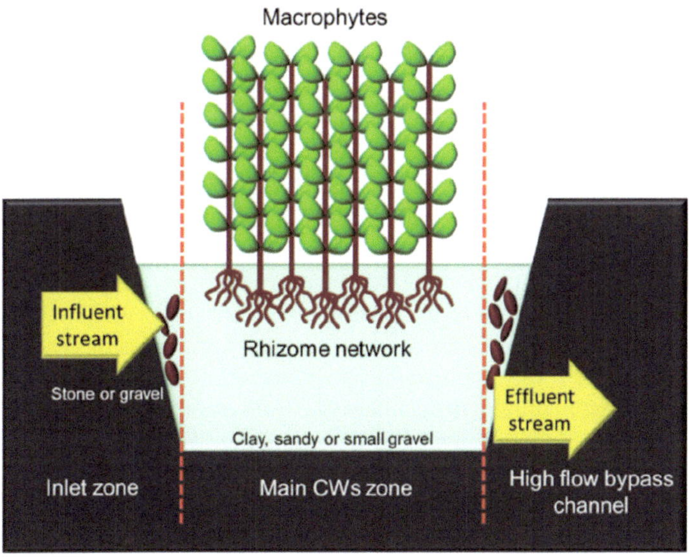

FIGURE 3.2 Wastewater within the CWs system.

for achieving 65.27% removal of total-P, 62.94% removal of total-N, and 61.85% removal of COD in eutrophic landscape river wastewater. Furthermore, the quality of the treated effluent, as measured by the reduction of turbidity, was improved in comparison to the initial influent after going through the CWs [88]. This finding is consistent with the study conducted by Pérez et al. [89] which has shown that horizontal-flow CWs are effective in reducing heavy metals, BOD, and COD in residential wastewater. However, it is necessary to use a combination of CWs systems to enhance nitrogen removal. In addition to their low energy requirements, CWs systems offer cost-effective operation and maintenance compared to conventional treatment methods. Furthermore, the implementation of CWs can also help with flood management and carbon sequestration in ecosystems [86].

3.5 CHALLENGES AND FUTURE PERSPECTIVES OF GREEN TECHNOLOGIES FOR WASTEWATER TREATMENT

The innovative clean technologies for treating wastewater have been interestingly focused to lessen the environmental concerns owing to the formation of by-product as well as the inefficiency of conventional processes to remove contaminated pollutants in released effluent. As illustrated in Section 3.3 and presented in Table 3.1, electricity, membrane, and natural associated processes are mainly applied to simplify the wastewater purification process as opposed to traditional procedures leading to shorter operating time, yet the efficiency of the treatment process and the quality of treated wastewater is still high. Table 3.2 indicates that the conventional treatment technique requires the further process in order to achieve high removal efficiency, e.g., coagulation followed by DAF process provided 39.2% of COD removal, on average, higher than the application of coagulation alone. The hazardous risk and the formation of undesired products from an exposure and reaction toward toxic chemicals, such as chlorine, which have high corrosive and toxic properties are reduced upon employing green technologies [90,91]. Jachimowski [92] reported that coagulation treatment caused an increase in chloride residual concentration in effluent. Even though the significant advantages of green treatment processes are inexpensive operating cost, requiring less space for installation, and high capability for pollutant removal, electricity is importantly needed to supply the electro-treatment process. In addition, the membrane-associated process commonly faces the challenge of membrane fouling caused by the diffusive transport of particulates and accordingly contributes to high and frequent maintenance costs to reduce membrane resistance [93,94]. Hence, it is essential to understand the properties of wastewater, including pH, composition, and ionic conditions, before implementing membrane filtration or MBRs in order to select appropriate membrane materials for the purification process and to mitigate future fouling difficulties [94]. Both open and closed PBRs could be the most suitable alternative method as their limitation is the simplest to cope with. The closed PBRs system has a low possibility of being contaminated during the treatment process and attained high removal efficiency (over 96% mineral removal), while the raceway ponds use the natural light which could save the energy supply [84]. Lastly, although the CWs require a large land for installation, this technique is

TABLE 3.2
Removal Efficiency of Different Wastewater Treatment Processes

Treatment Process	Removal Efficiency	References
Boiling	~92% of halomethanes	[38]
	69% of THMs	[42]
Coagulation	57.5% of COD	[48]
	96.8% of total suspended particles	
	90.5% of color index	
Coagulation, followed by DAF	~94.5% of oil droplets, COD, BOD, and turbidity	[43]
Single-layer sand filtration	76% of BOD	[56]
Double-layer sand filtration	85% of BOD	
Gravel-sand filtration	85% of COD	[54]
	99% of BOD	
Coarse sand and fine sand filtration	Over 94% of TOC and BOD	[55]
Aeration activated sludge	~60%–85% of COD	[61]
	85%–98% of BOD	
	40%–65% of adsorbable organically bound halogens	
	40%–85% of phosphorus	
	20%–50% of nitrogen	
Electrocoagulation	95.28% of COD	[63]
	92.30% of color	
	83.33% of turbidity	
Microfiltration	Over 94% of total petroleum hydrocarbon	[68]
MBRs	Over 70% of COD	[71]
MFCs	~83% of COD	[75–77]
PBRs	98% of phosphorus	[84]
	96% of nitrogen	
Constructed wetlands	61.85% of COD	[88]
	65.27% of total-P	
	62.94% of total-N	

mostly a natural process and less harmful to the environment. Thus, the application of this method could be achieved in abandoned areas to sustain feasible and practical use in long term.

3.6 CONCLUSION

The properties of wastewater are diverse and contingent upon the origins of its generation. Nevertheless, the overall makeup includes perilous elements such as excessive amounts of nitrogen and phosphate, organic contaminants, heavy metals, or harmful microorganisms. Since a significant amount of waste is typically discharged into natural water sources like rivers and oceans, there is a risk of contaminating drinking water and harming the aquatic ecosystem. To prevent this, regulations are put in place to limit the number of pollutants that can be released, ensuring the protection of living organisms and the preservation of the environment. Green technologies, such as bioreactors and membrane-associated processes, are more energy efficient and use fewer chemicals for wastewater treatment. As a result, they have fewer negative effects on the environment, human health, and animal life compared to conventional methods. Nevertheless, it is necessary to ascertain the ideal circumstances for each procedure, and the utilization of several methodologies is currently under assessment to get the utmost efficiency in wastewater treatment.

REFERENCES

1. BTL Liners Geomembrane Systems. Which Industries Produce the Most Wastewater? Available from: https://www.btlliners.com/which-industries-produce-the-most-wastewater#:~:text=Power%20plants%20are%20often%20the,nuclear%20power%20systems%20running%20smoothly. Accessed on 2 December 2023.
2. Jones ER, Van Vliet MT, Qadir M, et al. Country-level and gridded estimates of wastewater production, collection, treatment and reuse. *Earth Syst Sci Data.* 2021;13(2):237–254.
3. Jamal H. What Is the Composition of Wastewater? Environmental Engineering. 2017. Available from: https://www.aboutcivil.org/waste-water-composition. Accessed on 1 December 2023.
4. School Operations. Wastewater Treatment Systems. 2023. Available from: https://www2.education.vic.gov.au/pal/wastewater-treatment-systems/guidance/risks-hazards#:~:text=Contact%20with%20wastewater%20or%20its,such%20as%20hepatitis%20(liver%20infections). Accessed on 1 December 2023.
5. Alumichem. 6 Wastewater Parameters and Dealing with Discharge Limits. 2023. Available from: https://alumichem.com/6-wastewater-parameters-and-dealing-with-discharge-limits/#:~:text=In%20today%27s%20environmental%20framework%20we,%2DP)%20and%20Heavy%20Metals. Accessed on 2 December 2023.
6. NetSol Water Solutions Pvt. Ltd. How Do You Know If Your Wastewater Is Not Exceeding BOD Limits? Available from: https://www.netsolwater.com/how-do-you-know-if-your-wastewater-is-not-exceeding-bod-limits.php?blog=1650#:~:text=For%20discharge%20to%20sewer%20systems,typically%2010%20mg%20per%20litre. Accessed on 2 December 2023.
7. Environmental Encyclopedia. Suspended Solid. Available from: https://www.encyclopedia.com/environment/encyclopedias-almanacs-transcripts-and-maps/suspended-solid#:~:text=A%20typical%20TSS%20limit%20for,example%2C%20500%20mg%2FL. Accessed on 2 December 2023.
8. Bulter T. Sewage Parameters 6 Part 1: Total Nitrogen. Butler Manufacturing Services. Available from: https://butlerms.com/education-blog/sewage-parameters-6-part-1-total-nitrogen. Accessed on 2 December 2023.
9. Environmental Protection Agency. Effluent Standards and Limitations for Phosphorus. Department of Natural Resources. Chapter NR 217. 2011. Available from: https://www.epa.gov/sites/default/files/2014-12/documents/wiwqs-nr217.pdf. January Accessed on 2 December 2023.
10. Kinuthia GK, Ngure V, Beti D, et al. Levels of heavy metals in wastewater and soil samples from open drainage channels in Nairobi, Kenya: community health implication. *Sci Rep.* 2020;10(1):8434.

11. LibreTexts Biology. 17.3B: Wastewater and Sewage Treatment. Available from: https://bio.libretexts.org/Bookshelves/Microbiology/Microbiology_(Boundless)/17%3A_Industrial_Microbiology/17.03%3A_Wastewater_Treatment_and_Water_Purification/17.3B%3A_Wastewater_and_Sewage_Treatment#:~:text=Wastewater%20is%20treated%20in%203,and%20tertiary%20(extra%20filtration). Accessed on 2 December 2023.
12. Environmental Protection. Four Effective Processes to Treat Wastewater. 2018. Available from:https://eponline.com/articles/2018/02/08/four-effective-processes-to-treat-wastewater.aspx#:~:text=Four%20common%20ways%20to%20treat,about%20these%20processes%20in%20detail. Accessed on 2 December 2023.
13. Al-Abri M, Al-Ghafri B, Bora T, et al. Chlorination disadvantages and alternative routes for biofouling control in reverse osmosis desalination. *NPJ Clean Water.* 2019;2(1):2.
14. Yaqoob AA, Guerrero–Barajas C, Ahmad A, et al. Advanced technologies for wastewater treatment. In Shahid-ul-Islam, A. H. Shalla, M Shahadat (eds.), *Green Chemistry for Sustainable Water Purification.* 2023;179–202. Scrivener Publishing.
15. EPA South Australia. Black and Grey Water Management. 2021. Available from: https://www.epa.sa.gov.au/environmental_info/water_quality/programs/grey_and_black_water_discharge#:~:text=Definitions,includes%20both%20blackwater%20and%20greywater. Accessed on 3 December 2023.
16. Ghani LA, Mahmood NZ. Modeling domestic wastewater pathways on household system using the socio-MFA techniques. *Ecol Modell.* 2023;480:110328.
17. NGO International. Domestic Wastewater Treatment Technology. Available from: https://ngoenvironment.com/en/Tin-tuc-n5-DOMESTIC-WASTEWATER-TREATMENT-TECHNOLOGY-d138.html. Accessed on 3 December 2023.
18. Huang MH, Li YM, Gu GW. Chemical composition of organic matters in domestic wastewater. *Desalination.* 2010;262(1–3):36–42.
19. Yang Y, Hu Y, Duan A, et al. Characterization of preconcentrated domestic wastewater toward efficient bioenergy recovery: applying size fractionation, chemical composition and biomethane potential assay. *Bioresour Technol.* 2021;319:124144.
20. Fluence News Team. Decentralized Versus Centralized Treatment. 2021. Available from: https://www.fluencecorp.com/decentralized-vs-centralized-treatment/. Accessed on 3 December 2023.
21. United States Environmental Protection Agency. Small and Rural Wastewater Systems. 2023. Available from: https://www.epa.gov/small-and-rural-wastewater-systems/about-small-wastewater-systems. Accessed on 3 December 2023.
22. Feng C, Zhang S, Wang Y, et al. Synchronous removal of ammonium and phosphate from swine wastewater by two agricultural waste based adsorbents: performance and mechanisms. *Bioresour Technol.* 2020;307:123231.
23. Liu Y, Wang J. Reduction of nitrate by zero valent iron (ZVI)-based materials: a review. *Sci Total Environ.* 2019;671:388–403.
24. National Oceanic and Atmospheric Administration. What Is Eutrophication? 2023. Available from: https://oceanservice.noaa.gov/facts/eutrophication.html#:~:text=the%20nation%27s%20estuaries.-,Harmful%20algal%20blooms%2C%20dead%20zones%2C%20and%20fish%20kills%20are%20the,to%20estuaries%20and%20coastal%20waters. Accessed on 3 December 2023.
25. United States Environmental Protection Agency. The Effects: Dead Zones and Harmful Algal Blooms. 2023. Available from: https://www.epa.gov/nutrientpollution/effects-dead-zones-and-harmful-algal-blooms#:~:text=in%20your%20browser.-,Dead%20Zones,excess%20nutrients%20from%20upstream%20sources. Accessed on 3 December 2023.

26. Şentürk E, Ince M, Engin GO. Treatment efficiency and VFA composition of a thermophilic anaerobic contact reactor treating food industry wastewater. *J Hazard Mater.* 2010;176(1–3):843–848.
27. Haq I, Kalamdhad AS. Enhanced biodegradation of toxic pollutants from paper industry wastewater using *Pseudomonas* sp. immobilized in composite biocarriers and its toxicity evaluation. *Bioresour Technol Rep.* 2023;24:101674.
28. Du Z, Song J, Du S, et al. Numerical modeling of geological sequestration of brine wastewater due to coal mining in the Ordos Basin, China. *Sci Total Environ.* 2023;908:168580.
29. Fluence News Team. Wastewater Treatment for the Mining Industry. 2020. Available from: https://www.fluencecorp.com/wastewater-treatment-for-the-mining-industry/. Accessed on 4 December 2023.
30. Barałkiewicz D, Chudzińska M, Szpakowska B, et al. Storm water contamination and its effect on the quality of urban surface waters. *Environ Monit Assess.* 2014;186:6789–6803.
31. Kaur R, Gupta K. Blue-Green Infrastructure (BGI) network in urban areas for sustainable storm water management: a geospatial approach. *City Environ Interact.* 2022;16:100087.
32. Suchowska-Kisielewicz M, Nowogoński I. Influence of storms on the emission of pollutants from sewage into waters. *Sci Rep.* 2021;11(1):18788.
33. Gulliver JS, Erickson AJ, Weiss PT (eds.). Stormwater Treatment: Assessment and Maintenance. University of Minnesota, St. Anthony Falls Laboratory. Minneapolis, MN. 2010. Available from: https://stormwaterbook.safl.umn.edu/. Accessed on 6 December 2023.
34. Wang Y, Peng F, Zhao R, et al. Removal and transformation of disinfection by-products in water during boiling treatment. *Chemosphere.* 2023;326:138426.
35. Drinking-Water. Drinking Water Treatment – Chlorination. 2019. Available from: https://drinking-water.extension.org/drinking-water-treatment-chlorination/#:~:text=The%20different%20types%20of%20chlorine,when%20the%20chlorine%20demand%20fluctuates. Accessed on 4 December 2023.
36. Lenntech. Disinfectants-Chlorine. Available from: https://www.lenntech.com/processes/disinfection/chemical/disinfectants-chlorine.htm. Accessed on 4 December 2023.
37. International Agency for Research on Cancer. Chlorinated drinking-water; chlorination by-products; some other halogenated compounds; cobalt and cobalt compounds. *IARC Monogr Eval Carcinog Risks Humans.* 1990;52:7–515.
38. Ma S, Gan Y, Chen B, et al. Understanding and exploring the potentials of household water treatment methods for volatile disinfection by-products control: kinetics, mechanisms, and influencing factors. *J Hazard Mater.* 2017;321:509–516.
39. Rook JJ. Formation of haloform during chlorination of natural water. *Water Treat Exam.* 1972;21:259.
40. New York State Health Department. The Facts about Chlorine. Department of Health. 2004. Available from: https://www.health.ny.gov/environmental/emergency/chemical_terrorism/chlorine_tech.htm#:~:text=What%20is%20chlorine%27s%20mechanism%20of,which%20produce%20corrosive%20tissue%20damage. Accessed on 4 December 2023.
41. Peterson RS, Hill Jr CG, Amundson CH. Effects of temperature on the hydrolysis of lactose by immobilized β-galactosidase in a capillary bed reactor. *Biotechnol Bioeng.* 1989;34(4):429–437.
42. Batterman S, Huang AT, Wang S, et al. Reduction of ingestion exposure to trihalomethanes due to volatilization. *Environ Sci Technol.* 2000;34(20):4418–4424.
43. Abo-El Ela SI, Nawar SS. Treatment of wastewater from an oil and soap factory via dissolved air flotation. *Environ Int.* 1980;4(1):47–52.

44. AOS Treatment Solutions. What Is Sedimentation in Water Treatment? 2018. Available from: https://aosts.com/what-is-sedimentation-in-water-treatment-types-settling-tanks/#:~:text=A%20sedimentation%20tank%20in%20wastewater,aid%20in%20the%20settling%20process. Accessed on 6 December 2023.
45. Nathanson JA. Water Treatment. Coagulation and Flocculation. Britannica. Available from: https://www.britannica.com/technology/water-supply-system/Chlorination. Accessed on 5 December 2023.
46. Greaves J. How are coagulants and flocculants used in water and wastewater treatment? WCS Group. 2022. Available from: https://www.wcs-group.co.uk/wcs-blog/coagulants-flocculants-wastewater-treatment#types. Accessed on 6 December 2023.
47. Bubakova P, Pivokonsky M, Filip P. Effect of shear rate on aggregate size and structure in the process of aggregation and at steady state. *Powder Technol.* 2013;235:540–549.
48. Guvenc SY, Can-Güven E, Çifçi Dİ, et al. Biodegradability enhancement and sequential treatment of real chemical industry wastewater by chemical coagulation and Electro-Fenton processes. *Chem Eng Process Intensif.* 2023;194:109598.
49. Leppinen DM. A kinetic model of dissolved air flotation including the effects of interparticle forces. *J Water Supply Res Technol-AQUA.* 2000;49(5):259–268.
50. Rajapakse N, Zargar M, Sen T, et al. Effects of influent physicochemical characteristics on air dissolution, bubble size and rise velocity in dissolved air flotation: a review. *Sep Purif Technol.* 2022;289:120772.
51. Resl F. Addition of lime to improve sludge settling helps defer capital spending. *Water New Zealand Annual Conf.* City Care Ltd, Graham Morris, New Plymouth District Council. 2012.
52. Bruni MA, Spuhler D. Slow Sand Filtration. Available from: https://sswm.info/sswm-university-course/module-6-disaster-situations-planning-and-preparedness/further-resources-0/slow-sand-filtration. Accessed on 7 December 2023.
53. Bruni M, Spuhler D. Rapid Sand Filtration. Available from: https://sswm.info/sswm-university-course/module-6-disaster-situations-planning-and-preparedness/further-resources-0/rapid-sand-filtration#:~:text=Rapid%20sand%20filtration%20is%20a,and%20rapid%20pressure%20sand%20filters. Accessed on 7 December 2023.
54. Xi J, Mancl KM, Tuovinen OH. December. Biological treatment of cheese-processing effluents with gravel/sand filtration. *In Proceedings of 8th International Symposium on Agricultural Food Process Waste.* ASAE, St. Joseph, MI. 2000;151–157.
55. Kang YW, Mancl KM, Tuovinen OH. Treatment of turkey processing wastewater with sand filtration. *Bioresour Technol.* 2007;98(7):1460–1466.
56. Liu Q, Mancl K, Tuovinen OH. High fat wastewater remediation using layered sand filter biofilm systems. *In Proceedings of International Symposium on Animal Agricultural Food Process Waste.* ASAE, St. Joseph, MI. 2000;242–248.
57. Almojjly A, Johnson D, Oatley-Radcliffe DL, et al. Removal of oil from oil-water emulsion by hybrid coagulation/sand filter as pre-treatment. *J Water Process Eng.* 2018;26:17–27.
58. Vissers Sales Corporation. Benefits and Drawbacks of Industrial Filtration Systems. 2023. Available from: https://visserssales.com/benefits-and-drawbacks-of-industrial-filtration-systems/#:~:text=Disadvantages%3A,all%20impurities%20from%20the%20liquid. Accessed on 5 December 2023.
59. Shankar R, Kumar S, Prasad AK, et al. Biological wastewater treatment plants (WWTPs) for industrial wastewater. In M. Shah, S. Rodriguez-Couto (eds.), *Microbial Ecology of Wastewater Treatment Plants.* 2021;193–216. Elsevier.
60. Yap CC, Chan YJ. Approaching zero liquid discharge concept using high-rate integrated pilot-scale bioreactor in the palm oil mill effluent (POME) treatment. In AB Mohammad, WL Ang (eds.), *Integrated and Hybrid Process Technology for Water and Wastewater Treatment.* 2021;641–667. Elsevier.

61. Bajpai P (ed.). External treatment technologies used for pulp and paper mill effluents. In *Pulp and Paper Industry*. 2017;71–92. Elsevier.
62. Tuser C. What Is Activated Sludge? 2021. Available from: https://www.wwdmag.com/sludge-and-biosolids/article/10939479/what-is-activated-sludge#:~:text=The%20activated%20sludge%20process%20is%20a%20means%20of%20treating%20both,from%20wastewater,%20producing%20quality%20effluent. Accessed on 16 December 2023.
63. Ebba M, Asaithambi P, Alemayehu E. Development of electrocoagulation process for wastewater treatment: optimization by response surface methodology. *Heliyon*. 2022;8(5):e09383.
64. Sillanpää M, Shestakova M (eds.). Emerging and combined electrochemical methods. In *Electrochemical Water Treatment Methods: Fundamentals, Methods and Full Scale Applications*. Butterworth-Heinemann. 2017;131–225.
65. Benjamin MM, Lawler DF. *Water Quality Engineering: Physical/Chemical Treatment Processes*. John Wiley & Sons. 2013.
66. Díaz-Reinoso B. Concentration and purification of seaweed extracts using membrane technologies. In MD. Torres, S Kraan, H Dominguez (eds.), *Sustainable Seaweed Technologies*. Elsevier. 2020;371–390.
67. Eccles H. Ion exchange-future challenges/opportunities in environmental clean-up. In A Dyer, MJ Hudson, PA Williams (eds.), *Progress in Ion Exchange*. Woodhead Publishing. 1997;245–259.
68. Bhattacharyya A, Liu L, Walsh M, et al. Membrane technology for treating decanted oily wastewater from marine oil spill operations: comparison between membrane filtration and membrane bioreactor. *Mar Pollut Bull*. 2023;194:115397.
69. Martin KJ, Nerenberg R. The membrane biofilm reactor (MBfR) for water and wastewater treatment: principles, applications, and recent developments. *Bioresour Technol*. 2012;122:83–94.
70. Ødegaard H. Innovations in wastewater treatment: the moving bed biofilm process. *Water Sci Technol*. 2006;53(9):17–33.
71. Di Bella G, Di Prima N, Di Trapani D, et al. Performance of membrane bioreactor (MBR) systems for the treatment of shipboard slops: assessment of hydrocarbon biodegradation and biomass activity under salinity variation. *J Hazard Mater*. 2015;300:765–778.
72. Soltani S, Mowla D, Vossoughi M, et al. Experimental investigation of oily water treatment by membrane bioreactor. *Desalination*. 2010;250(2):598–600.
73. Ameen M, Zafar M, Ahmad M, et al. Prospects of Bioenergy Development in Future. In MR Rahimpour (ed.), *Reference Module in Earth Systems and Environmental Sciences*. Elsevier. 2023;497–508.
74. Roy S, Marzorati S, Schievano A, et al. Microbial fuel cells. In MA Abraham (ed.). *Encyclopedia of Sustainable Technologies*. Elsevier, Amsterdam, The Netherlands. 2017;245–259.
75. Abourached C, English MJ, Liu H. Wastewater treatment by microbial fuel cell (MFC) prior irrigation water reuse. *J Clean Prod*. 2016;137:144–149.
76. Feng Y, He W, Liu J, et al. A horizontal plug flow and stackable pilot microbial fuel cell for municipal wastewater treatment. *Bioresour Technol*. 2014;156:132–138.
77. Nimje VR, Chen CY, Chen HR, et al. Comparative bioelectricity production from various wastewaters in microbial fuel cells using mixed cultures and a pure strain of Shewanella oneidensis. *Bioresour Technol*. 2012;104:315–323.
78. García-Garibay M, Gómez-Ruiz L, Cruz-Guerrero AE, et al. Single cell protein | The algae. In CA Batt, ML Tortorello (eds.), *Encyclopedia of Food Microbiology* (2nd ed.). Academic Press. 2014;425–430.
79. Ramírez-Mérida LG, Zepka LQ, Jacob-Lopes E. Why does the photobioreactors fail. *J Bioprocess Biotech*. 2015;5(7):239.

80. Satpati GG, Pal R. Microalgae-biomass to biodiesel: a review. *J Algal Biomass Util.* 2018;9(4):11–37.
81. Kumar V, Jain SM. Plants and algae species: promising renewable energy production source. *Emir J Food Agric.* 2014;26:679–692.
82. Masojídek J, Torzillo G (eds.). Mass cultivation of freshwater microalgae. In *Reference Module in Earth Systems and Environmental Sciences.* Elsevier. 2014; 2226–2235.
83. Razzak SA, Bahar K, Islam KO, et al. Microalgae cultivation in photobioreactors: sustainable solutions for a greener future. *Green Chem Eng.* 2023;5:418–439.
84. Yan H, Chen Z, Ngo HH, et al. Nitrogen and phosphorus removal performance of sequential batch operation for algal cultivation through suspended-solid phase photobioreactor. *Bioresour Technol.* 2023;393:130143.
85. Melbourne Water. Constructed Wetlands. 2022. Available from: https://www.melbournewater.com.au/building-and-works/stormwater-management/options-treating-stormwater/constructed-wetlands#:~:text=Constructed%20wetlands%20are%20a%20series,our%20creeks%2C%20rivers%20and%20oceans. Accessed on 12 December 2023.
86. Vymazal J. Constructed wetlands for wastewater treatment. *Water.* 2010;2(3):530–549.
87. Brix H, Schierup HH. The use of aquatic macrophytes in water-pollution control. *Ambio.* 1989;28(2):100–107.
88. Sheng-Bing HE, Li YAN, Hai-Nan KONG, et al. Treatment efficiencies of constructed wetlands for eutrophic landscape river water. *Pedosphere.* 2007;17(4):522–528.
89. Pérez Y, Vargas E, García-Cortés D, et al. Efficiency and effectiveness of systems for the treatment of domestic wastewater based on subsurface flow constructed wetlands in Jarabacoa, Dominican Republic. *Water Sci Eng.* 2023;17:118–128.
90. Chemtech International Inc. Advantages and Disadvantages of Electrocoagulation Water Treatment. Waste-Water Treatment. 2020. Available from: https://chemtech-us.com/advantages-and-disadvantages-of-electrocoagulation-water-treatment/#:~:text=Electrocoagulation%20treats%20water%20without%20the,of%20the%20operation%20up%20front. Accessed on 29 February 2024.
91. US Environmental Protection Agency. Wastewater Technology Fact Sheet: Chlorine Disinfection. 1999. Available from: https://www3.epa.gov/npdes/pubs/chlo.pdf. Accessed on 29 February 2024.
92. Jachimowski A. Efficiency of Pollutant Removal from Surface Water in a Conventional Water Treatment System. *In E3S Web Conf. EDP Sci.* 2019;86:00030.
93. Caloris Engineering LLC. Membrane Filtration Problems. 2021. Available from: https://caloris.com/blog/membrane-filtration-problems/#:~:text=Feed%20quality%2C%20membrane%20life%2C%20fouling,are%20common%20membrane%20filtration%20problems. Accessed on 29 February 2024.
94. Li H, Chen V. Membrane fouling and cleaning in food and bioprocessing. In ZF Cui, HS Muralidhara (ed.), *Membrane Technology.* Butterworth-Heinemann. 2010;213–254.

4 Harnessing Microalgae for Wastewater Treatment

Optimizing Cultivation, Pollutant Removal, and Photobioreactor Technologies for a Sustainable Future

Saadia Ijaz and Mehboob Ahmed

This title emphasizes the active role of microalgae, optimization efforts, and the forward-looking perspective of the technology.

4.1 INTRODUCTION

Water makes up nearly 71% of the surface of planet Earth and is essential to life (Issac and Kandasubramanian, 2021). Once, an English poet, W. H. Auden, said, "Thousands have lived without love, not one without water" (Blair, 2015). The generation of wastewater (WW) from the overuse of clean water has become one of the most serious environmental issues on a global scale owing to rapid urbanization, explosive population growth, and industrialization (Fida et al., 2023). WW is 99.9% water, while the remaining 0.1% are suspended solids, natural and synthetic organic compounds, inorganic compounds, nutrients (nitrates and phosphates), metals, salts, minerals, pathogens with high chemical oxygen demand (COD) and biological oxygen demand (BOD) (Kumari et al., 2023). These contaminants in WW have an adverse effect on the water quality because of their toxicity, propensity for bioaccumulation, persistence, and ability to be transported and deposited over long distances by the atmosphere (Whelan et al., 2022). Recently, growing worries have been expressed regarding the presence of emerging pollutants (natural or synthetic organic compounds) in waterbodies and the associated risks they pose to aquatic life as well as human health (Shafi et al., 2023).

According to the United Nations' estimate, nearly 56% of freshwater is being converted into 2,212 km^3 of WW annually, of which nearly 80% is released without treatment (Morais et al., 2021). It is therefore critical to find sustainable solutions

 DOI: 10.1201/9781003459286-4

for wastewater treatment (WWT) and reuse because it is predicted that by 2030, the world will have a 40% shortage of clean water (Koop et al., 2022). Among these, the cultivation of microalgae may offer a workable substitute for the current conventional WWT technologies.

4.2 CONVENTIONAL WWT

Conventionally, the WW is systematically treated with physical, chemical, and biological methods (Pariente et al., 2022). The treatment starts with the preliminary process that goes to primary, secondary, tertiary, and/or advanced treatment for special pollutants and pathogens removal (Koul et al., 2022). The physical WWT processes include the use of screens; clarifiers (settling tanks or sedimentation); grit chambers; floatation; filtration; and, sometimes, flow equalization techniques to remove large non-biodegradable and floating solids (Wang and Wang, 2021). Chemical treatment targets lowering the turbidity of WW by adjusting the pH and eliminating dissolved matter to improve the quality of the water (Abdelfattah et al., 2023). Chlorination/chloramination, ozonation, ultraviolet light treatment, and flocculation are the main techniques used in secondary chemical WWT for disinfection; however, these techniques have some limitations because of high cost, dewatering, and increased maintenance (Acharya et al., 2020). Contrarily, biological methods rely on the biodegradation, bleaching, and biotransformation of pollutants by microorganisms like bacteria, fungi, and microalgae into biomass and gaseous byproducts such as CO_2, N_2, CH_4, SO_2, etc., thereby decreasing the levels of BOD and COD (Sahni, Hans and Kumar, 2023).

4.3 MICROALGAE-BASED WWT

The importance of treating sewage and WW with microalgae has long been acknowledged as a sustainable alternative (Agersborg and Hatfield, 1929; Cotton, 1910). The process of treating WW using microalgae was initially done by Oswald and Golueke (1960). Since then, microalgae have been studied extensively for WW reclamation, which is presently becoming increasingly popular for circular bioeconomy methods (Plöhn et al., 2021). Microalgae have shown to be a green and sustainable alternative for high-energy-consuming traditional biological treatment methods (Yap et al., 2021). Utilizing microalgae in WWT is a viable and economical way to bio-fix CO_2, besides being a biomass source that is renewable (Khoo et al., 2020). The idea behind the application of microalgae in WWT rests in their potential to utilize carbon (inorganic/inorganic), nitrogen, and phosphorus for their growth, all the while decreasing the amount of these pollutants in the water (Mathew et al., 2022). Microalgae can assimilate both organic (urea) and inorganic nitrogen (NH_4^+/NH_3, NO_3, NO_2, etc.) (Mohsenpour et al., 2021). Due to their photosynthetic nature, microalgae can provide molecular O_2 in WWT system which is required for aerobic heterotrophic bacteria to bioremediate carbonaceous pollutants (Ali, Khalid and Ajarem, 2023). The removal of inorganic nitrogen (N) and phosphorus (P) by microalgal–bacterial consortia from WW only needs to take place once, eliminating the requirement for several treatment stages resulting in lower greenhouse gas emission since microalgae assimilate NH_3 and PO_4 directly (Verma et al., 2020). Studies have shown that the WWT by microalgae in conjunction with heterotrophic bacteria results in a minimal N_2O release with the emission factor of 0.0047% g N_2O-N g^{-1} N-input (Mohsenpour et al., 2021). According to Aqualia, one of Europe's largest WWT companies which treats up to 500 Mm^3 of urban WW annually, conventional treatment of WW uses an average of 0.5 kWh/m^3, with a price of 0.2 Euro/m^3 or half the price of the energy used (Ambrosio et al., 2022). Thus, annually, this company single-handedly removes up to 25,000 and 5,000 t/year of nitrogen and phosphorus from the environment, respectively (Kaloudas, Pavlova and Penchovsky 2021). Biomass of up to 0.5 Mt/year could be produced by growing microalgae with this much nitrogen and phosphorous, which is 20 times more than what is currently produced globally (Ali et al., 2022). Therefore, significant environmental advantages come from using microalgae to treat WW and recycle nitrogen and phosphorus while consuming less energy and releasing fewer greenhouse gases (Nagarajan et al., 2021). Because interest in microalgal WWT has risen rapidly, the transition from conventional WWT processes to a low-carbon microalgal biorefinery has shifted the dynamics towards a resource-diversified circular bioeconomy (Ali et al., 2022). Since the key challenges of nutritional absorption and the substantial microalgal population have been considerably addressed, microalgae-based WWT has become a significant competitor to traditional WWT (Abdelfattah et al., 2023). Some of the advantages of microalgal-based WWT over conventional treatment are given in Table 4.1.

TABLE 4.1
Advantages of Microalgae-Based WWT over Conventional WWT

Conventional Wastewater Treatment	Microalgal-Based Wastewater Treatment
• Highly complex, expensive, and requires organic carbon sources	• Sustainable and cost-effective approach. No organic carbon is required
• Suspended solids are treated with gravity settling	• Prior settling of suspended solids is required due to poor light diffusion
• Organic pollutant is biodegraded by aerobic and anaerobic bacteria	• Organic pollutants are aerobically degraded by microalgae-heterotrophic bacterial consortia
• Higher energy is required for activated sludge through mechanical diffusers and pumps for aerobic degradation and to keep dissolved oxygen within bounds for bacterial metabolism	• Microalgae release O_2 for aerobic degradation
• Less efficient to recover heavy metals. Requires chemical precipitation, flocculation, ion exchange, adsorption, ultrafiltration, reverse osmosis	• Efficient in recovering heavy metals by active uptake into the cytoplasm of microalgae as well as by adsorption on microalgal cells
• High generation of greenhouse gases such as CO_2	• Reduced release of greenhouse gases

(*Continued*)

TABLE 4.1 (*Continued*)
Advantages of Microalgae-Based WWT over Conventional WWT

Conventional Wastewater Treatment	Microalgal-Based Wastewater Treatment
• Nitrogen removal is done in the tertiary stage, nutrients are removed by nitrification where wastewater containing NH_3 is oxidized to nitrite by autotrophic ammonia-oxidizing bacteria while nitrite is oxidized to nitrate by nitrite-oxidizing bacteria. NO_3 and NO_2 are then converted into N_2 during denitrification	• Nitrogen removal from wastewater is done through photosynthesis by assimilating nutrients in the form of inorganic ions such as NH^{4+}, NO^{3-}
• Oxygenic conditions are required for nitrification, whereas anoxygenic conditions are required for denitrification	• Nitrification and denitrification processes occur simultaneously in facultative oxidation ditches, ponds, or lagoons
• Ammonia is removed by stripping it at alkaline pH and high temperature through air or steam	• Volatilization of ammonia is done by increasing pH and dissolving oxygen
• In the absence of nitrate and oxygen, phosphorus is removed by enhanced biological phosphorus removal (EBPR)	• Phosphorus is removed through uptake by microalgal cells
• Pathogens are removed by treating wastewater with UV light, ozonation, chlorination, chemical disinfection	• Pathogens are removed by UV radiations due to high surface-to-volume ratios. Disinfection of pathogen occurs by raising pH and dissolving oxygen
• Secondary pollution from highly toxic sludge generation	• No toxic sludge generation and subsequent secondary pollution
• Do not have any other purpose	• High-value product isolation from biomass. Co-products and biomass used as sustainable animal and aquaculture feed. Biofuel production and biofertilizers

4.4 TYPES OF MICROALGAE

Although the term "microalgae" is commonly used for eukaryotic unicellular algae, their prokaryotic counterparts, cyanobacteria, and the blue-green algae, are also included in this term as they share many of the same functional and metabolic characteristics (Thoré et al., 2023).

4.4.1 Eukaryotic Microalgae

Eukaryotic microalgae are polyphyletic, oxygenic, photosynthetic, unicellular organisms that are producing 40% of O_2 on earth (Hachicha et al., 2022). They have high CO_2 fixation rates that are substantially higher than those of land plants since they can efficiently capture solar energy to synthesize their biomass (Ma et al., 2022). They are also metabolically flexible than land plants as they can shift from an autotrophic mode of nutrition to a heterotrophic or mixotrophic mode depending upon the environmental conditions (Vidotti et al., 2020). They also have high growth rates compared to land plants as their cell physiology and stress-tolerance ability are robust (Sharma et al., 2022). Microalgae are a rich source of proteins, lipids, omega-3 fatty acids, sugars, vitamins, pigments, enzymes, minerals, and other secondary metabolites of importance (Nagarajan et al., 2021). These high-value bioactive products are intensively exploited not only in various biotechnology fields but in nutraceutical, pharmaceutical, cosmeceutical, biomedical, and agricultural fields as well (Ali et al., 2022). Additionally, microalgae offer economic advantages in WW reclamation due to their ability to capture CO_2, remove nutrients, recover resources, and produce oxygen (Mu et al., 2021). Microalgae are well recognized for biologically removing carbon and nutrients (N and P) from municipal, agricultural, brewing, refinery, and industrial WW effluents (Zafar et al., 2023). The consortia of microalgae, bacteria, and biofilms have the capability to remove heavy metals from industrial WW and acid mine drainage (Zhao et al., 2023). The primary benefit of employing microalgae for WWT is their ability to provide molecular oxygen to the aerobic heterotrophic bacteria through photosynthesis while, at the same time, utilizing CO_2 produced by these bacteria (Iqbal et al., 2022). In this way, microalgae bypass mechanical aeration, which is costly and requires high energy consumption, which usually accounts for 45%–75% in conventional WWT systems (Zhu et al., 2019). Microalgal strains *belonging to genera Chlorella, Scenedesmus, Phaeodactylum, Botryococcus, Pediastrum, Nitzschia, Cosmarium, Micractinium, Chlamydomonas, Actinastrum,* etc., have been used in WWT in order to reduce COD, BOD, total nitrogen, and phosphorus (Cuellar-Bermudez et al., 2017).

4.4.2 Prokaryotic Microalgae

Cyanobacteria, previously known as blue-green algae, are a group of gram-negative eubacteria that can perform oxygenic photosynthesis like that of eukaryotic algae and plants (Jassim, Awadh and Al-Amery, 2023). These organisms also produce accessory photosynthetic pigments like phycobiliproteins which give them their blue-green appearance (Richardson, 2022). Cyanobacteria can be found inhabiting all biotopes on earth even those deemed extremely hostile for life (Kaštovský, 2024). They are ecologically well characterized for recycling nutrients and minerals (Singh et al., 2016). Due to their long evolutionary history, they have acquired high plasticity, making them adaptable to varying physiological conditions (Willis and Woodhouse, 2020). Majority of the cyanobacteria are diazotrophs that can fix atmospheric nitrogen along with production of molecular O_2, playing an essential role in biogeochemical cycles on earth (Upadhyay et al., 2021). They can assimilate nutrients like nitrogen (NO_2, NO_3, NH_4^{2-}) and phosphate (PO_4^{3-}) which allow them to thrive more robustly in sewage, agricultural, or urban effluents (Fernandes et al., 2022). The use of cyanobacteria in wastewater treatment plants (WWTPs) opens up the possibility of producing clean water with commercially useful byproducts such as proteins, carbohydrates, lipids, vitamins, omega-3 fatty acids, enzymes, amino acids, etc., similar to eukaryotic microalgae (Gupta et al., 2021). Additionally, the ability of cyanobacteria to consume CO_2 during photosynthesis contributes to the development of WWTPs with a carbon-neutral production process (Brennan and Owende, 2010). Various types of strains, such as *Oscillatoria, Spirulina, Leptolyngbya,*

TABLE 4.2
Eukaryotic and Prokaryotic Microalgal Strains Used for Different Wastewater Treatment

Microalgal Strain	Wastewater Source	Nutrient Removal Efficiency	References
Eukaryotic			
Scenedesmus quadracauda	Agricultural	92.15% TN, 100% P	Daneshvar et al. (2019)
Chlorella vulgaris	Domestic	62% N, 100%	Rinna et al. (2017)
Parachlorella kessleri	Municipal	99% N, 82% P	Aketo et al. (2020)
Tribonema sp.	Swine	89% N, 73% P	Cheng et al. (2020)
Scenedesmus obliquus	Domestic	94% N, 99% P	Fan et al. (2020b)
Botryococcus braunii	Domestic	63% N, 100%	Rinna et al. (2017)
C. sorokiniana	Urban	94% N, 100%	Álvarez-Díaz et al. (2017)
Desmodesmus **sp.**	Sewage	96% N, 85%	Mehrabadi, Farid and Craggs (2017)
Chlorella vulgaris, Scenedesmus obliquus, and Nannochloropsis **sp.**	Saline Municipal	100% N, 100% P	Fallahi et al. (2020)
Didymogenes **sp.**	Municipal	>90% TN, 99% P	Tang et al. (2023)
Prokaryotic			
Synechocystis **sp.**	Swine	76% N, 71% P	Cheng et al. (2020)
Nostoc sp. *and Tolypothrix* **sp.**	Municipal	>95% N, 35%–78% P	Arias, Uggetti, and García (2020)
Spirulina platensis	Swine	91.24% N, 87.44% P	Lu et al. (2021)
Arthrospira platensis	Cassava processing	91% N, 55% P	Araujo et al. (2021)
Scytonema hyalinum	Municipal	91% TN, 97% P	Wu et al. (2020)
Acaryochloris marina Oscillatoria **sp.** *and Pleurocapsa* **sp.**	Urban	53% N, and 88% P	Shahid et al. (2021)
Synechocystis **sp.**	Urine	87% N, and 99% P	Canizales et al. (2023)

TN, Total nitrogen; P, Phosphorus.

Phormidium, are being used as axenic or mixed cultures for WW remediation to reduce COD, BOD, total nitrogen, and phosphorus (Koul, Sharma, and Shah, 2022). Various strains of microalgae and cyanobacteria for the treatment of different WWs are given in Table 4.2.

4.5 MICROALGAL WWT SYSTEM COMPOSITIONS

One may think of microalgal-based WWT systems as a collection of microalgal cells that can either occur as monoculture (single strain), polyculture (various strains), and/or in combination with heterotrophic bacteria (microalgae–bacteria consortia)

(Shahid et al., 2020). For the treatment of WW, a number of alternative microalgal system compositions have been investigated at the lab and pilot scale, taking into account various conditions with either real or synthetic WW.

4.5.1 Monocultures and Polycultures

Microalgal monocultures and well-defined microalgae–bacteria consortia are basically used at the lab scale in artificially defined, sterile WW in order to understand the mechanism of WWT and the effects of various biotic and abiotic factors on their performance (Amaro et al., 2023). In contrast, microalgal polycultures, with or without bacterial consortium, are among the most prevalent microalgal WWT systems researched for real-case scenarios, as it is exceedingly difficult to sustain a monoculture of microalgae in WW given the diversity of invasive microalgae, heterotrophic bacteria, and various other microbes in the WW (González-González and de-Bashan, 2021). When compared to monocultures, polycultures of microalgae present two key benefits, which include increased nutrient absorption efficiency resulting in higher biomass production and increased bioremediation capability due to metabolic diversity and abiotic stress adaptation by different species (Ray, Nayak and Ghosh, 2022). The interaction of the various microalgal species, however, should be taken into consideration as it may lead to nutritional competition and the release of allelochemicals under stressful environmental conditions, which may prevent their growth (Amaro et al., 2023). Thus, careful selection of strains should be considered.

4.5.2 Microalgal–Bacterial Consortia

Previously thought as contamination, the synergistic relationship between microalgae and heterotrophic bacteria has proven to be an effective way to treat WW (Chan et al., 2022). Microalgae provide bacteria molecular oxygen and organic carbon, and, in turn, bacteria supply CO_2, growth factors, and essential vitamins which boost microalgal growth (Lafi, Panu and Liao, 2023). Furthermore, microalgal biomass surface can serve as a niche for bacteria to get protection from stressful environmental conditions (Ashraf, Ahmed and Lu, 2023). Microalgae also provide extracellular metabolites to improve the growth of bacteria (Amaro et al., 2023). Bacteria can also make synergistic relationships with microalgae through releasing signalling molecules and hormones like *N*-acyl-homoserine lactones and indole-3-acetic acid, respectively (Zhang et al., 2020a). *N*-acyl-homoserine lactones help in the creation of ecological niche formation by stimulation of granule development with microalgae, whereas indole-3-acetic acid stimulates growth of microalgae (Astafyeva et al., 2022). Horizontal gene transfer is another interaction which occurs between microalgae–bacteria consortia. Bacteria that are in association with microalgae are often reported to contain active genes for vitamin production, host surface recognition and adhesion, biofilm development, polysaccharide degradation, and different defensive mechanisms such as antibiotic production (Mugnai, Derossi and Hendlin, 2023). By this process, microalgae acquire features which also give them the ability to survive under stress. Thus, the application of bacteria–microalgae consortia for the treatment of WW can lead to enhanced uptake of pollutants while reducing high levels

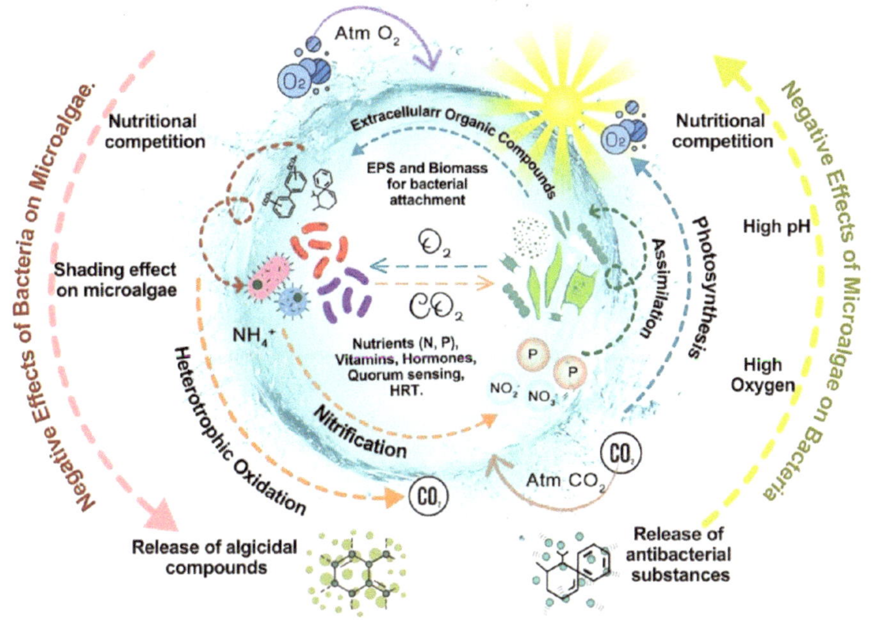

FIGURE 4.1 Schematic illustration of microalgae–bacteria interactions and exchange of essential molecules during WWT.

of contaminants that induce stress, increased production of microalgal biomass, ease of microalgal harvesting due to flocculation, robust acclimatization of the consortia to a fluctuating environment, and their ability to release extracellular polymeric substances (EPS) (Amaro et al., 2023). However, the negative effects caused by nutritional competition, bacterial shading on microalgae, bacterial algicidal compounds, biosynthesis of microalgal antibacterial compounds, and inhibition of bacteria due to the increase in pH and O_2 levels by microalgal photosynthetic process must be considered before establishing microalgal–bacterial consortia for WWT (Zhang et al., 2020a). A schematic illustration of microalgae–bacteria interactions is given in Figure 4.1.

4.6 FACTORS AFFECTING MICROALGAL WWT

Microalgal growth is not entirely based on nutrient availability but also on various biotic and abiotic factors. The most significant of the biotic factors is their high cell density which can cause self-shading from light sources thus leading to reduced photosynthesis and accumulation of auto-inhibitors (Dange et al., 2022). Other significant biotic factors include the correlation between heterotrophic bacteria and microalgae in a mix consortium, which relates to their stimulating or inhibiting effects on each other. These interactions can also change depending upon the type of strains, their growth phase, and the ratio of their cell densities (Fan et al., 2020a). There might erupt a competition for nutrients and space, especially when it comes to the microalgae–bacteria consortium, as discussed above, wherein one may suppress the growth of the other (Fallahi et al., 2021). Therefore, careful selection of strains should be considered before installing mix consortia for WWT. Similarly, several abiotic factors also influence the efficiency and effectiveness of microalgal-based WWT systems. These abiotic factors include light intensity for efficient photosynthesis, pH, temperature, WW composition, turbidity, and hydraulic retention time (HRT) (Inuwa, Pervez and Nazir, 2023).

4.6.1 Light

Light intensity is an important factor for microalgal growth as it drives the photosynthetic process. Raising the level of illumination up to a certain threshold, which differs for each species of microalgae, frequently speeds up the growth rates. Microalgae experience photoinhibition when light intensity exceeds the saturation point, whereas low-intensity light below this point slows their development (Khan, Shin and Kim, 2018). In a study done by Chávez-Fuentes, Ruiz-Marin, and Canedo-López (2018), the growth rate of *C. vulgaris* increased significantly when light intensity was increased from 70 to 140 μE/m^2/s). Conversely, in another study, *C. vulgaris* showed maximum growth at a saturation point of 80 μE/m^2/s; below this point, the growth was not sufficient, and above this point photoinhibition occurred (Khalili et al., 2015). Furthermore, modelling microalgal growth and metabolism also requires an accurate prediction of light distribution. The light quality must reside in photosynthetically active radiation (PAR) range of 400–750 nm on the electromagnetic spectrum as this is the range of radiation that microalgal cells most commonly utilize to drive photosynthesis (Montoya-Vallejo, Guzmán Duque, and Quintero Díaz, 2023). The research of Khalili et al. (2015) showed the highest growth rate of *C. vulgaris* under warm white light as opposed to blue and red light.

4.6.2 The pH Factor

The pH is another factor which significantly affects nutrient removal from WW, mostly due to ammonia stripping and phosphorus precipitation (Wu and Vaneeckhaute, 2022). By adding CO_2 (pure/flue gas), the pH is typically kept under control at 8.0 (Matito-Martos et al., 2021). High pH levels obtained during photosynthesis can increase the availability of free CO_2, facilitating its absorption by microalgae (Zhu et al., 2020); however, in the absence of pH regulation, an overly high pH, such as 11, can prevent microalgae from uptake and metabolism of nutrients (Amaro et al., 2023). Therefore, the optimum pH should be maintained between 7 and 9.

4.6.3 Temperature

Temperature is also very important when it comes to the optimized functioning of microalgae in WWT, as it can directly affect their enzymatic activity and stability. The microalgae cultivation for WW is usually done at shallow depths for proper light penetration, which can significantly cause the temperature to fluctuate between 5°C and 56°C (Inuwa, Pervez and Nazir, 2023). Microalgae optimally thrive between

15°C and 30°C. The growth rates of most microorganisms follow an asymmetric curve in response to temperatures lower or higher than the optimum temperature due to disturbances in the cellular physiology (Baker and Geider, 2021). Modelling and testing data show that a 10°C reduction in temperature would result in a halving of microalgae growth rates, while a rise in temperature over the optimum would cause a significant decline in growth rates (Kaur et al., 2022). High temperature affects the structure and stability of microalgal membranes and macromolecules, whereas low temperatures decrease the rates of biochemical reactions (Hajinajaf, Mehrabadi and Tavakoli, 2021). In their research work, Zhang, Zhan and Hong (2016) investigated how *Chlorella* sp. HQ's nutrient removal efficiency, growth, and lipid biosynthesis were affected by high and low temperatures, which showed that the microalgae could grow efficiently between the temperature range of 12°C and 25°C, but its growth was scarce at a high temperature of 38°C. That is why it is necessary to install a temperature regulation system at the WWT reactor that can maintain the temperature at its optimum for microalgal growth, but doing so would significantly raise investment and operating costs. This issue was addressed in the research of Xu et al. (2019) who cultivated *C. vulgaris* continuously for 15 days in municipal WW under low (4°C), high (35°C), and alternating night (4°C) and day (35°C) temperature regimes. They found the alternating day/night regime, which mimics a natural scenario, to produce the highest biomass (99.21 mg/L/day) and nutrient removal efficiency [96.5% total nitrogen (TN), 97.8% NH_3-N, 99.2% total phosphorus (TP), and 83.0% COD] without the installation of a high-end temperature control system.

4.6.4 Mixing and HRT

The most crucial operating aspect of a photobioreactor is mixing because it maintains microalgae in suspension and prevents nutrient, gas, or temperature gradients from developing. Additionally, mixing encourages microalgae to rotate alternately between light and dark zones with an interval of as little as 10 ms for decent production of biomass in order to perform photosynthesis in the light zone and to prevent cells from light saturation and photoinhibition in the dark zone Xu et al. (2019). Additionally, it increases the rates of mass transfer between the liquid media, the environment, and the cells to facilitate effective cell nutrition absorption, the removal of excessive dissolved oxygen (DO) by stripping, and the facilitation of atmospheric CO_2 in culture media (Masojídek et al., 2021). Another factor that also contributes to the efficacy of a microalgal-based WWT system is the HRT, which affects the microalgal strains efficiency, algal-bacterial ratio, and nutrient removal ability (Oruganti et al., 2022). HRT is the connection between the volume of the bioreactor and the flow rate of the feed, which typically indicates the duration of time that the substrates and cells spend within the bioreactor (Goswami, Agrawal, and Verma, 2023). Anbalagan et al. (2016) showed in their research that the TN removal efficiency increased from 64.6% ± 16.2% (4-day HRT) to 81.5% ± 5.1% (6-day HRT), whereas the TP also increased from 56% ± 5% (4-day HRT) to 80% ± 12% (6-day HRT). Thus, it is important to optimize HRT and avoid making it longer or shorter than normal, since longer HRT can cause nutrient deficiency and also slow down the growth of microalgae because of the shading effect. Conversely, a short HRT causes the washout of microalgal cells and prevents them from removing nutrients from WW (Goswami, Agrawal, and Verma, 2023). It has been suggested that HRT of 2–8 days is ideal for microalgae WWT systems especially when it comes to municipal WW (Nguyen, Ngo and Nguyen, 2022).

4.6.5 WW Composition

The nature of WW and its composition such as types of contaminants and their concentrations, pH, suspended solids (turbidity), and colour greatly affects the WWT efficiency of microalgae. The unbalanced ratio and fluctuating concentrations of the main nutrients, such as phosphorus, nitrogen, and carbon, provide the most obstacles to microalgae-based WW cleanup. The synthesis of proteins necessary for the assimilation of phosphorus depends on nitrogen, and insufficient nitrogen levels can prevent this process from occurring (Oliveira et al., 2023). As a consequence, phosphorus deficiency can reduce the production of ATP, which is crucial for the metabolism of microalgae. The N/P mass ratios between 5 and 30 are considered optimum (Ahmed et al., 2022). Conversely, in some microalgal species, the photosynthetic activity of some microalgal species can be inhibited by NH_3 toxicity due to high levels of N-NH_4^+ concentration (100 mg N-NH_4^+ L^{-1}) at pH more than 8 (Su, 2021). Inefficient nutrient removal from WW can also be the result of the presence of some toxic or recalcitrant contaminants such as heavy metals, polyaromatic hydrocarbons (PAHs), personal care products, phenolic compounds, etc. These contaminants have the ability to prevent microalgal growth and even change the morphology of the microalgal cells (Posadas et al., 2017).

High turbidity (suspended solids) and soluble coloured compounds are another limiting factors that can affect the production of biomass, as they can absorb light and hinder its penetration into the microalgal cultures for photosynthesis. Although longer wavelengths of light, such as orange-red (>600 nm), could be used for photosynthesis under these circumstances for deeper penetration into WW as compared with shorter wavelengths (green-blue-violet <500 nm); highly coloured and turbid WW still limit the penetration of light and subsequently stop photosynthesis (Hena, Gutierrez, and Croué, 2021). To overcome this problem, WWs need to be pre-treated using techniques like filtration, gravity settling, flocculation, or oxidation for efficient phycoremediation when suspended solids (turbidity/colour) and ammonia are present in high concentrations in the WW, as these can adversely impede microalgal growth (Nitsos et al., 2020).

The adaptation of microalgae in WW has also been made easier by the development of several strategies like inoculating mixed cultures of microalgae; selecting naturally or genetically modified tolerant strains; and mixing different types of WWs together or with synthetic media such as BG-11, BBM, acidified BBM, half-strength Chu 10, modified Hoagland medium, etc., or their modified forms (Markou et al., 2021). In their research, Leite, Hoffmann, and Daniel (2019) cultured *Chlorella sorokiniana* in municipal and piggery WW mixtures and obtained a high organic matter removal efficiency of more than 90%, whereas removal efficiencies of 46%–56%, 40%–60%, and 100% were recorded for inorganic carbon, orthophosphate, and ammonia, respectively.

4.7 MICROALGAL WWT MECHANISM

4.7.1 Pre-Treatment of WW

WW comes in many forms, from simple municipal WW to intricate industrial effluent. However, the majority of them have high total suspended solids (TSS), which contribute to turbidity, and high COD and/or BOD, both of which indicate the presence of organic materials (Ng et al., 2022). When compared to the traditional activated sludge technique, microalgae are often less effective in eliminating organic compounds from WW, despite the fact that many of them may grow mixotrophically (Lee et al., 2019). Therefore, it is expected that microalgal WWT will be suitable for WWs with low BOD levels, such as those from landfill leachate, anaerobic digesters, or secondary aerobic treatment (activated sludge). In microalgae-based WWT systems, anaerobic digestion (AD) is a useful pre-treatment technique because it may reduce the load of organic waste, including both COD and BOD, and transform organic materials (animal and human waste, etc.) into methane that can be used as fuel to produce CO_2 (Bele, Rajagopal, and Goyette, 2023). Stoichiometric equations (Eqs. 4.1 and 4.2) describing these conversions are given as follows:

$$C_{10}H_{20}O_6N_2 \text{ (approx. formula of WW)} + 3H_2O \rightarrow 5.5CH_4 + 4.5CO_2 + 2NH_3 \quad (4.1)$$

$$CH_4 + 2O_2 \rightarrow CO_2 + 2H_2O \quad (4.2)$$

Important nutrients and a residual COD are still present in the secondary effluent or digestate following the AD process such as phosphate and ammonium that can be easily assimilated by microalgae (de Carvalho et al., 2022). The subsequent generation of tertiary effluent can then be disposed of in an ecofriendly manner.

4.7.2 Removal of Carbon

Microalgae use CO_2 as their main source of carbon by bio-converting it via the light-dependent water photolysis process into biomass as shown in the Eqs. 4.3 and 4.4 [98].

$$CO_{2(aq)} + H_2O \leftrightarrow HCO^-_{3\,(aq)} \quad (4.3)$$

$$H_2O + HCO_3^- + \text{Nutrients}^{(hv)} \rightarrow \text{Biomass}(CH_2O) + 1/2O_2 + 3OH \rightarrow pH \uparrow \quad (4.4)$$

Depending on the pH of the solution, CO_2 exists as four distinct inorganic carbon (Ci) forms: bicarbonate (HCO_3^-), carbonate (CO_3^{2-}), carbonic acid (H_2CO_3), and dissolved CO_2 (Li et al., 2023). Microalgae in reality grow in the medium with pH range of 7.5–8.4 at which approximately 90% is HCO_3^-, whereas the concentrations of CO_3^{2-} and H_2CO_3 are low (Kong et al., 2021). As a result, the produced microalgal biomass contains 43%–56% carbon. Consequently, 1.8 kg of CO_2 is required to produce 1 kg of microalgae (Li et al., 2023). Different microalgae have varying degrees of CO_2 tolerance. In a study done on *Nannochloropsis oceanica*, CO_2-tolerating genes were reported to activate in the presence of 60% CO_2 (v/v) (Wang et al., 2022a). Even in the absence of photosynthesis, certain microalgae can still grow by utilizing organic substrates, resulting in their ability to adapt to mixotrophic and heterotrophic modes of living (Vidotti et al., 2020). Differences in pH that depend on the stoichiometry of microalgal growth can be used to identify these types of metabolisms as given in the following Eqs. 4.5 and 4.6 (Park et al., 2021).

4.8 HETEROTROPHIC GROWTH

$$CH_2O \rightarrow C(\text{biomass}) + CO_2 + H_2O \rightarrow \text{pH} \downarrow \tag{4.5}$$

4.9 MIXOTROPHIC GROWTH

$$HCO_3 + CH_2O \overset{(hv)}{\rightarrow} C(\text{biomass}) + 3OH^- + CO_2 \rightarrow \text{pH (no change)} \tag{4.6}$$

4.9.1 Removal of Nutrients

Nutrient removal is the major process that contributes to the microalgal-based WWT systems. Ammonium (NH_4^+), nitrate (NO_3^-), nitrite (NO_2^-), and organic nitrogen are the main types of nitrogenous contaminants usually found in WW; of these, ammonium and nitrates are the most prevalent elements that microalgae can assimilate for their growth (Bhatti et al., 2023). Phosphorus usually exists in its inorganic and organic forms in the WW, with soluble orthophosphate (PO_4^{3-}) frequently serving as the sole assimilative compound for microalgal growth (Sanz-Luque, Bhaya, and Grossman, 2020). In microalgae–bacteria WWT systems, nutrient removal is primarily mediated by three methods, including assimilatory, abiotic, and dissimilatory nutrient removal. The enzymes nitrate and nitrite reductases aid in the reduction of nitrate and nitrite during assimilation, eventually resulting in the production of ammonium after which glutamate (Glu) and ATP help NH_4^+ incorporate into the intracellular amino acid glutamine, as shown in Eq. 4.5 (Kumar and Bera, 2020). Various microalgal strains accumulate phosphates (PO_4^{3-}) as granules within their cells with or without phosphorus starvation, enabling their luxurious uptake and removal from WW (Lia et al., 2021).

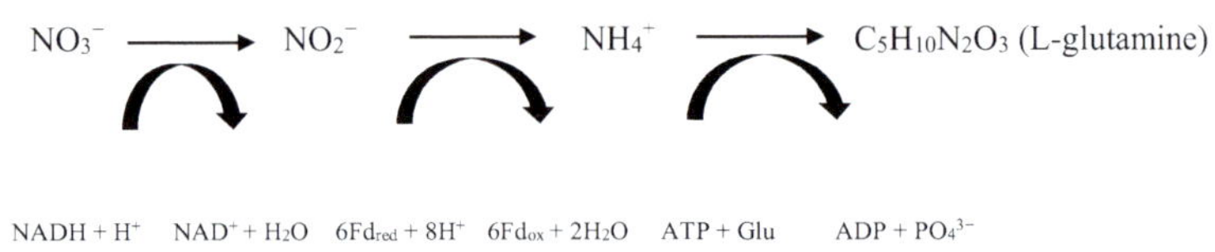

In the abiotic nutrient removal mechanism, the pH of the medium increases due to CO_2 removal during microalgal photosynthesis, as shown in Eq. 4.7 (Posadas et al., 2017).

$$CO_2(l) + H_2O(l) \leftrightarrow H_2CO_3 \leftrightarrow HCO_3^- + H^+ \leftrightarrow CO_3^{2-} + 2H^+ + \text{pH} \uparrow \tag{4.7}$$

Furthermore, ammonia stripping occurs at an elevated pH of more than 8.5 in the presence of Ca^{2+} and Mg^{2+} and dissolves oxygen from photosynthesis, whereas phosphorus is removed through precipitation as hydroxyapatite $[Ca_5(OH)(PO_4)^3]$ and struvite $[NH_4MgPO{\cdot}6H_2O]$ in addition to microalgal assimilation and luxurious uptake (Li et al., 2022a), as shown in the following equations (Posadas et al., 2017):

$$NH_4^+ \leftrightarrow NH_3(l) + H^+ \leftrightarrow NH_3(g) \tag{4.8}$$

$$3HPO_4^{2-} + 5Ca^{2+} + 4OH^- \rightarrow Ca_5(OH)(PO_4)_3(s)(\text{precipitates}) + 3H_2O \tag{4.9a}$$

$$Mg^{2+} + NH_4^+ + HPO_4^{2-} + OH^- + 5H_2O \rightarrow MgNH_4PO_4.6H_2O(s)(\text{precipitates}) \tag{4.9b}$$

Dissimilatory removal of nutrients in typical activated sludge systems involves nitrification and denitrification procedures, as shown in Eqs. 4.10 and 4.11 (Posadas et al., 2017).

$$NH_4^+ + 2O_2 \rightarrow NO_3^- + 2H^+ + H_2O \tag{4.10}$$

$$8NO_3^- + 5CH_3COOH \rightarrow 8HCO_3^- + 6H_2O + 2CO_2 + 4N_2 \tag{4.11}$$

Interestingly, this sequential process can happen concurrently in microalgal–bacterial photobioreactors (PBRs) because the growth medium and the interior of the algal-bacterial flocs or biofilms have diffusional gradients (Fallahi et al., 2021).

4.10 MICROALGAL-BASED WWT SYSTEMS

The treatment of WW by microalgae has no doubt certain merits; however, to establish a microalgal WWT system, several challenges must be addressed that range from microalgal cultivation settings to the composition and quantity of WW that needs to be treated. Various microalgal cultivation models have been engineered to optimize the growth of microalgae that can efficiently remove pollutants from WW, as described in the study of Bekirogullari et al. (2020). Generally, microalgae can be cultivated in batches and/or in continuous or semi-continuous cultivation in suspended or immobilized settings. These configurations can be further divided into open or closed systems under stirred or non-stirred conditions. Following WWT, the biomass of microalgae can provide a number of commercially valuable bioprospective products, including biofuels (biodiesel, biohydrogen, bioethanol, and biogas), bioelectricity, pigments, enzymes, polyunsaturated fatty acids (PUFA), biomass for food and feed, biofertilizer, etc. Moreover, the treated water can further be used in various industries (pharmaceuticals, cosmeceuticals, nutraceuticals, etc.), agriculture, gardening, etc. A schematic classification of microalgal cultivation settings is given in Figure 4.2.

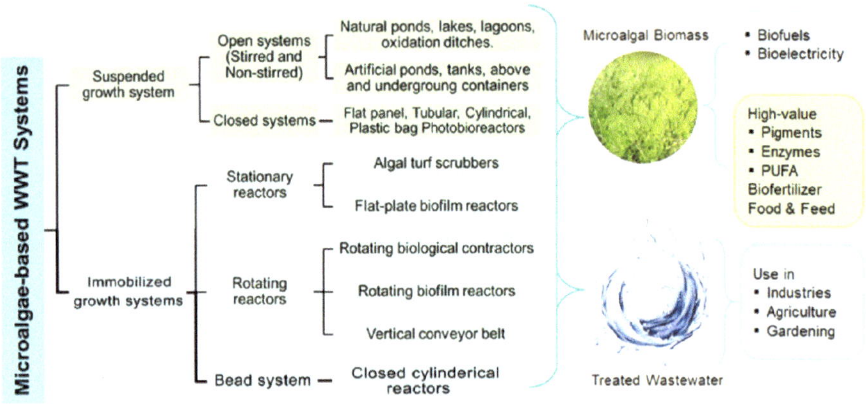

FIGURE 4.2 Schematic classification of microalgal cultivation settings in WWT. Water image designed by Vectorpouch/Freepik.

4.11 SUSPENDED MICROALGAL SYSTEMS

A microalgal suspended growth system is the most widely used method of treating WW because of low energy consumption, affordable construction, scale-up ease and ease of cleaning (Hena, Gutierrez, and Croué, 2021). It is the traditional microalgal cultivation that has been developed over many years which enables unrestricted cell movement in the aqueous medium (Ummalyma et al., 2023). Suspension cultures of microalgae can be cultured in batch, semi-batch, or continuous conditions (Rosmahadi et al., 2021). In general, these cultures can be engineered as open systems or closed systems (PBRs). The degree to which these two reactors are exposed to their surroundings differs. PBRs have very little exposure to the atmosphere compared to open systems, which are completely exposed. The main difficulty in suspension cultures is the harvest of microalgal biomass, which can be done through sedimentation, auto-flocculation, bio-flocculation, coagulation-flocculation, electro-coagulation, microalgal floatation, and filtration. These techniques make the downstream process intensive and costly. Some advantages and disadvantages of these two systems are given in Table 4.3.

4.12 OPEN CULTIVATION SYSTEMS

Open ponds have traditionally been used to produce microalgae on a large scale in commercial settings because of their affordability and simplicity of use (Al-Jabri et al., 2020). They are classified as natural and artificial systems. Natural systems are composed of microalgal growth in various ponds, ditches, or lagoons. Artificial systems consist of customized tanks, ponds, ditches, and various types of surface and subsurface containers which are environmentally controlled (Tan et al., 2020). Non-stirred and stirred ponds are additional categories for open systems. Stirred ponds encourage microalgal growth more readily due to mechanical aeration

TABLE 4.3
Advantages and Disadvantages of Open and Closed Wastewater Treatment Systems

	Advantages	Disadvantages
Open-pond system	Easier to build	Difficulty in maintaining monocultures due to contamination
	Relatively cheap and maintenance cost is low	Difficulty in controlling the process parameters such as pH, temperature, light
	Low energy input	High CO_2 loss due to very short HRT
	Easy operation and scalability	Seasonal variation in biomass productivity and nutrient removal rates
	Ease of cleaning	Increased evaporation rates
	Don't require arable land to grow	High rate of contamination from invasive microalgal species
	Higher production volume	Poor mixing and light availability for efficient photosynthesis
	High durability	Large land area requirement
Closed photobioreactor system	Ease in controlling process conditions (pH, temperature)	Expensive with high maintenance cost
	Small area requirement	Biofouling of reactors leading to inhibition of photosynthesis and low biomass production
	Less prone to contamination	Difficult to clean up
	Beneficial for sensitive strains	Relatively small illuminated-area-to-volume ratio
	Low hydrodynamic stress	Difficult to scale up
	Low water and CO_2 losses	Internal cooling is required
	High biomass yield	Difficult to construct and operate

(typically 10–40 rpm, depending on abiotic and biotic factors), efficient illumination, and nutrient dispersion, whereas non-stirred ponds are more frugal and easier to manage as they operate under natural conditions with surface aeration with wind (Sultan and Pei, 2023).

Open systems function by using sunlight and atmospheric air as energy and carbon sources for the microalgae, respectively (Chia et al., 2021). Because the microalgal cells have a negatively charged surface, the growth medium maintains the suspension of the cells, thus avoiding cell agglomeration (Grira et al., 2023). The microalgae primarily receive energy from sunlight through photosynthesis; however, they can also shift to heterotrophic nutritional strategy in the absence of light (Hong et al., 2022). Given the high chlorophyll content of the suspended microalgal growth in WW, a lot of oxygen is produced during photosynthesis in the open system (Sarwer et al., 2022). Furthermore, microalgal growth can benefit greatly from the presence of CO_2 produced by bacteria, which can speed up photosynthesis and produce more biomass and oxygen (Thanigaivel et al., 2022). As a result of the beneficial gaseous exchange between microalgae and bacteria, a mixed consortium of both are formed leading to the conclusion that this system might make aeration in conventional WW processes less expensive and sustainable (Iqbal et al., 2022).

4.12.1 Unstirred Ponds: Waste Stabilization Ponds

The treatment of WW in oxidation ponds, also referred to as waste stabilization (WS) ponds, lagoons, or oxidation ditches, involves the interaction of sunlight, bacteria, and microalgae. These ponds are an example of unstirred ponds, which are among the earliest known biological WWT methods, dating back as far as 3,000 years (Aditya et al., 2022). Air is brought into these ponds either by wind or by molecular oxygen produced by algal blooms during photosynthesis which accepts terminal electrons and acts as an oxidizing agent (Ho and Goethals, 2020). The waste stabilization pond (WSP) is stratified according to aerobic, anaerobic, and/or facultative zones depending on the type of bacteria and how quickly they grow there (Mahapatra, Samal, and Dash, 2022). The bottom portion of the pond is dominated by anaerobic microorganisms, whereas the upper portion is aerobic owing to the presence of algae and air-liquid interface (Gupta et al., 2021). Aerobic microbes transform waste into phosphates, carbon dioxide, and ammonia which can be utilized by algae present in the aerobic zone of the pond to give off oxygen (Molitor and Schnoor, 2022). Moreover, the presence of oxygen in the upper layer minimizes the odorous and volatile compounds by oxidation, thereby reducing the foul smell caused by the anaerobic process (Gupta et al., 2021). The bottom portion of the pond is deficient in oxygen and thus is dominated by anaerobic microorganisms. These microbes convert organic compounds present in waste into gaseous byproducts such as H_2S, CO_2, and CH_4 (Bartosiewicz et al., 2021). Conversely, facultative zone provides the biodegradation of organic nutrients employing both anaerobic and aerobic processes. This zone exhibits a synergistic relationship between microalgae and bacteria. While microalgae produce oxygen when exposed to sunlight, bacteria use anaerobic and aerobic metabolism for the oxidation of organic compounds in WW; however, when the sunlight is absent, microalgae utilize the oxygen for heterotrophic metabolism and oxidize biodegradable organic matter (Abdelfattah et al., 2023). That is why, the depth of the facultative ponds is kept to 2–3 m in order for sunlight to penetrate efficiently thereby balancing the degradation of organic compounds through HRT and loading of organic matter (Ranjit, Jhansi, and Reddy, 2021). WSP is also advantageous in removing pathogenic microorganisms from WW without the addition of hazardous chemicals. This is mainly done by the autotrophic and heterotrophic conditions generated during day and night by microalgae and bacteria. Microalgae utilize CO_2, generated by heterotrophic bacteria, for photosynthesis throughout the day; the pH of the aerobic zone can reach as high as 10; however, when it is dark, both algae and bacteria release CO_2, which causes the pH to go below 7 (Wang et al., 2020). This change in pH through day and night together with high HRT yields good pathogen eradication. Facultative ponds are also used for nutrient removal, as both nitrification and denitrification processes are encouraged simultaneously due to the formation of oxygenic and anoxygenic zones in the pond (Dias, Passos, and von Sperling, 2017). Alkaline pH in the aerobic zone of the pond aids in ammonia stripping where

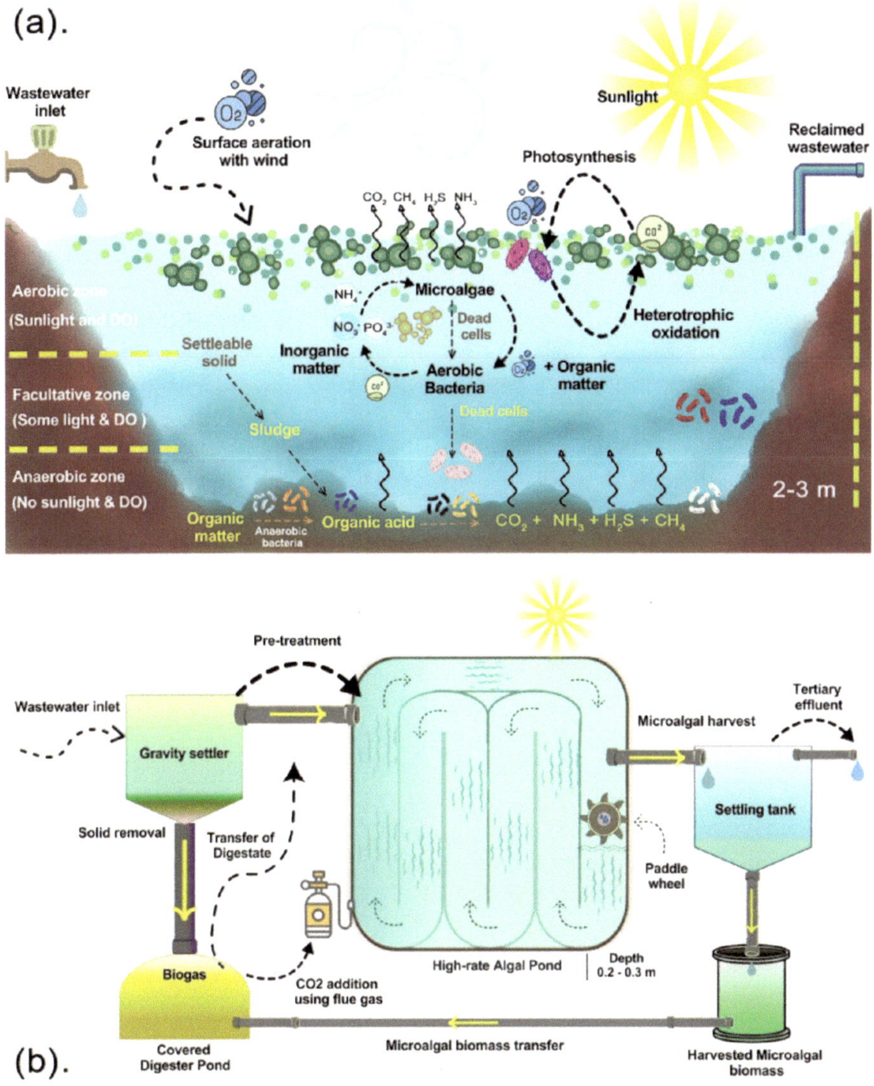

FIGURE 4.3 Designs of open-stirred and unstirred microalgal cultivation WWT systems. (a) The working mechanism of WS pond. (b) High-rate microalgal ponds (HRAP) supplemented with CO_2-enriched flue gas formed in a primary sludge in the anaerobic digester.

NH_4^+ ions are transformed into free NH_3 (Chuka-Ogwude et al., 2020). The working mechanism of WSP is illustrated in Figure 4.3a.

4.12.2 Stirred Ponds: High-Rate Algal Ponds (HRAP)

These types of ponds, which are shallower (0.2–0.4 m) and more advanced variants of raceway ponds, are the most prevalent kind of open-stirred ponds (Luo et al., 2017).

A typical rate of organic load for HRAPs is 100–150 kg BOD_5/ha day, depending on the environmental conditions (Randrianarison and Ashraf, 2017). At moderate latitudes, the yearly microalgal biomass production through WWT by HRAP is around 40–70 tons/ha/year (Mehrabadi, Craggs, and Farid, 2015). These ponds are different from conventional pond systems in that they strive to enhance microalgal growth for high-rate WWT. HRAP system is also more cost-effective than conventional pond systems since they tend to take up less area, are expandable, don't emit any smells or fumes, consistently remove the maximum amount of nutrients, and can even eradicate pathogens due to fluctuations in the pH (Chambonniere, Bronlund, and Guieysse, 2021). The principle of HRAP is based on mutualistic interactions between the microalgae and bacteria, whereby the microalgal photosynthetic process provides bacteria the oxygen for aerobic degradation of organic matter, and bacteria provide nutrients (N, P) to microalgae (Lafi, Panu, and Liao, 2023). A typical HRAP design consists of various rectangular grids with oval-shaped channels as shown in Figure 4.3b. A paddlewheel is often used for gentle mixing of WW with a mean horizontal flow rate set at 0.15–0.3 m/s for higher biomass productivity (Torres-Franco et al., 2021). Microalgal cell proliferation is also ensured by the vertical mixing in HRAP all the while preserving nutrient diffusion by limiting the sedimentation and flocculation of microalgae–bacteria biomass (Nagarajan et al., 2023). Extensively long channels can lower the efficiency of HRAP due to the formation of dead-dark zones (Gupta et al., 2021). Therefore, to prevent the buildup of these zones, it is essential to balance the water velocity and mass transfer. An ideal balance of the organic matter loading rate, HRT, depth of pond, supply of CO_2, horizontal mixing velocity, and pH must continually be maintained for the HRAP to function properly (Mishra et al., 2023). Prior to being transferred to HRAP for treatment, the WW is often pre-treated in anaerobic covered digester ponds (CDP) and/or gravity settlers, which is essential for removing the solids from the WW (Singh et al., 2020). After pre-treatment, the microalgae uptake nutrients from digestate from AD for their growth and thus help to provide secondary and tertiary treatment to the WW (Oruganti et al., 2022).

4.12.2.1 HRAP with CO_2 Augmentation from Biogas

The HRAP can be further modified to enhance its performance by injecting CO_2 as flue gas, powered by solar energy and collected from CDP, into the WW due to carbon limitation during the day due to volatilization of ammonia and precipitation of phosphate, while the microalgae are conducting photosynthesis. The harvested microalgal biomass from the microalgal settling tank and/or WW solids from a gravity settler (primary clarifier), when stabilized in a CDP, yield biogas (flue gas) with an average yield comparable to that of high-temperature mixed digesters that are 0.30–0.45 m^3 CH_4/kg with 50%–60% microalgal volatile solid conversion (Benemann et al., 2012). Biogas is generally composed of CH_4 (50%–75%), CO_2 (25%–50%) and other miscellaneous gases like H_2S, N_2, and NH_3 in trace amounts (2%–8%) (Li et al., 2019). The supernatant digestate (liquid fertilizer) can then be added to clarifiers and subsequently HRAP for secondary WWT and biomass harvesting, ultimately resulting in clarified tertiary effluent (Figure 4.3b). It is possible to recycle the resulting biomass to the CDP for co-digestion with the raw WW. Due to their single closed-loop working mechanism, co-digesting microalgal biomass can sustainably enhance the AD process and subsequent biogas output (Leong and Chang, 2022). If enough area is available, the HRAP+CO_2 system might remove tertiary nutrients with minimal additional energy use than an HRAP designed to remove BOD (Molazadeh et al., 2019). Theoretically, WWs with a wide variety of nutrient content may be used to completely assimilate both nitrogen and phosphorus into microalgal biomass using the HRAP+CO_2 system. Nutrient absorption rates can approach 16 kg N/ha day (nitrogen) and 2 kg P/ha day (phosphorus) based on the typical microalgal nitrogen (8%) and phosphorus (1%) and microalgal biomass productivity of 20 g/m^{2}day (Razaviarani et al., 2023).

4.13 CLOSED CULTIVATION SYSTEM

This system consists of various types of PBRs made up of glass, plexiglass, polycarbonate, and plastic bags. In these reactors, cultures of microalgae are suspended in a growth medium or WW, which circulates in an illuminated, transparent array of tubes and plates (Chai et al., 2021). The rationale behind the design of a closed cultivation system is to transform an open pond into a greenhouse by covering it with a transparent barrier (Leong, Huang, and Chang, 2021). These reactors can be placed vertically, horizontally, or inclined, depending on the design, to receive maximum light either from a natural or artificial energy source. By heating or cooling the tube, temperature may be easily regulated. Through direct CO_2 injection or the addition of acid, the pH fluctuations are minimized. Furthermore, they show higher productivity, lower harvesting costs, and ease of cultivation. Various factors affect the efficiency of closed PBRs such as optimum illumination, homogenous supply of CO_2 with minimum gaseous loss, efficient removal of O_2 generated by microalgae during photosynthesis that can negatively affect cell growth, and scalable nature of the reactor design (Assunção and Malcata, 2020). The closed PBR system offers several advantages. It has a larger surface area-to-volume ratio compared to open pond systems, which enhances two key aspects: light utilization rates, allowing for more efficient use of light, and mass-transfer efficiency, facilitating better transfer of gases, nutrients, and other substances (Chanquia et. al., 2022). Higher irradiance promotes increased photosynthetic rates and subsequent cellular growth (Paul et al., 2021). Moreover, since closed systems are not exposed to the environment, they operate with reduced risk of contamination, water loss, and CO_2 emissions (Assunção and Malcata, 2020). Therefore, more sensitive microalgal strains can be grown under optimum conditions. Despite these benefits, additional mechanisms are still needed to provide microalgae the proper assistance to carry out intensive operations. These mechanisms, which are required to manage gaseous (CO_2 and O_2) concentrations, pH regulation, steady liquid velocity, and stirring conditions, include aeration, pumping, and degassing, which raise capital and operating costs and cause difficulties during use (Valchev and Ribarova, 2022). Several different PBR designs have been developed and patented for WWT and microalgae cultivation, which can be grouped into three basic types: tubular, flat plate, and fermenter type.

4.13.1 Tubular PBR

Tubular photobioreactors (TRs) are formed by a series of parallel glass or plastic tubes (10–60 mm in diameter and length of more than 100 m), which can be arranged in loops or systems that are horizontal, vertical, or tilted at an angle. These reactors are mostly designed for continuous outdoor cultivation of microalgae, where the reactors are placed strategically at an angle to the sun to capture maximum solar radiation, ensuring high rates of photosynthesis. Due to the fact that they are impervious to microbial contamination, such reactors are designed to hold substantial amounts of culture media or WW (Priyadharshini et al., 2021). In a typical TR, an airlift or mechanical pump is used to continually move the microalgae and growth medium through the tubes to a reservoir (Al-Dailami et al., 2022). Many pilot-scale TBs have been designed for the cultivation of a variety of microalgae. Despite their advantages, TRs frequently experience issues such as excessive concentrations of DO, which prevents photosynthesis and causes photobleaching, unfavourable pH gradients, and negative CO_2 gradients (Banerjee et al., 2023). Additionally, when dealing with large-scale processes, photoinhibition occurs as a result of dense growth of microalgae on the reactor walls, which prevents light from penetrating the reactor (Dange et al., 2022). Different configurations of tubular reactors have been designed, of which vertical, horizontal, and helical are the most noteworthy.

4.13.1.1 Vertical Tubular Reactor (VTR)

A typical VTR is made up of a vertical transparent reactor that enables the light to penetrate through it for the growth of autotrophic microalgae. At the base of the reactor, a gas sparger device is installed that creates tiny bubbles that act as a propellant for stirring, gaseous exchange of CO_2 and O_2 as illustrated in Figure 4.4a. Normally, physical agitation is not required in these reactors. VTR can be designed as bubble columns or airlift reactors.

The containers used to create bubble columns are transparent, cylinder shaped, and have an interior diameter of at least 20 cm (Mehariya et al., 2021). This photobioreactor consists of a bubble column, a gas chamber, an air sparger, an illumination source, and an aerator. These reactors benefit from minimal initial investment cost, a high surface-area-to-volume ratio, the absence of mobile components, adequate heat and mass transfer, a reasonably homogenous culture environment, and effective gaseous (O_2 and CO_2) exchange (Sirohi et al., 2022). In bubble column reactors, the suspension culture of microalgae is agitated by gas phase, which is injected into the culture medium as tiny bubbles generated by a distributor or sparger installed at the base of the column (Mehariya et al., 2021). Therefore, the design of the distributor is critical to the performance of a bubble column. In tall bubble columns, perforated plates serve as the sparger, breaking apart and redistributing merged bubbles (Dange et al., 2022). An artificial illumination covering PAR is usually provided from the outside of the column; however, an interior illumination design can also be utilized since it has a better rate of light penetration and more even lighting (Sirohi et al., 2022).

The design of the airlift PBRs consists of internally linked, transparent cylindrical containers. The airlift system separates the fluid volume into two zones to enhance circulation, oxygen transmission, and balance forces in the reactor (Li et al., 2022b).

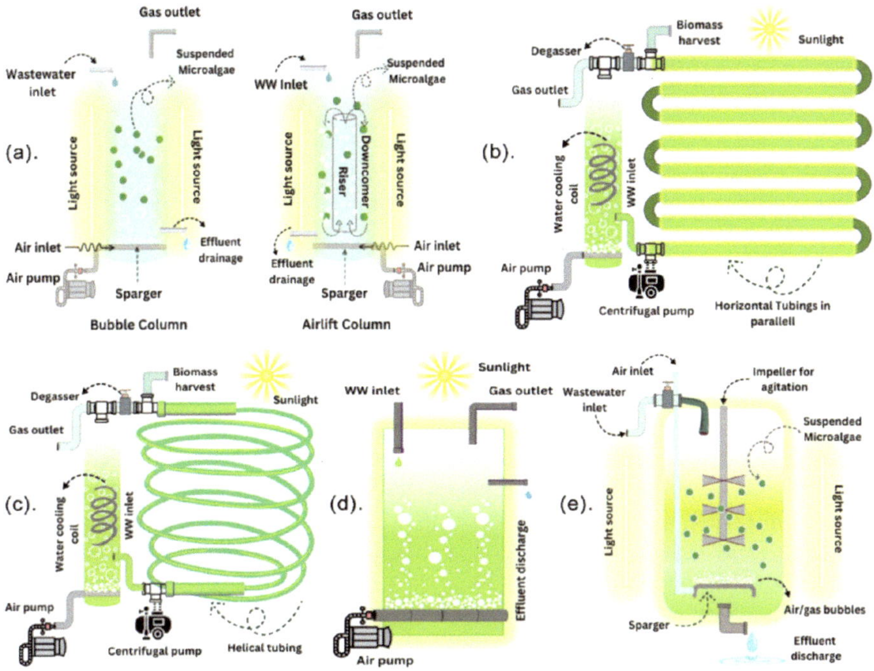

FIGURE 4.4 Designs of closed photobioreactor systems for suspended microalgal cultures. (a) Tubular reactors, VTRs (bubble column and airlift), horizontal tubular reactors, and helical tubular rectors. (b) Flat-plate reactor. (c) FTR.

One of the two zones receives gas from a sparger and is known as the riser, whereas the other zone that is not sparged is the downcomer (Wang et al., 2022b). The liquid inside the reactor is circulated based on the fluid density gradient. Since the gas is introduced from the riser zone, it is denser than the downcomer zone. As a result, the microalgal suspension rises in the aeration zone and flows down in the downcomer zone. An airlift reactor has the benefit of generating flow circulation where the culture continuously alternates between light and dark phases, giving the microalgal cells the effects of flashing lights (Borowiak and Krzywonos, 2022). The performance of the photobioreactor is governed by the residence time of gas in different zones, which has an impact on variables including mass transfer between gases and liquids, heat transfer, mixing, and turbulence (Rezvani and Rostami, 2023). Despite their benefits, the bubble and airlift VTRs are costly, which is mainly due to the amount of gas used in the reactors.

4.13.1.2 Horizontal Tubular Reactor

To maximize light penetration, translucent polypropylene acrylic or polyvinyl chloride tubes with tiny internal diameters are used to construct horizontal tubular photobioreactors (HTR) which can be arranged in several parallel horizontal rows of tubes. An air pump is used to keep the culture in motion by agitating and mixing it. Due to their orientation towards the sun and excellent light conversion efficiency, the varied designs are advantageous in outdoor culture. The enhancement of air-residence time to provide increased amounts of dissolved CO_2 is the most important feature of an HTR system that distinguishes it from a vertical column bioreactor. In addition to being able to use natural sunlight, these devices could also employ artificial light. The high energy consumption of HTR, which is over 2,000 W/m^3, is a major disadvantage when compared to bubble column and flat-plate PBRs, which use 50 W/m^3 (Zhang, 2015). To achieve stirred conditions with adequate light-dark cycles, a high linear liquid flow rate of around 20–50 m/s are required (Singh and Sharma, 2012).

4.13.1.3 Helical Tubular Reactor

A unique kind of horizontal bioreactor is a photobioreactor of the helical type, often known as a coil type. It is made up of a degassing device for O_2 release that can be installed independently or attached to a thin transparent tube (Rezvani and Rostami, 2023). A centrifugal pump moves the culture along the lengthy tube to the degassing device. Although either direction may be used to circulate the culture medium and CO_2 gas mixture in the reactor, infusion from the bottom leads to increased photosynthetic efficiency (Banerjee et al., 2023). Most coil-type systems offer an increased surface-area-to-culture-volume ratio that allows for efficient lighting as well as simple temperature and contamination control (Sirohi et al., 2022). Additionally, they also benefit from having a good balance between energy input and photosynthetic efficiency. This kind of technology uses less energy to function and places less mechanical strain on the microalgal cells. Due to the tiny internal tube size, cleaning issues with tubular systems are challenging to solve; there is currently no suitable mechanical method to clean the interior of a lengthy tube.

4.13.2 Flat-Plate PBR

Conceptually, flat-plate reactors (FPRs) are built to utilize sunlight effectively. They are made of two transparent, adhesive sheets (glass, plexiglass, or polycarbonate) that range in length up to 70 m in terms of the length of the light path (*d*) that passes through them (Nagarajan et al., 2023) as shown in Figure 4.4b. It features a high surface-area-to-volume ratio in addition to a gaseous exchange system (Sirohi et al., 2022). Microalgal cultures are evenly exposed to sunlight or a PAR light source due to this large and uniform surface of the reactor. Air is bubbled through a perforated tube from one side to agitate the culture, or it is rotated manually by a motor. FPRs are perfect for high output in both indoor and outdoor settings, with potential ease of design scale-up due to several advantages, such as an increased level of light penetrating the surface of the plate, leading to a decreased buildup of dissolved oxygen (Zheng et al., 2022). While flat-plate systems have these advantages, they may also present some issues related to cleaning the reactor; high energy inputs; a large area for installation; and, moreover, poor mass production per unit area efficiency (Sirohi et al., 2022).

4.13.3 Fermenter-Type PBRs

The fermenter-type (FTR) PBR are made of glass, plexiglass, or plastics in a steel framework. These PBRs have circular cross-sections, as shown in Figure 4.4c, and can be designed in a variety of shapes, such as tubular, cuboidal, rectangular, etc.

The sunlight harvesting ability of FTR-PBRs during photosynthesis is not very efficient due to the drawback of having a low surface-to-volume ratio (Siddiki et al., 2022). Consequently, only a small number of these PBRs have been engineered over time. However, this drawback can be overcome by internal illumination giving homogeneous distribution of light (Banerjee et al., 2023). Furthermore, under proper organic carbon sources, microalgae can also be grown heterotrophically in FTR (Siddiki et al., 2022). Due to these feats, this PBR type is beginning to gain interest, especially from the industries, as it holds the potential of higher biomass output per land area. The operating modes for FTR include batch, semi-continuous, and continuous. Axenic cultures can be maintained for a long time in this type of reactor, since the physiological parameters such as light, pH, temperature, HTR, etc., can be completely regulated (Siddiki et al., 2022).

4.13.4 Photo Sequencing Batch Reactor (PSBR)

This type of reactor is a modification of sequencing batch reactors (SBRs). For a certain time period, the reactor is filled with culture medium or WW while being constantly exposed to an illumination source with PAR efficiency followed by brief aeration resulting in higher growth of microalgae, as indicated by increased oxygen production (Liu et al., 2022). The development of heterotrophic microorganisms that consume dissolved organic carbon in WW is eventually aided by this process. Additionally, oxygen also helps the nitrification process, which produces nutrients like NO_2^- and NO_3^- that the microalgal cells actively take up (Jin et al., 2023). Following the aeration process, the resultant sludge settles at the base of the reactor, while the treated water is discharged. A minute amount of sludge is saved for the following batch, which is composed of rich consortia of microalgae and bacteria, thereby decreasing the amount of time needed to treat the following batch.

4.14 IMMOBILIZED MICROALGAL SYSTEM

Microalgae cell immobilization or attachment technology was basically developed to solve the difficulties of harvesting microalgal biomass produced in the suspension cultures (Plöhn et al., 2021). Immobilized microalgae take up nutrients from WW and utilize them to grow their biomass. The metabolic conversion of WW constituents and the easy and inexpensive harvesting of the generated biomass are two of the processing goals that may be achieved using microalgae immobilization (Zhang, Zhan, and Hong, 2016). Immobilization methods work by entrapping or anchoring the microorganism inside an inert carrier (natural or synthetic) to make it more stable and resilient (Tong and Derek, 2023). The microalgal immobilization technique offers a promising alternative to traditional WWT due to enhanced cell density and enzymatic activity, resilience to extreme environments, easier harvesting of the biomass, and enriched bioproduct generation (Gong et al., 2022). Microalgae can be immobilized through adsorption on a solid matrix, entrapment in natural or synthetic polymers, and liquid-phase emulsions, amassing behind a semi-permeable membrane, as well as through affinity and covalent coupling (Caldwell et al., 2021). The immobilized microalgae not only produce a high amount of biomass that can be harvested and used as a lucrative byproduct, but are also highly resistant to harmful pollutants found in WW. Immobilized microalgae-based WWT systems are usually used to treat low-concentration WW such as municipal, sewage, and domestic WW as well as secondary and tertiary effluents (Song et al., 2022). Compared to using normal suspended culture WWT methods for WWT, immobilized microalgae applications offer a number of benefits such as flexibility in the designing of PBRs, high rates of reaction and uptake, generating high cell density, ease of cultivation and harvesting, efficient process stability, circumventing cell washout problems, ease in the replacement of microalgal strains, protection from harsh abiotic factors, and sustainability of microalgae. Despite the advantages, the main problems with immobilized microalgae systems include light utilization, gel stability, immobilization process costs, and substrate penetration through the gel matrix, which can reduce the effectiveness of the systems required to remove pollutants (Nguyen et al., 2021).

4.15 APPROACHES IN MICROALGAL IMMOBILIZATION

4.15.1 Adsorption onto a Matrix (Passive Immobilization)

Microalgae and other microbes possess a natural propensity to reversibly attach and proliferate on solid surfaces, which Moreno Osorio et al. (2021) described as "passive immobilization" (Walther et al., 2022). Natural or artificial materials can be used as adsorbent materials or carriers for passive immobilization. The application of attaching microalgae to a solid matrix for growth is an intriguing alternative, where microalgae are continually harvested. This sort of immobility falls within the category of non-enclosure techniques, according to Katarzyna, Sai, and Singh (2015). The microbes that accumulate at a solid surface are termed as "aggregates", whereas those developed at water/solid interfaces are defined as "biofilms" [Moreno Osorio et al. (2021)]. Aggregates of microorganisms can also entrap themselves in an extracellularly produced EPS so as to adhere together or on a solid surface; this property is known as auto-flocculation or granular biofilms (Wang and Hong, 2022).

4.15.2 Entrapment in Porous Matrix (Active Immobilization)

The active immobilization techniques of entrapment of microalgae are done through covalent bonding, which involves flocculant agents, chemical attachment, or gel trapping (Rahul et al., 2021). In this technique, the biomass grows around or within a granule whose size gradually increases until it typically reaches several millimetres in diameter. The granule is made of a natural or synthetic polymer (alginate, carrageenan, and polyacrylamide gel) that is mixed with the microalgae cells, resulting in the formation of immobilized microalgae beads after being stabilized with divalent ions, such as Ca^{2+} (Yarkent et al., 2022). Since the immobilized beads are substantially larger than free cell cultures, they may be separated from water using a straightforward filtering technique such as sieving without the input of energy. Therefore, microalgal biomass can easily be employed on a commercial scale.

4.16 BIOFILM-BASED PBRS: PASSIVE IMMOBILIZATION

Biofilm-based microalgal PBRs are grabbing the attention recently as a promising alternative to suspension cultures. Torpey et al. (1971) originally designed the microalgal biofilm reactors in the 1970s using a revolving disk made of aluminium for the removal of nitrogen and phosphorus from municipal WW. In an attached growth-based WWT system, the microalgal biomass integrates an attachment matrix, whereas the influent WW supplies substrates in sufficient volumes. Microalgal biomass is readily scraped off from the matrix by regulating the thickness of the biofilm, leading to a consistent effluent and biomass ratio (Zhuang et al., 2018). Currently, biofilm-based reactors for WWT under optimal conditions are classified into two major groups: stationary and rotating reactors.

4.16.1 Stationary Biofilm PBRs

In stationary biofilm reactors, there are no moving elements in the configuration; however, mats of polymers are used as matrix for the attachment of microalgal cells. To ensure that all algae cells receive light, the matrix is passed over by a thin layer of WW effluent. Moreover, by scraping, the biofilm's thickness is regulated. Algae turf scrubbers (ATS) and flat-panel biofilm reactors, as shown in Figures 4.5a and b, are the two primary reactor designs mentioned in the published literature.

4.16.1.1 Algae Turf Scrubbers (ATS)

The ATS is a 0.1–0.2 m-deep reactor that is used for the sustainable cultivation of microalgae. It resembles an open-pond system with the exception of a thin polymer matrix installed at the bottom (Martini et al., 2019). Due to their lower depth, these systems operate more simply yet need a tremendous amount of surface area. This reactor usually treats municipal and agricultural WW with the help of periphyton, which consists of mixed cultures of microalgae and other heterotrophic bacteria. The ATS reactor works by involving the use of a substrate liner for periphyton in a raceway pond with a gentle slope (Wollmann et al., 2019). Following that, the contaminants are either filtered or decomposed when the WW is channelled through the microalgal biomass in the form of a biofilm.

4.16.1.2 Flat-Panel Biofilm (FPB) Reactors

The microalgal biofilm is immersed in the WW reservoir that is stirred or circulated by a pump in a standard flat-panel biofilm reactor (Gupta et al., 2021). This reactor looks like the flat-plate photobioreactor for suspension cultures, with the exception that a supportive attachment material is inserted there to encourage the formation of biofilms instead. The reservoir of the FPB reactor may also be sealed off from the outside environment, which can also help reduce contamination, but at the expense of increased cost. Furthermore, the cost of operation is also made more expensive by the additional energy required for bubbling air or the circulation of liquid.

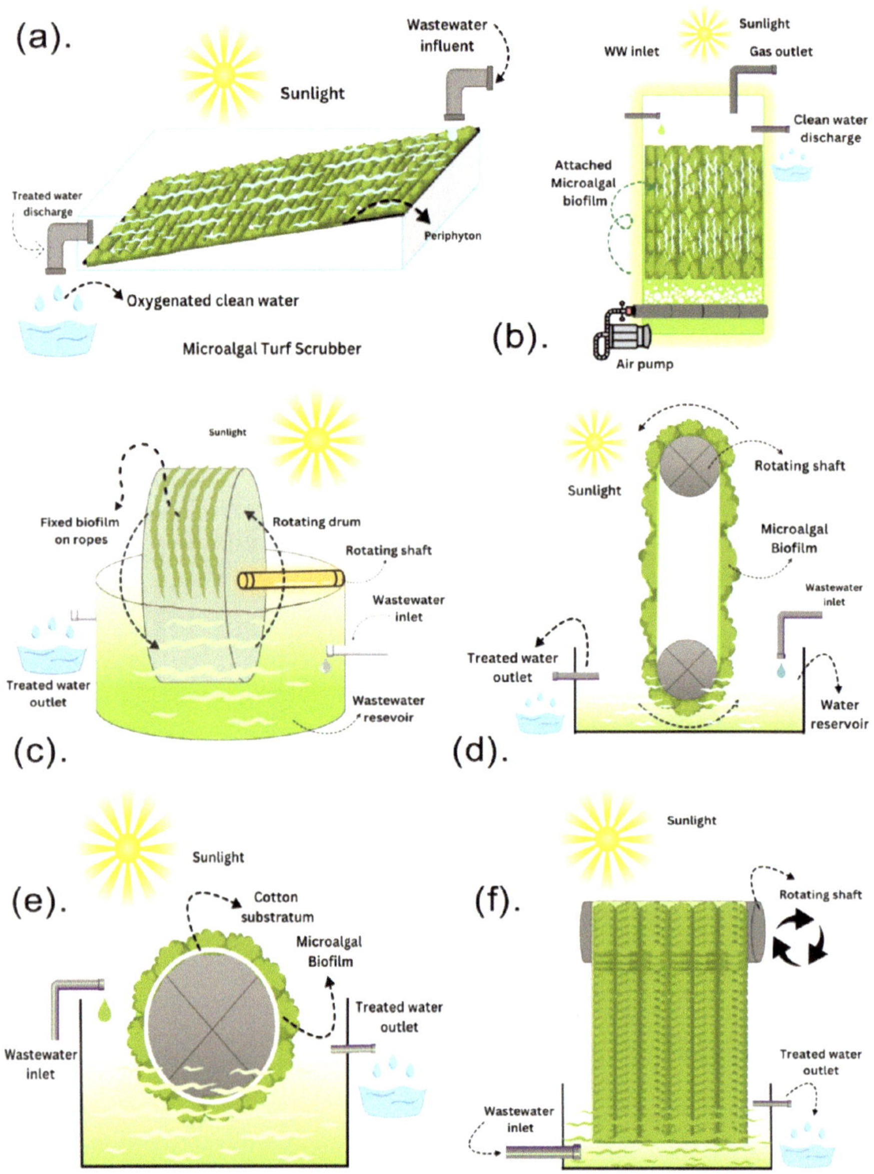

FIGURE 4.5 Stationary and rotating microalgal biofilm-based PBRs. (a) ATS. (b) (FPB) reactors. (c) RBC. (d) Revolving RAB. (e) RABR. (f) VCBD.

4.16.2 Rotating Biofilm PBRs

The PBRs with rotating biofilm bases have been designed to efficiently use sunlight and gaseous exchange, while also reducing the need for land area and growth substrate. The microalgae-biofilm matrix is fixed to an electrically powered rotating solid support in the form of a shaft, drum, belt, etc., which can periodically rotate the attached microalgae between the air and liquid (WW) phases (Maurya, Kumar, and Raj, 2023). Many designs regarding this technology have been proposed and developed; however, the three, rotating biological contactors (RBC) and vertical conveyor belt design (VCBD) are the most common, as shown in Figure 4.5c–f.

4.16.2.1 RBC

One of the most commonly employed rotating biofilm PBRs for WWT is the RBC-type PBRs as shown in Figure 4.5c. Microalgal cells are attached to a drum in RBC systems, which rotates continually between liquid and air (Zhang et al., 2021). In this way, the microalgal biofilm that forms on the drum is periodically exposed to WW while removing nutrients from it.

4.16.2.2 Revolving Algal Biofilm Reactor (RAB)

This reactor is a modification of RBC which employs vertical spinning disks and belts that are partially immersed in the nutrient-rich liquid reservoir as shown in Figure 4.5d. The microalgae cells are immobilized on the inert matrix [plastics like polyvinyl chloride (PVC)] on the conveyor belt in order to increase the surface area. The rotating action of conveyer belts causes aeration, which results in the gaseous exchange of CO_2 and O_2 when the WW-contact-induced degradation of the medium occurs following exposure to air and light (Wang et al., 2022c). The surface area of the inert matrix and the concentration, type, and quality of contaminants are all factors that affect how well RAB removes nutrients from WW. EPS formed by the activity of microalgae–bacterial consortia serve as adsorbent materials and help in the formation of a diffusion barrier (Krishnan, Patil, and Nancharaiah, 2023). Additionally, EPS build microbial communities, facilitate interactions, offer a favourable microenvironment, act as a carbon and energy reserve during times of water and nutrient stress, as well as a barrier against the harsh external environment (Jones, 2023). The symbiotic relationship between various autotrophic and heterotrophic bacteria is created as soon as the microalgal biofilm develops which leads to the breakdown of organic compounds and the elimination of nutrients from WW. RAB reactor is quite versatile since, in a single integrated system, the rotation of the conveyor belt also produces aerobic, anaerobic, and anoxic microenvironment gradients, leading to a variety of removal bioprocesses (Gupta et al., 2021).

4.16.2.3 Rotating Algal Biofilm Reactors (RABR)

In RABRs, a spinning cylindrical drum is wrapped in a matrix of natural (such as cotton) or artificial (like plastic or polymers) substrates that are used as ropes for the attachment of microalgal biomass as shown in Figure 4.5e. By pulling the rope through a hole of a pre-designed diameter, excessive microalgal biomass is harvested, enabling the biofilm to be kept at a consistent thickness (Ennaceri et al., 2023). These systems outperform RBCs and are currently being developed further.

4.16.2.4 VCBD

Like RABRs, VCBDs operate on the same principles but with smaller footprints due to the narrower canal that houses the conveyor belt as shown in Figure 4.5f. This is accomplished by using larger conveyor belts, which provide a similar amount of surface for adhesion to a few rotations of a narrow rope around the RABR drum (Hu et al., 2021a). The biomass can then be scraped off from the conveyor belt easily. This reactor design is arguably the most promising patented technology regarding immobilized microalgal-based WWT systems.

4.17 MICROALGAL BEADS SYSTEM PBRS: *ACTIVE IMMOBILIZATION*

The active immobilized algal bead systems are yet another reactor design for WWT that has received much attention. The microalgal cells in these reactors mirror those in enclosed suspended growth systems, with the exception that they are contained within distinct beads that have a specified biomass concentration rather than floating freely in the aqueous medium (Zhang et al., 2020b). The initial fabrication of the beads takes place in a laboratory setting using the aforementioned covalent bonding, adsorption, encapsulation in a polymer (natural, synthetic) or semi-permeable membrane (Rahul et al., 2021). The most common and easy approach for fixing microalgae alive inside the gel matrix is immobilization in polymers. Alginate, carrageenan, agar, agarose, and chitosan are examples of natural polymers that have been used in experiments along with a number of synthetic polymers such as acrylamide, resins, polyvinyl, polyurethane, etc. (Yarkent et al., 2022). Regardless of the polymers utilized, the material must be hydrophilic in order for WW to permeate into the beads. This technology was basically developed in order to bypass the laborious and expensive microalgal harvesting procedures and to tighten the regulations for microalgal monocultures to prevent contamination. Additionally, these systems enable the use of larger microalgal biomass production, which results in processes that are more intense. The immobilized algal bead reactors also have some limitations such as the requirement for the same supplemental equipment package as the enclosed suspended growth systems, plus a further phase of preparatory encapsulation. Furthermore, to sustain the high clearance rates, the long-term stability of bead-immobilized algal cultures still has to be enhanced, necessitating an understanding of biofilm development (Janpum et al., 2022). Similar to suspended cultures, a microalgae–bacteria consortia can increase clearance rates, especially for chemical pollutants. In these immobilized consortia, several synergistic benefits have been observed, including enhanced growth rates, increased pigment production, improved nutrient removal (nitrogen and phosphorus), and altered physiological responses. (Priya et al., 2022). Immobilization in polymers have shown to give protection to delicate microalgae against the harshness of WW environment (Zhang et al., 2020b). Microalgal cells within the gel matrix are guarded by bacterial biofilm surrounding the beads against not only the grazers and microbial fauna but also from toxic pollutants present in the WW (Tong, Yang, and Du, 2020).

4.17.1 Fluidized Bed PBRs

To solve the challenge of the microalgal biomass harvesting procedure, fluidized bed PBRs were developed. These PBRs consist of packed beds of floating substratum like carrageen, chitosan, or alginate beads that are used to immobilize microalgal biomass (Siddiqui et al., 2023) as shown in Figure 4.6a. These microalgal beads are then laid on the bottom of the reactor, containing WW coming in from

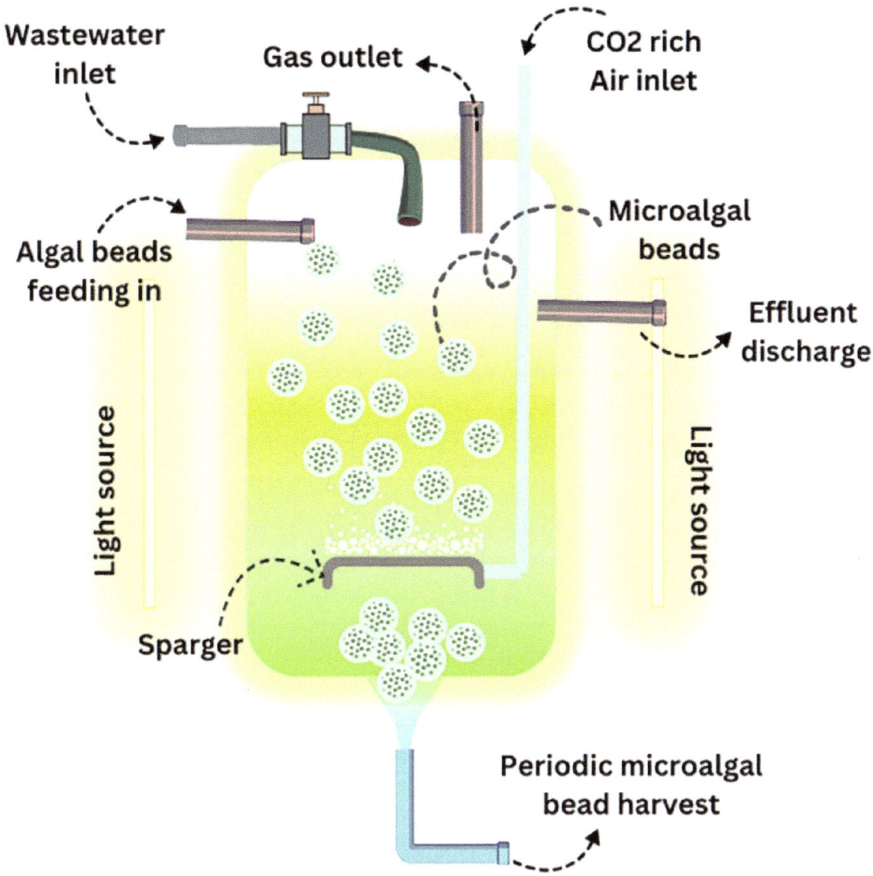

FIGURE 4.6 Immobilized bead system. A schematic illustration of fluidized bed photobioreactor (FBPB) containing immobilized algal beads for WWT.

a secondary effluent, after which CO_2 is pumped into the reactor through a distributor to make the bed fluidized (Hu et al., 2021a). The light source is provided with artificial illumination with optimized PAR efficiency. As a result, the cells receive improved nutrition, oxygen transmission, mass transfer, and efficient light. Following their immobilization in the beads, the microalgae used the nutrients present in the WW, concurrently reducing the levels of COD, nitrogen, and phosphorus (Rodríguez López et al., 2023). It is also possible to combine fluidized bed systems with other PBRs, such as stirred reactors and bubble columns, to provide synergistic benefits in terms of growth and treatment.

4.18 MICROALGAL BIOMASS YIELD

Microalgae adapt their metabolic systems and development patterns to control their composition and growth characteristics under various culture circumstances. In addition to the chosen microalgae strain, growing conditions and the particular photobioreactor design—all affect the biomass and product yields. Additionally, the size of the reactor affects the process yield; the bigger the reactor capacity, the more challenging it is to maintain the conditions necessary for microalgal growth, which lowers the system output (Siddiki et al., 2022). Moreover, by controlling carbon metabolism, a metabolic network alternative that is theoretically carbon neutral may improve biomass and product yields. Microalgae were found to have a potential biomass production of 100–200 g DW/m^2/day, whereas their actual productivity rate was 15–30 g DW/m^2/day (Sun et al., 2018).

4.19 RELIABILITY AND TECHNOLOGY READINESS OF MICROALGAE WWT SYSTEMS

The process reliability of the microalgae-based WWT systems for nutrient removal from secondary treated effluent can be classified on the basis of four major aspects, which include WW quality (depicting how sensitive is the system regarding variable WW quality at the inlet of the microalgae reactors), operational parameters and process control (how stable and controlled the process is regarding quality of effluent), microalgal harvesting methods (depicting the applicability of harvesting methods for microalgal biomass), and produced biomass (showing how safe the biomass is for reuse). In Table 4.4, a visual representation of the estimated reliability of the configurations of the microalgal WWT systems with respect to these parameters and the variation in WW quality is shown. These estimations are based on the reported research compiled in the study of Valchev and Ribarova (2022). By comparing these parameters, it has been concluded that the highest reliability can be achieved by rotating biofilm systems and microalgal beads systems with a reliability score of 75% each, followed by closed-suspended systems with a reliability score of 63%. Conversely, stationary biofilm systems present a low reliability of 13%, whereas open suspended systems have the lowest score with 0% reliability. According to the internationally established and recognized NASA or European Commission (EC) criteria, the technological maturity of any system depends upon nine technological readiness levels (TRL) (Buchner et al., 2019). As for microalgal WWT technology, no reports have been published that indicate TRL reaching beyond 6–7, which has been achieved by the open-pond system despite having the lowest reliability score and also by closed-suspended systems (Valchev and Ribarova, 2022). This shows that the design and development of any type of reactor is not only driven by its reliability but also by its economic feasibility and ease of component assembly. Conversely, the use of closed-suspended WWT technology appears to be more appropriate in the area of culturing microalgal monoculture for the generation of valuable compounds and fuels. Higher parameter control yields more consistent results, but the complex design of the reactor, the challenging operational procedures, and the higher prices appear to hinder the use of microalgae-based WWT technology. The most promising reactor types for microalgal WWT are the rotating biofilm systems. Some businesses try to commercialize rotating biofilms using various patent approaches and design solutions, despite the scarcity of published data (Hu et al., 2021b). In conclusion, VCBDs and RABRs, which are based on microalgae rotating biofilm PBRs, have shown promising potential when it comes to scale-up and obtaining the validation of technology in relevant environments (TRL 5) (Valchev and Ribarova, 2022).

TABLE 4.4
Reliability and Technology Readiness Status of Microalgae-based WWT Systems

	Reliability Parameters					Technology Readiness
Photobioreactor Systems	**Inlet WW Quality**	**Operating Conditions/ Process Control**	**Harvesting Efficiency**	**Produced Biomass**	**% Reliability**	**TRL Score**
Open suspended systems (WSP, HRAPS)	● (yellow)	● (yellow)	● (yellow)	● (yellow)	0	6–7
Closed-suspended systems (Photobioreactors)	● (green)	● (green)	● (yellow)	● (blue)	63	6–7
Stationary biofilm systems (Passive immobilization)	● (yellow)	● (yellow)	● (blue)	● (yellow)	13	4–5
Rotating biofilm systems (Passive immobilization)	● (green)	● (blue)	● (green)	● (blue)	75	5–7
Microalgal beads systems (Active immobilization)	● (blue)	● (green)	● (blue)	● (green)	75	3–5

Source: Adapted from Valchev and Ribarova (2022).
● (green), Higher reliability; ● (blue), Moderate reliability; ● (yellow), Low reliability; TRL, technology readiness level.

4.20 CHALLENGES AND FUTURE PERSPECTIVE

Even though microalgal WWT has been a subject of discussion in many publications as a viable and renewable alternative to traditional WWT methods, its application in large-scale WWT could only be accomplished after a number of significant obstacles have been surmounted. Due to these obstacles, microalgae-based WWT technology is primarily restricted to laboratory and pilot-scale settings, despite being a viable alternative to traditional WWT procedures. Consequently, there are few modest-sized facilities using this technology that have been developed on an industrial scale. In order to make it efficient and scalable at the industrial level, microalgal-based processes must be improved. These improvements can be made by implementing certain strategies, such as making the process robust regardless of abiotic conditions; improving the microalgal strains through genetic and metabolic engineering so they can withstand unfavourable conditions; adapting and improving the efficiency of strains for bioproduct yields and pollution removal; lowering the HRT, which is between 2 and 8 days, to amounts that are roughly equivalent to the 0.3 days achieved by traditional techniques; lowering the energy usage to less than 0.5 kWh/m^3 of WW; and making certain that the released water complies with rules in terms of quality and safety. Another factor that hinders microalgae from being applied on a large scale is the dark colour of WW and hazardous pollutants. It is advised to pretreat the WW influent before adding microalgae as a solution to this problem. Additionally, it is believed that a number of pretreatments are required to create the appropriate nutritional balance for microalgal growth in order to overcome toxicity due to the vast diversity of nutrients and their concentration in WW. It may be possible to employ optimized combinations of several WW sources as a sole and balanced nutrition medium for microalgae. Therefore, to achieve these objectives, it is necessary to reassess both the biological and technical aspects of WWT procedures.

4.21 CONCLUSION

The sustainability and environmental advantages of microalgae-based WWT systems have garnered a lot of interest recently. In addition to acting as CO_2 sinks and facilitating the recovery of nutrients from WW, these systems lessen the greenhouse effect. This enables the WWT industry to adopt the ideas of the circular economy. For the treatment of WW, microalgae can be grown as suspensions or they can be fixed or immobilized in and on a solid matrix in open or closed reactors. These reactors come in many designs and have shown efficient performance; however, they all have their own advantages and disadvantages. The key is to design an efficient, easy, and cost-effective photobioreactor for the treatment of WW where the microalgae can grow sustainably with minimum stress.

REFERENCES

Abdelfattah, Abdallah, Sameh Samir Ali, Hassan Ramadan, Eslam Ibrahim El-Aswar, Reham Eltawab, Shih-Hsin Ho, Tamer Elsamahy et al. "Microalgae-based wastewater treatment: Mechanisms, challenges, recent advances, and future prospects." *Environmental Science and* Ecotechnology 13 (2023): 100205.

Acharya, Diptikanta, Sagarika Satapathy, D. Tapas Kumar Dora, Sushma Dave, and Jayashankar Das. "Advancement in treatment technologies of biopharmaceutical industrial effluents." In Maulin P. Shah (ed.), *Microbial Bioremediation & Biodegradation* (2020), pp. 467–484. Springer.

Aditya, Lisa, T. M. Indra Mahlia, Luong N. Nguyen, Hang P. Vu, and Long D. Nghiem. "Microalgae-bacteria consortium for wastewater treatment and biomass production." *Science of the Total Environment* 838 (2022): 155871.

Agersborg, H. P. K., and William D. Hatfield. "The biology of a sewage treatment plant: A preliminary survey: Decatur, Illinois." *Sewage Works Journal* 1, no. 4 (1929): 411–424.

Ahmed, Shams Forruque, M. Mofijur, Tahlil Ahmed Parisa, Nafisa Islam, F. Kusumo, Abrar Inayat, Irfan Anjum Badruddin, T. M. Yunus Khan, and Hwai Chyuan Ong. "Progress and challenges of contaminate removal from wastewater using microalgae biomass." *Chemosphere* 286 (2022): 131656.

Aketo, Tsuyoshi, Yumiko Hoshikawa, Daisuke Nojima, Yusuke Yabu, Yoshiaki Maeda, Tomoko Yoshino, Hiroyuki Takano, and Tsuyoshi Tanaka. "Selection and characterization of microalgae with potential for nutrient removal from municipal wastewater and simultaneous lipid production." *Journal of Bioscience and Bioengineering* 129, no. 5 (2020): 565–572.

Al-Dailami, Anas, Iwamoto Koji, Imran Ahmad, and Masafumi Goto. "Potential of photobioreactors (PBRs) in cultivation of microalgae." *Journal of Advanced Research in Applied Sciences and Engineering Technology* 27, no. 1 (2022): 32–44.

Ali, Atif, Zunera Khalid, and Jamaan S. Ajarem. "Wastewater treatment by using microalgae: Insights into fate, transport, and associated challenges." *Chemosphere* 338 (2023): 139501.

Ali, Sameh Samir, Mostafa El-Sheekh, Alessandro Manni, Héctor A. Ruiz, Tamer Elsamahy, Jianzhong Sun, and Michael Schagerl. "Microalgae-mediated wastewater treatment for biofuels production: A comprehensive review." *Microbiological Research* 265 (2022): 127187.

Al-Jabri, Hareb, Probir Das, Shoyeb Khan, Mahmoud Thaher, and Mohammed AbdulQuadir. "Treatment of wastewaters by microalgae and the potential applications of the produced biomass: A review." *Water* 13, no. 1 (2020): 27.

Álvarez-Díaz, Pablo, Jesús Ruiz González, Zouhayr Arbib, Jesus Barragan, Maria DelCarmen Garrido-Pérez, and José Antonio Perales-Vargas-Machuca. "Freshwater microalgae selection for simultaneous wastewater nutrient removal and lipid production." *Algal Research* 24 (2017): 477–485.

Amaro, Helena M., Eva M. Salgado, Olga C. Nunes, José C. M. Pires, and Ana F. Esteves. "Microalgae systems-environmental agents for wastewater treatment and further potential biomass valorisation." *Journal of Environmental Management* 337 (2023): 117678.

Ambrosio, Rafael, Lara Sanchez Rizza, Mauro Do Nascimento, Hugo Guillermo Jimenez Pacheco, Lilia Mary Miranda Ramos, Jose A. Hernandez, and Leonardo Curatti. "Promises and challenges for expanding the use of N_2-fixing cyanobacteria as a fertilizer for sustainable agriculture." In Prashant Singh, Maria Fillat, and Ajay Kumar (eds.), *Cyanobacterial Lifestyle and Its Applications in Biotechnology* (2022), pp. 99–158. Academic Press.

Anbalagan, Anbarasan, Sebastian Schwede, Carl-Fredrik Lindberg, and Emma Nehrenheim. "Influence of hydraulic retention time on indigenous microalgae and activated sludge process." *Water Research* 91 (2016): 277–284.

Araujo, Glacio S., Clarice S. Santiago, Renato T. Moreira, Maximiano P. Dantas Neto, and Fabiano A. N. Fernandes. "Nutrient removal by Arthrospira platensis cyanobacteria in cassava processing wastewater." *Journal of Water Process Engineering* 40 (2021): 101826.

Arias, Dulce María, Enrica Uggetti, and Joan García. "Assessing the potential of soil cyanobacteria for simultaneous wastewater treatment and carbohydrate-enriched biomass production." *Algal Research* 51 (2020): 102042.

Ashraf, Noreen, Fiaz Ahmad, and Yandu Lu. "Synergy between microalgae and microbiome in polluted waters." *Trends in Microbiology* 31, no. 1 (2023): 9–21.

Assunção, Joana, and F. Xavier Malcata. "Enclosed "non-conventional" photobioreactors for microalga production: A review." *Algal Research* 52 (2020): 102107.

Astafyeva, Yekaterina, Marno Gurschke, Minyue Qi, Lutgardis Bergmann, Daniela Indenbirken, Imke de Grahl, Elena Katzowitsch et al. "Microalgae and bacteria interaction—Evidence for division of diligence in the alga microbiota." *Microbiology Spectrum* 10, no. 4 (2022): e00633–22.

Baker, Kirralee G., and Richard J. Geider. "Phytoplankton mortality in a changing thermal seascape." *Global Change Biology* 27, no. 20 (2021): 5253–5261.

Banerjee, Srijoni, Dipankar Ghosh, Chetan Pandit, Sagnik Saha, Anwesha Mohapatra, Soumya Pandit, Minaxi Sharma, Kandi Sridhar, Baskaran Stephen Inbaraj, and Ram Prasad. "Microalgal pandora for potent bioenergy production: A way forward?" *Fuel* 333 (2023): 126253.

Bartosiewicz, Maciej, Liah X. Coggins, Patricia Glaz, Alicia Cortés, Sebastien Bourget, Elke S. Reichwaldt, Sally MacIntyre, Anas Ghadouani, and Isabelle Laurion. "Integrated approach towards quantifying carbon dioxide and methane release from waste stabilization ponds." *Water Research* 202 (2021): 117389.

Bekirogullari, Mesut, Gonzalo M. Figueroa-Torres, Jon K. Pittman, and Constantinos Theodoropoulos. "Models of microalgal cultivation for added-value products: A review." *Biotechnology Advances* 44 (2020): 107609.

Bele, Vaibhavi, Rajinikanth Rajagopal, and Bernard Goyette. "Closed loop bioeconomy opportunities through the integration of microalgae cultivation with anaerobic digestion: A critical review." *Bioresource Technology Reports* 21 (2023): 101336.

Benemann, Rupert Craggs Tryg Lundquist John, John H. Todd, Anthony McInnis, Sandeep Kumar, and Anju Dahiya. "Part IV: From waste water to fuel production." *The Science of Algal Fuels: Phycology, Geology, Biophotonics, Genomics and* Nanotechnology 25 (2012): 424.

Bhatti, Shabana, Robert Richards, Cheryl L. Wall, Margaret J. MacPherson, Kevin Stemmler, Crystal G. Lalonde, Shane J. J. Patelakis, Sean M. Tibbetts, and Patrick J. McGinn. "Phycoremediation and simultaneous production of protein-rich algal biomass from aquaculture and agriculture wastewaters." *Journal of Chemical Technology & Biotechnology* 98, no. 8 (2023): 1918–1935.

Blair, John G. "The poetics of parable." In Blair, J.G. (ed.), *The Poetic Art of WH Auden* (2015), Vol. 2302, p. 54. Princeton University Press.

Borowiak, Daniel, and Małgorzata Krzywonos. "Bioenergy, biofuels, lipids and pigments—Research trends in the use of microalgae grown in photobioreactors." *Energies* 15, no. 15 (2022): 5357.

Brennan, Liam, and Philip Owende. "Biofuels from microalgae: A review of technologies for production, processing, and extractions of biofuels and co-products." *Renewable and Sustainable Energy Reviews* 14, no. 2 (2010): 557–577.

Buchner, Georg A., Kai J. Stepputat, Arno W. Zimmermann, and Reinhard Schomäcker. "Specifying technology readiness levels for the chemical industry." *Industrial & Engineering Chemistry Research* 58, no. 17 (2019): 6957–6969.

Caldwell, Gary S., Pichaya In-Na, Rachel Hart, Elliot Sharp, Assia Stefanova, Matthew Pickersgill, Matthew Walker, Matthew Unthank, Justin Perry, and Jonathan G. M. Lee. "Immobilising microalgae and cyanobacteria as biocomposites: New opportunities to intensify algae biotechnology and bioprocessing." *Energies* 14, no. 9 (2021): 2566.

Canizales, Sebastian, Pei-Hsuan Chen, Hardy Temmink, René H. Wijffels, and Marcel Janssen. "Cyanobacteria cultivation on human urine for nutrients recovery." *Algal Research* 71 (2023): 103064.

Chai, Wai Siong, Wee Gee Tan, Heli Siti Halimatul Munawaroh, Vijai Kumar Gupta, Shih-Hsin Ho, and Pau Loke Show. "Multifaceted roles of microalgae in the application of wastewater biotreatment: A review." *Environmental Pollution* 269 (2021): 116236.

Chambonniere, Paul, John Bronlund, and Benoit Guieysse. "Pathogen removal in high-rate algae pond: State of the art and opportunities." *Journal of Applied Phycology* 33, no. 3 (2021): 1501–1511.

Chan, Sook Sin, Kuan Shiong Khoo, Kit Wayne Chew, Tau Chuan Ling, and Pau Loke Show. "Recent advances biodegradation and biosorption of organic compounds from wastewater: Microalgae-bacteria consortium: A review." *Bioresource Technology* 344 (2022): 126159.

Chanquia, Santiago N., Guillem Vernet, and Selin Kara. "Photobioreactors for cultivation and synthesis: Specifications, challenges, and perspectives." *Engineering in Life Sciences* 22, no. 12 (2022): 712–724.

Chávez-Fuentes, Paolah, Alejandro Ruiz-Marin, and Yunuen Canedo-López. "Biodiesel synthesis from *Chlorella vulgaris* under effect of nitrogen limitation, intensity and quality light: Estimation on the based fatty acids profiles." *Molecular Biology Reports* 45 (2018): 1145–1154.

Cheng, Pengfei, Dongjie Chen, Wei Liu, Kirk Cobb, Nan Zhou, Yuhuan Liu, Hui Liu et al. "Auto-flocculation microalgae species *Tribonema* sp. and *Synechocystis* sp. with T-IPL pretreatment to improve swine wastewater nutrient removal." *Science of the Total Environment* 725 (2020): 138263.

Chia, Shir Reen, Kit Wayne Chew, Hui Yi Leong, Shih-Hsin Ho, Heli Siti Halimatul Munawaroh, and Pau Loke Show. "CO_2 mitigation and phycoremediation of industrial flue gas and wastewater via microalgae-bacteria consortium: Possibilities and challenges." *Chemical Engineering Journal* 425 (2021): 131436.

Chuka-Ogwude, David, James Ogbonna, Michael A. Borowitzka, and Navid R. Moheimani. "Screening, acclimation and ammonia tolerance of microalgae grown in food waste digestate." *Journal of Applied Phycology* 32 (2020): 3775–3785.

Cotton, Arthur D.. "On the growth of *Ulva latissima, L.* in water polluted by sewage." *Bulletin of Miscellaneous Information (Royal Botanic Gardens, Kew)* 1910, no. 1 (1910): 15–19.

Cuellar-Bermudez, Sara P., Gibran S. Aleman-Nava, Rashmi Chandra, J. Saul Garcia-Perez, Jose R. Contreras-Angulo, Giorgos Markou, Koenraad Muylaert, Bruce E. Rittmann, and Roberto Parra-Saldivar. "Nutrients utilization and contaminants removal. A review of two approaches of algae and cyanobacteria in wastewater." *Algal Research* 24 (2017): 438–449.

Daneshvar, Ehsan, Mohammad Javad Zarrinmehr, Eleni Koutra, Michael Kornaros, Omidvar Farhadian, and Amit Bhatnagar. "Sequential cultivation of microalgae in raw and recycled dairy wastewater: Microalgal growth, wastewater treatment and biochemical composition." *Bioresource Technology* 273 (2019): 556–564.

Dange, Pujaa, Shreya Gawas, Soumya Pandit, Lukhanyo Mekuto, Piyush K. Gupta, Poojhaa Shanmugam, Ravishankar Patil, and Srijoni Banerjee. "Trends in photobioreactor technology for microalgal biomass production along with wastewater treatment: Bottlenecks and breakthroughs." In Maulin P. Shah, Susana Rodriguez-Couto, Celia Bertha Vargas De La Cruz, and Jayanta Kumar Biswas (eds.), *An Integration of Phycoremediation Processes in Wastewater Treatment* (2022), pp. 135–154. Elsevier.

de Carvalho, Júlio Cesar, Denisse Tatiana Molina-Aulestia, Walter José Martinez-Burgos, Susan Grace Karp, Maria Clara Manzoki, Adriane Bianchi Pedroni Medeiros, Cristine Rodrigues et al. "Agro-industrial wastewaters for algal biomass production, bio-based products, and biofuels in a circular bioeconomy". *Fermentation* 8, no. 12 (2022): 728.

Dias, Daniel F. C., Ricardo GomesPassos, and Marcos von Sperling. "A review of bacterial indicator disinfection mechanisms in waste stabilisation ponds." *Reviews in Environmental Science and Bio/Technology* 16 (2017): 517–539.

Ennaceri, Houda, Tasneema Ishika, Victor Okorie Mkpuma, and Navid Reza Moheimani. "Microalgal biofilms: Towards a sustainable biomass production." *Algal Research* (2023): 103124.

Fallahi, Alireza, Fariba Rezvani, Hashem Asgharnejad, Ehsan Khorshidi Nazloo, Nima Hajinajaf, and Brendan Higgins. "Interactions of microalgae-bacteria consortia for nutrient removal from wastewater: A review." *Chemosphere* 272 (2021): 129878.

Fallahi, Alireza, Nima Hajinajaf, Omid Tavakoli, and Mohammad Hossein Sarrafzadeh. "Cultivation of mixed microalgae using municipal wastewater: Biomass productivity, nutrient removal, and biochemical content." *Iranian Journal of Biotechnology* 18, no. 4 (2020): e2586.

Fan, Honghui, Kai Wang, Chuan Wang, Fangyuan Yu, Xinxia He, Jun Ma, and Xiaoting Li. "A comparative study on growth characters and nutrients removal from wastewater by two microalgae under optimized light regimes." *Environmental Technology & Innovation* 19 (2020a): 100849.

Fan, Jie, Yue Chen, Tian C. Zhang, Bin Ji, and Liang Cao. "Performance of *Chlorella sorokiniana*-activated sludge consortium treating wastewater under light-limited heterotrophic condition." *Chemical Engineering Journal* 382 (2020b): 122799.

Fernandes, Tânia V., Lukas M. Trebuch, and René H. Wijffels. "Microalgae-based technologies for circular wastewater treatment." In Goksel N. Demirer, and Sibel Uludag-Demirer (eds.), *Integrated Wastewater Management and Valorization Using Algal Cultures* (2022), pp. 81–112. Elsevier.

Fida, Misbah, Peiyue Li, Yuanhang Wang, S. M. Khorshed Alam, and Abel Nsabimana. "Water contamination and human health risks in Pakistan: A review." *Exposure and Health* 15, no. 3 (2023): 619–639.

Gong, You-Zi, Qiu-Ya Niu, Yun-Guo Liu, Jie Dong, and Meng-Meng Xia. "Development of multifarious carrier materials and impact conditions of immobilised microbial technology for environmental remediation: A review." *Environmental Pollution* 314 (2022): 120232.

González-González, Lina Maria, and Luz E. De-Bashan. "Toward the enhancement of microalgal metabolite production through microalgae–bacteria consortia." *Biology* 10, no. 4 (2021): 282.

Goswami, Rahul Kumar, Komal Agrawal, and Pradeep Verma. "Multifaceted role of microalgae for municipal wastewater treatment: A futuristic outlook toward wastewater management." *CLEAN–Soil, Air,* Water 51, no. 3 (2023): 2100286.

Grira, Soumaya, Hadil Abu Khalifeh, Mohammad Alkhedher, and Mohamad Ramadan. "The conventional microalgal biofuel production process and the alternative milking pathway: A review." *Energy* 277 (2023): 127547.

Gupta, Prabuddha, Ashok Kumar Bishoyi, Mahendrapal Singh Rajput, Ujwal Trivedi, and Gaurav Sanghvi. "Biotechnological advances for utilization of algae, microalgae, and cyanobacteria for wastewater treatment and resource recovery." In Pradeep Verma, and Maulin P. Shah (eds.), *Phycology-Based Approaches for Wastewater Treatment and Resource Recovery* (2021), pp. 1–24. CRC Press.

Hachicha, Rihab, Fatma Elleuch, Hajer Ben Hlima, Pascal Dubessay, Helene de Baynast, Cedric Delattre, Guillaume Pierre et al. "Biomolecules from microalgae and cyanobacteria: Applications and market survey." *Applied* Sciences 12, no. 4 (2022): 1924.

Hajinajaf, Nima, Abbas Mehrabadi, and Omid Tavakoli. "Practical strategies to improve harvestable biomass energy yield in microalgal culture: A review." *Biomass and Bioenergy* 145 (2021): 105941.

Hena, Sufia, Leonardo Gutierrez, and Jean-Philippe Croué. "Removal of pharmaceutical and personal care products (PPCPs) from wastewater using microalgae: A review." *Journal of Hazardous Materials* 403 (2021): 124041.

Ho, Long, and Peter L. M. Goethals. "Municipal wastewater treatment with pond technology: Historical review and future outlook." *Ecological Engineering* 148 (2020): 105791.

Hong, Joung Sook, Wonsik Shin, Hyorin Nam, Jin-Ho Yun, Hee-Sik Kim, and Kyung Hyun Ahn. "Sedimentation and rheological study of microalgal cell (*Chlorella sp.* HS_2) suspension." *Biotechnology and Bioprocess Engineering* 27, no. 3 (2022): 451–460.

Hu, Xinjuan, Yulie E. Meneses, Ashraf Aly Hassan, Jayne Stratton, and Shuhao Huo. "Application of alginate immobilized microalgae in treating real food industrial wastewater and design of annular photobioreactor: A proof-of-concept study." *Algal Research* 60 (2021a): 102524.

Hu, Yingbing, Yu Xiao, Kang Liao, Yunyue Leng, and Qian Lu. "Development of microalgal biofilm for wastewater remediation: From mechanism to practical application." *Journal of Chemical Technology & Biotechnology* 96, no. 11 (2021b): 2993–3008.

Inuwa, A. B., A. Pervez, and R. Nazir. "Microalgae-based wastewater treatment system: Current state, antibiotic resistant bacteria and antibiotic resistance genes reduction potentials." *International Journal of Environmental Science and Technology* 20, no. 12 (2023): 14053–14072.

Iqbal, Khushboo, Abhishek Saxena, Priyanshi Pande, Archana Tiwari, Naveen Chandra Joshi, Ajit Varma, and Arti Mishra. "Microalgae-bacterial granular consortium: Striding towards sustainable production of biohydrogen coupled with wastewater treatment." *Bioresource Technology* 354 (2022): 127203.

Issac, Merlin N., and Balasubramanian Kandasubramanian. "Effect of microplastics in water and aquatic systems." *Environmental Science and Pollution Research* 28 (2021): 19544–19562.

Janpum, Chalampol, Nuttapon Pombubpa, Tanakarn Monshupanee, Aran Incharoensakdi, and Pichaya In-Na. "Advancement on mixed microalgal-bacterial cultivation systems for nitrogen and phosphorus recoveries from wastewater to promote sustainable bioeconomy." *Journal of Biotechnology* 360 (2022): 198–210.

Jassim, Yazi Abdullah, Eman Fadhel Abbas Awadh, and Shaemaa Muhi Hasoon Al-Amery. "A review of general properties of blue-green algae (Cyanobacteria)." *Biomedicine and Chemical Sciences* 2, no. 2 (2023): 143–148.

Jin, Yaruo, Wei Zhan, Rui Wu, Yahong Han, Shanshan Yang, Jie Ding, and Nanqi Ren. "Insight into the roles of microalgae on simultaneous nitrification and denitrification in microalgal-bacterial sequencing batch reactors: Nitrogen removal, extracellular polymeric substances, and microbial communities." *Bioresource Technology* 379 (2023): 129038.

Jones, Gerald Benjamin. "An Integrated Experimental and Modeling Approach to Design Rotating Algae Biofilm Reactors (RABRs) Via Optimizing Algae Biofilm Productivity, Nutrient Recovery, and Energy Efficiency." (2023). *All Graduate Plan B and other Reports, Spring 1920 to Spring 2023*, 1719.

Kaloudas, Dimitrios, Nikolet Pavlova, and Robert Penchovsky. "Phycoremediation of wastewater by microalgae: A review." *Environmental Chemistry Letters* 19 (2021): 2905–2920.

Kaštovský, Jan. "Welcome to the jungle! An overview of modern taxonomy of cyanobacteria." *Hydrobiologia* 851, no. 4 (2024): 1063–1077.

Katarzyna, Ledwoch, Gu Sai, and Oinam Avijeet Singh. "Non-enclosure methods for non-suspended microalgae cultivation: Literature review and research needs." *Renewable and Sustainable Energy Reviews* 42 (2015): 1418–1427.

Kaur, Manpreet, Khem Chand Saini, Hiramoni Ojah, Rajalakshmi Sahoo, Kriti Gupta, Adesh Kumar, and Felix Bast. "Abiotic stress in algae: Response, signalling and transgenic approaches." *Journal of Applied Phycology* 34, no. 4 (2022): 1843–1869.

Khalili, Arezoo, Ghasem D. Najafpour, Ghazaleh Amini, and Faezeh Samkhaniyani. "Influence of nutrients and LED light intensities on biomass production of microalgae *Chlorella vulgaris*." *Biotechnology and Bioprocess Engineering* 20 (2015): 284–290.

Khan, Muhammad Imran, Jin Hyuk Shin, and Jong Deog Kim. "The promising future of microalgae: Current status, challenges, and optimization of a sustainable and renewable industry for biofuels, feed, and other products." *Microbial Cell Factories* 17 (2018): 1–21.

Khoo, Choon Gek, Man Kee Lam, Abdul Rahman Mohamed, and Keat Teong Lee. "Hydrochar production from high-ash low-lipid microalgal biomass via hydrothermal carbonization: Effects of operational parameters and products characterization." *Environmental Research* 188 (2020): 109828.

Kong, Wenwen, Boxiong Shen, Honghong Lyu, Jia Kong, Jiao Ma, Zhuozhi Wang, and Shuo Feng. "Review on carbon dioxide fixation coupled with nutrients removal from wastewater by microalgae." *Journal of Cleaner Production* 292 (2021): 125975.

Koop, Stef H. A., Chloé Grison, Steven J. Eisenreich, Jan Hofman, and Kees van Leeuwen. "Integrated water resources management in cities in the world: Global solutions." *Sustainable Cities and Society* 86 (2022): 104137.

Koul, Bhupendra, Dhananjay Yadav, Swati Singh, Manoj Kumar, and Minseok Song. "Insights into the domestic wastewater treatment (DWWT) regimes: A review." *Water* 14, no. 21 (2022): 3542.

Koul, Bhupendra, Komal Sharma, and Maulin P. Shah. "Phycoremediation: A sustainable alternative in wastewater treatment (WWT) regime." *Environmental Technology & Innovation* 25 (2022): 102040.

Krishnan, Sivakumar, Sunil A. Patil, and Y. V. Nancharaiah. "Environmental microbial biofilms: Formation, characteristics, and biotechnological applications." In Nabin Aryal, Yifeng Zhang, Sunil A. Patil, and Deepak Pant (eds.), *Material-Microbes Interactions* (2023), pp. 3–45. Academic Press.

Kumar, Ajay, and Surojit Bera. "Revisiting nitrogen utilization in algae: A review on the process of regulation and assimilation." *Bioresource Technology Reports* 12 (2020): 100584.

Kumari, Astha, Nityanand Singh Maurya, Abhishek Kumar, Rajanee Kant Yadav, and Amit Kumar. "Options for the disposal and reuse of wastewater sludge, associated benefit, and environmental risk." In Basak Kiliç Taseli (ed.), *Sewage Management* (2023). IntechOpen, 1–15.

Lafi, Hana, Umed Panu, and Baoqiang Liao. "Effect of the organic carbon to nutrient (N and P) ratio on the biological performance of a microalgal–bacterial membrane photobioreactor." *Environmental Science: Water Research &* Technology 9, no. 8 (2023): 2021–2030.

Lee, Sang-Ah, Nakyeong Lee, Hee-Mock Oh, and Chi-Yong Ahn. "Enhanced and balanced microalgal wastewater treatment (COD, N, and P) by interval inoculation of activated sludge." *Journal of Microbiology and Biotechnology* 29 (2019): 1434–1443.

Leite, Luan de Souza, Maria Teresa Hoffmann, and Luiz Antonio Daniel. "Microalgae cultivation for municipal and piggery wastewater treatment in Brazil." *Journal of Water Process Engineering* 31 (2019): 100821.

Leong, Yoong Kit, and Jo-Shu Chang. "Integrated role of algae in the closed-loop circular economy of anaerobic digestion." *Bioresource Technology* 360 (2022): 127618.

Leong, Yoong Kit, Chi-Yu Huang, and Jo-Shu Chang. "Pollution prevention and waste phycoremediation by algal-based wastewater treatment technologies: The applications of high-rate algal ponds (HRAPs) and algal turf scrubber (ATS)." *Journal of Environmental Management* 296 (2021): 113193.

Li, Gang, Wenbo Xiao, Tenglun Yang, and Tao Lyu. "Optimization and process effect for microalgae carbon dioxide fixation technology applications based on carbon capture: A comprehensive review." *C* 9, no. 1 (2023): 35.

Li, Lifeng, Xiaoyun Xu, Wujun Wang, Raymond Lau, and Chi-Hwa Wang. "Hydrodynamics and mass transfer of concentric-tube internal loop airlift reactors: A review." *Bioresource Technology* 359 (2022b): 127451.

Li, Yin, Christopher P. Alaimo, Minji Kim, Norman Y. Kado, Joshua Peppers, Jian Xue, Chao Wan et al. "Composition and toxicity of biogas produced from different feedstocks in California." *Environmental Science &* Technology 53, no. 19 (2019): 11569–11579.

Li, Zejiao, Jixiang Wang, Xingyu Chen, Zhongfang Lei, Tian Yuan, Kazuya Shimizu, Zhenya Zhang, and Duu-Jong Lee. "Insight into aerobic phosphorus removal from wastewater in algal-bacterial aerobic granular sludge system." *Bioresource Technology* 352 (2022a): 127104.

Lia, Weiran, Sidi Penga, Huaiyong Jiaoa, Jiahao Zhanga, Yucai Huanga, and Lianpeng Suna. "Performance assessment of microalgae-based wastewater treatment (MBWT) system in response to operation modes, hydraulic retention time (HRT) and cyclical light." *Desalination and Water Treatment* 229 (2021): 79–92.

Liu, Jiajing, Qi Zhang, Guangjing Xu, and Fan Gao. "Simultaneous nitritation, denitritation and phosphorus removal in an algal-bacterial consortium system treating low-strength mariculture wastewater." *Journal of Water Process Engineering* 49 (2022): 103056.

Lu, Weidong, Shijie Liu, Zhiyan Lin, and Manjia Lin. "Enhanced microalgae growth for biodiesel production and nutrients removal in raw swine wastewater by carbon sources supplementation." *Waste and Biomass Valorization* 12 (2021): 1991–1999.

Luo, Shuai, John A. Berges, Zhen He, and Erica B. Young. "Algal-microbial community collaboration for energy recovery and nutrient remediation from wastewater in integrated photobioelectrochemical systems." *Algal Research* 24 (2017): 527–539.

Ma, Zengling, Wai Yan Cheah, I-Son Ng, Jo-Shu Chang, Min Zhao, and Pau Loke Show. "Microalgae-based biotechnological sequestration of carbon dioxide for net zero emissions." *Trends in Biotechnology* 40, no. 12 (2022): 1439–1453.

Mahapatra, Saswat, Kundan Samal, and Rajesh Roshan Dash. "Waste stabilization pond (WSP) for wastewater treatment: A review on factors, modelling and cost analysis." *Journal of Environmental Management* 308 (2022): 114668.

Markou, Giorgos, Alexandros Diamantis, Dimitris Arapoglou, Dimitris Mitrogiannis, Cristina González-Fernández, and Adrian Unc. "Growing *Spirulina* (*Arthrospira platensis*) in seawater supplemented with digestate: Trade-offs between increased salinity, nutrient and light availability." *Biochemical Engineering* Journal 165 (2021): 107815.

Martini, Felipe Augusto, Aline Rubert, Maiara Priscilla de Souza, Lourdes Teresinha Kist, Michele Hoeltz, Lisianne Brittes Benitez, Tiele Medianeira Rizzetti, Pablo Diego Gressler, and Rosana de Cassia de Souza Schneider. "Periphytic biomass composition and exploitation from algae turf scrubber system." *SN Applied Sciences* 1 (2019): 1–9.

Masojídek, Jiří, Karolina Ranglova, Gergely Ernö Lakatos, Ana Margarita Silva Benavides, and Giuseppe Torzillo. "Variables governing photosynthesis and growth in microalgae mass cultures." *Processes* 9, no. 5 (2021): 820.

Mathew, Merwin Mammen, Kanchan Khatana, Vaidehi Vats, Raunak Dhanker, Ram Kumar, Hans-Uwe Dahms, and Jiang-Shiou Hwang. "Biological approaches integrating algae and bacteria for the degradation of wastewater contaminants—A review." *Frontiers in Microbiology* 12 (2022): 801051.

Matito-Martos, Ismael, Claudia Sepúlveda, Cintia Gómez, Gabriel Acien, Julio Perez-Carbajo, Jose A. Delgado, V. I. Águeda et al. "Potential of CO_2 capture from flue gases by physicochemical and biological methods: A comparative study." *Chemical Engineering Journal* 417 (2021): 128020.

Maurya, Annapurna, Rajesh Kumar, and Abhay Raj. "Biofilm-based technology for industrial wastewater treatment: Current technology, applications and future perspectives." *World Journal of Microbiology and Biotechnology* 39, no. 5 (2023): 112.

Mehariya, Sanjeet, Rahul Kumar Goswami, Pradeep Verma, Roberto Lavecchia, and Antonio Zuorro. "Integrated approach for wastewater treatment and biofuel production in microalgae biorefineries." *Energies* 14, no. 8 (2021): 2282.

Mehrabadi, Abbas, Mohammed M. Farid, and Rupert Craggs. "Potential of five different isolated colonial algal species for wastewater treatment and biomass energy production." *Algal Research* 21 (2017): 1–8.

Mehrabadi, Abbas, Rupert Craggs, and Mohammed M. Farid. "Wastewater treatment high-rate algal ponds (WWT HRAP) for low-cost biofuel production." *Bioresource* Technology 184 (2015): 202–214.

Mishra, Vandana, Nikhil Mudgal, Deepak Rawat, Pankaj Poria, Paromita Mukherjee, Udita Sharma, Poonam Kumria et al. "Integrating microalgae into textile wastewater treatment processes: Advancements and opportunities." *Journal of Water Process Engineering* 55 (2023): 104128.

Mohsenpour, Seyedeh Fatemeh, Sebastian Hennige, Nicholas Willoughby, Adebayo Adeloye, and Tony Gutierrez. "Integrating micro-algae into wastewater treatment: A review." *Science of the Total Environment* 752 (2021): 142168.

Molazadeh, Marziyeh, Hossein Ahmadzadeh, Hamid R. Pourianfar, Stephen Lyon, and Pabulo Henrique Rampelotto. "The use of microalgae for coupling wastewater treatment with CO_2 biofixation." *Frontiers in Bioengineering and Biotechnology* 7 (2019): 42.

Molitor, Hannah R., and Jerald L. Schnoor. "Natural treatment systems and integrated watershed management for decarbonization." In Zhiyong Jason Ren, and Krishna Pagilla (eds.), *Pathways to Water Sector Decarbonization, Carbon Capture and Utilization* (2022): 207. IWA Publishing.

Montoya-Vallejo, Carolina, Fernando León Guzmán Duque, and Juan Carlos Quintero Díaz. "Biomass and lipid production by the native green microalgae *Chlorella sorokiniana* in response to nutrients, light intensity, and carbon dioxide: Experimental and modelling approach." *Frontiers in Bioengineering and Biotechnology* 11 (2023): 1149762.

Morais, Etiele G., Nathana L. Cristofoli, Inês B. Maia, Tânia Magina, Paulo R. Cerqueira, Margarida Ribau Teixeira, João Varela, Luísa Barreira, and Luísa Gouveia. "Microalgal systems for wastewater treatment: Technological trends and challenges towards waste recovery." *Energies* 14, no. 23 (2021): 8112.

Moreno Osorio, Jairo Hernán, Antonino Pollio, Luigi Frunzo, Piet Nicolaas Luc Lens, and Giovanni Esposito. "A review of microalgal biofilm technologies: Definition, applications, settings and analysis." *Frontiers in Chemical Engineering* 3 (2021): 737710.

Mu, Ruimin, Yantian Jia, Guixia Ma, Leran Liu, Kaixuan Hao, Feng Qi, and Yuanyuan Shao. "Advances in the use of microalgal–bacterial consortia for wastewater treatment: Community structures, interactions, economic resource reclamation, and study techniques." *Water Environment Research* 93, no. 8 (2021): 1217–1230.

Mugnai, Sergio, Natalia Derossi, and Yogi Hendlin. "Algae communication, conspecific and interspecific: The concepts of phycosphere and algal-bacteria consortia in a photobioreactor (PBR)." *Plant Signaling & Behavior* 18, no. 1 (2023): 2148371.

Nagarajan, Dillirani, Chun-Yen Chen, Duu-Jong Lee, and Jo-Shu Chang. "Photobioreactors for microalgae-based wastewater treatment." In Rakesh Kumar, Sunil Kumar, and Ashok Pandey (eds.), *Current Developments in Biotechnology and Bioengineering* (2023), pp. 121–152. Elsevier.

Nagarajan, Dillirani, Sunita Varjani, Duu-Jong Lee, and Jo-Shu Chang. "Sustainable aquaculture and animal feed from microalgae–nutritive value and techno-functional components." *Renewable and Sustainable Energy Reviews* 150 (2021): 111549.

Ng, Mary, Sadou Dalhatou, Jessica Wilson, Boniface Pone Kamdem, Mercy Bankole Temitope, Hugues Kamdem Paumo, Hayet Djelal, Aymen Amine Assadi, Phuong Nguyen-Tri, and Abdoulaye Kane. "Characterization of slaughterhouse wastewater and development of treatment techniques: A review." *Processes* 10, no. 7 (2022): 1300.

Nguyen, Hoang Tam, Yeomin Yoon, Huu Hao Ngo, and Am Jang. "The application of microalgae in removing organic micropollutants in wastewater." *Critical Reviews in Environmental Science and Technology* 51, no. 12 (2021): 1187–1220.

Nguyen, Thi An Hang, Huu Hao Ngo, and Manh Khai Nguyen. "Case studies of algae-based biomaterials for environmental remediation." In Sunita J. Varjani (ed.), *Biomass, Biofuels, and Biochemicals* (2022): 291–320. Elsevier.

Nitsos, Christos, Rayen Filali, Behnam Taidi, and Julien Lemaire. "Current and novel approaches to downstream processing of microalgae: A review." *Biotechnology* Advances 45 (2020): 107650.

Oliveira, Adriana Paulo de Sousa, Paula Assemany, Lidiane Covell, Gabriella Peterlini Tavares, and Maria Lúcia Calijuri. "Microalgae-based wastewater treatment for micropollutant removal in swine effluent: High-rate algal ponds performance under different zinc concentrations." *Algal Research* 69 (2023): 102930.

Oruganti, Raj Kumar, Keerthi Katam, Pau Loke Show, Venkataramana Gadhamshetty, Venkata Krishna Kumar Upadhyayula, and Debraj Bhattacharyya. "A comprehensive review on the use of algal-bacterial systems for wastewater treatment with emphasis on nutrient and micropollutant removal." *Bioengineered* 13, no. 4 (2022): 10412–10453.

Oswald, William J., and Claeence G. Golueke. "Biological transformation of solar energy." In Geoffrey M. Gadd, and Sima Sariaslani (eds.), *Advances in Applied Microbiology*, vol. 2 (1960): 223–262. Academic Press.

Pariente, María Isabel, Yolanda Segura, Silvia Álvarez-Torrellas, J. A. Casas, Z. M. de Pedro, E. Diaz, J. García et al. "Critical review of technologies for the on-site treatment of hospital wastewater: From conventional to combined advanced processes." *Journal of Environmental Management* 320 (2022): 115769.

Park, Jeong-Eun, Shan Zhang, Thi Hiep Han, and Sun-Jin Hwang. "The contribution ratio of autotrophic and heterotrophic metabolism during a mixotrophic culture of *Chlorella sorokiniana*." *International Journal of Environmental Research and Public* Health 18, no. 3 (2021): 1353.

Paul, Swastik, Souhardya Bera, Rishav Dasgupta, Shibsankar Mondal, and Subhasis Roy. "Review on the recent structural advances in open and closed systems for carbon capture through algae." *Energy Nexus* 4 (2021): 100032.

Plöhn, Martin, Olivia Spain, Sema Sirin, Mario Silva, Carlos Escudero-Oñate, Laura Ferrando-Climent, Yagut Allahverdiyeva, and Christiane Funk. "Wastewater treatment by microalgae." *Physiologia Plantarum* 173, no. 2 (2021): 568–578.

Posadas, E., Cynthia Alcántara, P. A. García-Encina, Luisa Gouveia, Benoit Guieysse, Zane Norvill, F. G. Acién et al. "Microalgae cultivation in wastewater." In Cristina Gonzalez-Fernandez, and Raúl Muñoz (eds.), *Microalgae-Based Biofuels and Bioproducts* (2017): 67–91. Woodhead Publishing.

Priya, A. K., A. A. Jalil, Sethumathavan Vadivel, Kingshuk Dutta, Saravanan Rajendran, Manabu Fujii, and Matias Soto-Moscoso. "Heavy metal remediation from wastewater using microalgae: Recent advances and future trends." *Chemosphere* 305 (2022): 135375.

Priyadharshini, Stephen Dayana, Palanisamy Suresh Babu, Sivasubramanian Manikandan, Ramasamy Subbaiya, Muthusamy Govarthanan, and Natchimuthu Karmegam. "Phycoremediation of wastewater for pollutant removal: A green approach to environmental protection and long-term remediation." *Environmental Pollution* 290 (2021): 117989.

Rahul, Soosai, Michael, M. A. Sundaramahalingam, C. S. Shivamthi, P. Varalakshmi, S. Karthikumar, S. Sabarathinam, and I. Ganesh Moorthy. "Insights about sustainable biodiesel production from microalgae biomass: A review." *International Journal of Energy* Research 45, no. 12 (2021): 17028–17056.

Randrianarison, Gilbert, and Muhammad Aqeel Ashraf. "Microalgae: A potential plant for energy production." *Geology, Ecology, and Landscapes* 1, no. 2 (2017): 104–120.

Ranjit, Pabbati, Vulise Jhansi, and Kondakindi Venkateswar Reddy. "Conventional wastewater treatment processes." In Naga Raju Maddela, Luz C. García Cruzatty, and Sagnik Chakraborty (eds.), *Advances in the Domain of Environmental Biotechnology: Microbiological Developments in Industries, Wastewater Treatment and Agriculture* (2021), pp. 455–479. Springer.

Ray, Ayusmita, Manoranjan Nayak, and Amit Ghosh. "A review on co-culturing of microalgae: A greener strategy towards sustainable biofuels production." *Science of the Total Environment* 802 (2022): 149765.

Razaviarani, V., G. Arab, N. Lerdwanawattana, and Y. Gadia. "Algal biomass dual roles in phycoremediation of wastewater and production of bioenergy and value-added products." *International Journal of Environmental Science and* Technology 20, no. 7 (2023): 8199–8216.

Rezvani, Fariba, and Khosrow Rostami. "Photobioreactors for utility-scale applications: Effect of gas–liquid mass transfer coefficient and other critical parameters." *Environmental Science and Pollution Research* 30, no. 31 (2023): 76263–76282.

Richardson, Tammi L. "The colorful world of cryptophyte phycobiliproteins." *Journal of Plankton Research* 44, no. 6 (2022): 806–818.

Rinna, Francesca, Silvia Buono, Iago Teles Dominguez Cabanelas, Iracema Andrade Nascimento, Giovanni Sansone, and Carmela Maria Assunta Barone. "Wastewater treatment by microalgae can generate high quality biodiesel feedstock." *Journal of Water Process Engineering* 18 (2017): 144–149.

Rodríguez López, Nuria, José Valeriano Alvarez Cabal, Marcos Cueto Cuiñas, and Francisco Ortega Fernndez. "Applicability of technology maturity level evaluation methodologies within small-and medium-sized organizations: Prospects and proposals." *Systems* 11, no. 8 (2023): 387.

Rosmahadi, Nurulfarah Adilah, Wai-Hong Leong, Hemamalini Rawindran, Yeek-Chia Ho, Mardawani Mohamad, Noraini A. Ghani, Mohammed J. K. Bashir, Anwar Usman, Man-Kee Lam, and Jun-Wei Lim. "Assuaging microalgal harvesting woes via attached growth: A critical review to produce sustainable microalgal feedstock." *Sustainability* 13, no. 20 (2021): 11159.

Sahni, Nidhi, Meenu Hans, and Sachin Kumar. "Fermentative approaches in wastewater treatment for harnessing renewable energy." In Pratibha Dheeran and Sachin Kumar (eds.), *Extremophiles* (2023), pp. 61–79. CRC Press.

Sanz-Luque, Emanuel, Devaki Bhaya, and Arthur R. Grossman. "Polyphosphate: A multifunctional metabolite in cyanobacteria and algae." *Frontiers in Plant Science* 11 (2020): 557161.

Sarwer, Asma, Seham M. Hamed, Ahmed I. Osman, Farrukh Jamil, Ala'A. H. Al-Muhtaseb, Nawaf S. Alhajeri, and David W. Rooney. "Algal biomass valorization for biofuel production and carbon sequestration: A review." *Environmental Chemistry Letters* 20, no. 5 (2022): 2797–2851.

Shafi, Mozim, Ruby Jan, and Khalid Muzamil Gani. "Selection of priority emerging contaminants in surface waters of India, Pakistan, Bangladesh, and Sri Lanka." *Chemosphere* 341 (2023): 139976.

Shahid, Ayesha, Muhammad Usman, Zahida Atta, Syed Ghulam Musharraf, Sana Malik, Ali Elkamel, Muhammad Shahid, Nuha Abdulhamid Alkhattabi, Munazza Gull, and Muhammad Aamer Mehmood. "Impact of wastewater cultivation on pollutant removal, biomass production, metabolite biosynthesis, and carbon dioxide fixation of newly isolated cyanobacteria in a multiproduct biorefinery paradigm." *Bioresource Technology* 333 (2021): 125194.

Shahid, Ayesha, Sana Malik, Hui Zhu, Jianren Xu, Muhammad Zohaib Nawaz, Shahid Nawaz, Md Asraful Alam, and Muhammad Aamer Mehmood. "Cultivating microalgae in wastewater for biomass production, pollutant removal, and atmospheric carbon mitigation: A review." *Science of the Total Environment* 704 (2020): 135303.

Sharma, Monika, El-Sayed Salama, Peng Zhang, Lihong Zhang, Xiaohong Xing, Jianwei Yue, Zhongzhong Song, Lan Nan, Su Yujun, and Xiangkai Li. "Microalgae-assisted microbial fuel cells for electricity generation coupled with wastewater treatment: Biotechnological perspective." *Journal of Water Process Engineering* 49 (2022): 102966.

Siddiki, Sk Yasir Arafat, M. Mofijur, P. Senthil Kumar, Shams Forruque Ahmed, Abrar Inayat, F. Kusumo, Irfan Anjum Badruddin et al. "Microalgae biomass as a sustainable source for biofuel, biochemical and biobased value-added products: An integrated biorefinery concept." *Fuel* 307 (2022): 121782.

Siddiqui, Shahida Anusha, Yuan Seng Wu, Trideep Saikia, İlknur Ucak, Maliha Afreen, Mohd Asif Shah, and Raphael D. Ayivi. "Production and growth of microalgae in urine and wastewater: A review." *Environmental Chemistry Letters* 21, no. 5 (2023): 2789–2823.

Singh, Dig Vijay, Atul Kumar Upadhyay, Ranjan Singh, and D. P. Singh. "Eco-friendly and eco technological approaches in treatment of wastewater by different algae and cyanobacteria." In Atul Kumar Upadhyay, and D. P. Singh (eds.), *Algae and Sustainable Technologies* (2020), pp. 43–64. CRC Press.

Singh, Jay Shankar, Arun Kumar, Amar N. Rai, and Devendra P. Singh. "Cyanobacteria: A precious bio-resource in agriculture, ecosystem, and environmental sustainability." *Frontiers in Microbiology* 7 (2016): 186282.

Singh, Ram Narayan., and Shaishav Sharma. "Development of suitable photobioreactor for algae production–A review." *Renewable and Sustainable Energy Reviews* 16, no. 4 (2012): 2347–2353.

Sirohi, Ranjna, Ashutosh Kumar Pandey, Panneerselvam Ranganathan, Shikhangi Singh, Aswathy Udayan, Mukesh Kumar Awasthi, Anh Tuan Hoang, Chaitanya Reddy Chilakamarry, Sang Hyoun Kim, and Sang Jun Sim. "Design and applications of photobioreactors: A review." *Bioresource Technology* 349 (2022): 126858.

Song, Yuling, Lijun Wang, Xi Qiang, Wenhui Gu, Zengling Ma, and Guangce Wang. "The promising way to treat wastewater by microalgae: Approaches, mechanisms, applications and challenges." *Journal of Water Process Engineering* 49 (2022): 103012.

Su, Yanyan. "Revisiting carbon, nitrogen, and phosphorus metabolisms in microalgae for wastewater treatment." *Science of the Total Environment* 762 (2021): 144590.

Sultan, Marriya, and De-Sheng Pei. "Mesocosm study in the reservoir ecosystem." In De-Sheng Pei, Naima Hamid, Marriya Sultan, and Thodhal Yoganandham Suman (eds.), *Reservoir Ecotoxicology* (2023), pp. 137–149. Cham: Springer International Publishing.

Sun, Han, Weiyang Zhao, Xuemei Mao, Yuelian Li, Tao Wu, and Feng Chen. "High-value biomass from microalgae production platforms: Strategies and progress based on carbon metabolism and energy conversion." *Biotechnology for Biofuels* 11 (2018): 1–23.

Tan, Jia Sen, Sze Ying Lee, Kit Wayne Chew, Man Kee Lam, Jun Wei Lim, Shih-Hsin Ho, and Pau Loke Show. "A review on microalgae cultivation and harvesting, and their biomass extraction processing using ionic liquids." *Bioengineered* 11, no. 1 (2020): 116–129.

Tang, Jialing, Xiangjiang Qu, Si Chen, Yunhui Pu, Xinrui He, Zhihui Zhou, Huijun Wang et al. "Microalgae cultivation using municipal wastewater and anaerobic membrane effluent: Lipid production and nutrient removal." *Water* 15, no. 13 (2023): 2388.

Thanigaivel, Sundaram, Sundaram Vickram, Sivasubramanian Manikandan, Santhana Raj Deena, Ramasamy Subbaiya, Natchimuthu Karmegam, Muthusamy Govarthanan, and Woong Kim. "Sustainability and carbon neutralization trends in microalgae bioenergy production from wastewater treatment: A review." *Bioresource Technology* 364 (2022): 128057.

Thoré, Eli S. J., Koenraad Muylaert, Michael G. Bertram, and Tomas Brodin. "Microalgae." *Current Biology* 33, no. 3 (2023): R91–R95.

Tong, Cheah Yi, and Chan Juinn ChiehDerek. "Bio-coatings as immobilized microalgae cultivation enhancement: A review." *Science of the Total Environment* (2023): 163857.

Tong, Kai, Lijun Yang, and Xiaoze Du. "Modelling of TiO_2-based packing bed photocatalytic reactor with Raschig rings for phenol degradation by coupled CFD and DEM." *Chemical Engineering Journal* 400 (2020): 125988.

Torpey, Wilbur N., H. Heukelekian, A. Joel Kaplovsky, and R. Epstein. "Rotating disks with biological growths prepare wastewater for disposal or reuse." *Journal (Water Pollution Control Federation)* 43 (1971): 2181–2188.

Torres-Franco, Andres Felipe, Cleber Cunha Figueredo, Lucas Barros, Björn Gücker, Iola GoncalvesBoëchat, Raul Muñoz, and Cesar Rossas Mota. "Assessment of a deep, LED-enhanced high-rate algal pond for the treatment of digestate." *Algal Research* 59 (2021): 102444.

Ummalyma, Sabeela Beevi, Ranjna Sirohi, Aswathy Udayan, Pooja Yadav, Abhay Raj, Sang Jun Sim, and Ashok Pandey. "Sustainable microalgal biomass production in food industry wastewater for low-cost biorefinery products: A review." *Phytochemistry Reviews* 22, no. 4 (2023): 969–991.

Upadhyay, Divya, Kartikeya Shukla, Arti Mishra, Tanu Jindal, Shikha Sharma, and Smriti Shukla. "Molecular aspects and oxygen relations of nitrogen fixation in cyanobacteria." In Cristina Cruz, Kanchan Vishwakarma, Devendra Kumar Choudhary, and Ajit Varma (eds.), *Soil Nitrogen Ecology* (2021), pp. 521–568. Springer International Publishing.

Valchev, Dobril, and Irina Ribarova. "A review on the reliability and the readiness level of microalgae-based nutrient recovery technologies for secondary treated effluent in municipal wastewater treatment plants." *Processes* 10, no. 2 (2022): 399.

Verma, Kavita, Kiran Kumar Panga, Vijaya Krishna Saranga, and Himabindu Vurimindi. "Phycoremediation of sewage-contaminated lake water using microalgae–bacteria co-culture." *Water, Air, & Soil Pollution* 231 (2020): 1–16.

Vidotti, Annamaria D. S., Diego M. Riaño-Pachón, Lucia Mattiello, Laís Albuquerque Giraldi, Flavia V. Winck, and Telma T. Franco. "Analysis of autotrophic, mixotrophic and heterotrophic phenotypes in the microalgae Chlorella vulgaris using time-resolved proteomics and transcriptomics approaches." *Algal Research* 51 (2020): 102060.

Walther, Jakob, Niklas Erdmann, Michael Stoffel, Katharina Wastian, Anna Schwarz, Dorina Strieth, Kai Muffler, and Roland Ulber. "Passively immobilized cyanobacteria *Nostoc* species BB 92.2 in a moving bed photobioreactor (MBPBR): Design, cultivation, and characterization." *Biotechnology and Bioengineering* 119, no. 6 (2022): 1467–1482.

Wang, Chun, Yinghong Tan, Lifang Zhu, Chengxu Zhou, Xiaojun Yan, Qingshan Xu, Roger Ruan, and Pengfei Cheng. "The intrinsic characteristics of microalgae biofilm and their potential applications in pollutants removal: A review." *Algal Research* 68 (2022c): 102849.

Wang, Mu-Hao Sung, and Lawrence K. Wang. "Glossary of natural resources and environmental pollution control." In Lawrence K. Wang, Mu Hao Sung Wang, Yung-Tse Hung, and Nazih K. Shammas (eds.), *Environmental and Natural Resources Engineering* (2021), pp. 421–494. Springer Nature.

Wang, Xiaoyan, and Yu Hong. "Microalgae biofilm and bacteria symbiosis in nutrient removal and carbon fixation from wastewater: A review." *Current Pollution Reports* 8, no. 2 (2022): 128–146.

Wang, Yufei, Linhua Fan, Nick Crosbie, and Felicity A. Roddick. "Photodegradation of emerging contaminants in a sunlit wastewater lagoon, seasonal measurements, environmental impacts and modelling." *Environmental Science: Water Research &* Technology 6, no. 12 (2020): 3380–3390.

Wang, Zhenyi, Jun Cheng, Yixuan Sun, Xuanxiang You, Feifei Chu, and Weijuan Yang. "Mutation adaptation and genotoxicity of microalgae induced by long-term high CO_2 stress." *Chemical Engineering* Journal 445 (2022a): 136745.

Wang, Zongliang, Hongshan Guo, Tong Zhou, Zhenmin Cheng, and Zibin Huang. "Influence of sparger type on mass transfer in a pilot-scale internal loop airlift reactor." *Processes* 10, no. 2 (2022b): 429.

Whelan, Michael J., and Jaeshin Kim. "Application of multimedia models for understanding the environmental behavior of volatile methylsiloxanes: Fate, transport, and bioaccumulation." *Integrated Environmental Assessment and Management* 18, no. 3 (2022): 599–621.

Willis, Anusuya, and Jason N. Woodhouse. "Defining cyanobacterial species: Diversity and description through genomics." *Critical Reviews in Plant Sciences* 39, no. 2 (2020): 101–124.

Wollmann, Felix, Stefan Dietze, Jörg-Uwe Ackermann, Thomas Bley, Thomas Walther, Juliane Steingroewer, and Felix Krujatz. "Microalgae wastewater treatment: Biological and technological approaches." *Engineering in Life Sciences* 19, no. 12 (2019): 860–871.

Wu, Haotian, and Céline Vaneeckhaute. "Nutrient recovery from wastewater: A review on the integrated physicochemical technologies of ammonia stripping, adsorption and struvite precipitation." *Chemical Engineering* Journal 433 (2022): 133664.

Wu, Li, Long Qian, Zhikang Deng, Xiaolong Zhou, Bolin Li, Shubin Lan, Lie Yang, and Zulin Zhang. "Temperature modulating sand-consolidating cyanobacterial biomass, nutrients removal and bacterial community dynamics in municipal wastewater." *Bioresource Technology* 301 (2020): 122758.

Xu, Kaiwei, Xiaotong Zou, Hao Wen, Yating Xue, Yanhui Qu, and Yanpeng Li. "Effects of multi-temperature regimes on cultivation of microalgae in municipal wastewater to simultaneously remove nutrients and produce biomass." *Applied Microbiology and Biotechnology* 103 (2019): 8255–8265.

Yap, Jiunn Kwok, Revathy Sankaran, Kit Wayne Chew, Heli Siti Halimatul Munawaroh, Shih-Hsin Ho, J. Rajesh Banu, and Pau Loke Show. "Advancement of green technologies: A comprehensive review on the potential application of microalgae biomass." *Chemosphere* 281 (2021): 130886.

Yarkent, Cagla, Bahar Aslanbay Guler, Ceren Gurlek, Yaprak Sahin, Ayse Kose, Suphi S. Oncel, and Esra Imamoglu. "Algal alginate in biotechnology: Biosynthesis and applications." In Irem Deniz, Esra Imamoglu, and Tugba Keskin Gundogdu (eds.), *Properties and Applications of Alginates* (2022), pp. 1–26. IntechOpen.

Zafar, Mohd, Avnish Pareek, Taqi Ahmed Khan, Ramkumar Lakshminarayanan, and Naveen Dwivedi. "Smart microalgae wastewater treatment: IoT and edge computing applications with LCA and technoeconomic analysis." *Next-Generation Algae* 1 (2023): 1–47.

Zhang, Bing, Wei Li, Yuan Guo, Zhiqiang Zhang, Wenxin Shi, Fuyi Cui, Piet N. L. Lens, and Joo Hwa Tay. "Microalgal-bacterial consortia: From interspecies interactions to biotechnological applications." *Renewable and Sustainable Energy Reviews* 118 (2020a): 109563.

Zhang, Huichao, Xin Li, Zhongyi An, Zhiwei Liu, Chunxiao Tang, and Xiaodong Zhao. "Treatment of polyacrylamide-polluted wastewater using a revolving algae biofilm reactor: Pollutant removal performance and microbial community characterization." *Bioresource Technology* 332 (2021): 125132.

Zhang, Qiao, Jing-jing Zhan, and Yu Hong. "The effects of temperature on the growth, lipid accumulation and nutrient removal characteristics of *Chlorella* sp. HQ." *Desalination and Water* Treatment 57, no. 22 (2016): 10403–10408.

Zhang, Xing. *Microalgae Removal of CO_2 from Flue Gas*, Vol. 14. IEA Clean Coal Centre, 2015.

Zhang, Yixin, Huan-Hsuan Hsu, John J. Wheeler, Sishuang Tang, and Xiaocheng Jiang. "Emerging investigator series: Emerging biotechnologies in wastewater treatment: From biomolecular engineering to multiscale integration." *Environmental Science: Water Research & Technology* 6, no. 8 (2020b): 1967–1985.

Zhao, Dehua, Wai Yan Cheah, Sai Hin Lai, Eng-Poh Ng, Kuan Shiong Khoo, Pau Loke Show, and Tau Chuan Ling. "Symbiosis of microalgae and bacteria consortium for heavy metal remediation in wastewater." *Journal of Environmental Chemical Engineering* 11, no. 3 (2023): 109943.

Zheng, Heshan, Yu Wang, Shuo Li, Dillirani Nagarajan, Sunita Varjani, Duu-Jong Lee, and Jo-Shu Chang. "Recent advances in lutein production from microalgae." *Renewable and Sustainable Energy Reviews* 153 (2022): 111795.

Zhu, Chenba, Xiaoqian Zhai, Yimei Xi, Jinghan Wang, Fantao Kong, Yunpeng Zhao, and Zhanyou Chi. "Efficient CO_2 capture from the air for high microalgal biomass production by a bicarbonate pool." *Journal of CO_2 Utilization* 37 (2020): 320–327.

Zhu, Shunni, Lei Qin, Pingzhong Feng, Changhua Shang, Zhongming Wang, and Zhenhong Yuan. "Treatment of low C/N ratio wastewater and biomass production using co-culture of *Chlorella vulgaris* and activated sludge in a batch photobioreactor." *Bioresource Technology* 274 (2019): 313–320.

Zhuang, Lin-Lan, Dawei Yu, Jian Zhang, Fei-fei Liu, Yin-Hu Wu, Tian-Yuan Zhang, Guo-Hua Dao, and Hong-Ying Hu. "The characteristics and influencing factors of the attached microalgae cultivation: A review." *Renewable and Sustainable Energy Reviews* 94 (2018): 1110–1119.

5 Microalgal Treatment of Wastewater by Suspended and Immobilized Algal Systems

An Approach Contributing to Circular Bioeconomy

L. H. S. Hassan

5.1 INTRODUCTION: BACKGROUND AND DRIVING FORCES

Water pollution has increased as a result of urbanization, industrialization, and human activity, releasing heavy metals into the environment that are not biodegradable or thermally degradable (Chen et al., 2023). Some of the common techniques used to treat wastewater include chemical precipitation (Zhang and Duan, 2023), adsorption (Syeda et al., 2022), electrodialysis (Min et al., 2021), ion exchange (Nekouei at al., 2019), membrane filtration (Efome at al., 2019), coagulation/flocculation (El Gaayda et al., 2023; El Gaayda et al., 2021), and electrochemical precipitation (Liu et al., 2019a). Each of these approaches has benefits and drawbacks of its own. The development of novel, environmentally friendly, and long-lasting techniques that can guarantee the complete removal of heavy metals is of great interest in light of the promotion of environmental sustainability (Shahrokhi-Shahraki et al., 2021).

Bioremediation is an environmentally friendly technology using microorganisms' metabolic potential to remove pollutants (Chen et al., 2023). Microalgae "as photosynthetic organisms" are essential in the aquatic ecosystem. Recently, microalgae have gained significant interest as a source of bioenergy production and potential for wastewater treatment (e.g., domestic, agricultural, and industrial wastewater) (Ahmad et al., 2019a; Srimongkol et al., 2022; Kathi et al., 2023; Amaro et al., 2023) due to their ability to biosorb pollutants which are required for their biomass formation (Dasgupta et al., 2019). Wastewater is a nutrient-rich media used for high productivity of microalgae biomass, including nitrogen; phosphorus; heavy metals; total suspended solids (TSS); total dissolved solids (TDS); total volatile solids (TVS); total fixed solids (TFS); fat, oil, and grease (FOG); biochemical oxygen demand (BOD); and chemical oxygen demand (COD) (López-Sánchez etal., 2022; Hashmi et al., 2023). Wastewater bioremediation by microalgae resembles cost-effective,

 DOI: 10.1201/9781003459286-5

renewable, and ecofriendly process. This process was developed to be occurred in open and closed cultivation systems to enhance microalgal biomass productivity. The produced microalgal cellular contents of polysaccharides, lipids, and protein can be converted toward biofuel production (Sukla et al., 2019). By combining wastewater treatment (WWT) with microalgae cultivation, a low-cost ecofriendly approach can be achieved, retaining large amounts of natural hydrocarbons suitable for biofuel production. Factors such as nutrient quantity, quality, light, carbon dioxide, temperature, pH, turbulence, and salinity are considered for effective microalgae cultivation (Sukla et al., 2019; Borowiak and Krzywonos, 2022). Biological treatment is a secondary method for eliminating biodegradable pollutants. Microalgae genera like *Chlorella*, *Chlamydomonas*, *Scenedesmus*, *Ankistrodesmus*, *Botryococcus*, *Euglena*, *Limnospira*, *Oscillatoria*, and *Phormidium* are microalgae genera that exhibit as effective bioremediation agents, as they can grow and tolerate toxins (Hosseinzadeh et al., 2022; Hashmi et al., 2023).

Algal cells can attach to particular surfaces, producing biomembranes that reduce the cost of separating algal biomass. The typical harvesting and dewatering procedure can be simplified and made more energy effective by using larger immobilized microalgae beads. These beads prevent the release of immobilized microorganisms into wastewater, reducing disruption to native ecosystems caused by foreign microorganisms (Chen et al., 2023).

This chapter aims to provide an overview of microalgae in wastewater pollution treatment, emphasizing the use of both immobilized and free-suspended microalgal cells. It explores the mechanisms of pollutant removal by immobilized algae, the factors influencing biosorbent efficiency, and presents examples from prior research in this field. In addition, it highlights applications of the biomass produced through these processes.

5.2 POTENTIAL PHYCOREMEDIATION IN WWT USING SUSPENDED FREE CELLS AND IMMOBILIZED MICROALGAL BEADS

Three major groups of treatment systems can be used in WWT by microalgae, including suspended cell system, algal bead (active immobilization), and attached (passive immobilization) growth (Valchev and Ribarova, 2022).

5.2.1 Suspended Cell System

Suspended cells culture is the common microalgal cultivation system, which can be performed either in open or closed cultivation systems (Borowiak and Krzywonos, 2022). Using natural or artificial open ponds is a popular choice for WWT. High-rate algal ponds (HRAP) is an open suspended algal growth reactor, which is easy and inexpensive to construction and operation (González-Camejo et al., 2021). Microalgal cells are freely floating into the wastewater medium through shallow ponds, using natural solar illumination as a light source. However, monitoring algal biomass is crucial to prevent internal shading inhibition if hyper concentration of algae is reached (Leong et al., 2021; Valchev and Ribarova, 2022). In closed photobioreactors, barriers made of transparent or semi-transparent materials, such as glass, plexiglass, polycarbonate, and plastic bags, are used to keep the algae-wastewater suspension away from the environment. Systems for treating wastewater that are frequently used have cylinder, flat panel, and tubular designs. Algal insulation enhances the overall management and light utilization. To provide the necessary medium for algae to achieve intense processes like aeration, pumping, and degassing, additional systems must be put in place. In addition to complicating operations and impeding the development of enclosed suspended-growth algae-based WWT systems, this raises capital and operating costs (Paddock, 2019; Valchev and Ribarova, 2022).

5.2.2 Immobilized Microalgal Beads (Active Immobilization)

Immobilized cell system is a green technology that involves attaching microalgal cells to a carrier to enhance stability. The process involves encapsulating target cells in a porous polymer layer, allowing substrate diffusion. Immobilized systems improve resistance to cell growth disruption, prevent photoinhibition, minimize cytotoxicity, and aid microalgal cells in adapting to environmental stress or toxicity. They increase volumetric output, substrate usage, and resistance to harmful elements (Chen et al., 2023). Immobilized cells increase nutrient uptake and help their harvesting and recovery more than suspended free cells, and allow the reuse of the beads for another removal (Agarwal et al., 2019; Kumar and Eswari-Jujjavarappu, 2023). Immobilized systems are more resistant to microbial contamination and less sensitive to environmental factors, such as pH and temperature, in addition to reducing the effect of the shading layer (López-Sánchez et al., 2022). Immobilizing algal cells is primarily used to maintain living cells with restricted mobility while they function within a matrix and to maintain metabolic activity for as long as possible in the cells (Hejna et al., 2022). Several chemical and organic materials, including polyacrylamide gel, alginians, agar, and carrageens, can be used to immobilize algal cells (Zabochnicka, 2022). The immobilization matrix should offer chemical and mechanical stability (Plöhn et al., 2021). Chen et al. (2020) used sponge as a solid immobilized surface. Paper was used as biofilm support material for the cultivation of *Chlorella* sp. for the treatment of anaerobic digestate food effluent, the microalgal biomass productivity reached 9.96 g (DW)/m^2/day and nutrient removal rate 100% NH_4^+-N, 85% PO_4^{3+}-P, and 40.2% COD (Mkpuma et al., 2023).

5.2.2.1 Immobilization of Algal Cells by Adsorption

Adsorption forms physical bonds between water-insoluble carriers and immobilized algae, causing cell leakage during use due to weak molecular forces and gentle bonding (Han et al., 2023). Surface immobilized algae reduce pollutants (heavy metals) burden in effluent through adsorption and passage through algal cells, reducing recovery costs due to easier performance (Vasilieva et al., 2021a).

5.2.2.2 Immobilization of Algal Cells by Encapsulation

Encapsulation is a permanent immobilization technique where cells are confined in a capsule space, allowing them to float freely (Whitton et al., 2019). The ability to encapsulate microalgae for nutrient remediation with calcium alginate beads has been demonstrated in studies, but there are drawbacks, including poor swelling, poor mechanical properties, and potential mass loss during adsorption (Chen et al., 2023).

5.2.2.3 Immobilization of Algal Cells by Entrapping

Entrapment of algal cells occurs by using a fiber or organic gel polymer as a support matrix (Chen et al., 2023). Algal cell inactivation is a major problem for entrapment immobilization. Natural polymers used to immobilize algal cells have low porosity, which can limit nutrient diffusion and reduce the effectiveness of immobilized cells' bio-removal processes (Giese et al., 2020).

5.2.2.4 Immobilization of Algal Cells by Self-Immobilization

Resembling to the action of filamentous fungi, channels between functional groups on the outer cell membrane are created by multifunctional reagents and cross-linking immobilization. Through electrostatic interactions, crosslinked polyethyleneimine polymer successfully immobilized *C. vulgaris* cells. Engaging without a carrier lowers material costs and replaces time-consuming technologies. Conditions that are artificially created promote cell growth and density. However, exposure to chemicals and environments can change cellular makeup, which could potentially harm the cell surface (Vasilieva et al., 2021a).

5.2.3 Attached Growth (Passive Immobilization)

In WWT systems, algal biomass is treated by fusing a media matrix with a biochemical substrate. Algal harvesting is made simple by controlling biofilm thickness, resulting in consistent effluent and biomass quality. Stationary and rotating reactors are preferred to achieve the best conditions for algal growth and WWT performance (Valchev and Ribarova, 2022).

5.3 FACTORS INFLUENCING IMMOBILIZATION

Not only the biological characteristics of the immobilized microalgae influence their bioremediation activities, but also other physico-chemical factors of wastewater, and the process parameters can affect this including pH, temperature, contact time, carrier type, light intensity and molecular weight, ionic radius, and oxidation state (Chu and Phang, 2019; Dasgupta et al., 2019; Zhuang et al., 2020; Chen et al., 2023).

5.3.1 *pH* Value

pH significantly impacts site dissociation, metal ion solution chemistry, hydrolysis, complexation, redox reactions, precipitation, and metal ion speciation and biosorption affinity, affecting protonation and ionization of functional groups in adsorbates (Chu and Phang, 2019; Chen et al., 2023). As pH rises, more functional groups will separate from the biosorbent surface. pH 5 is considered the optimal pH for eliminating Cu ions, as it impacts metal chemistry and binding on algal cell surfaces. Immobilized fresh algal cells adsorbed the most Fe (II), Mn (II), and Zn (II) at pH 6, 40.98 mg/g Mn (II), and 37.43 mg/g Zn (II) (Ahmad et al., 2019a). Contrarily, in alkaline pH conditions, protons must compete with OH for suitable adsorption sites in interstitial spaces or on the adsorbent surface, preventing Cr(VI) adsorption (Wang et al., 2022). Low pH levels lead to biosorbent surface protonation, which lowers the immobilized algal biomass adsorption capacity and lowers the anticipated zero charge point (Fawzy et al., 2022).

5.3.2 Algal Species

Algae are a crucial sorbent material, but their toxicity levels may be strain specific, affecting their potential remediation capacity for various metals (Chen et al., 2023). *Chlorella vulgaris* is excellent for WWT and biomass production due to its robustness and resistance to invaders and harsh conditions (Taufikurahman et al., 2020).

5.3.3 Temperature

The temperature has a significant implication on the microorganisms' metabolism and cell death, which also affects the adsorption of nutrients and heavy metals (Chen et al., 2023). Enzymes that transfer ions operate at their peak efficiency at room temperature, suggesting that rising temperatures reduce the effectiveness of heavy metal adsorption (Ahmad et al., 2019a). Raising temperature increases the surface activity and kinetic energy of an adsorbate, improving biosorption. However, it may also disrupt the biosorbent's physical structure. This increases active sites, enlarging pores, reducing diffusion boundary layer thickness, and increasing ion mobility and surface activity (Chen et al., 2023). Algae biosorb ions through heat-absorbing processes, increasing efficiency with temperature when optimal temperature for immobilized algae bioreaction is limited (Fawzy et al., 2022).

5.3.4 Immobilized Carriers

Different carriers have an effect on how well algae are immobilized, and using fresh materials made from agricultural or industrial waste can increase sustainability. The naturally occurring polymer sodium alginate, which is derived from the cell walls of coastal brown algae, has the advantages of biocompatibility, low cost, binding ability, and hydrophilicity. It offers a promising alternative for immobilizing microalgae due to its biocompatibility, affordability, and hydrophilicity (Chen et al., 2023). A deacetylated form of chitin called chitosan is created when chitin is subjected to chemical and enzymatic deacetylation (Kuczajowska-Zadrożna et al., 2020).

Alginate is a natural macroalgal polymer which can form microalgal polymeric matrix beads in the cross-linking solution (Wu et al., 2020). The use of alginate beads for treating municipal wastewater reverse osmosis concentrate (ROC) may be constrained by the potential for alginate beads to disintegrate in the presence of high ion concentrations. The combined impact of the ROC's pH, alkalinity, and salinity on the physical stability, chemical properties, biomass production, and nutrient removal performance of microalgal cells entrapped in alginate was examined. The initial reason for the decrease in algal bead stability was water adsorption brought on by the loss of calcium from the alginate matrix. When alkalinity >400 mg/L and pH is 9.5, the compressive strength was reduced by >65%, which caused the beads to disintegrate during ROC treatment. Algal beads' performance for various ROC compositions could be predicted using empirical models that were created and validated. With respect to the viability of using an alginate-immobilized microalgal system to treat ROC streams of municipal wastewater, this study improves understanding and decision-making (Mohseni et al., 2021). Three microalgae species—*S. obliquus, Auxenochlorella protothecoides* and *C. vulgaris* and a cyanobacterium *Arthrospira maxima*—were investigated in the bioremediation of agro-industrial winery wastewater under heterotrophic and mixotrophic conditions for 8 days; the best removal performance was for *C. vulgaris* under mixotrophic conditions with removal efficiency 91% for total nitrogen, 40% for phosphorus, 92% for COD, and 49% for polyphenols, and maximum biomass 147.5 mg/L/day and carotenoid content reached 8.7 mg/g (Marchão et al., 2021). Chitosan uses chemical bonds to immobilize microalgae, increasing their stability. Van der Waals forces and weak interactions like hydrogen bonding, on the other hand, lead to detachment during cleaning. Chitosan interacts with cross-linking substances like glutaraldehyde to create highly porous hydrophilic polymers that are more stable and have greater mechanical strength (Vasilieva et al., 2021a). Sodium alginate, carrageenan, and polyvinyl alcohol showed high immobilization efficiency on algae, despite not being able to adsorb metal ions (Chen et al., 2023).

5.3.5 Contact Time

Contact time determines the kinetic rate of biosorption, and as the process moves along, the biosorbent becomes saturated. Biosorption and desorption rates are equal at the point of equilibrium. Since immobilized algae have slower kinetics than free algae, more heavy metal adsorption sites can be reached during pre-exposure. The adsorption rate decreases as a result of increased repulsive interactions between adsorbent and adsorbing ions and a reduction in vacant sites as the biosorbent loses binding to biomass after equilibrium (Chen et al., 2023). It takes 450 minutes for algal cell equilibrium during calcium alginate immobilization (Ahmad et al., 2019a). In 180 minutes of contact time, immobilized algae can remove 96.4% of Cu (II) ions, according to research (Lieswito et al., 2019). Reactive groups have an impact on the adsorption of heavy metals onto Fe_2O_3@Microalgae composite surfaces, achieving the best binding equilibrium for Cr (VI), Pb (II), and Cd (II) in less than 30 minutes (Shen et al., 2020). Cd (II) biosorption is faster when empty binding sites contact Cd (II) ions, with rapid removal within 15 minutes and gradual improvement until equilibrium at 90 minutes (Fawzy et al., 2022). Monolayer active spots biosorb, decreasing the efficiency of removal. Due to binding within the immobilized matrix, immobilized biomass has slower kinetics than pristine algae, whereas the binding sites in pristine biomass are accessible to Fe^{3+} ions (Chen et al., 2023).

5.3.6 Adsorbent Dosage

Biosorption effectiveness depends on sorbent amount, with higher dosages causing cellular aggregates, reducing effective surface area, and causing decreased efficiency (Ahmad et al., 2019a). Adsorbent surface functional groups are determined by adsorbent amount. Increased algal dosage decreases ions' biosorption ability due to increased unsaturated active sites on biosorbents. Adsorbent size increases due to increased surface area, enhancing metal adsorption efficiency and increasing functional site mass of adsorbed metal (Chen et al., 2023). Increased adsorbent amount can significantly improve heavy metal removal effectiveness, but it doesn't significantly impact the adsorption capacity due to the reaction between heavy metal ions and the adsorbent's active sites. The concentration gradient between the biosorbent and heavy metal ions may also reduce the adsorption capacity (Wang et al., 2022).

5.4 POLLUTANT REMOVAL EFFICIENCY BY SUSPENDED MICROALGAL CELLS AND IMMOBILIZED BEADS

5.4.1 Removal of Nutrients

Hydrogen bonding significantly broadens the band and reduces the mean adsorption frequency. The NH_4^+ ion bonds weakly with the carbonyl oxygen of the free COO^- ions in the polymeric matrix's electrostatic environment. According to Etter's rule, the best hydrogen bond acceptor (COO^-) forms a hydrogen bond with the best hydrogen pair donor (NH_4^+). Electrostatic attraction between the protonated amino group and the negatively charged COO^- ions is the basis for the NH_4^+ ion adsorption. The correlation coefficients obtained from the Langmuir isotherm agree with this interaction's constant and equal adsorption energy. Microalgae assimilate weakly bonded ions (NH_4^+ and PO_4^{3-}) into the alginate structure, causing vacant COO^- and OH^- groups. This increases nutrient uptake from the environment, and the cycle repeats until microalgae restrict nutrient uptake, potentially due to stagnant phases, saturating the bead with adsorbed nutrients (Banerjee et al., 2019; Kube et al., 2020).

Many previous studies assessing microalgal efficiency in WWT have been mentioned in Table 5.1. Benítez et al. (2019) used a native Ecuadorian microalgal strain in the removal of nitrogen and phosphorus by using agitation and aeration photobioreactors. The results reported removal efficiency of ammonia reached 52.6%–55.6% and 67.0% for phosphorus, while NO_3^-N production rate reached 87.0%–93.1%. Liu et al. (2019b) compared the uptake of different ammonia concentrations by using free suspended and immobilized *Scenedesmus obliquus*, with respect to mixotrophy and heterotrophy effects. The results indicated the higher performance of the immobilized *S. obliquus* beads with high tolerance to high ammonia concentrations and higher uptake 96.6% of 50 mg/L under mixotrophic cultivation. A comparison between alginate-immobilized and suspended free cells of *C. vulgaris* was investigated in the treatment of lagoon municipal wastewater; this study reported that there was not any difference in nutrients uptake (N and P) between the immobilized cells and the suspended cells (Kube et al., 2020). *Chlorella vulgaris* was used to study the effect of alginate concentration on the removal of nutrients from wastewater; the results informed that 3% of alginate is optimal for nutrient removal in addition to the ability of alginate to adsorb NH_4^+-N (95%) and PO_4^{3-}-P (99%) (Banerjee et al., 2019). Suspended free cells and immobilized fresh cells of *Scenedesmus vacuolatus* ACUF_053 and *Chlorella vulgaris* ACUF_809 were used for wastewater bioremediation in biofilm PBR. The results exhibited high nutrients removal by *S. vacuolatus*

TABLE 5.1
Immobilized Microalgae Removal Efficiency of Nutrients from Wastewater

Microalgal Strain	Microalgal Status	Wastewater Type	Removal Efficiency				Productivity		Cultivation Conditions	References
			N	P	COD	BOD	Biomass	Other		
Nannochloropsis sp.	Immobilized	TPOME	—	—	55%	—	—	—	110% POME after 9 days	Emparan et al. (2019)
Phormidium fragile (cyanobacterium)	- Suspended free cells	WI	51.56%	66.67%	51.74%	37.05%	—	—	pH 7.3–9.1, 2,500 lux, 25°C, 40 mL algal dose, 10 days	Shabana et al. (2019)
		WII	58.89%	61.23%	54.55%	59.85%	—	—		
	- Immobilized	WI	100%	100%	72.56%	—	—	—		
		WII	87.28%	100%	71.98%	—	—	—		
Leptolyngbya sp. (cyanobacterium)	Suspended free cells	Poultry wastewater	88.2%	97.4%	94.0%	—	—	79.3% total sugars	Laboratory scale	Patrinou et al. (2020)
			69.4%	81.0%	82.0%	—	335.3 mg/L/day	83.8% of total sugars, 19.6% (w/w) lipid	Pilot scale	
Chlorella sorokiniana AK-1 (microalga)	Suspended free cells	50% swine wastewater	—	—	—	—	5.45 g/L	0.27 g/L/day protein	—	Chen et al. (2020)
	Immobilized beads		97.0%	92.8%	90.1%	—	8.08	0.272 g/L/day protein		
	Reuse immobilized beads		—	—	—	—	6.51 g/L	0.15 g/L/day of protein		

(Continued)

TABLE 5.1 (*Continued*)
Immobilized Microalgae Removal Efficiency of Nutrients from Wastewater

Microalgal Strain	Microalgal Status	Wastewater Type	Removal Efficiency				Productivity		Cultivation Conditions	References
			N	P	COD	BOD	Biomass	Other		
C. sorokiniana AK-1 (microalga)	Sponge-immobilized beads	Piggery wastewater	94.1%	96.9%	95.7%	99.0%	4.56 mg/g	0.48 g/L/day protein, 3.56 mg/L/day lutein	—	Chun-Yen et al. (2021)
A Nordic microalga *Scenedesmaceae* family	Suspended free cells	Greenhouse-recirculating hydroponic effluent	18%–35%	40%–98%	—	—	—	—	pH 7.5–9, 36 days	João et al. (2021)
Ascochlorissps ADW007		Dairy wastewater	72%–80%	80%–97%	94%–96%	—	—	—	—	Kumar et al. (2019)
Nannochloropsis oculata	Alginate-immobilized beads	POME	80%	83%	73%	—	—	—	—	Minturo et al. (2022)
Chlorococcum oleofaciens	Alginate-immobilized beads		93.51%	90.43%	50.72%	—	—	—	—	Tan et al. (2021)
Synechococcus pevalekii	Alginate-immobilized beads	Industrial wastewater	91%	—	—	—	—	—	10 days of cultivation	Colusse et al. (2021)
Chlorococcum sp.	Suspended free cells	Municipal wastewater	—	90.1%	—	—	82 mg/L/day	Lipid 25 mg/L/day	50% MWW	
Chlorella vulgaris	Suspended free cells Immobilized	Pharmaceutical municipal wastewater	65% 81.8%	0% 86.2%	—	—	—	—	Total removal of fluoxetine within 6 days	Silva et al. (2022)

which was 60–240 mg/L for nitrate and 20–40 mg/L for phosphate. While *C. vulgaris* recorded about 90% of nutrient removal efficiency (Moreno-Osorio et al., 2019), Chen et al. (2020) investigated the biorefinery performance of three *Chlorella sorokiniana* strains including *C. sorokiniana* AK-1, *C. sorokiniana* MS-C1, and *C. sorokiniana* TJ5 in 50% swine wastewater. The best results were obtained by *C. sorokiniana* AK-1, which recorded 5.45 g/L and 0.27 g/L/day biomass and protein content, respectively. The immobilized cells increased the biomass and protein contents to 8.08 g/L and 0.272 g/L/day, respectively, with nutrients removal rate reaching 97.0% of nitrogen, 92.8% phosphorus, and 90.1% COD. This immobilization technique allowed the reuse of the beads for another recovery and allowed the reuse of the beads with high efficiency reaching 6.51 g/L of biomass productivity and 0.15 g/L/day of protein contents. Sponge-immobilized *C. sorokiniana* AK-1 showed great nutrient removal efficiency from piggery wastewater, with removal rate reaching 94.1%, 96.9%, 95.7%, and 99.0% for total N, P, COD, and BOD, respectively, in addition to protein content and productivity reaching 61.1%, 0.48 g/L/day, respectively (Chun-Yen et al., 2021). In another study, immobilized fresh *C. vulgaris* cells were investigated in the bioremediation of pharmaceuticals (fluoxetine) wastewaters; the results showed high removal efficiency of 100%, 86.2%, and 81.8% for fluoxetine, phosphorus, and nitrogen, respectively (Silva et al., 2022). Alazaiza et al. (2023) reported the efficiency of *C. vulgaris* in the removal of nutrients from 50% wastewater with removal rate reaching 95%, 97%, and 84% of nitrogen, phosphorus, and COD, respectively, and biomass productivity 400 mg/L after 14 days. In down-flow hanging sponge reactors, immobilized *Nannochloropsis oculata* was used to treat palm oil mill effluent (POME), which showed color removal, and high removal efficiencies reached 73% of COD, 80% of ammonia, and 83% of phosphate (Minturo et al., 2022).

Free and immobilized cyanobacterium *Phormidium fragile* were used in the treatment of pharmaceutical laboratory wastewater (WI) and production wastewater (WII); the results showed higher removal of immobilized cells of ammonia, phosphorus, and COD, which recorded 100%, 100%, and 72.56%, respectively, for WI and 87.28%, 100%, and 71.98%, respectively, for WII (Shabana et al., 2019). A study compared the performance of suspended and immobilized cyanobacterium *Microcystis aeruginosa* by cultivation in wastewater culture using anaerobic digestion effluent (ADE). The results reported the ability of immobilized beads to resist the high pollutant concentration, as they recorded 86.4% cell vitality higher than that of the free suspended cells which recorded 75.2%, with a high efficiency in nutrient removal 93.05% and 100% for ammonia (NH4+-N) and phosphorus (PO_4^{3-}) (Gong et al., 2019). When the filamentous cyanobacterium *Leptolyngbya* sp. suspension was investigated in poultry waste biorefinery under laboratory and pilot scale, the results of the laboratory scale tests showed high removal reached 88.2% of nitrogen, 97.4% of phosphorus, 94.0% of COD, 79.3% of total sugars, while the removal efficiency of pilot scale reached 69.4% of nitrogen, 81.0% of phosphorus, 82.0% of COD, 83.8% of total sugars, with high biomass content recording 335.3 mg/L/day and lipid content 19.6% (w/w) (Patrinou et al., 2020).

5.4.2 Removal of Heavy Metals

Heavy metals are a group of about 53 transition elements, which pollute water resources and have very serious biohazard effect on the ecosystem (Ahmad et al., 2019b). The hazardous effects of heavy metal stress can accumulate in the food chain, which endangers crops by lowering plant growth and yield parameters, extending their environmental persistence, and reducing plant growth and yield parameters. In aquatic systems, these non-essential metal ions are very mobile and are thought to be extremely dangerous even in minute amounts. Through biosorption and accumulation in the aquatic food chain, humans may be exposed to heavy metal ions, which can have serious health effects (Chen et al., 2023; Hashmi et al., 2023). Heavy metals resources that pollute water are various such as fertilizers, dyeing, mining, paper, paint, and plastic industries (Chu and Phang, 2019). So, there is a need for innovation of ecofriendly economical absorbents for heavy metals. Microalgae are efficient in heavy metals uptake by various cellular mechanisms, which differ with species, and some factors. These factors are metal ion features, pH value, temperature, ionic strength, time of contact, concentration and nature of the reactants (Ahmad et al., 2018; Chu and Phang, 2019).

Some researchers hypothesize that sorption occurs in two phases: initial adsorption based on heavy metals' affinity toward functional groups and a second phase involving ion exchange with calcium alginate (Pinto et al., 2023). Biological methods are cost effective, efficient in waste removal, and have potential for regeneration. Organisms, species, growth conditions, physiological states, and cell ages impact metal binding to them. Biomaterials exhibit strong removal, binding, and metal selectivity, relying on lipids, carbohydrates, and proteins in cell walls for efficient metal sorption (Ahmad et al., 2019b; Chen et al., 2023).

In Table 5.2 are some examples of immobilized microalgal cells for heavy metals bioremediation. Ahmad et al. (2018) used freshwater microalga *C. vulgaris* in two ways, free suspended cells and alginate-immobilized beads in the removing of some metal ions from industrial wastewater (Fe^{+2}, Mn^{+2}, and Zn^{+2}). The immobilized microalgal beads showed better biosorption results than suspended cells under growth conditions of pH, temperature, metal ions concentration, and contact time (differ according to the metal ions removed). The biosorption capacity of freely suspended *C. vulgaris* biomass recorded 74.54 for Fe (II), 69.19 for Zn (II), and 65.14 mg/g for Mn (II), while immobilized algal beads recorded 129.83, 115.90, and 105.29 mg/g, respectively. *Selenastrum capricornutum* and *Microcystis aeruginosa* have been used in the removal of cadmium ions (Cd) and produced CdSe nanoparticles which were established as a nano-sensor to detect mercury (Hg^{+2}) (Zhang et al., 2019). Alginate-immobilized *Galdieria sulphuraria* has been used in the treatment of mining wastewater to remove rare earth elements (REEs) lanthanum (La), samarium (Sm), and yttrium (Y); the removal efficiency reached 97.19% for La, 98.87% for Sm, and 96.19% for Y (Sun et al., 2022). It was indicated that alginate-immobilized *Chlamydomonas reinhardtii* showed high adsorption capacity of Hg^{+2} and lead (Pb^{+2}) which related to the presence of more heavy-metal-binding sites of functional groups on the cell wall of *C. reinhardtii* including hydroxyl, glycoprotein, protein, and sulfhydryl (Spain et al., 2021).

TABLE 5.2
Immobilized Microalgae Removal Efficiency of Heavy Metals from Wastewater

Microalgae	Biomass Used	Wastewater Type	Adsorption Efficiency	Cultivation Conditions	References
Chlorella vulgaris (microalga)	Suspension Alginate immobilized	Industrial wastewater POME	— Fe (II) 129.83, Mn (II) 115.90, Zn (II) 105.29 mg/g	pH of 6.0, biomass dosage 0.4 g/L, contact time of 300 minutes, 25°C	Ahmad et al. (2018)
Galdieria sulphuraria (microalga)	Alginate immobilized	Mining wastewater	La 97.19%, Sm 98.87%, Y 96.19%	Acidified pH 2.5	Sun et al. (2022)
Chlamydomonas reinhardtii (microalga)	Alginate immobilized	Sewage wastewater	Cd (II)	Weak acidic	Spain et al. (2021)
Mixed culture of *Chlorella sorokiniana, Monoraphidium* sp. and *Scenedesmus obliquus*	Alginate immobilized	Artificial wastewater	Cu (II) 96.4%	0.5 g/g biosorbent concentration, pH 4, contact time 180 minutes, 35°C	Lieswito et al. (2019)
Spirulina maxima	Suspended free cells	Steel hot-rolling wastewater	Fe ions 97.9%	25°C, 16:8 hours, 80–100 μmol/s/m^2 continuous illumination, 100 L/h aeration flow	Blanco-Vieites et al. (2022)
Chlorella vulgaris, Scenedesmus spinosus	Suspended free cells	Metal mine tailings wastewater	Cu (II) 64.7% Mo (III) 99.9%	96 hours, 20°C 60 μmol/m^2/s	Urrutia et al. (2019)

5.4.3 Removal of Dyes

Alginate-immobilized *Chlorella* sp. beads showed efficient results when cultivated in textile wastewater at pH 10 for 7 days; the removing rate reached 80.2% for $NH_4^+ - N$, 70.8% for COD, and 77.9% dye removal, and produced fatty acid methyl esters by 10% (Wu et al., 2020). A team of researchers developed an algal textile material consisting of *Chlorella sorokiniana* IPPAS C-1 alginate-immobilized beads with cotton textile. This algal textile beads showed removal efficiency of phosphorus and nitrogen by 99% and 76%, respectively (Melnikova et al., 2022).

5.5 MECHANISMS OF MICROALGAL ELIMINATION OF POLLUTANTS

Contrarily, an exogenous CO_2 source is provided to enclosed algae-based WWT systems (suspended and immobilized algal growth). When the C:N ratios at the inlet are lower, the enclosed reactors are more stable and dependable because the aeration system using CO_2 enriched air (1%–5% CO_2) regulates the inorganic carbon content in wastewater through a sensory system (Xiaogang et al., 2022). Therefore, in terms of C:N inlet ratio variation, enclosed suspended-growth reactors achieve the highest reliability. The enclosed immobilized algal bead systems also profit from the aeration system, but the nutrient transfer is less effective due to the insulation of the granules. Furthermore, since shell tears can occur, reducing their reliability to an average level, the amount of aeration and algal growth inside the granules should be closely monitored (Hu et al., 2021). Microalgae cultivation to eliminate pollutants utilizes four main mechanisms: bioaccumulation, bioadsorption, biodegradation, and bioaugmentation (De-Carvalho et al., 2022).

5.5.1 Bioaccumulation

Bioaccumulation occurs when pollutants build up inside cells, altering their structure (De-Carvalho et al., 2022). It involves transporting pollutants across the cell membrane into the cell, requiring energy for active compartmentalization (Zohoorian et al., 2020; Chen et al., 2023). Biomass accumulates or metabolizes substances by absorbing chemicals and nutrients through its surface. The process from metal ion absorption to movement within the cell or organelle takes time, and ion-selective transport proteins in the cell membrane are essential (Chugh et al., 2022; Chen et al., 2023).

5.5.2 Bioadsorption

In bioadsorption, the contaminant adheres extracellularly to the surface of the cell by electrostatic forces of the negatively charged microalgae cell walls due to the presence of carboxyl, phosphoryl functional groups and various polymers, facilitating adsorption processes (Xiong et al., 2018). Biosorption is the extracellular process of sequestering pollutants like heavy metals and dyes through passive binding in living or dead biomass. Microalgae have a high surface-to-volume ratio and binding affinity, making them suitable for biosorption. There are a number of mechanisms at play, including complexation, ion exchange, precipitation, and physical adsorption. Microalgae's cell walls contain receptors of functional groups, including carboxyl, amino, phosphate, hydroxyl, imidazole, and sulfate, which attach the pollutants by adsorption, chelation/complexation, electrostatic interactions, surface precipitation, and ion exchange. (Chu and Phang, 2019; Chen et al., 2023). The protonation of the functional groups on algal biomass particles and the amino and hydroxyl groups on carriers may affect the electrostatic repulsion between positively charged surfaces and metal cations. Pollutants such as heavy metals can easily bind to the surfaces, lipids, and proteins of algae thanks to polysaccharides. When the pH of the solution rapidly changes during biosorption or when the concentration of the metals increases to saturation, algae bind to heavy metals by precipitating and accumulating on the cell surface. The immobilization of microalgae produces a smoother algal cell surface (Chen et al., 2023).

5.5.3 Biodegradation

Biodegradation, which can take place intracellularly or extracellularly, involves decomposition reactions that lessen the toxicity of the pollutant (Xiong et al., 2018; De-Carvalho et al., 2022). Enzymatic and biochemical transformation convert pollutants to less harmful compounds (Chen et al., 2023).

5.5.4 Bioaugmentation

Bioaugmentation involves combining microalgae and other microorganisms to broaden pollutants' bioremediation range and create synergy. One method involves CO_2 and O_2 gas exchange in the effluent, but the understanding of population interactions is limited (Sforza et al., 2018).

5.6 HARVESTING AND RECOVERY OF MICROALGAL BIOMASS AFTER WWT

5.6.1 Harvesting and Recovery of Suspended Biomass

Suspended-growth systems are the most challenging to harvest algal biomass after treatment due to their small size, density, and net negative charge. These factors make microalgae resistant to free sedimentation, making it difficult to integrate modern technology into large-scale WWT facilities. The small size of microphytes, density, and net negative charge make it difficult to harvest algal biomass (Dasgupta et al., 2019; Valchev and Ribarova, 2022). Current technologies for harvesting suspended algal growth, such as centrifugation, membrane filtration, and coagulation/flocculation, achieve high efficiency. However, their commercial use for large-scale WWTs is limited due to their reliance on precise dosages of external reagents or excessive energy, reducing the overall reliability of suspended-growth reactors (Valchev and Ribarova, 2022; Xiaogang et al., 2022).

5.6.2 Harvesting and Recovery of Immobilized Algal Beads from Growth System

Harvesting is made easier and more flexible by immobilizing microalgae in a matrix. It has been demonstrated that this technology is more effective in removing nutrients than cells that are suspended freely (Dasgupta et al., 2019). The post-treatment harvesting of free algal suspension is negatively impacted by factors neutralized by the initial separation of algal biomass into distinct beads. High concentrations of algae result in denser biomass agglomerates with larger diameters, affecting the harvesting process; additionally, harvesting is no longer affected by individual cell charge due to large groups of cells (1.5×10^6 cells in a bead) (Valchev and Ribarova, 2022). Preparing algal beads eliminates drawbacks, simplifying biomass separation after treatment using free gravitational sedimentation or coarse sieving (Kube et al., 2020). Immobilized bead algal technology enables facilitated algal harvesting, increasing their dependability for downstream processing. Controlling aeration intensity and agitation in the reactor and monitoring wastewater flow through the sieve is crucial to prevent shear stress on beads. This reduces the likelihood of biomass leaking outside individual beads (Valchev and Ribarova, 2022).

5.7 BENEFITS OF MICROALGAL TREATMENT OVER CONVENTIONAL WWT TECHNOLOGY

Microalgae produced in excess during the WWT process can be very valuable and have applications in the production of fertilizer and biofuel, as shown in Table 5.3, the extraction of pigment for paints and colorants, the production of bioplastics, etc. (Al-Jabri et al., 2021). Algal biomass in WWT is susceptible to contamination and prolonged contact with bacteria and protozoa. The main post-treatment applications include fertilizer production and third-generation biofuels like ethanol, methanol, propane, and biodiesel. Algae's high nutrient content, nitrogen fixation ability, high calorific value, and lipid storage capacity make it a promising resource for these applications. The risk of algae contamination is related to the safe application of surplus biomass in WWT processes. More stable nutrient and lipid content is a result of lower contamination risk. But if a harsh native mix of algal strains is used as fertilizer, invasive algae can release poisonous extracellular allelochemicals, increasing the risk of soil toxicity magnification (Valchev and Ribarova, 2022). Enclosed immobilized growth reactors are systems that double isolate biomass from the environment by enclosing the system to the atmosphere and separating algal cells within individual beads. These systems achieve stable operation, constant effluent results, and enhanced control over parameters. Although not as reliable as immobilized algal beads, they still achieve a high-reliability rating.

Mixed microalgae culture was used to treat dairy wastewater (DWW); after the cultivation, the removal rate of COD reached 90% and the biomass increased to 1.4 g/L with biochemical contents recorded 38%, 22%, 15%, and 54.12% for carbohydrates, lipids, proteins, and reducing sugars, respectively (Hemalatha et al., 2019). Then *Saccharomyces cerevisiae* was used to ferment the reducing sugars to bioethanol (116.2 mg/g) (Hemalatha et al., 2019). Katam and Bhattacharyya (2019) used mixed microalgal culture in suspended free cells and immobilized form in the treatment of domestic wastewater toward biofuel production. The results showed higher performance of immobilized microalgae than suspended cells with higher nutrient uptake reaching 91% for nitrogen and 93% for phosphorus; further, about 40% lipid content more than the suspended cells which reached only 16%. Immobilized Scenedesmus sp. was cultivated on palm oil mill wastewater under 25% v/v cell concentration, 128 μmol·proton/m^2/s, which achieved high CO_2 uptake of 4.63 kg/day/m^3, 2.98 g/L biomass, and 35.92% lipid content. The produced lipid content was upgraded to biogas (Srinuanpan et al., 2019). Immobilized *Clostridium butyricum* achieved maximum biohydrogen production by feeding it in a fixed bed reactor with acid algae hydrolysate and partially packed hybrid-immobilized beads, resulting in

TABLE 5.3
Application of Algal Cultivations in Wastewater to Produced Biofuel

Microalgal Strain	Microalgal Status	Wastewater Type	Produced Biomass				Produced Biofuel	Cultivation Conditions	References
			Biomass	Carbohydrates	Lipids	Protein			
Mixed microalgae culture	Suspended free cells	Dairy wastewater	1.4 g/L	38%, 54.12% reducing sugars	22%	15%	Bioethanol	—	Hemalatha et al. (2019)
Scenedesmus sp.	Immobilized	Palm oil mill wastewater	2.98 g/L	—	35.92%	—	Biogas (>95% methane)	25% v/v of 10^6 cells/mL microalgal cell, 20% effluent, 128 μmol·proton/m²/s light intensity	Srinuanpan et al. (2019)
S. obliquus, C. vulgaris	Immobilized beads	Artificial wastewater	—	—	—	—	Biohydrogen	pH 7.5 and 30°C and 140 μE/m²/s of light intensity	Ruiz-Marin et al. (2020)
Chlorella sp., *Nannochloropsis* sp., *Scenedesmus* sp.	Suspended cells	Sewage sludge	—	—	—	—	Biogas	Anaerobic digestion of solid microalgae residues	Torres et al. (2023)
Nannochloropsis oceanica	Suspended cells	*Saccharomyces cerevisiae* yeast wastewater	—	21.19%	41.97%	—	Biodiesel	—	Senousy et al. (2023)
Monoraphidium braunii	Suspended cells	Sewage water	—	—	5.26 mg/L/day	—	Biodiesel	—	El-Sheekh et al. (2023)

biohydrogen yield of 2.3 mol H_2/mol hexose (Anburajan et al., 2019). The goal of the study of Ruiz-Marin et al. (2020) is to produce biohydrogen from immobilized *C. vulgaris* and *S. obliquus* microalgae in urban wastewater. Prior to anaerobic conditions at pH 7.5 and 30°C, the researchers used two light conditions at 140 μE/m²/s of light intensity and sulfur deprivation. Purple light eliminated organic carbon and produced the most hydrogen at 128 and 60.4 mL H_2/L for *S. obliquus* and *C. vulgaris*, respectively, while blue light promoted more algal growth. A study on *Tetraspora* sp. CU2551 found that cells immobilized on a calcium alginate matrix showed enhanced H_2 production under anaerobic conditions. The study found successful aerobic H_2 production in trapped cells under optimal immobilization conditions. When incubated in sulfur-deprived media, H_2 production increased, and the process could be repeated up to six times using refreshed media. After six cycles, H_2 production reached 12.8±0.9 mL H_2/25 mL^{-1} of medium, significantly higher than the initial rate (Maswanna et al., 2020).

REFERENCES

Agarwal P., Gupta, R., and Agarwal N. (2019) Advances in synthesis and applications of microalgal nanoparticles for wastewater treatment. *J. Nanotechnol.,* 2019, 1–9.

Ahmad A., Bhat A.H., and Buang A. (2018) Biosorption of transition metals by freely suspended and Ca-alginate immobilised with *Chlorella vulgaris*: kinetic and equilibrium modeling. *J. Clean. Prod., 171,* 1361–1375.

Ahmad A., Bhat A.H., and Buang, A. (2019a) Enhanced biosorption of transition metals by living *Chlorella vulgaris* immobilized in Ca-alginate beads. *Environ. Technol., 40,* 1793–1809.

Ahmad A., Bhat A.H., Buang A., Shah S.M.U., and Afzal M. (2019b) Biotechnological application of microalgae for integrated palm oil mill effluent (POME) remediation: a review. *Int. J. Environ. Sci. Technol., 16,* 1763.

Alazaiza M.Y.D., He S., Su D., Abu Amr S.S., Toh P.Y., and Bashir M.J.K. (2023) Sewage water treatment using *Chlorella vulgaris* microalgae for simultaneous nutrient separation and biomass production. *Separations, 10,* 229.

Al-Jabri H., Das P., Khan S., Thaher M., and Abdulquadir M. (2021) Treatment of wastewaters by microalgae and the potential applications of the produced biomass: a review. *Water, 13,* 27.

Amaro H.M., Salgado E.M., Nunes O.C., Pires J.C.M., and Esteves A.F. (2023) Microalgae systems - environmental agents for wastewater treatment and further potential biomass valorisation. *J. Environ. Manag., 337,* 117678.

Anburajan P., Yoon J.-J., Kumar G., Park J.-H., and Kim S.-H. (2019) Evaluation of process performance on biohydrogen production in continuous fixed bed reactor (C-FBR) using acid algae hydrolysate (AAH) as feedstock. *Int. J. Hydrog. Energy., 44,* 2164–2169.

Banerjee S., Tiwade P., Sambhav K., Banerjee C., and Bhaumik S.K. (2019) Effect of alginate concentration in wastewater nutrient removal using alginate-immobilized microalgae beads: uptake kinetics and adsorption studies. *Biochem. Eng. J., 149,* 107241.

Benítez M.B., Champagne P., Ramos A., Torres A.F., and Ochoa-Herrera V. (2019) Wastewater treatment for nutrient removal with Ecuadorian native microalgae. *Environ. Technol.,* 40(22), 2977–2985. doi:10.1080/09593330.2018.1459874

Blanco-Vieites M., Suárez-Montes D., Delgado F., Álvarez-Gil M., Battez A.H., and Rodríguez E. (2022) Removal of heavy metals and hydrocarbons by microalgae from wastewater in the steel industry. *Algal Res., 64,* 102700.

Borowiak D., and Krzywonos M. (2022) Bioenergy, biofuels, lipids and pigments—research trends in the use of microalgae grown in photobioreactors. *Energies, 15,* 5357.

Chen C.-Y., Kuo E.-W., Nagarajan D., Ho S.-H., Dong C.-D., Lee D.-J., and Chang J.-S. (2020) Cultivating *Chlorella sorokiniana* AK-1 with swine wastewater for simultaneous wastewater treatment and algal biomass production. *Bioresour. Technol., 302,* 122814.

Chen Z., Osman A.I., Rooney D.W., Oh W.-D., and Yap P.-S. (2023) Remediation of heavy metals in polluted water by immobilized algae: current applications and future perspectives. *Sustainability, 15*, 5128.

Chu W.-L., and Phang S.-M. (2019) Biosorption of heavy metals and dyes from industrial effluents by Microalgae. In Alam, M.A., Wang, Z., Eds., *Microalgae Biotechnology for Development of Biofuel and Wastewater Treatment*, pp. 599–634. Springer Nature Singapore Pte Ltd.

Chugh M., Kumar L., Shah M.P., and Bharadvaja N. (2022) Algal bioremediation of heavy metals: an insight into removal mechanisms, recovery of by-products, challenges, and future opportunities. *Energy Nexus, 7*, 100129.

Chun-Yen C., En-Wei K., Dillirani N., Cheng-Di D., Duu-Jong L., Sunita V., Su S.L., and Jo-Shu C. (2021) Semi-batch cultivation of *Chlorella sorokiniana* AK-1 with dual carriers for the effective treatment of fullstrength piggery wastewater treatment. *Bioresour. Technol., 326*, 124773.

Colusse G.A., Santos A.O., Rodrigues J.M., Barga M.C., Duarte M.E.R., De-Carvalho J.C., and Noseda M.D. (2021) Rice vinasse treatment by immobilized *Synechococcus pevalekii* and its effect on *Dunaliella Salina* cultivation. *Bioprocess Biosyst. Eng., 44*, 1477–1490.

Dasgupta C.N., Toppo K., Nayaka S., and Singh A.K. (2019) Bioremediation of municipal sewage using potential microalgae. In Gupta, S.K., Bux, F. Eds., *Application of Microalgae in Wastewater Treatment*, pp. 121–144. Springer Nature.

De-Carvalho J.C., Molina-Aulestia D.T., Martinez-Burgos W.J., Karp S.G., Manzoki M.C., Medeiros A.B.P., Rodrigues C., Scapini T., Vandenberghe L.P.S., Vieira S., Woiciechowski A.L., Soccol V.T., and Soccol C.R. (2022) Agro-industrial wastewaters for algal biomass production, bio-based products, and biofuels in a circular bioeconomy. *Fermentation, 8*, 728.

Efome J.E., Rana D., Matsuura T., and Lan C.Q. (2019) Effects of operating parameters and coexisting ions on the efficiency of heavy metal ions removal by nano-fibrous metal-organic framework membrane filtration process. *Sci. Total Environ., 674*, 355–362.

El-Gaayda J., Rachid Y., Titchou F.E., Barra I., Hsini A., Yap P.-S., Oh W.-D., Swanson C., Hamdani M., and Akbour R.A. (2023) Optimizing removal of chromium (VI) ions from water by coagulation process using central composite design: effectiveness of grape seed as a green coagulant. *Sep. Purif. Technol., 307*, 122805.

El-Gaayda J., Titchou F.E., Oukhrib R., Yap P.-S., Liu T., Hamdani M., and Ait-Akbour R. (2021) Natural flocculants for the treatment of wastewaters containing dyes or heavy metals: a state-of-the-art review. *J. Environ. Chem. Eng., 9*, 106060.

El-Sheekh M.M., Galal H.R., Mousa A.S.H., and Farghl A.A.M. (2023) Coupling wastewater treatment, biomass, lipids, and biodiesel production of some green microalgae. *Environ. Sci. Pollut. Res., 30*, 35492–35504.

Emparan Q., Harun R., and Jye Y.S. (2019) Phycoremediation of treated palm oil mill effluent (TPOME) using *Nannochloropsis* sp. cells immobilized in the biological sodium alginate beads: effect of POME concentration. Immobile phycoremediation. *Biorescources, 14*(4), 9429–9443.

Fawzy M.A., Darwish H., Alharthi S., Al-Zaban M.I., Noureldeen A., and Hassan S.H.A. (2022) Process optimization and modeling of Cd^{2+} biosorption onto the free and immobilized Turbinaria ornata using Box–Behnken experimental design. *Sci. Rep., 12*, 3256.

Giese E.C., Silva D.D.V., Costa A.F.M., Almeida S.G.C., and Dussán K.J. (2020) Immobilized microbial nanoparticles for biosorption. *Crit. Rev. Biotechnol., 40*, 653–666.

Gong W., Xie B., Deng S., Fan Y., Tang X., and Liang H. (2019) Enhancement of anaerobic digestion effluent treatment by microalgae immobilization: characterized by fluorescence excitation-emission matrix coupled with parallel factor analysis in the photobioreactor. *Sci. Total Environ., 678,* 105–113.

González-Camejo J., Ferrer J., Seco A., and Barat R. (2021) Outdoor microalgae-based urban wastewater treatment: recent advances, applications, and future perspectives. *Wiley Interdiscip. Rev. Water, 8*, 12.

Han M., Zhang C., and Ho S.-H. (2023) Immobilized microalgal system: an achievable idea for upgrading current microalgal wastewater treatment. *Environ. Sci. Ecotechnol., 14,* 100227.

Hashmi Z., Bilad M.R., Fahrurrozi-Zaini J., Lim J.W., and Wibisono Y. (2023) Recent progress in microalgae-based technologies for industrial wastewater treatment. *Fermentation, 9,* 311.

Hejna M., Kapuścińska D., and Aksmann A. (2022) Pharmaceuticals in the aquatic environment: a review on eco-toxicology and the remediation potential of algae. *Int. J. Environ. Res. Public Health, 19*, 7717.

Hemalatha M., Sravan J.S., Min B., and Venkata Mohan S. (2019) Microalgae-biorefinery with cascading resource recovery design associated to dairy wastewater treatment. *Bioresour. Technol.*, 284, 424–429.

Hosseinzadeh A., Zhou J.L., Li X., Afsari M., and Altaee A. (2022) Techno-economic and environmental impact assessment of hydrogen production processes using bio-waste as renewable energy resource. *Renew. Sustain. Energy Rev., 156*, 111991.

Hu X., Meneses Y.E., Hassan A.A., Stratton J., and Huo S. (2021) Application of alginate immobilized microalgae in treating real food industrial wastewater and design of annular photobioreactor: a proof-of-concept study. *Algal Res., 60*, 102524.

João S., Dimitar V., Juha N., Esa T., Sema S., and Yagut A. (2021) Nutrient removal from hydroponic effluent by Nordic microalgae: from screening to a greenhouse photobioreactor operation. *Algal Res., 55,* 102247.

Katam K., and Bhattacharyya D. (2019) Simultaneous treatment of domestic wastewater and bio-lipid synthesis using immobilized and suspended cultures of microalgae and activated sludge. *J. Ind. Eng. Chem., 69*, 295–303.

Kathi S., Singh S., Yadav R., Singh A.N., and Mahmoud A.E.D. (2023) Wastewater and sludge valorisation: a novel approach for treatment and resource recovery to achieve circular economy concept. *Front. Chem. Eng., 5,* 1129783.

Kube M., Spedding B., Gao L., Fan L., and Roddick F. (2020) Nutrient removal by alginate-immobilised Chlorella vulgaris: response to different wastewater matrices. *J. Chem. Technol. Biotechnol., 95*(6), 1790–1799.

Kuczajowska-Zadrożna M., Filipkowska U., and Jóźwiak T. (2020) Adsorption of Cu (II) and Cd (II) from aqueous solutions by chitosan immobilized in alginate beads. *J. Environ. Chem. Eng., 8,* 103878.

Kumar A.K., Sharma S., Patel A., Dixit G., and Shah E. (2019) Comprehensive evaluation of microalgal based dairy effluent treatment process for clean water generation and other value-added products. *Int. J. Phytorem., 21*(6), 519–530.

Kumar T., and Eswari-Jujjavarappu S. (2023) A critical review on an advanced bioelectrochemical system for carbon dioxide sequestration and wastewater treatment. *Total Environ. Res. Themes, 5*, 100023.

Leong Y.K., Huang C.-Y., and Chang J.-S. (2021) Pollution prevention and waste phycoremediation by algal-based wastewater treatment technologies: the applications of high-rate algal ponds (HRAPs) and algal turf scrubber (ATS). *J. Environ. Manag., 296*, 113193.

Lieswito N.A., Rinanti A., and Fachrul M.F. (2019) Removal of heavy metal (Cu^{2+}) by immobilized microalgae biosorbent with effect of temperature and contact time. *J. Phys. Conf. Ser., 1402*, 022106.

Liu C., Wu T., Hsu P.-C., Xie J., Zhao J., Liu K., Sun J., Xu J., Tang J., Ye Z., Lin D., and Cui Y. (2019a) Direct/alternating current electrochemical method for removing and recovering heavy metal from water using graphene oxide electrode. *ACS Nano, 13*, 6431–6437.

Liu X., Wang K., Zhang J., Wang J., Wu J., and Peng F. (2019b) Ammonium removal potential and its conversion pathways by free and immobilized *Scenedesmus obliquus* from wastewater. *Bioresour. Technol., 283,* 184–190.

López-Sánchez A., Silva-Gálvez A.L., Aguilar-Juárez Ó., Senés-Guerrero C., Orozco-Nunnelly D.A., Carrillo-Nieves D., and Gradilla-Hernández M.S. (2022) Microalgae-based livestock wastewater treatment (MbWT) as a circular bioeconomy approach: enhancement of biomass productivity, pollutant removal and high-value compound production. *J. Environ. Manag., 308,* 114612.

Marchão L., Fernandes J.R., Sampaio A., Peres J.A., Tavares P.B., and Lucas M.S. (2021) Microalgae and immobilized TiO_2/UV-A LEDs as a sustainable alternative for winery wastewater treatment. *Water Res., 203,* 117464.

Maswanna T., Lindblad P., and Maneeruttanarungroj C. (2020) Improved biohydrogen production by immobilized cells of the green alga Tetraspora sp. CU2551 incubated under aerobic condition. *J. Appl. Phycol.,* 2020, 10–25.

Melnikova A., Namsaraev Z., Komova A., Meuser I., Roeb M., Ackermann B., Klose H., and Kuchendorf C.M. (2022) AlgalTextile - a new biohybrid material for wastewater treatment. *Biotechnol. Rep., 33,* e00698.

Min K.J., Kim J.M., and Park K.Y. (2021) Characteristics of heavy metal separation and determination of limiting current density in a pilot-scale electrodialysis process for plating wastewater treatment. *Sci. Total Environ., 757,* 143762.

Minturo G.J., Noorain R., Hitam S.M.S., Shoiful A., Azni M.E., Ernawati L., Abdullah R., and Mohamad R. (2022) Immobilized *Nannochloropsis oculata* in a down-flow hanging sponge (DHS) reactor for the treatment of palm oil mill effluent (POME). In *IOP Conference Series: Earth and Environmental Science*; Institute of Physics: London, UK, Volume 1017.

Mkpuma V.O., Moheimani N.R., and Ennaceri H. (2023) Commercial paper as a promising carrier for biofilm cultivation of Chlorella sp. for the treatment of anaerobic digestate food effluent (ADFE): effect on the photosynthetic efficiency. *Sci. Total Environ., 898,* 165439.

Mohseni A., Kube M., Fan L., and Roddick F.A. (2021) Treatment of wastewater reverse osmosis concentrate using alginate-immobilised microalgae: integrated impact of solution conditions on algal bead performance. *Chemosphere, 276,* 130028.

Moreno-Osorio J.H., Pinto G., Pollio A., Frunzo L., Lens Piet N.L., and Esposito G. (2019) Start-up of a nutrient removal system using *Scenedesmus vacuolatus* and *Chlorella vulgaris* biofilms. *Bioresour. Bioprocess., 6,* 27.

Nekouei R.K., Pahlevani F., Assefi M., Maroufi S., and Sahajwalla V. (2019) Selective isolation of heavy metals from spent electronic waste solution by macroporous ion-exchange resins. *J. Hazard. Mater., 371,* 389–396.

Paddock M.B. (2019) *Microalgae Wastewater Treatment: A Brief History. Preprints,* 2019120377.

Patrinou V., Tsolcha O.N., Tatoulis T.I., Stefanidou N., Dourou M., Moustaka-Gouni M., Aggelis G., and Tekerlekopoulou A.G. (2020) Biotreatment of poultry waste coupled with biodiesel production using suspended and attached growth microalgal-based systems. *Sustainability, 12*(12), 5024.

Pinto J., Colónia J., Abdolvaseei A., Vale C., Henriques B., and Pereira E. (2023) Algal sorbents and prospects for their application in the sustainable recovery of rare earth elements from E-waste. *Environ. Sci. Pollut. Res., 30,* 74521–74543.

Plöhn M., Spain O., Sirin S., Silva M., Escudero-Oñate C., Ferrando-Climent L., Allahverdiyeva Y., and Funk C. (2021) Wastewater treatment by microalgae. *Physiol. Plant.,* 173(2), 568–578.

Ruiz-Marin A., Canedo-López Y., and Chávez-Fuentes P. (2020) Biohydrogen production by *Chlorella vulgaris* and *Scenedesmus obliquus* immobilized cultivated in artificial wastewater under different light quality. *AMB Expr., 10,* 191.

Sforza E., Pastore M., Santeufemia-Sanchez S., and Bertucco A. (2018) Bioaugmentation as a strategy to enhance nutrient removal: symbiosis between *Chlorella protothecoides* and *Brevundimonas diminuta*. *Bioresour. Technol. Rep., 4*, 153–158.

Shabana E.F., Senousy H.H., and Khourshid E.B. (2019) Pharmaceutical wastewater treatment using free and immobilized Cyanobacteria. *Egyptian J. Phycol., 20*(1), 123–154.

Shahrokhi-Shahraki R., Benally C., El-Din M.G., and Park J. (2021) High efficiency removal of heavy metals using tire-derived activated carbon vs commercial activated carbon: insights into the adsorption mechanisms. *Chemosphere, 264*, 128455.

Senousy H.H., El-Sheekh M.M., Khairy H.M., El-Sayed H.S., Mahmoud G.A.-E., and Hamed A.A. (2023) Biodiesel production from the marine alga *Nannochloropsis oceanica* grown on yeast wastewater and the effect on its biochemical composition and gene expression. *Plants, 12*(16), 2898.

Shen L., Wang J., Li Z., Fan L., Chen R., Wu X., Li J., and Zeng W. (2020) A high-efficiency Fe_2O_3@Microalgae composite for heavy metal removal from aqueous solution. *J. Water Process Eng., 33*, 101026.

Silva A.D.M., Fernandes D.F., Figueiredo S.A., Freitas O.M., and Delerue-Matos C. (2022) Fluoxetine and nutrients removal from aqueous solutions by phycoremediation. *Int. J. Environ. Res. Public Health, 19*, 6081.

Spain O., Plohn M., and Funk C. (2021) The cell wall of green microalgae and its role in heavy metal removal. *Physiol. Plant., 10*, 13405.

Srimongkol P., Sangtanoo P., Songserm P., Watsuntorn W., and Karnchanatat A. (2022) Microalgae-based wastewater treatment for developing economic and environmental sustainability: current status and future prospects. *Front. Bioeng. Biotechnol., 10*, 904046.

Srinuanpan S., Cheirsilp B., Boonsawang P., and Prasertsan P. (2019) Immobilized oleaginous microalgae as effective two-phase purify unit for biogas and anaerobic digester effluent coupling with lipid production. *Bioresour. Technol., 281*, 149–157.

Sukla L.B., Pradhan D., and Subbaiah T. (2019) Future prospects of microalgae in wastewater treatment. In Sukla, L.B., et al. Eds., *The Role of Microalgae in Wastewater Treatment*, pp. 129–135. Springer Nature Singapore Pte Ltd.

Sun Y., Lu T., Pan Y., Shi M., Ding D., Ma Z., Liu J., Yuan Y., Fei L., and Sun Y. (2022) Recovering rare earth elements via immobilized red algae from ammonium-rich wastewater. *Environ. Sci. Ecotechnol., 12*, 100204.

Syeda H.I., Sultan I., Razavi K.S., and Yap P.-S. (2022) Biosorption of heavy metals from aqueous solution by various chemically modified agricultural wastes: a review. *J. Water Process Eng., 46*, 102446.

Tan K.A., Lalung J., Morad N., Ismail N., Omar W.M.W., Khan M.A., Sillanpää M., and Rafatullah M. (2021) Post-treatment of palm oil mill effluent using immobilised green microalgae *Chlorococcum oleofaciens*. *Sustainability, 13*, 1562.

Taufikurahman T., Ardiansyah M.A., Astutiningsih N.T., and Suyono E.A. (2020) Cultivation of *Chlorella vulgaris* in anaerobically digested dairy manure wastewater (ADDMW) for protein and chlorophyll production. *Sains Malaysiana, 49*, 2035–2042.

Torres A., Padrino S., Brito A., and Díaz L. (2023) Biogas production from anaerobic digestion of solid microalgae residues generated on different processes of microalgae-to-biofuel production. *Biomass Conv. Bioref., 13*, 4659–4672.

Urrutia C., Yañez-mansilla E., and Jeison D. (2019) Bioremoval of heavy metals from metal mine tailings water using microalgae biomass. *Algal Res., 43*, 101659.

Valchev D., and Ribarova I. (2022) A review on the reliability and the readiness level of microalgae-based nutrient recovery technologies for secondary treated effluent in municipal wastewater treatment plants. *Processes, 10*, 399.

Vasilieva S., Lobakova E., Grigoriev T., Selyakh I., Semenova L., Chivkunova O., Gotovtsev P., Antipova C., Zagoskin Y., Scherbakov P., Lukyanov A., Lukanina K., and Solovchenko A. (2021a) Bio-inspired materials for nutrient biocapture from wastewater: microalgal cells immobilized on chitosan-based carriers. *J. Water Process Eng., 40*, 101774.

Vasilieva S., Lobakova E., and Solovchenko A. (2021b) Biotechnological applications of immobilized microalgae. In Gothandam, K.M., Ranjan, S., Dasgupta, N., Lichtfouse, E., Eds., *Environmental Biotechnology*, Volume 3, pp. 193–220. Springer International Publishing: Cham.

Wang J., Liang Q., Cui L., Wu X., Li J., Zeng W., and Shen L. (2022) Adsorption characteristics of Cr (VI) on microalgae immobilized by different carriers. *Int. J. Phytoremediat., 24,* 704–720.

Whitton R., Ometto F., Villa R., Pidou M., and Jefferson B. (2019) Influence of light regime on the performance of an immobilized microalgae reactor for wastewater nutrient removal. *Algal Res., 44*, 101648.

Wu J.-Y., Lay C.-H., Chen C.-C., Wu S.-Y., Zhou D., and Mohamed-Abdula P. (2020) Textile wastewater bioremediation using immobilized *Chlorella* sp. Wu-G23 with continuous culture. *Clean Technol. Environ. Policy, 23*, 153 161.

Xiaogang H., Jalalah M., Jingyan W., Zheng Y., Li X., and Salama E.-S. (2022) Microalgal growth coupled with wastewater treatment in open and closed systems for advanced biofuel generation. *Biomass Convers. Biorefinery., 12*, 1939–1958.

Xiong J., Kurade M.B., and Jeon B. (2018) Can microalgae remove pharmaceutical contaminants from water? *Trends Biotechnol., 36*, 30–44.

Zabochnicka M. (2022) Industrial wastewater as a growth medium for microalgal biomass for a sustainable circular bioeconomy. *Appl. Sci., 12*, 10299.

Zhang Y., and Duan X. (2023) Chemical precipitation of heavy metals from wastewater by using the synthetical magnesium hydroxy carbonate. *Water Sci. Technol., 81*, 1130–1136.

Zhang Z., Yan K., Zhang L., Wang Q., Guo R., Yan Z., and Chen J. (2019) A novel cadmium-containing wastewater treatment method: bio-immobilization by microalgae cell and their mechanism. *J. Hazard. Mater., 374*, 420–427.

Zhuang L-L., Li M., and Hao-Ngo H. (2020) Non-suspended microalgae cultivation for wastewater refinery and biomass production. *Bioresour. Technol., 308*, 123320.

Zohoorian H., Ahmadzadeh H., Molazadeh M., Shourian M., and Lyon S. (2020) Microalgal bioremediation of heavy metals and dyes. In Konur, O., Ed., *Handbook of Algal Science, Technology and Medicine*, pp. 659–674. Academic Press: Cambridge, MA.

6 Microalgae-Bacteria Symbiotic System for Wastewater Treatment

Sameh S. Ali, Rania Al-Tohamy, Majid Al-Zahrani, Michael Kornaros, and Michael Schagerl

6.1 INTRODUCTION

Wastewater treatment (WWT) is a critical process that ensures the protection of water resources and the environment (Ali et al., 2024; Chan et al., 2022). It involves the removal of various pollutants, including organic matter, nutrients (e.g., nitrogen and phosphorus), heavy metals, and pathogens, from wastewater before its discharge (Ahmed et al., 2022). The conventional methods of WWT, such as activated sludge processes and chemical treatments, have limitations in terms of energy consumption, high costs, and incomplete removal of pollutants (Ahmed et al., 2021). Recently, the integration of MBSS has gained attention as a potential solution to address the shortcomings of traditional WWT methods (Fallahi et al., 2021; Sahu et al., 2023). Microalgae are unicellular photosynthetic organisms that can utilize light energy and nutrients present in wastewater for growth and biomass production. They have a high affinity for nutrients, especially nitrogen (N) and phosphorus (P), and can efficiently remove them from wastewater through a process called assimilation (Nagarajan et al., 2022). Additionally, microalgae can capture carbon dioxide (CO_2) during photosynthesis, making them effective in mitigating greenhouse gas emissions (Reetu et al., 2023).

On the other hand, bacteria play a crucial role in the breakdown and degradation of organic matter. They break down complex organic compounds into simpler forms that can be readily utilized by microalgae (Nagarajan et al., 2022, 2023; Sahu et al., 2023). This mutualistic relationship between microalgae and bacteria creates a synergistic effect, where microalgae provide oxygen and organic carbon to the bacteria, while bacteria supply necessary nutrients and growth factors to the microalgae (Fallahi et al., 2021). This symbiotic interaction enhances the efficiency of nutrient removal and organic matter degradation compared to individual microalgae or bacteria cultures in WWT.

The advantages of MBSS in WWT are manifold. First, they offer a sustainable and energy-efficient approach as microalgae utilize sunlight as their energy source and do not require external carbon sources. This reduces the energy consumption and operational costs associated with WWT. Second, these systems have the potential for resource recovery (Johnson et al., 2020; Zhao et al., 2023). The biomass produced by microalgae can be harvested and used for various purposes, such as biofuel

 DOI: 10.1201/9781003459286-6

production, animal feed, or as a source of valuable compounds (Ali et al., 2022a–c; Khan et al., 2022). Moreover, the removal of nutrients by microalgae helps mitigate eutrophication, a major environmental concern associated with excessive nutrient discharge into water bodies (Pechsiri et al., 2023).

However, the application of MBSS in WWT faces several challenges. The selection of suitable microalgae and bacteria strains is crucial for achieving optimal performance. Factors such as temperature, light intensity, pH, and nutrient availability need to be carefully controlled to ensure the growth and activity of both microalgae and bacteria (Zhou et al., 2023). Biomass harvesting from the symbiotic system can be challenging, as microalgae can form aggregates or biofilms, which may require specialized techniques for efficient separation (Huang et al., 2023).

To this end, this chapter aims to delve into the mechanisms and interactions involved in MBSS for WWT and explore their performance and efficiency in nutrient removal, organic matter degradation, and biomass production. Additionally, this chapter will discuss the factors influencing the performance of these systems and the challenges associated with their implementation. By critically evaluating the current knowledge and identifying research gaps, this chapter aims to contribute to the advancement and wider adoption of MBSS as a sustainable and cost-effective approach for WWT.

6.2 MBSS: MECHANISMS AND INTERACTIONS

6.2.1 Mutualistic Relationship between Microalgae and Bacteria

The mutualistic relationship between microalgae and bacteria in the MBSS is a key factor in its effectiveness in WWT (Johnson et al., 2020). Therefore, understanding the mechanisms and interactions involved in this mutualistic relationship and highlighting the metabolic exchange and spatial arrangements that facilitate nutrient removal and organic matter degradation is essential.

As shown in Figure 6.1, the metabolic exchange between microalgae and bacteria is based on the release and utilization of oxygen, organic carbon, and nutrients. Microalgae, through photosynthesis, produce oxygen as a byproduct. This oxygen is crucial for the survival and growth of bacteria, as it is required for the breakdown of organic matter through aerobic respiration (Dai and Wang, 2024). The presence of microalgae in the symbiotic system ensures a continuous supply of oxygen, creating an oxygen-rich environment that supports the growth and activity of bacteria. In addition to oxygen, microalgae also release organic carbon compounds, such as sugars and amino acids, into the surrounding environment during photosynthesis. These organic carbon compounds serve as an energy source for bacteria (González-González and de-Bashan, 2021). Bacteria possess a wide range of metabolic capabilities that allow them to break down complex organic compounds (e.g., proteins and lipids) into low molecular weight and inorganic compounds (e.g., ammonia and nitrate). These simple forms can be readily assimilated by microalgae for growth and biomass production (Ali et al., 2022c; Fallahi et al., 2021). Therefore, the metabolic exchange of organic carbon between microalgae and bacteria enhances the overall efficiency of organic matter degradation in the symbiotic system.

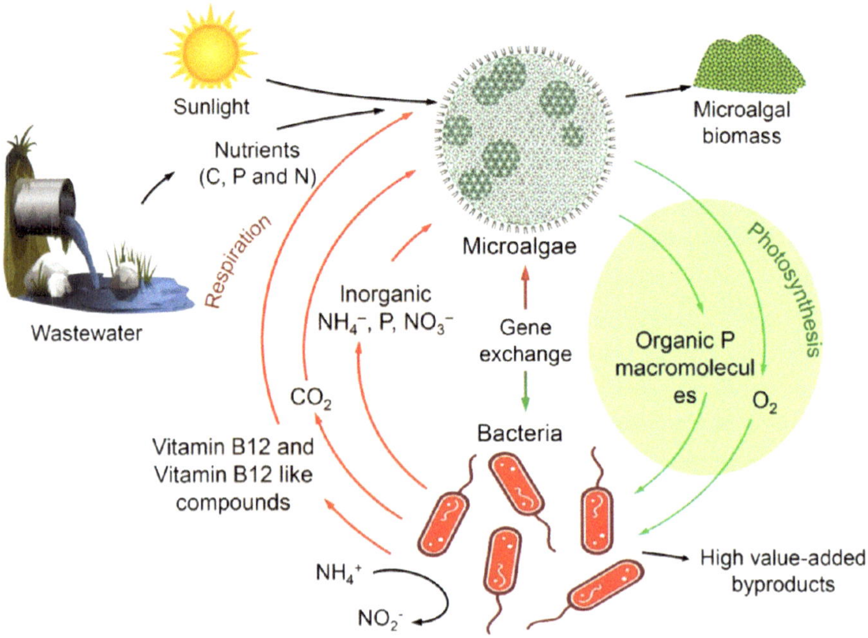

FIGURE 6.1 Microalgae-bacteria symbiotic systems and their mutualistic relationship.

Furthermore, bacteria play a crucial role in the provision of growth factors and nutrients to microalgae. Bacteria are capable of synthesizing and releasing essential growth factors, such as vitamins B12 and B12-like compounds that are required for the growth and metabolism of microalgae (Helliwell et al., 2016). These growth factors are often limiting in wastewater environments, and the presence of bacteria in the symbiotic system ensures their availability to microalgae. Additionally, bacteria provide microalgae with necessary nutrients (e.g., N and P) by breaking down complex organic compounds into simpler forms that can be assimilated by microalgae (Ahmed et al., 2022). This nutrient supply from bacteria enhances the nutrient removal efficiency of microalgae and reduces the concentration of pollutants in the wastewater.

In this regard, Cassan et al. (2021) found that *Azospirillum brasilense* T6SS is required for *Chlorella sorokiniana-Azospirillum* synthetic mutualism to promote the production of lipids, carbohydrates, and photosynthetic pigments by the microalgae due to the production of indole acetic acid (IAA) phytohormones. Additionally, *A. brasilense* T6SS provides antagonistic activities against several pathogens (e.g., *Agrobacterium* spp., *Dickeya* spp., *Pectobacterium* spp., and *Ralstonia* spp.) in vitro, suggesting that in addition to promoting growth, *A. brasilense* might provide bio-control protection to microalgae against pathogenic microorganisms (Cassan et al., 2021).

The spatial arrangements within the MBSS also contribute to the mutualistic relationship between microalgae and bacteria. Microalgae can form aggregates or biofilms, which provide a surface for the attachment and growth of bacteria (Cheah and Chan, 2021). These aggregates or biofilms can be composed of both microalgae and bacteria, creating a close physical association between the two organisms. This physical association allows for a more intimate metabolic exchange and enhances the efficiency of nutrient removal and organic matter degradation. The proximity of microalgae and bacteria in the aggregates or biofilms facilitates the transfer of oxygen, organic carbon, and nutrients between the two organisms, further strengthening their mutualistic relationship (Wicker et al., 2023).

The spatial arrangements within the symbiotic system also play a role in protecting microalgae from detrimental factors, such as grazing by zooplankton or adverse environmental conditions (Aditya et al., 2022). The formation of aggregates or biofilms provides a physical barrier that can shield microalgae from predation and create a more stable microenvironment. Additionally, the presence of bacteria in the aggregates or biofilms can protect microalgae by producing antimicrobial compounds or competing with harmful microorganisms for resources. This protection allows microalgae to thrive and contribute to the overall performance of the symbiotic system (Penesyan et al., 2021).

Conclusively, the mutualistic relationship between microalgae and bacteria in the MBSS is based on a metabolic exchange of oxygen, organic carbon, and nutrients. Microalgae provide oxygen and organic carbon compounds to bacteria, supporting their growth and organic matter degradation capabilities. Bacteria, in return, supply microalgae with essential growth factors and nutrients, enhancing their nutrient removal efficiency. The spatial arrangements within the symbiotic system further facilitate the metabolic exchange and protect microalgae. Understanding and harnessing these mechanisms and interactions between microalgae and bacteria are crucial for optimizing the performance of the MBSS in WWT.

6.2.2 Nutrient Cycling and Pollutant Removal Mechanisms

The MBSS is highly efficient in nutrient cycling and pollutant removal due to the combined metabolic capabilities of microalgae and bacteria (Fallahi et al., 2021). One of the key benefits of the MBSS is its ability to remove nutrients, particularly N and P, from wastewater (Qv et al., 2023). The metabolic exchange between microalgae and bacteria enhances the overall efficiency of nutrient removal in the symbiotic system (Fallahi et al., 2021). In the case of N removal, bacteria in the symbiotic system play a crucial role in the nitrogen cycle (Ayre et al., 2021). They are capable of converting organic nitrogen (e.g., proteins and amino acids) into ammonia through biochemical processes (e.g., ammonification and proteolysis). As shown in Figure 6.2, ammonia can then be further transformed by nitrifying bacteria into nitrate through nitrification (Ayre et al., 2021; Fallahi et al., 2021). Microalgae can assimilate nitrate through a process called assimilatory nitrate reduction, converting it into organic nitrogen compounds and incorporating it into their biomass (Rezvani et al., 2020). This nitrogen cycling between microalgae and bacteria allows for the efficient removal of nitrogen from wastewater, preventing its release into the environment and reducing the risk of eutrophication.

Similarly, as shown in Figure 6.3, the symbiotic system also contributes to the removal of phosphorus from wastewater. Bacteria in the system can hydrolyze organic phosphorus compounds, such as phospholipids and nucleic acids, into

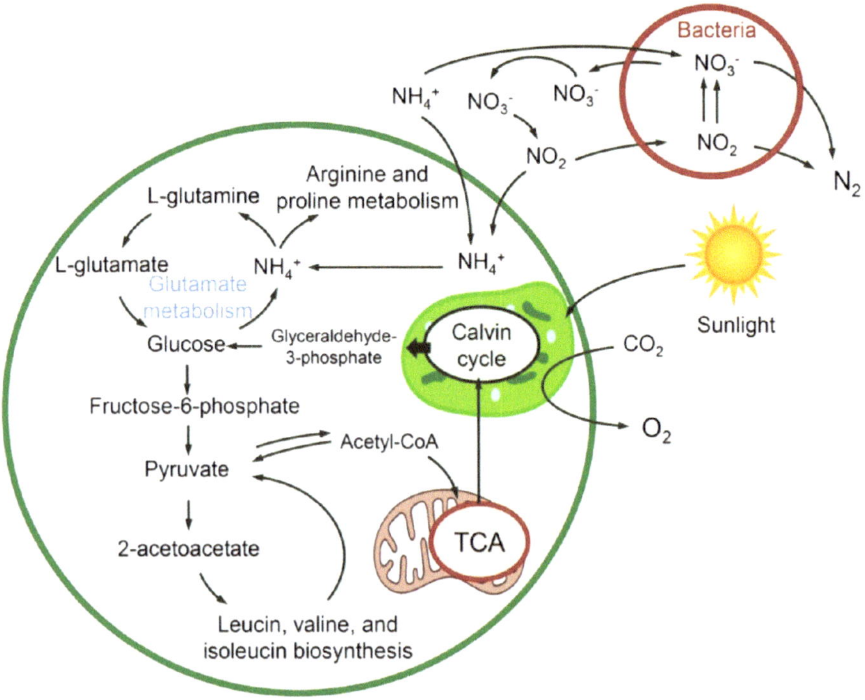

FIGURE 6.2 Nitrogen cycling between microalgae and bacteria during nutrient removal from wastewater.

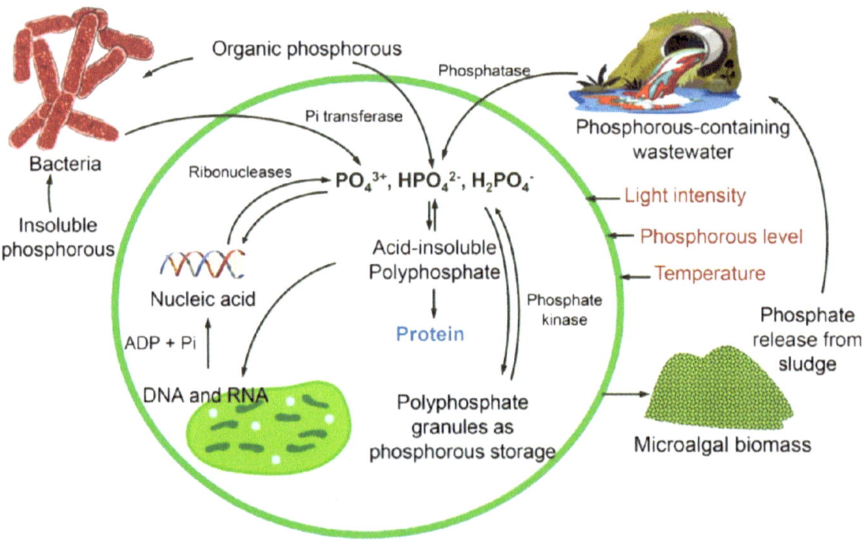

FIGURE 6.3 Phosphorous assimilation between microalgae and bacteria during nutrient removal from wastewater.

inorganic phosphate through processes like phosphatase activity (Billah et al., 2019). Microalgae can then assimilate this inorganic phosphate through a process called phosphate assimilation, incorporating it into their biomass (Su, 2021). This cycling of phosphorus between microalgae and bacteria ensures efficient removal of phosphorus from wastewater, reducing its concentration and minimizing the risk of eutrophication. Apart from nutrient cycling, the MBSS also exhibits efficient pollutant removal mechanisms (Nagarajan et al., 2022). Microalgae can remove a wide range of pollutants, including heavy metals, organic pollutants, and emerging contaminants (Zhao et al., 2023). Mustafa et al. (2021) stated that microalgae can accumulate these pollutants in their biomass through various mechanisms, such as adsorption, absorption, and biosorption. Once accumulated in the microalgae biomass, these pollutants can be effectively removed from the wastewater by harvesting and disposing of the microalgae.

Bacteria in the symbiotic system also contribute to pollutant removal through their metabolic capabilities. They can degrade and mineralize organic pollutants (e.g., pharmaceuticals, pesticides, and petroleum hydrocarbons) into simpler and less toxic forms (Bhatt et al., 2022). Bacteria possess a wide range of enzymes (e.g., dehydrogenases, oxidases, and hydrolases) which enable them to break down complex organic compounds (Kumari and Das, 2023). This degradation of organic pollutants by bacteria reduces their concentration in the wastewater, improving water quality and minimizing the environmental impact (Ahmed et al., 2021). Furthermore, the spatial arrangements within the MBSS also play a role in pollutant removal (Aditya et al., 2022; Fallahi et al., 2021). The formation of aggregates or biofilms by microalgae and bacteria provides a surface area for the attachment and accumulation of pollutants. This physical association between microalgae and bacteria enhances the removal of pollutants from the wastewater, as the pollutants can be efficiently adsorbed or absorbed by the biomass (Cheah and Chan, 2021; Moreno Osorio et al., 2021; Wicker et al., 2023). The proximity between microalgae and bacteria in the aggregates or biofilms also facilitates the metabolic exchange and cooperation in pollutant removal (Fallahi et al., 2021).

Thus, the MBSS exhibits efficient nutrient cycling and pollutant removal mechanisms. The metabolic capabilities of microalgae and bacteria enable the removal of nutrients from wastewater through a series of metabolic exchanges. This nutrient cycling minimizes the risk of eutrophication and helps maintain water quality. Additionally, the symbiotic system contributes to the removal of pollutants through the accumulation and degradation of pollutants by microalgae and bacteria. The spatial arrangements within the system further enhance pollutant removal by providing a surface for attachment and facilitating the metabolic exchange between microalgae and bacteria. Understanding and optimizing these nutrient cycling and pollutant removal mechanisms are crucial for harnessing the full potential of the MBSS in WWT.

6.2.3 Synergistic Interactions and Coexistence in the Symbiotic System

The MBSS is characterized by synergistic interactions and coexistence between microalgae and bacteria (Cassan et al., 2021; Sahu et al., 2023). One of the key factors contributing to the synergistic interactions in the MBSS is the exchange of metabolites and signaling molecules (González-González and de-Bashan, 2021). Microalgae and bacteria release a variety of metabolites, such as sugars, amino acids, organic acids, and vitamins, into the surrounding environment. These metabolites act as signaling molecules, facilitating communication between microalgae and bacteria. For example, Awdhesh Kumar Mishra and Kodiveri Muthukaliannan (2022) reported that microalgae can release quorum-sensing molecules, such as *N*-acyl homoserine lactones, which can induce the expression of specific genes in bacteria. In turn, bacteria can release growth-stimulating substances, such as auxins and cytokinins, which promote the growth and development of microalgae. For instance, Baggesen (2014) isolated three bacterial species, *Halomonas* sp., *Marinobacter* sp., and *Pelagibaca* sp. from one xenic culture of *Dunaliella salina*. These bacteria were found to enhance *D. salina* growth under iron-limited conditions. This exchange of metabolites and signaling molecules allows for a coordinated response and mutual growth enhancement in the symbiotic system (Awdhesh Kumar Mishra and Kodiveri Muthukaliannan, 2022; González-González and de-Bashan, 2021).

Another important aspect of the synergistic interactions in the symbiotic system is the complementary metabolic capabilities of microalgae and bacteria (Tong et al., 2023). These complementary metabolic capabilities of microalgae and bacteria create a mutually beneficial relationship, where microalgae provide organic carbon compounds and oxygen to bacteria, while bacteria supply microalgae with essential growth factors and nutrients. This metabolic complementarity enhances the overall efficiency of nutrient cycling and pollutant removal in the symbiotic system (Ali et al., 2022c; Fallahi et al., 2021; Tong et al., 2023).

The spatial arrangements within the MBSS also contribute to the synergistic interactions and coexistence between microalgae and bacteria (Aditya et al., 2022). The formation of aggregates or biofilms by microalgae and bacteria creates a microenvironment with distinct physical and chemical conditions (Sauer et al., 2022). These microenvironments can have different oxygen and nutrient gradients, pH levels, and light availability, which can favor the growth and activity of specific microorganisms. The close physical association between microalgae and bacteria in the aggregates or biofilms allows for a more intimate metabolic exchange and efficient transfer of metabolites and signaling molecules (Awdhesh Kumar Mishra and Kodiveri Muthukaliannan, 2022; Wicker et al., 2023). This physical association also protects against external stressors and competition from other microorganisms, promoting the coexistence and long-term stability of the symbiotic system.

Furthermore, the presence of microalgae in the symbiotic system can also influence the microbial community structure and diversity (Hu et al., 2023). Microalgae release a variety of organic compounds into the environment, which can act as a carbon source for other microorganisms. These organic compounds can promote the growth and activity of specific bacterial groups, leading to shifts in the microbial community composition (Ferro et al., 2020). Additionally, microalgae can release antimicrobial compounds or compete with harmful microorganisms for resources, suppressing their growth and promoting the dominance of beneficial bacteria (Stirk and van Staden, 2022). This influence of microalgae on the microbial community structure can further enhance the synergistic interactions and coexistence within the symbiotic system.

Therefore, the MBSS are characterized by synergistic interactions and coexistence between microalgae and bacteria. The exchange of metabolites and signaling molecules facilitates communication and coordination between microalgae and bacteria, promoting mutual growth enhancement. The complementary metabolic capabilities of microalgae and bacteria enable efficient nutrient cycling and pollutant removal in the symbiotic system. The spatial arrangements within the system provide a physical association that enhances the metabolic exchange and protects microorganisms from external stressors. The presence of microalgae also influences the microbial community structure, promoting the dominance of beneficial bacteria. Understanding and harnessing these synergistic interactions and coexistence mechanisms are crucial for optimizing the performance and stability of the MBSS.

6.3 PERFORMANCE AND EFFICIENCY OF MBSS

6.3.1 Nutrient Removal Efficiency

The performance and efficiency of MBSS in nutrient removal are crucial factors in evaluating their potential for WWT. Table 6.1 shows the efficiency of several MBSS to remove nutrients from different wastewater. MBSS have been shown to exhibit high nitrogen removal efficiency from wastewater. The combined metabolic capabilities of microalgae and bacteria allow for efficient nitrogen cycling and assimilation into biomass. Microalgae can assimilate nitrogen in the form of nitrate, which is converted into organic nitrogen compounds, while bacteria can mineralize organic nitrogen compounds, releasing ammonia (Mujtaba and Lee, 2016).

The efficiency of nitrogen removal in MBSS can be influenced by various factors, including the initial nitrogen concentration in wastewater, the ratio of microalgae to bacteria, and the availability of light and nutrients (Jia and Yuan, 2016; Rezvani et al., 2020). Higher initial nitrogen concentrations in wastewater can lead to higher nitrogen removal rates, as microalgae have a higher demand for nitrogen under these conditions (Morán-Valencia et al., 2023). Additionally, maintaining an optimal ratio of microalgae to bacteria is crucial for efficient nitrogen removal (Rezvani et al., 2020), as excessive bacterial growth can result in the consumption of nitrogen compounds by bacteria before they can be assimilated by microalgae. Providing sufficient light and nutrients can also enhance nitrogen removal efficiency by promoting microalgae growth and photosynthetic activity (Rezvani et al., 2020).

MBSS also demonstrate high efficiency in phosphorus removal from wastewater. Microalgae can assimilate phosphorus in the form of inorganic phosphate, which is incorporated into their biomass through phosphate assimilation (Fallahi et al., 2021). On the other hand, bacteria contribute to phosphorus removal by hydrolyzing organic phosphorus compounds, such as phospholipids and nucleic acids, into inorganic phosphate through phosphatase activity (Billah et al., 2019). The efficiency of phosphorus removal in MBSS can be influenced by factors such as the initial phosphorus concentration in wastewater, the availability of light and nutrients, and the presence of competing microorganisms (Dölle and Röhlen, 2023; González-Camejo et al., 2022). Higher initial phosphorus concentrations in wastewater can lead to higher phosphorus removal rates, as microalgae have a higher

TABLE 6.1
Nutrient Removal Efficiency from Different Wastewater Using Microalgae-Bacteria Symbiotic System

Wastewater Type	Reactor Type	Microalgae Species	Bacterial Species	Nutrient Removal (%)			Remarks	References
				Nitrogen	Phosphorous	COD		
ND	ND	*Chlorella sorokiniana*	*Brevundimonas* sp.	65.3	42.6	ND	The microalgae-bacteria symbiosis system (MBSS) can effectively remove pollutants from wastewater. *Brevundimonas promotes* microalgal biomass growth and microalgae facilitate nutrient removal	(Qv et al., 2023)
Dairy wastewater was used in the study	ND	*Tetraselmis indica*	*Pseudomonas aeruginosa*	83.76	79.83	87.49	MBSS showed higher biomass growth and nutrient removal. Chlorophyll-a content of consortium was higher than pure microalgae culture	(Talapatra et al., 2023)
Secondary treated municipal wastewater	ND	*Tetraselmis indica* and *Chlorella protothecoide*	*Pseudomonas* sp. and *Bacillus pumilus*	86.76	81.35	83.59	*C. protothecoide-Pseudomonas* sp. system showed the highest lipid productivity and nutrient removal	(Talapatra et al., 2021)
Pre-treated domestic wastewater	Photobioreactor working under sequencing batch mode (SBR)	ND	ND	>40	>90	>60	Algal-bacterial symbiosis enhances the removal of pollutants. Algal photosynthesis replaces mechanical aeration for greenhouse gas reduction	(Tahseen Sayara et al., 2021)

(Continued)

TABLE 6.1 (*Continued*)
Nutrient Removal Efficiency from Different Wastewater Using Microalgae-Bacteria Symbiotic System

Wastewater Type	Reactor Type	Microalgae Species	Bacterial Species	Nutrient Removal (%) Nitrogen	Phosphorous	COD	Remarks	References
Swine wastewater	ND	*Chlorella* sp. MA1 and *Coelastrella* sp. KE4	Bacterial consortium	12.79–99.52	35.66–96.06	ND	The ABS is effective for swine wastewater treatment. The symbiotic system promotes pollutant removal and pathogen inactivation.	(Qu et al., 2021)
Real pre-treated domestic wastewater	Photobioreactor of 200L	ND	ND	55	91	70	The algal-bacterial photobioreactor demonstrated efficient removal of nutrients and organic pollutants. The system showed promising results in improving the quality of treated wastewater	(T Sayara et al., 2021)
Raw municipal wastewater.	Lab-scale bioreactor	*Chlorella vulgaris* and *Scenedesmus obliquus*	The main bacterial phyla were *Proteobacteria, Firmicutes, Bacteroidetes, Chloroflexi*	2.34/d	1.67/day	ND	Nutrient removal and biomass yield varied with seasonal conditions.	(Xu et al., 2021)
Domestic wastewater	ND	ND	ND	94.85	97.4–99.79	85.36–88.96	Microalgal-bacterial consortia effectively treat domestic wastewater and showed a high reduction in physicochemical parameters	(Moondra et al., 2020)
Swine wastewater (SW) with high-strength ammonium.	Quartz reactor	*Chlorella vulgaris*	Four phyla of bacteria mainly *Proteobacteria* and *Firmicutes*	95	96	99	The biochar-mediated absorption coupled with MBSS enhanced nutrient recovery from SW and achieved high nutrient recovery efficiencies	(Yu et al., 2020)

ND, not detected; COD, chemical oxygen demand.

demand for phosphorus under these conditions (Dölle and Röhlen, 2023). Providing sufficient light and nutrients (e.g., CO_2 and N) can enhance phosphorus removal efficiency by promoting microalgae growth and photosynthetic activity (Li et al., 2023). However, the presence of competing microorganisms, such as algae or bacteria that are not part of the symbiotic system, can decrease the phosphorus removal efficiency by consuming phosphorus compounds before they can be assimilated by microalgae (González-Camejo et al., 2022).

In general, MBSS have demonstrated high nutrient removal efficiency, particularly for nitrogen and phosphorus. The combined metabolic capabilities of microalgae and bacteria, along with the synergistic interactions between them, allow for efficient nutrient cycling and assimilation into biomass. However, it is important to optimize various factors, such as the initial nutrient concentration, the microalgae-to-bacteria ratio, and the availability of light and nutrients, to maximize the nutrient removal efficiency of these systems. Additionally, the presence of competing microorganisms in the wastewater can also affect the nutrient removal efficiency and should be carefully managed. Understanding and optimizing these factors are crucial for harnessing the full potential of MBSS in WWT and nutrient recovery.

6.3.2 Organic Matter Degradation and Removal

In addition to nutrient removal, the performance and efficiency of MBSS in WWT also rely on their ability to degrade and remove organic matter (Chan et al., 2022). Organic matter is a major component of wastewater and can contribute to water pollution if not effectively treated (Touliabah et al., 2022). Microalgae and bacteria in the symbiotic system play complementary roles in the degradation of organic matter. Microalgae can utilize light energy through photosynthesis to convert CO_2 into organic carbon compounds. These organic carbon compounds can serve as a carbon source for bacteria, promoting their growth and metabolic activity. Bacteria, in turn, possess a wide range of metabolic capabilities for the breakdown and mineralization of organic matter (Chan et al., 2022). They can produce extracellular enzymes, such as proteases, amylases, and lipases, which hydrolyze proteins, carbohydrates, and lipids into smaller, more easily assimilable compounds. The breakdown of complex organic matter by bacteria releases simpler organic compounds that can be taken up and assimilated by microalgae, completing the cycle of organic matter degradation and removal (Kumari and Das, 2023).

The efficiency of organic matter degradation and removal in MBSS can be influenced by various factors. One important factor is the availability of light and nutrients (Vergara et al., 2016). Insufficient light or nutrient availability can limit the degradation and assimilation of organic matter, leading to lower removal efficiency.

Another factor that can affect organic matter degradation and removal efficiency is the composition and complexity of the organic matter present in the wastewater (Chan et al., 2022). Different types of organic compounds require specific enzymatic activities for their breakdown (Touliabah et al., 2022). The presence of a diverse microbial community in the symbiotic system increases the likelihood of having a wide range of enzymatic activities, thereby enhancing the degradation of different organic compounds (Blair et al., 2021). However, the efficiency of organic matter degradation can be affected by the presence of inhibitory substances, such as heavy metals or toxic organic compounds, which can inhibit enzyme activity and microbial growth (Zdarta et al., 2022).

The spatial arrangement within the MBSS also plays a role in organic matter degradation and removal efficiency (Fallahi et al., 2021). The formation of biofilms by microalgae and bacteria creates microenvironments with different conditions with different oxygen and nutrient gradients, and light availability, which can influence the growth and activity of microorganisms involved in organic matter degradation (Awdhesh Kumar Mishra and Kodiveri Muthukaliannan, 2022; Wicker et al., 2023). For example, aerobic bacteria that require oxygen for their metabolic activity can be found at the surface of the aggregates or biofilms, where oxygen is more readily available. Anaerobic bacteria that can degrade complex organic matter under oxygen-limited conditions can be found in the deeper layers of the aggregates or biofilms. Fallahi et al. (2021) stated that this spatial arrangement allows for more efficient utilization of oxygen and nutrients and enhances the overall organic matter degradation and removal efficiency.

Therefore, MBSS exhibit efficient organic matter degradation and removal capabilities in WWT. The complementary roles of microalgae and bacteria in the degradation and assimilation of organic matter contribute to the overall efficiency of the system. Factors such as light and nutrient availability, the composition and complexity of organic matter, and the spatial arrangement within the system can influence the organic matter degradation and removal efficiency. Optimizing these factors is crucial for maximizing the performance and efficiency of MBSS in the treatment of organic-rich wastewater. Further research is needed to better understand and optimize these factors and to explore the potential of MBSS in organic matter recovery and resource recycling.

6.3.3 Biomass Production and Resource Recovery

As shown in Figure 6.4, one of the key advantages of MBSS is their ability to produce biomass, which can be further utilized for various purposes, such as bioenergy production, nutraceuticals, pharmaceuticals, biofertilizers, animal feed, and nutrient recovery (Khan et al., 2022). Microalgae are known for their rapid growth rates and high biomass production capabilities. They can convert CO_2 and nutrients into biomass through photosynthesis, utilizing light energy to fix carbon and synthesize organic compounds (Li et al., 2022). In MBSS, the coupling of microalgae with bacteria can further enhance biomass production (González-González and de-Bashan, 2021). Bacteria can provide additional organic carbon sources and nutrients to microalgae through their metabolic activities. For example, bacteria can produce and release organic compounds, such as volatile fatty acids and amino acids, through the degradation of organic matter. These compounds can serve as additional carbon sources for microalgae, promoting their growth and biomass production (Fuentes et al., 2016).

The efficiency of biomass production in MBSS can be influenced by various factors. One important factor is the availability of light and nutrients (Fuentes et al., 2016). Sufficient light energy is crucial for microalgae to carry out photosynthesis and fix CO_2. Adequate nutrient availability (e.g., nitrogen and phosphorus) is essential

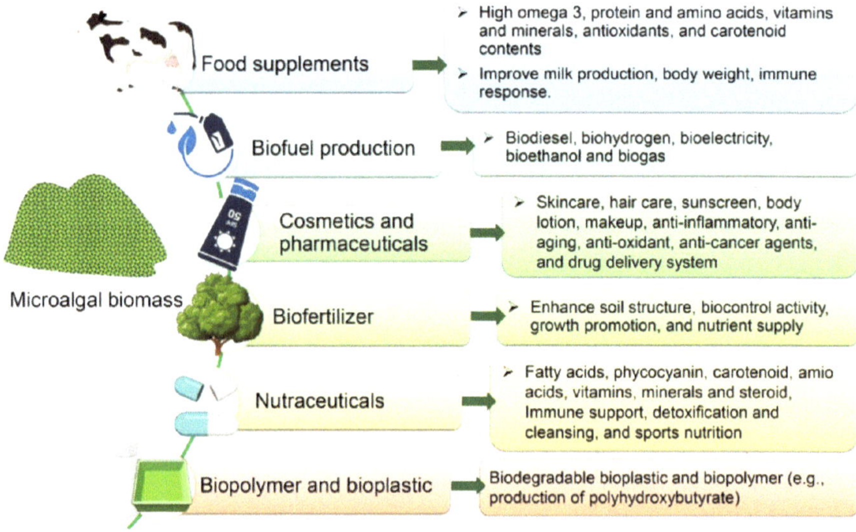

FIGURE 6.4 Microalgal biomass for the production of biofuel and high value-added products.

for the growth and metabolic activity of microalgae and bacteria. Insufficient light or nutrient availability can limit biomass production (Vergara et al., 2016). Another factor that can affect biomass production efficiency is the composition and complexity of the wastewater (Singh and Mishra, 2021). Different types of wastewater can contain different concentrations and compositions of organic matter, nutrients, and other compounds. The presence of inhibitory substances (e.g., heavy metals or toxic organic compounds) can also affect biomass production by inhibiting the growth and metabolic activity of microorganisms (Chan et al., 2022). Therefore, it is crucial to consider the characteristics of the wastewater and optimize the system design and operation accordingly to maximize biomass production.

MBSS not only produce biomass but also have the potential for resource recovery (Moondra et al., 2020; Xu et al., 2021; Yu et al., 2020). The biomass produced in these systems can be utilized for various purposes, such as bioenergy production, animal feed, and nutrient recovery (Ali et al., 2024). Bioenergy production is one of the key applications of microalgae biomass. Microalgae have high lipid content, which can be converted into biodiesel through the process of transesterification (Reetu et al., 2023). The lipids can also be converted into bioenergy (e.g., bioethanol and biogas) through various fermentation processes. Additionally, microalgae biomass can be used as a feedstock for anaerobic digestion, producing biogas rich in methane (Vargas-Estrada et al., 2021).

Microalgae biomass can also be used as a source of animal feed. It contains high levels of proteins, amino acids, and essential fatty acids, making it a valuable nutritional resource for livestock and aquaculture industries (Dineshbabu et al., 2019). The biomass can be processed into different forms, such as pellets or powders, and incorporated into animal feed formulations. Furthermore, MBSS have the potential for nutrient recovery from wastewater (Qu et al., 2021; Qv et al., 2023; Talapatra et al., 2023). The biomass produced in these systems can be rich in nutrients. These nutrients can be recovered through various methods, such as drying and ashing the biomass, or through the extraction of specific compounds (Leng et al., 2021). The recovered nutrients can then be used as fertilizers or recycled back into the WWT process.

Generally, MBSS demonstrate high potential for biomass production and resource recovery. The coupling of microalgae with bacteria enhances biomass production capabilities, and the biomass produced can be utilized for various purposes, such as bioenergy production, animal feed, and nutrient recovery. Factors such as light and nutrient availability, wastewater characteristics, and system design and operation need to be optimized to maximize biomass production and resource recovery. Further research is needed to explore and optimize the potential of MBSS in these applications, as well as to assess the economic feasibility and environmental sustainability of these systems.

6.3.4 Energy Requirements and Carbon Footprint

The performance and efficiency of MBSS are not only evaluated based on their treatment capabilities and biomass production but also on their energy requirements and carbon footprint (Kim et al., 2019). It is crucial to assess the energy inputs and outputs of these systems to determine their sustainability and environmental impact. MBSS require energy inputs for various processes, such as mixing, aeration, lighting, and harvesting. The energy requirements of these systems can vary depending on factors such as the scale of the system, wastewater characteristics, and the desired treatment efficiency (Wang and Hong, 2022). The main energy-consuming processes in these systems are aeration and lighting.

Aeration is essential to provide oxygen for the aerobic bacteria involved in organic matter degradation and nutrient removal. The oxygen transfer is typically achieved through mechanical mixing and the use of diffusers or aerators (Chen et al., 2023). The energy consumption for aeration can be significant, especially in large-scale systems. Various strategies can be employed to optimize aeration efficiency and reduce energy consumption, such as using energy-efficient aerators, optimizing oxygen transfer rates, and implementing process control algorithms (Yusoff et al., 2019).

Lighting is another energy-consuming process in MBSS. Microalgae require light for photosynthesis, which is essential for biomass production and nutrient uptake. Artificial lighting is often used in indoor or enclosed systems to provide the necessary light energy. The energy consumption for lighting can be significant, especially in large-scale systems (Al-Ghaili et al., 2020). To optimize energy efficiency, various lighting technologies can be employed, such as light-emitting diode (LED) lights, which have higher energy efficiency and longer lifespans compared to traditional lighting sources.

The carbon footprint of MBSS is determined by the greenhouse gas emissions associated with their energy requirements, as well as the carbon sequestration potential of the biomass produced (Guo et al., 2023). The main greenhouse gas emissions in these systems come from energy consumption, particularly from electricity generation. Electricity generation in many regions is still heavily reliant on fossil fuels, which results in significant CO_2 emissions (Dasan et al., 2019). Therefore, reducing the energy requirements and increasing the energy efficiency of MBSS can help to reduce their carbon footprint. This can be achieved through the use of energy-efficient equipment, renewable energy sources, and process optimization.

On the other hand, the biomass produced in MBSS can potentially sequester CO_2 from the atmosphere. Microalgae fix CO_2 through photosynthesis, converting it into organic carbon compounds (Dai and Wang, 2024; Li et al., 2022). The biomass can be used for various purposes, such as bioenergy production or as a feedstock for other industries (Khan et al., 2022). If biomass is used for bioenergy production, the CO_2 released during combustion can be considered as part of a closed carbon cycle, as the CO_2 emitted is equal to the amount of CO_2 fixed during photosynthesis (Saleh and Hassan, 2023). This can help to offset the carbon footprint of the system.

Therefore, the energy requirements and carbon footprint of MBSS are important considerations for their sustainability and environmental impact. Strategies to optimize energy efficiency, such as using energy-efficient equipment and renewable energy sources, can help to reduce energy consumption and greenhouse gas emissions. Additionally, the biomass produced in these systems can potentially sequester CO_2, contributing to the offset of greenhouse gas emissions. Further research is needed to assess and optimize the energy requirements and carbon footprint of MBSS, as well as to evaluate their overall sustainability and environmental benefits compared to conventional WWT methods.

6.4 FACTORS INFLUENCING THE PERFORMANCE OF MBSS

6.4.1 Selection of Microalgae and Bacteria Strains

The selection of appropriate microalgae and bacteria strains is crucial for the performance and efficiency of MBSS. The choice of strains can significantly impact the treatment capabilities, biomass production, and overall stability of the system (González-González and de-Bashan, 2021).

6.4.1.1 Microalgae Strains

When selecting microalgae strains, several factors need to be considered. One important factor is the growth rate and biomass productivity of the microalgae. Fast-growing microalgae strains with high biomass productivity are desirable for efficient biomass production (Aditya et al., 2022). The growth rate can be influenced by factors such as light intensity, nutrient availability, and temperature (Borella et al., 2021; Chen et al., 2023). Some commonly used microalgae species with high growth rates include *Chlorella* spp., *Scenedesmus* spp., and *Nannochloropsis* spp. (Metsoviti et al., 2019).

Another factor to consider is the nutrient uptake capability of the microalgae. Different microalgae strains have varying abilities to take up and utilize nutrients, such as nitrogen and phosphorus (Qu et al., 2021; T Sayara et al., 2021; Tahseen Sayara et al., 2021). Nutrient removal is a key function of microalgae in symbiotic systems, as they can uptake and assimilate nutrients from the wastewater, helping to

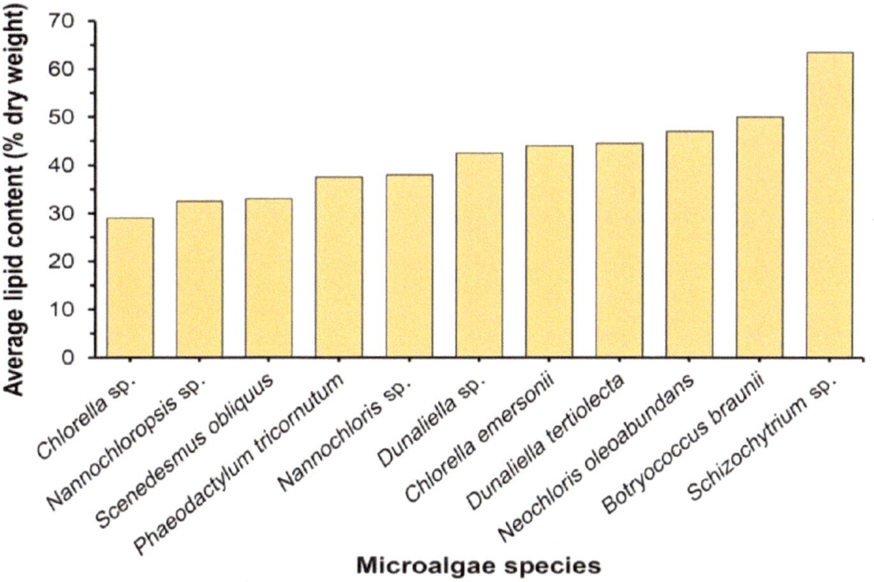

FIGURE 6.5 Lipid contents of different microalgal species.

reduce the nutrient load. Therefore, selecting microalgae strains with high nutrient uptake capabilities can enhance the treatment efficiency of the system.

Furthermore, the lipid content of microalgae is an important consideration, especially if the system aims to produce biomass for bioenergy production (Brindhadevi et al., 2021; Reetu et al., 2023). Microalgae strains with high lipid content are desirable, as lipids can be converted into biodiesel or other biofuels. As shown in Figure 6.5, some microalgae species known for their high lipid content include *Chlorella spp.*, *Nannochloropsis* spp., and *Botryococcus braunii* (Saifullah et al., 2014).

6.4.1.2 Bacteria Strains

The selection of bacteria strains in MBSS is equally important. Bacteria play a crucial role in organic matter degradation, nutrient cycling, and overall system stability (González-González and de-Bashan, 2021). Some factors to consider when selecting bacteria strains include their metabolic activities, growth rates, and compatibility with the selected microalgae strains. Bacteria strains with diverse metabolic capabilities are desirable in symbiotic systems (Ge et al., 2023). This allows for the degradation of a wide range of organic compounds present in the wastewater. For instance, Lutzu et al. (2021) stated that bacteria capable of degrading complex organic matter, such as proteins, cellulose, and fats, can facilitate the breakdown of these compounds into simpler forms that can be readily utilized by microalgae. Bacteria strains known for their metabolic versatility include species from the genera, such as *Bacillus* sp., *Pseudomonas* sp., and *Rhodopseudomonas* sp. (Ge et al., 2023).

The growth rate and biomass production capability of bacteria strains are also important considerations. Bacteria with fast growth rates can help maintain a stable symbiotic relationship with microalgae and prevent the overgrowth of undesirable bacteria (Tong et al., 2023). Additionally, bacteria strains with high biomass production can contribute to the overall biomass productivity of the system. Compatibility between microalgae and bacteria strains is crucial for the establishment and maintenance of a symbiotic relationship (González-González and de-Bashan, 2021). Some microalgae and bacteria strains have been shown to have mutualistic interactions, where the bacteria provide additional nutrients or organic carbon sources to the microalgae, enhancing their growth and biomass production. It is important to select bacteria strains that can coexist and support the growth of the selected microalgae strains without causing any negative effects, such as competition or inhibition (González-González and de-Bashan, 2021).

Hence, the selection of microalgae and bacteria strains is a critical factor in the performance and efficiency of MBSS. The choice of strains should consider factors such as growth rate, biomass productivity, nutrient uptake capability, lipid content, metabolic activities, and compatibility. By selecting appropriate strains, it is possible to optimize the treatment capabilities, biomass production, and overall stability of the system. Further research is needed to explore and identify the most suitable microalgae and bacteria strains for different applications and wastewater types, as well as to understand the mechanisms and interactions underlying the symbiotic relationship between microalgae and bacteria.

6.4.2 Optimization of Operational Parameters

The optimization of operational parameters is a crucial aspect of maximizing the performance and efficiency of MBSS. Various factors, including light, temperature, pH, nutrient concentration, and hydraulic retention time, can significantly influence the growth, metabolism, and overall performance of microalgae and bacteria during the WWT process (Oruganti et al., 2022).

6.4.2.1 Light

Light is a vital parameter in MBSS as it is necessary for photosynthesis, biomass production, and nutrient uptake by microalgae. The intensity, duration, and spectral quality of light can impact the growth rate and lipid content of microalgae (Nwoba et al., 2019). Therefore, Brindhadevi et al. (2021) reported that optimizing light conditions is important to maximize biomass productivity and lipid accumulation. One approach to optimizing light is by selecting the appropriate light source. Natural sunlight is often the most cost-effective option, especially for outdoor systems. However, sunlight availability can be variable, and the quality and intensity of light can vary depending on the geographical location and weather conditions (Peter et al., 2022). In indoor systems, artificial lighting can be used to provide a consistent and controllable light source. LEDs are commonly used in indoor systems due to their energy efficiency, spectral control, and long lifespan (Borella et al., 2021).

Another approach to optimizing light is by controlling the light intensity and duration. The light intensity can be adjusted by varying the distance between the light source and the culture, using shading devices, or by controlling the power input (Nwoba et al., 2019). The duration of light exposure can be controlled by regulating the photoperiod, which is the duration of light and darkness in a day. Different microalgae species have specific light requirements, and optimizing the light intensity and duration based on the specific strain can enhance their growth and performance (Maltsev et al., 2021).

6.4.2.2 Temperature

Temperature is a critical parameter that directly affects the growth, metabolism, and overall performance of microalgae and bacteria in symbiotic systems. Different microalgae and bacteria strains have specific temperature ranges in which they thrive (Xu et al., 2021). Optimizing the temperature conditions can promote the growth and metabolic activities of the microorganisms and maximize the treatment capabilities of the system (Li et al., 2020). The temperature of the system can be controlled through various means, such as adjusting the ambient temperature in outdoor systems or using temperature control systems in indoor systems. It is important to maintain the temperature within the optimal range for both microalgae and bacteria. For example, the optimal temperature for many microalgae species is typically between 20°C and 30°C (Khan et al., 2018), while some bacteria strains may have different temperature requirements. Maintaining a stable temperature is crucial to prevent fluctuations that can negatively impact the performance of the symbiotic system. Temperature fluctuations can lead to stress on the microorganisms, affecting their growth, metabolism, and nutrient uptake. Additionally, extreme temperatures can also promote the growth of undesirable microorganisms, leading to competition or inhibition of the targeted microalgae and bacteria strains.

6.4.2.3 pH Value

pH is an important parameter that influences the growth, metabolism, and nutrient uptake of microalgae and bacteria in symbiotic systems. Different microorganisms have specific pH ranges in which they can thrive (Wang and Hong, 2022). Therefore, optimizing the pH conditions is essential to promote their growth and enhance the treatment capabilities of the system. The pH of the system can be controlled by adding acid or base solutions to adjust the pH or by using pH control systems (Choi et al., 2017). The optimal pH range for most microalgae species is typically between 7 and 9, while bacteria strains may have different pH requirements (Sales and Abreu, 2014). Maintaining the pH within the optimal range is crucial to prevent adverse effects on the microorganisms, such as cell lysis or inhibition.

6.4.2.4 Nutrient Concentration

Optimizing nutrient concentration is essential to ensure that microalgae and bacteria have access to the necessary nutrients for growth and metabolism (Figueroa-Torres et al., 2021). The nutrient requirements vary depending on the strain and the desired treatment objectives. Nutrient concentrations (e.g., N and P) need to be maintained within the appropriate range to avoid nutrient limitation or excessive nutrient concentrations that can lead to eutrophication (Tiwari and Pal, 2022). The nutrient concentration can be controlled by adjusting the influent wastewater composition or by adding nutrient solutions to the system. Monitoring and controlling the nutrient concentration is crucial to prevent imbalances that can negatively impact the growth and performance of the microorganisms (Magwaza et al., 2020). Additionally, nutrient cycling within the system, such as the release of nutrients from biomass during harvesting and recycling, can also contribute to optimizing nutrient availability and reducing nutrient requirements (Farooq, 2021).

6.4.2.5 Hydraulic Retention Time

The hydraulic retention time (HRT), which is the time that the wastewater spends in the system, is an important operational parameter that influences the treatment capabilities of MBSS (Ozcelik et al., 2022). The HRT affects the contact time between the microorganisms and the wastewater, allowing for sufficient nutrient uptake, organic matter degradation, and biomass production (Zhang et al., 2020). The HRT needs to be optimized based on the specific wastewater characteristics, treatment objectives, and desired treatment efficiency (Barbera et al., 2022). A longer HRT allows for more contact time and can enhance nutrient removal and biomass production. However, a longer HRT can also lead to increased reactor volume and higher energy requirements. On the other hand, a shorter HRT may result in lower treatment efficiency and biomass productivity. Therefore, finding the right balance between treatment performance and system operational requirements is crucial.

6.4.2.6 Other Operational Parameters

In addition to the parameters discussed above, several other operational parameters can influence the performance of MBSS. These include dissolved oxygen concentration, mixing or agitation, CO_2 concentration, and trace element availability (Yang and Xin, 2023). Maintaining an adequate dissolved oxygen concentration is essential for the aerobic metabolism of both microalgae and bacteria (Sánchez Zurano et al., 2020). Insufficient oxygen levels can lead to anaerobic conditions, affecting the treatment efficiency and promoting the growth of anaerobic microorganisms. Mixing or agitation is often necessary to ensure proper oxygen transfer and to prevent sedimentation or settling of microalgae and bacteria (Banerjee and Mukhopadhyay, 2021). CO_2 concentration is another parameter that needs to be optimized, especially for microalgae. CO_2 is a crucial substrate for photosynthesis, and an adequate supply of CO_2 can enhance the growth and biomass production of microalgae. CO_2 can be supplied through various means, such as direct injection, sparging, or utilizing flue gases from industrial processes (Nguyen et al., 2022). Trace elements, such as iron, manganese, and zinc, are essential micronutrients for microorganisms. Their availability in the system needs to be optimized to ensure proper growth and metabolic activities. Some trace elements can be supplemented through nutrient solutions or by utilizing trace element-rich wastewater sources (Abdelfattah et al., 2023).

Therefore, the optimization of operational parameters is crucial for maximizing the performance and efficiency of MBSS. Parameters such as light, temperature, pH, nutrient concentration, HRT, dissolved oxygen concentration, mixing or agitation, CO_2 concentration, and trace element availability need to be carefully considered and controlled. By optimizing these parameters, it is possible to enhance the growth, metabolism, nutrient uptake, biomass production, and overall treatment capabilities of microalgae and bacteria in symbiotic systems. Further research is needed to explore and understand the interactions and synergies between these parameters and their effects on the performance of MBSS.

6.4.3 Nutrient Availability and Concentration

Nutrient availability is a crucial factor that directly impacts the performance and efficiency of MBSS. Proper nutrient dosing strategies and the availability of essential nutrients play a vital role in the growth, metabolism, and overall performance of microalgae and bacteria (Fallahi et al., 2021). Microalgae and bacteria require essential nutrients, including nitrogen (N), phosphorus (P), and trace elements, for their growth and metabolic activities (Barbera et al., 2022; Farooq, 2021; Figueroa-Torres et al., 2021). The availability of these nutrients in the system is crucial to ensure optimal performance. Insufficient nutrient availability can limit growth and biomass production, while excessive nutrient concentrations can lead to eutrophication and undesirable algal blooms (Tiwari and Pal, 2022). The primary sources of nutrients in MBSS are the influent wastewater and biomass recycling. Wastewater contains varying concentrations of nutrients, depending on its origin and composition (Gupta et al., 2019). However, the nutrient composition may not always be in the optimal range for microalgae and bacteria growth. Therefore, it is often necessary to optimize nutrient availability through dosing strategies and nutrient supplementation.

Nutrient concentration involves the controlled addition of nutrients to the system to maintain optimal nutrient concentrations. Several dosing strategies can be employed to enhance nutrient availability in MBSS (Sahu et al., 2023). These strategies include batch culture, continuous culture, and feedback control systems. In batch culture, nutrients are added to the system in a single dose or multiple doses at specific time intervals. This strategy is commonly used when there is limited control over the nutrient composition of the influent wastewater or when the wastewater has low nutrient concentrations (Liyanaarachchi et al., 2021). Batch culture allows for the supplementation of nutrients to ensure that the microorganisms have access to the required nutrients for growth and metabolism (Di Caprio et al., 2021). However, this strategy may result in fluctuations in nutrient concentrations, and careful monitoring is necessary to prevent nutrient imbalances.

Continuous culture involves the continuous addition of nutrients to the system at a constant rate. This loading strategy ensures a more stable nutrient supply, reducing fluctuations in nutrient concentrations (Han et al., 2022). Continuous loading is often used when the wastewater has relatively stable nutrient concentrations or when nutrient uptake rates by microalgae and bacteria are well understood (Fallahi et al., 2021). This strategy allows for the maintenance of nutrient concentrations within the optimal range, promoting consistent growth and performance of the microorganisms. Feedback control systems involve monitoring the nutrient concentrations in the system and adjusting the nutrient dosing rates based on the measured values (Wolf et al., 2023). This strategy enables real-time adjustments in nutrient addition, ensuring that the microorganisms have access to nutrients in the required amounts. Feedback control systems can be automated, utilizing sensors and controllers to monitor and regulate nutrient concentrations (Nwoba et al., 2022). This strategy allows for precise control over nutrient availability, optimizing the growth and performance of microalgae and bacteria.

The nutrient concentration in MBSS needs to consider several factors, including the nutrient requirements of the microorganisms, the nutrient composition of the influent wastewater, and the desired treatment objectives (Fallahi et al., 2021).

The nutrient loading rate and frequency need to be optimized to ensure sufficient nutrient availability without causing nutrient imbalances or excessive nutrient concentrations (Tiwari and Pal, 2022). The nutrient loading rate should be determined based on the specific nutrient requirements of the microorganisms. Different microalgae and bacteria strains have varying nutrient uptake rates and preferences (Díaz et al., 2022). It is essential to understand the nutrient uptake dynamics of the selected strains to determine the appropriate loading rate. Nutrient uptake rates can be influenced by factors such as light intensity, temperature, pH, and the presence of other nutrients (Goswami et al., 2022). Therefore, optimizing these parameters in conjunction with nutrient dosing is crucial for maximizing nutrient uptake and utilization.

The nutrient concentration and loading need to be determined based on the nutrient consumption rate of the microorganisms and the nutrient retention time in the system (Solmaz and Işık, 2020). Microalgae and bacteria can rapidly take up and utilize nutrients, especially during periods of active growth. Therefore, frequent dosing may be necessary to maintain nutrient concentrations within the optimal range. However, dosing frequency should also consider the nutrient retention time in the system to prevent excessive nutrient accumulation or wastage. Hence, nutrient availability and dosing strategies are crucial factors that influence the performance and efficiency of MBSS. Proper nutrient availability is necessary to support the growth, metabolism, and overall performance of microalgae and bacteria. Nutrient dosing strategies, such as batch dosing, continuous dosing, and feedback control systems, are employed to optimize nutrient uptake and utilization.

The availability of essential nutrients, including nitrogen, phosphorus, and trace elements, is critical for the growth and metabolic activities of microalgae and bacteria (Nagarajan et al., 2022). While wastewater is a primary source of nutrients in these systems, the nutrient composition may not always be within the optimal range. Insufficient nutrient availability can limit growth and biomass production, while excessive nutrient concentrations can lead to eutrophication (Tiwari and Pal, 2022). Therefore, it is necessary to optimize nutrient availability through dosing strategies and nutrient supplementation. Batch culture involves adding nutrients to the system in a single dose or multiple doses at specific time intervals. This strategy is used when there is limited control over the nutrient composition of the influent wastewater or when the nutrient concentrations are low. Batch culture allows for the supplementation of nutrients to ensure microorganisms have access to the required nutrients. However, careful monitoring is necessary to prevent nutrient imbalances (Havlik et al., 2015).

On the other hand, continuous culture involves the continuous addition of nutrients to the system at a constant rate. Benvenuti et al. (2016) stated that this strategy provides a more stable nutrient supply, reducing fluctuations in nutrient concentrations. It is often used when the wastewater has relatively stable nutrient concentrations or when nutrient uptake rates are well understood. Continuous culture ensures that nutrient concentrations remain within the optimal range, promoting consistent growth and performance of microorganisms (Tavares et al., 2023). Feedback control systems are another dosing strategy that monitors nutrient concentrations in real time and adjusts dosing rates accordingly. This strategy allows for precise control over nutrient availability, optimizing the growth and performance of microalgae and bacteria. By utilizing sensors and controllers, feedback control systems automate the adjustment of nutrient addition based on measured values (Wolf et al., 2023).

When dosing nutrients, several factors need to be considered, including the nutrient requirements of the microorganisms, the nutrient composition of the influent wastewater, and the treatment objectives (Fallahi et al., 2021; Zetterlund et al., 2023). The dosing rate should be determined based on the specific nutrient requirements of the microorganisms. Different strains have varying nutrient uptake rates and preferences, and understanding these dynamics is crucial for determining the appropriate dosing rate (Zhang et al., 2020). Therefore, frequent dosing may be necessary to maintain nutrient concentrations within the optimal range. However, the nutrient retention time should also be considered to prevent excessive accumulation or wastage. Whitton et al. (2016) stated that the nutrient composition of the influent wastewater should be considered when determining the dosing strategy. Raper et al. (2018) reported that the wastewater may contain varying concentrations of nutrients, and these should be taken into account to avoid over- or under-dosing. Nutrient analysis of the wastewater can help identify any nutrient deficiencies or excesses and guide the nutrient dosing strategy. Furthermore, consideration should be given to the stoichiometric ratios of nutrients, such as the nitrogen to phosphorus (N:P) ratio. Microalgae and bacteria have specific N:P ratio requirements for optimal growth and biomass production (Arcila et al., 2021). Maintaining the N:P ratio within the appropriate range can prevent nutrient limitation or imbalance. Adjusting nutrient loading rates and ratios can help maintain the desired N:P ratio and promote balanced growth.

Thus, nutrient availability and concentration strategies are essential for optimizing the performance of MBSS. Proper nutrient availability and concentration ensure that microorganisms have access to essential nutrients without causing imbalances or excessive concentrations. By employing batch culture, continuous culture, or feedback control systems, nutrient uptake and utilization can be optimized, leading to improved growth and performance in these systems.

6.4.4 Design Considerations for Reactor Configuration

The design of the reactor configuration plays a crucial role in the performance and efficiency of MBSS. The reactor configuration determines the spatial arrangement and interaction between microalgae and bacteria, as well as the mass transfer of nutrients, gases, and light (Fallahi et al., 2021). Hence, it is important to discuss the key design considerations for reactor configuration, including the choice of reactor type, reactor geometry, and mixing and aeration strategies.

6.4.4.1 Reactor Type

Several types of reactors can be used for MBSS, including open ponds, raceway ponds, photobioreactors (PBRs), and membrane-based systems (Nagarajan et al., 2023). The selection of the reactor type depends on various factors, such as the scale of operation, available space, and desired treatment objectives (Sathinathan et al., 2023). Open ponds are the simplest and most cost-effective reactor type. They are shallow, earthen basins that allow for the cultivation of microalgae and bacteria in natural sunlight. Open ponds have large surface areas, which facilitate the absorption of sunlight and promote photosynthesis. However, they have limited control over environmental conditions, such as temperature and light intensity, and are susceptible to contamination and evaporation (Qin et al., 2019).

Raceway ponds are an improved version of open ponds that feature a circular or oval shape with a paddlewheel or other flow-inducing devices. These ponds provide better mixing and circulation of the culture medium, enhancing mass transfer and reducing light attenuation. Raceway ponds offer higher productivity and efficiency compared to open ponds but still have limited control over environmental conditions (Neethu et al., 2020). However, PBRs are closed systems that provide better control over environmental conditions, such as light intensity, temperature, and pH. PBRs can be tubular, flat-panel, or airlift reactors, and they are often equipped with artificial lighting systems to ensure a consistent light supply (Banerjee and Mukhopadhyay, 2021; Qin et al., 2019). PBRs offer higher biomass productivity and better scalability compared to open ponds and raceway ponds. However, they are more expensive to construct and operate (Banerjee and Mukhopadhyay, 2021).

Membrane-based systems, such as membrane photobioreactors (MPBRs) and membrane bioreactors (MBRs), utilize membranes to separate microalgae and bacteria from the culture medium (Luo et al., 2019). Membrane-based systems offer better control over biomass retention and provide higher biomass concentrations. They also allow for the recovery of biomass and the reuse of culture medium, reducing water and nutrient requirements. However, membrane fouling and cleaning can be significant challenges in these systems (Castro-Muñoz and García-Depraect, 2021).

6.4.4.2 Reactor Geometry

The geometry of the reactor plays a crucial role in the light distribution, mixing, and mass transfer in MBSS. The choice of reactor geometry depends on various factors, including the reactor type, biomass concentration, and the desired treatment objectives (Legrand et al., 2021). In open ponds and raceway ponds, the geometry is typically rectangular or circular. Rectangular ponds provide a larger surface area for light absorption and better mixing (Molinuevo-Salces et al., 2019). However, circular ponds offer better hydraulic flow characteristics, minimizing dead zones and improving mass transfer (Rayen et al., 2019). The choice of geometry should consider the balance between light absorption, mixing, and mass transfer efficiency. In PBRs, the reactor geometry can be tubular, flat panel, or airlift. Tubular reactors are cylindrical and provide a high surface-to-volume ratio, allowing for efficient light absorption and gas exchange (Banerjee and Mukhopadhyay, 2021; Qin et al., 2019). Flat-panel reactors consist of transparent panels arranged in a parallel or inclined configuration, maximizing light penetration. Airlift reactors utilize air bubbles to circulate the culture medium, enhancing mixing and gas exchange (Carvalho et al., 2006). The choice of PBR geometry depends on factors such as light availability, biomass concentration, and the desired level of control over environmental conditions.

6.4.4.3 Mixing and Aeration Strategies

Proper mixing and aeration are essential for maintaining homogeneity in MBSS. Mixing promotes the distribution of nutrients, gases, and light, ensuring equal exposure for microalgae and bacteria. Aeration enhances gas exchange, providing oxygen for respiration and CO_2 removal (Chen et al., 2023). In open ponds and raceway ponds, mixing is primarily achieved through paddlewheels or other flow-inducing devices (Saxena et al., 2023). These devices create flow patterns that circulate the culture medium, minimizing the formation of dead zones and promoting mass transfer. Aeration in open ponds and raceway ponds is usually passive, relying on natural air-water interfaces and wind-induced turbulence (Sur et al., 2016).

In PBRs, mixing is typically achieved using mechanical agitation, such as the rotation of tubular reactors or the recirculation of the culture medium in flat-panel reactors (Sirohi et al., 2022). Mechanical agitation ensures homogeneity and improves mass transfer. Aeration in PBRs can be achieved through the injection of air or CO_2, depending on the requirements of microalgae and bacteria (Chanquia et al., 2021). Aeration systems should be designed to provide sufficient oxygen supply and prevent oxygen limitation. Additionally, membrane-based systems rely on the pumping of the culture medium through membranes to achieve mixing (Luo et al., 2019). The pressure generated during membrane filtration helps in the circulation of the culture medium, promoting mass transfer and preventing biomass settling. Aeration in membrane-based systems can be achieved through the injection of air or oxygen into the culture medium or by utilizing membrane aerators that introduce air directly into the culture medium through the membranes (Castro-Muñoz and García-Depraect, 2021). The choice of mixing and aeration strategies depend on the specific requirements of the microalgae and bacteria, as well as the characteristics of the reactor type and geometry.

Hence, the design of the reactor configuration is a critical factor in the performance of MBSS. The selection of the reactor type, such as open ponds, raceway ponds, PBRs, or membrane-based systems, depends on factors such as scale, available space, and treatment objectives. The reactor geometry, whether rectangular, circular, tubular, flat panel, or airlift, should be chosen to optimize light distribution, mixing, and mass transfer efficiency. Additionally, the mixing and aeration strategies, whether through paddlewheels, mechanical agitation, or membrane filtration, should be designed to ensure homogeneity, proper nutrient and gas distribution, and sufficient oxygen supply. By carefully considering these design considerations, the performance and efficiency of MBSS can be significantly improved, leading to enhanced biomass productivity and WWT capabilities.

6.5 CHALLENGES AND LIMITATIONS

6.5.1 Contaminant Tolerance and Inhibitory Substances

One of the challenges in MBSS is the tolerance of the organisms to various contaminants present in wastewater. Microalgae and bacteria are known to have different sensitivities to different contaminants, and their growth can be inhibited or even completely halted under certain conditions (Molina et al., 2019). It is crucial to understand the specific tolerance limits and sensitivities of the microalgae and bacteria species used in the symbiotic system to optimize their performance.

Contaminants commonly found in wastewater include heavy metals, organic pollutants, and nutrients such as nitrogen and phosphorus. Heavy metals, such as lead, cadmium, and mercury, can be toxic to microalgae and bacteria, affecting their growth and metabolic processes (Manikandan et al., 2021). Organic pollutants, such as phenols, pesticides, and hydrocarbons, can also have inhibitory effects on microalgae and bacteria (Touliabah et al., 2022). Additionally, high concentrations of nutrients, particularly nitrogen and phosphorus, can lead to eutrophication and disrupt the balance of the symbiotic system (Tiwari and Pal, 2022).

To overcome these challenges, several strategies can be employed. First, the selection of microalgae and bacteria species with high tolerance to specific contaminants can be advantageous. Screening and selecting strains that are naturally resistant or can metabolize and remove specific contaminants can enhance the system's performance (González-González and de-Bashan, 2021). Genetic engineering techniques can also be utilized to enhance the tolerance of microalgae and bacteria to contaminants (Ranjbar and Malcata, 2022).

Another approach is the use of pre-treatment methods to remove or reduce the contaminants before they enter the symbiotic system (Rąza et al., 2022). Physical, chemical, and biological pre-treatment methods, such as filtration, coagulation, adsorption, and activated carbon, can be employed to remove or reduce contaminants (Rohers et al., 2021). This reduces the load on the microalgae and bacteria, allowing them to thrive in a less inhibitory environment.

6.5.2 Biomass Harvesting and Dewatering Methods

Biomass harvesting and dewatering are critical steps in MBSS, as the harvested biomass can be used for various applications, including biofuel production, animal feed, and fertilizer (Ali et al., 2016; Khan et al., 2022). However, Barros et al. (2015) stated that the efficient and cost-effective harvesting and dewatering of microalgae and bacteria are challenging due to their small size and low settling velocities. Microalgae are typically harvested using physical, chemical, or biological methods (Li et al., 2020; Moreno Osorio et al., 2021; Najjar and Abu-Shamleh, 2020); as shown in Table 6.2, each technique has its advantages and disadvantages.

Physical methods include sedimentation, centrifugation, and filtration. Sedimentation is the simplest and least expensive method but is only suitable for larger microalgae species with higher settling velocities (Najjar and Abu-Shamleh, 2020). Centrifugation offers higher separation efficiency but is energy intensive and expensive. Filtration methods (e.g., microfiltration and ultrafiltration) can achieve high separation efficiencies but are limited by membrane fouling and require frequent cleaning (Ennaceri et al., 2022), while chemical methods (e.g., flocculation and coagulation) can enhance the settling of microalgae by aggregating them into larger flocs. Various chemicals, (e.g., polymers, aluminum salts, and cationic surfactants) can be used as flocculants and coagulants. However, the use of chemicals may introduce additional costs and environmental concerns (Li et al., 2020). However, biological methods (e.g., bioflocculation and biofilm formation) utilize the natural abilities of microalgae and bacteria to form aggregates. Bioflocculation involves the cultivation of specific microalgae or bacteria strains that can produce extracellular polymeric substances (EPS) to promote flocculation (Choi et al., 2020; Mao et al., 2019). Biofilm formation involves the attachment of microalgae and bacteria to a solid support, such as a mesh or a membrane, forming a biofilm that can be easily harvested (Moreno Osorio et al., 2021).

Dewatering of the harvested biomass is also a crucial step in the process. Dewatering methods include mechanical dewatering, such as centrifugation and

TABLE 6.2
Advantages and Disadvantages of Different Physical, Chemical, and Biological Harvesting Techniques

Method Category	Harvesting Technique	Principle	Advantages	Disadvantages
Physical	Sedimentation	Gravity settling of microalgae	• Simple and cost-effective: Does not require specialized equipment or chemicals • Suitable for high-density microalgae with relatively large cell size • Can be combined with other methods for improved efficiency	• Requires long settling time, which may lead to increased water loss and contamination • Lower separation efficiency for low-density and small-sized microalgae • Difficult to scale up for large-scale production
Physical	Centrifugation	Centrifugal force to separate microalgae	• High separation efficiency: Can achieve rapid and effective separation • Versatile: Can handle a wide range of microalgal species and biomass concentrations • Allows for control over separation parameters	• Energy-intensive and expensive: Requires significant energy input • Requires specialized equipment and skilled operation • Maintenance and operational costs are relatively high
Physical	Filtration	Sieving or membrane filtration	• High biomass recovery rate: Can achieve efficient separation • Scalable: Can be used for both small and large-scale operations • Can be combined with other methods for improved efficiency	• Clogging issues with small microalgae: Fine meshes or small pore size may lead to blockages • Membrane fouling and reduced permeability over time • Additional pre-treatment steps may be required to remove debris and prevent fouling
Physical	Flotation	Air bubble attachment to microalgae	• Fast separation and high biomass recovery rate • Effective for a wide range of microalgal species • Can be integrated with other methods for improved efficiency	• Chemicals required for bubble generation: May introduce additional costs and environmental concerns • Requires careful control of operating parameters to avoid cell damage • Lower separation efficiency for low-density microalgae

(Continued)

TABLE 6.2 (*Continued*)
Advantages and Disadvantages of Different Physical, Chemical, and Biological Harvesting Techniques

Method Category	Harvesting Technique	Principle	Advantages	Disadvantages
Chemical	Flocculation	Aggregation of microalgae using chemicals	• High separation efficiency: Allows for rapid sedimentation or flotation • Effective for a wide range of microalgal species • Can be combined with other methods for improved efficiency	• Chemical addition and potential environmental impact: Requires the use of flocculants/coagulants • Additional steps for chemical dosing, mixing, and pH adjustment • Residual chemicals may require treatment or disposal
Chemical	Coagulation	Formation of larger flocs using chemicals	• Effective for small-sized microalgae: Enhances their removal efficiency • Can be combined with other methods for improved efficiency • Can remove other impurities like suspended solids and metals	• Chemical addition and potential environmental impact: Requires the use of coagulants • Additional steps for chemical dosing, mixing, and pH adjustment • Residual chemicals may require treatment or disposal
Biological	Bioflocculation	Microbial-induced aggregation of microalgae	• Environmentally friendly: Relies on microbial activity and natural processes • Can be applied to a wide range of microalgal species • Potential for simultaneous wastewater treatment and biomass harvesting	• Longer harvesting time compared to other methods: Requires cultivation of flocculating microorganisms • More suitable for larger-scale operations due to longer settling times • Process optimization and control are challenging
Biological	Biofilm Formation	Microalgae attachment to solid surfaces	• Continuous operation and high biomass yield: Provides a stable and continuous harvest • Can be integrated with various substrates (e.g., textiles, meshes) for biofilm growth	• Requires additional infrastructure for biofilm growth: Substrate materials, maintenance, and cleaning • Potential biofilm detachment and subsequent contamination • Limited to specific microalgal strains that exhibit attachment properties

pressing, and thermal drying methods, such as spray drying and drum drying. Mechanical dewatering methods are effective but require high energy input and can damage the biomass. Thermal drying methods are energy intensive but can achieve high moisture removal and produce a stable dry biomass (Min et al., 2022; Muhammad et al., 2020).

In recent years, novel technologies such as electroflocculation, acoustic separation, and microbubble-based flotation have been explored for biomass harvesting and dewatering (Ghernaout, 2019; Gilmour and Zimmerman, 2020). These technologies show promising results in terms of efficiency, energy consumption, and product quality. However, further research and development are needed to optimize these techniques for large-scale applications in MBSS.

6.5.3 Scale-Up and Implementation on a Larger Scale

One of the challenges in MBSS is the successful scale-up and implementation on a larger scale (Zhang et al., 2020). Many studies and pilot projects have demonstrated the feasibility and potential of these systems at a laboratory or small-scale level. However, scaling up these systems to a commercial scale presents several technical, operational, and economic challenges. At larger scales, maintaining the stability and performance of the MBSS becomes more challenging (Zhang et al., 2020). Chang et al. (2017) stated that factors such as light distribution, mixing efficiency, and mass transfer become more complex and require careful engineering and design considerations. The choice of reactor type, geometry, and mixing and aeration strategies becomes critical in ensuring the success of the system.

Another challenge in scaling up MBSS is the availability and sourcing of suitable microalgae and bacteria strains. Identifying and cultivating strains that are highly productive, adaptable to varying environmental conditions, and resistant to contaminants can be a time-consuming and labor-intensive process (Zhang et al., 2020). Furthermore, ensuring the consistency and quality of the strains at a larger scale can be challenging.

Operational challenges also arise when scaling up MBSS. The management and control of environmental parameters, such as temperature, light intensity, and nutrient concentrations, become more complex. Monitoring and maintaining optimal conditions throughout the system requires advanced sensing and automation technologies (Sahu et al., 2023). Additionally, the handling and processing of larger volumes of wastewater and biomass present logistical challenges that need to be addressed.

Economic feasibility and cost analysis are fundamental considerations when scaling up MBSS. The capital and operational costs associated with constructing and operating large-scale systems can be significant. Factors such as land requirements, energy consumption, labor costs, and maintenance expenses need to be carefully evaluated to ensure the economic viability of the system (Yang et al., 2020). Cost-effective strategies, such as utilizing waste heat or integrating the system with other industrial processes, should be explored to improve the economic feasibility.

To overcome these challenges, a holistic and integrated approach is needed. Collaboration between researchers, engineers, and industry stakeholders is essential to exchange knowledge, share experiences, and develop innovative solutions.

Pilot projects and demonstration plants can provide valuable insights and data for optimizing the system design, operation, and economic feasibility (Hellsmark et al., 2016). Additionally, continuous research and development efforts are needed to improve the understanding of microalgae-bacteria interactions, develop new technologies, and enhance the overall performance and efficiency of MBSS.

Hence, scaling up and implementing MBSS on a larger scale presents several challenges. Understanding the tolerance of microalgae and bacteria to contaminants, developing efficient biomass harvesting and dewatering methods, addressing scale-up and implementation challenges, and conducting economic feasibility and cost analysis are crucial considerations. Overcoming these challenges requires a multidisciplinary approach, collaboration among stakeholders, and continuous research and development efforts. By addressing these challenges, MBSS have the potential to become a sustainable and cost-effective solution for WWT and biomass production.

6.6 FUTURE PERSPECTIVES AND RESEARCH DIRECTIONS

6.6.1 Genetic Engineering and Strain Improvement

Genetic engineering holds great potential for improving the performance and resilience of microalgae and bacteria in symbiotic systems (Li et al., 2022). By manipulating the genes responsible for tolerance to contaminants, nutrient uptake, and biomass productivity, researchers can develop strains that are more robust and efficient in WWT and biomass production (Ali et al., 2022b). Genetic engineering can be used to enhance the tolerance of microalgae and bacteria to specific contaminants present in wastewater (El-Sheekh et al., 2021). This can be achieved by introducing genes encoding detoxification enzymes or transporters that enable the organisms to metabolize or sequester the contaminants more effectively. By selecting and engineering strains with improved tolerance, the inhibitory effects of contaminants can be minimized, leading to better overall system performance.

Another area of focus is enhancing the nutrient uptake and utilization capabilities of microalgae and bacteria. By engineering the genes involved in nutrient assimilation and metabolism (Sun et al., 2019), researchers can develop strains that are more efficient in removing nitrogen and phosphorus from wastewater. This can help reduce the eutrophication potential of the treated effluent and improve the nutrient removal efficiency of the system. Furthermore, genetic engineering can be used to improve biomass productivity and lipid accumulation in microalgae. By manipulating the genes responsible for photosynthesis, carbon fixation, and lipid synthesis, researchers can develop strains that have higher biomass and lipid yields. This can enhance the economic feasibility of MBSS for biofuel production and other value-added applications.

6.6.2 Integration with Other Treatment Technologies

To further enhance the efficiency and effectiveness of MBSS, integration with other treatment technologies should be explored. Combining symbiotic systems with conventional WWT processes, such as activated sludge and anaerobic digestion, can lead to synergistic effects and improved overall performance. Activated sludge processes can be integrated with MBSS to enhance nutrient removal and reduce the energy consumption of aeration. The microalgae can utilize the excess nutrients in the activated sludge effluent, reducing the nutrient load and improving the water quality. Additionally, the oxygen produced by the microalgae during photosynthesis can supplement the oxygen requirements of the aerobic bacteria in the activated sludge process, reducing the need for mechanical aeration.

Anaerobic digestion is another technology that can be integrated with MBSS (Yan et al., 2022). The biomass produced in the symbiotic system can be used as a substrate for anaerobic digestion, generating biogas and further reducing the organic load of the wastewater. The anaerobic digestion process can also help in the removal of residual contaminants and pathogens, enhancing the overall treatment efficiency. The integration of MBSS with other treatment technologies requires careful design and optimization to ensure compatibility and maximize the benefits. It is important to consider factors such as reactor configuration, HRT, and nutrient cycling to achieve optimal performance and synergy between different processes.

6.6.3 Techno-Economic Analysis and Life Cycle Assessment

Techno-economic analysis and life cycle assessment (LCA) are essential tools for evaluating the economic feasibility and environmental sustainability of MBSS. These analyses provide insights into the costs, benefits, and potential environmental impacts associated with the implementation of the technology. The techno-economic analysis involves assessing the capital and operational costs, revenue streams, and profitability of the system. It considers factors such as equipment costs, energy consumption, labor costs, and maintenance expenses. By quantifying these costs and revenue streams, techno-economic analysis can identify the key drivers of the system's economics and help optimize the design and operation to improve cost effectiveness.

LCA is a comprehensive tool that evaluates the environmental impacts of a technology throughout its life cycle, from raw material extraction to end-of-life disposal. LCA takes into account various environmental indicators, such as greenhouse gas emissions, energy consumption, water use, and land occupation (Ali et al., 2024). By conducting LCA, the environmental hotspots and trade-offs of MBSS can be identified, and strategies for improving environmental performance can be developed. The integration of techno-economic analysis and LCA can provide a holistic understanding of the economic and environmental implications of MBSS. It can guide decision-making processes, help identify cost-saving opportunities, and inform policy and investment decisions. Additionally, techno-economic analysis and LCA can facilitate comparisons between different WWT technologies and support the development of sustainable and economically viable solutions.

6.6.4 Regulatory Framework and Policy Implications

The successful implementation of MBSS on a larger scale requires a supportive regulatory framework and policy environment (Sahu et al., 2023). Regulations and policies play a crucial role in facilitating technology adoption, ensuring environmental protection, and promoting market development. Regulatory frameworks need to be developed to address the potential risks and benefits associated with MBSS. Clear guidelines and standards should be established for the treatment and use of treated wastewater, biomass products, and biofuels. Regulatory agencies need to consider factors such as water quality standards, nutrient management, and the potential release of genetically modified organisms into the environment.

Policies and incentives can also play a significant role in promoting the adoption of MBSS. Governments can provide financial incentives, such as grants, tax credits, and subsidies, to support research and development, pilot projects, and commercial deployment. Additionally, policies that promote the use of renewable energy and biofuels can create market demand for the biomass products generated from symbiotic systems. Furthermore, public awareness and education are crucial in gaining public acceptance and support for MBSS. Governments and research institutions should invest in outreach and education programs to inform the public about the benefits, safety, and potential applications of the technology. This can help address concerns and misconceptions and build trust in the technology. Collaboration between stakeholders, including researchers, regulators, policymakers, industry representatives, and environmental groups, is essential in shaping the regulatory framework and policy environment. Regular dialog and knowledge exchange can help identify challenges, explore solutions, and ensure that the regulations and policies are practical, effective, and aligned with the goals of sustainable development.

Therefore, future research should focus on genetic engineering and strain improvement to enhance the performance and resilience of microalgae and bacteria in symbiotic systems. Integration with other treatment technologies, such as activated sludge and anaerobic digestion, should be explored to achieve synergistic effects and improve overall performance. Techno-economic analysis and LCA are necessary to evaluate the economic feasibility and environmental sustainability of MBSS. Finally, a supportive regulatory framework and policy environment are essential for the successful implementation and widespread adoption of the technology.

6.7 CONCLUSION

MBSS offer numerous advantages for WWT and biomass production. They excel in efficient nutrient removal, biomass production, and the potential for biofuel generation. The symbiotic relationship between microalgae and bacteria enhances overall system performance, while nutrient removal reduces eutrophication potential. However, challenges remain, such as the hindrance caused by contaminants in wastewater and the need for scalability. Strategies like genetic engineering and strain improvement can enhance tolerance and resilience. Careful design and optimization are required for scaling up to commercial operations. Future research should focus on addressing these challenges and exploring potential applications of MBSS. Genetic engineering, integration with other treatment technologies, and techno-economic analysis are essential for enhancing tolerance, achieving synergistic effects, evaluating feasibility, and supporting sustainable solutions. MBSS hold significant potential for sustainable WWT and biomass production. With further research and development, they can contribute to a more sustainable and circular approach to wastewater management, reducing environmental impact and providing valuable resources for various applications.

REFERENCES

Abdelfattah, A., Ali, S.S., Ramadan, H., El-Aswar, E.I., Eltawab, R., Ho, S.-H., Elsamahy, T., Li, S., El-Sheekh, M.M., Schagerl, M., Kornaros, M., Sun, J., 2023. Microalgae-based wastewater treatment: Mechanisms, challenges, recent advances, and future prospects. *Environ. Sci. Ecotechnol.* 13, 100205. https://doi.org/10.1016/j.ese.2022.100205

Aditya, L., Mahlia, T.M.I., Nguyen, L.N., Vu, H.P., Nghiem, L.D., 2022. Microalgae-bacteria consortium for wastewater treatment and biomass production. *Sci. Total Environ.* 838, 155871. https://doi.org/10.1016/j.scitotenv.2022.155871

Ahmed, S.F., Mofijur, M., Nuzhat, S., Chowdhury, A.T., Rafa, N., Uddin, M.A., Inayat, A., Mahlia, T.M.I., Ong, H.C., Chia, W.Y., Show, P.L., 2021. Recent developments in physical, biological, chemical, and hybrid treatment techniques for removing emerging contaminants from wastewater. *J. Hazard. Mater.* 416, 125912. https://doi.org/10.1016/j.jhazmat.2021.125912

Ahmed, S.F., Mofijur, M., Parisa, T.A., Islam, N., Kusumo, F., Inayat, A., Le, V.G., Badruddin, I.A., Khan, T.M.Y., Ong, H.C., 2022. Progress and challenges of contaminate removal from wastewater using microalgae biomass. *Chemosphere* 286, 131656. https://doi.org/10.1016/j.chemosphere.2021.131656

Al-Ghaili, A.M., Kasim, H., Al-Hada, N.M., Othman, M., Saleh, M.A., 2020. A review: Buildings energy savings: Lighting systems performance. *IEEE Access* 8, 76108–76119. https://doi.org/10.1109/access.2020.2989237

Ali, S.S., Dar, M.A., El-Sheekh, M., Elsamahy, T., Abdelfattah, A., Abdelkarim, E., Pispas, K., Mastropetros, S.G., Zagklis, D., Kornaros, M., Sun, J., 2022a. Microalgae as a Renewable Resource for Bioplastic Production. In M.M. El-Sheekh, N. Abdullah, I. Ahmad (eds.), *Handbook of Research on Algae as a Sustainable Solution for Food, Energy, and the Environment*. IGI Global. https://doi.org/10.4018/978-1-6684-2438-4.ch019

Ali, S.S., Elsamahy, T., Abdelkarim, E.A., Sobhy, M., Srinivasan, G.R., Zahoor, Madadi, M., Sun, J., 2024. Life cycle assessment of bioenergy production from biomass residue. In S. Jafarinejad, B. Beckingham (eds.), *The Renewable Energy-Water-Environment Nexus*, pp. 81–112. Megan R. Ball. https://doi.org/10.1016/b978-0-443-13439-5.00003-x

Ali, S.S., El-Sheekh, M., Manni, A., Ruiz, H.A., Elsamahy, T., Sun, J., Schagerl, M., 2022b. Microalgae-mediated wastewater treatment for biofuels production: A comprehensive review. *Microbiol. Res.* 265, 127187. https://doi.org/10.1016/j.micres.2022.127187

Ali, S.S., Mastropetros, S.G., Schagerl, M., Sakarika, M., Elsamahy, T., El-Sheekh, M., Sun, J., Kornaros, M., 2022c. Recent advances in wastewater microalgae-based biofuels production: A state-of-the-art review. *Energy Rep.* 8, 13253–13280. https://doi.org/10.1016/j.egyr.2022.09.143

Ali, S.S., Shaaban, M.T., Abomohra, A.E.F. and El-Safity, K., 2016. Macroalgal activity against multiple drug resistant Aeromonas hydrophila: A novel treatment study towards enhancement of fish growth performance. *Microb. Pathog.* 101, 89–95.

Arcila, J.S., Céspedes, D., Buitrón, G., 2021. Influence of wavelength photoperiods and N/P ratio on wastewater treatment with microalgae–bacteria. *Water Sci. Technol.* 84, 712–724. https://doi.org/10.2166/wst.2021.257

Awdhesh Kumar Mishra, R., Kodiveri Muthukaliannan, G., 2022. Role of microalgal metabolites in controlling quorum-sensing-regulated biofilm. *Arch. Microbiol.* 204. https://doi.org/10.1007/s00203-022-02776-2

Ayre, J.M., Mickan, B.S., Jenkins, S.N., Moheimani, N.R., 2021. Batch cultivation of microalgae in anaerobic digestate exhibits functional changes in bacterial communities impacting nitrogen removal and wastewater treatment. *Algal Res.* 57, 102338. https://doi.org/10.1016/j.algal.2021.102338

Baggesen, C., 2014. A Study of Microalgal Symbiotic Communities with the Aim to Increase Biomass and Biodiesel Production. Technical University of Denmark, Department of Chemical and Biochemical Engineering.

Banerjee, P., Mukhopadhyay, A., 2021. Microalgal photobioreactors as an integrated approach for simultaneous wastewater treatment, carbon sequestration, and recovery of valuable resources. In A.K. Rathoure (ed.), *Handbook of Research on Waste Diversion and Minimization Technologies for the Industrial Sector*, pp. 174–195. IDI Global. https://doi.org/10.4018/978-1-7998-4921-6.ch008

Barbera, E., Turetta, M., Sforza, E., 2022. The effect of the internal nutrient quota accumulation on algal-based wastewater treatment: Decoupling HRT and SRT to improve the process. *J. Water Process Eng.* 49, 103112. https://doi.org/10.1016/j.jwpe.2022.103112

Barros, A.I., Gonçalves, A.L., Simões, M., Pires, J.C.M., 2015. Harvesting techniques applied to microalgae: A review. *Renew. Sustain. Energy Rev.* 41, 1489–1500. https://doi.org/10.1016/j.rser.2014.09.037

Benvenuti, G., Bosma, R., Ji, F., Lamers, P., Barbosa, M.J., Wijffels, R.H., 2016. Batch and semi-continuous microalgal TAG production in lab-scale and outdoor photobioreactors. *J. Appl. Phycol.* 28, 3167–3177. https://doi.org/10.1007/s10811-016-0897-1

Bhatt, P., Bhandari, G., Bhatt, K., Simsek, H., 2022. Microalgae-based removal of pollutants from wastewaters: Occurrence, toxicity and circular economy. *Chemosphere* 306, 135576. https://doi.org/10.1016/j.chemosphere.2022.135576

Billah, M., Khan, M., Bano, A., Hassan, T.U., Munir, A., Gurmani, A.R., 2019. Phosphorus and phosphate solubilizing bacteria: Keys for sustainable agriculture. *Geomicrobiol. J.* 36, 904–916. https://doi.org/10.1080/01490451.2019.1654043

Blair, E.M., Dickson, K.L., O'Malley, M.A., 2021. Microbial communities and their enzymes facilitate degradation of recalcitrant polymers in anaerobic digestion. *Curr. Opin. Microbiol.* 64, 100–108. https://doi.org/10.1016/j.mib.2021.09.008

Borella, L., Ortolan, D., Barbera, E., Trivellin, N., Sforza, E., 2021. A multiwavelength model to improve microalgal productivity and energetic conversion in a red-blue light emitting diodes (LEDs) continuous photobioreactor. *Energy Convers. Manag.* 243, 114330. https://doi.org/10.1016/j.enconman.2021.114330

Brindhadevi, K., Mathimani, T., Rene, E.R., Shanmugam, S., Chi, N.T.L., Pugazhendhi, A., 2021. Impact of cultivation conditions on the biomass and lipid in microalgae with an emphasis on biodiesel. *Fuel* 284, 119058. https://doi.org/10.1016/j.fuel.2020.119058

Carvalho, A.P., Meireles, L.A., Malcata, F.X., 2006. Microalgal reactors: A review of enclosed system designs and performances. *Biotechnol. Prog.* 22, 1490–1506. https://doi.org/10.1002/bp060065r

Cassan, F.D., Coniglio, A., Amavizca, E., Maroniche, G., Cascales, E., Bashan, Y., de-Bashan, L.E., 2021. The *Azospirillum brasilense* type VI secretion system promotes cell aggregation, biocontrol protection against phytopathogens and attachment to the microalgae *Chlorella sorokiniana*. *Environ. Microbiol.* 23, 6257–6274. https://doi.org/10.1111/1462-2920.15749

Castro-Muñoz, R., García-Depraect, O., 2021. Membrane-based harvesting processes for microalgae and their valuable-related molecules: A review. *Membranes (Basel)* 11, 585. https://doi.org/10.3390/membranes11080585

Chan, S.S., Khoo, K.S., Chew, K.W., Ling, T.C., Show, P.L., 2022. Recent advances biodegradation and biosorption of organic compounds from wastewater: Microalgae-bacteria consortium: A review. *Bioresour. Technol.* 344, 126159. https://doi.org/10.1016/j.biortech.2021.126159

Chang, J.-S., Show, P.-L., Ling, T.-C., Chen, C.-Y., Ho, S.-H., Tan, C.-H., Nagarajan, D., Phong, W.-N., 2017. Photobioreactors. In C. Larroche, M.Á. Sanromán, G. Du, A. Pandey (eds.), *Current Developments in Biotechnology and Bioengineering*, pp. 313–352. Elsevier. https://doi.org/10.1016/b978-0-444-63663-8.00011-2

Chanquia, S.N., Vernet, G., Kara, S., 2021. Photobioreactors for cultivation and synthesis: Specifications, challenges, and perspectives. *Eng. Life Sci.* 22, 712–724. https://doi.org/10.1002/elsc.202100070

Cheah, Y.T., Chan, D.J.C., 2021. Physiology of microalgal biofilm: A review on prediction of adhesion on substrates. *Bioengineered* 12, 7577–7599. https://doi.org/10.1080/21655979.2021.1980671

Chen, S., Li, X., Ma, X., Qing, R., Chen, Y., Zhou, H., Yu, Y., Li, J., Tan, Z., 2023. Lighting the way to sustainable development: Physiological response and light control strategy in microalgae-based wastewater treatment under illumination. *Sci. Total Environ.* 903, 166298. https://doi.org/10.1016/j.scitotenv.2023.166298

Choi, O.K., Hendren, Z., Kim, G.D., Dong, D., Lee, J.W., 2020. Influence of activated sludge derived-extracellular polymeric substance (ASD-EPS) as bio-flocculation of microalgae for biofuel recovery. *Algal Res.* 45, 101736. https://doi.org/10.1016/j.algal.2019.101736

Choi, Y.Y., Joun, J.M., Lee, J., Hong, M.E., Pham, H.-M., Chang, W.S., Sim, S.J., 2017. Development of large-scale and economic pH control system for outdoor cultivation of microalgae Haematococcus pluvialis using industrial flue gas. *Bioresour. Technol.* 244, 1235–1244. https://doi.org/10.1016/j.biortech.2017.05.147

Dai, C., Wang, F., 2024. Potential applications of microalgae–bacteria consortia in wastewater treatment and biorefinery. *Bioresour. Technol.* 393, 130019. https://doi.org/10.1016/j.biortech.2023.130019

Dasan, Y.K., Lam, M.K., Yusup, S., Lim, J.W., Lee, K.T., 2019. Life cycle evaluation of microalgae biofuels production: Effect of cultivation system on energy, carbon emission and cost balance analysis. *Sci. Total Environ.* 688, 112–128. https://doi.org/10.1016/j.scitotenv.2019.06.181

Di Caprio, F., Tayou Nguemna, L., Stoller, M., Giona, M., Pagnanelli, F., 2021. Microalgae cultivation by uncoupled nutrient supply in sequencing batch reactor (SBR) integrated with olive mill wastewater treatment. *Chem. Eng. J.* 410, 128417. https://doi.org/10.1016/j.cej.2021.128417

Díaz, V., Leyva-Díaz, J.C., Almécija, M.C., Poyatos, J.M., del Mar Muñío, M., Martín-Pascual, J., 2022. Microalgae bioreactor for nutrient removal and resource recovery from wastewater in the paradigm of circular economy. *Bioresour. Technol.* 363, 127968. https://doi.org/10.1016/j.biortech.2022.127968

Dineshbabu, G., Goswami, G., Kumar, R., Sinha, A., Das, D., 2019. Microalgae–nutritious, sustainable aqua- and animal feed source. *J. Funct. Foods* 62, 103545. https://doi.org/10.1016/j.jff.2019.103545

Dölle, K., Röhlen, P., 2023. Determination of the feasibility of a renewed approach to the mechanical cleaning stage in wastewater treatment plant: A laboratory study. *Asian J. Adv. Res. Rep.* 17, 126–134. https://doi.org/10.9734/ajarr/2023/v17i9525

El-Sheekh, M., El-Dalatony, M.M., Thakur, N., Zheng, Y., Salama, E.-S., 2021. Role of microalgae and cyanobacteria in wastewater treatment: Genetic engineering and omics approaches. *Int. J. Environ. Sci. Technol.* 19, 2173–2194. https://doi.org/10.1007/s13762-021-03270-w

Ennaceri, H., Fischer, K., Schulze, A., Moheimani, N.R., 2022. Membrane fouling control for sustainable microalgal biodiesel production: A review. *Renew. Sustain. Energy Rev.* 161, 112335. https://doi.org/10.1016/j.rser.2022.112335

Fallahi, A., Rezvani, F., Asgharnejad, H., Khorshidi Nazloo, E., Hajinajaf, N., Higgins, B., 2021. Interactions of microalgae-bacteria consortia for nutrient removal from wastewater: A review. *Chemosphere* 272, 129878. https://doi.org/10.1016/j.chemosphere.2021.129878

Farooq, W., 2021. Sustainable production of microalgae biomass for biofuel and chemicals through recycling of water and nutrient within the biorefinery context: A review. *GCB Bioenergy* 13, 914–940. https://doi.org/10.1111/gcbb.12822

Ferro, L., Hu, Y.O.O., Gentili, F.G., Andersson, A.F., Funk, C., 2020. DNA metabarcoding reveals microbial community dynamics in a microalgae-based municipal wastewater treatment open photobioreactor. *Algal Res.* 51, 102043. https://doi.org/10.1016/j.algal.2020.102043

Figueroa-Torres, G.M., Pittman, J.K., Theodoropoulos, C., 2021. Optimisation of microalgal cultivation via nutrient-enhanced strategies: The biorefinery paradigm. *Biotechnol. Biofuels* 14, 64. https://doi.org/10.1186/s13068-021-01912-2

Fuentes, J.L., Garbayo, I., Cuaresma, M., Montero, Z., González-Del-Valle, M., Vílchez, C., 2016. Impact of microalgae-bacteria interactions on the production of algal biomass and associated compounds. *Mar. Drugs* 14, 100. https://doi.org/10.3390/md14050100

Ge, Z.-B., Zhai, Z.-Q., Xie, W.-Y., Dai, J., Huang, K., Johnson, D.R., Zhao, F.-J., Wang, P., 2023. Two-tiered mutualism improves survival and competitiveness of cross-feeding soil bacteria. *ISME J.* 17, 2090–2102. https://doi.org/10.1038/s41396-023-01519-5

Ghernaout, D., 2019. Electrocoagulation process: A mechanistic review at the dawn of its modeling. *J. Environ. Sci. Allied Res.* 2, 22–38. https://doi.org/10.29199/2637-7063/esar-201019

Gilmour, D.J., Zimmerman, W.B., 2020. Microbubble intensification of bioprocessing. *Adv. Microb. Physiol.* https://doi.org/10.1016/bs.ampbs.2020.07.001

González-Camejo, J., Aparicio, S., Pachés, M., Borrás, L., Seco, A., 2022. Comprehensive assessment of the microalgae-nitrifying bacteria competition in microalgae-based wastewater treatment systems: Relevant factors, evaluation methods and control strategies. *Algal Res.* 61, 102563. https://doi.org/10.1016/j.algal.2021.102563

González-González, L.M., de-Bashan, L.E., 2021. Toward the enhancement of microalgal metabolite production through microalgae-bacteria consortia. *Biology (Basel)* 10, 282. https://doi.org/10.3390/biology10040282

Goswami, R.K., Agrawal, K., Verma, P., 2022. Multifaceted role of microalgae for municipal wastewater treatment: A futuristic outlook toward wastewater management. *Clean Soil Air Water* 51, 2100286. https://doi.org/10.1002/clen.202100286

Guo, Q., Qi, F., Mu, R., Yu, G., Ma, G., Meng, Q., 2023. Advances in sustainable wastewater treatment: Microalgal–bacterial consortia process, greenhouse gas reduction and energy recovery technologies. *Water Environ. J.* 37, 192–205. https://doi.org/10.1111/wej.12839

Gupta, S., Pawar, S.B., Pandey, R.A., 2019. Current practices and challenges in using microalgae for treatment of nutrient rich wastewater from agro-based industries. *Sci. Total Environ.* 687, 1107–1126. https://doi.org/10.1016/j.scitotenv.2019.06.115

Han, M., Zhang, C., Ho, S.-H., 2022. Immobilized microalgal system: An achievable idea for upgrading current microalgal wastewater treatment. *Environ. Sci. Ecotechnol.* 14, 100227. https://doi.org/10.1016/j.ese.2022.100227

Havlik, I., Scheper, T., Reardon, K.F., 2015. Monitoring of microalgal processes. *Adv. Biochem. Eng. Biotechnol.* 153, 89–142. https://doi.org/10.1007/10_2015_328

Helliwell, K.E., Lawrence, A.D., Holzer, A., Kudahl, U.J., Sasso, S., Kräutler, B., Scanlan, D.J., Warren, M.J., Smith, A.G., 2016. Cyanobacteria and eukaryotic algae use different chemical variants of vitamin B12. *Curr. Biol.* 26, 999–1008. https://doi.org/10.1016/j.cub.2016.02.041

Hellsmark, H., Frishammar, J., Söderholm, P., Ylinenpää, H., 2016. The role of pilot and demonstration plants in technology development and innovation policy. *Res. Policy* 45, 1743–1761. https://doi.org/10.1016/j.respol.2016.05.005

Hu, H., Wu, D.-D., Yu, L., Hu, Y., Meng, F.-L., Wei, D., 2023. Pollutants removal, microbial community shift and oleic acid production in symbiotic microalgae–bacteria system. *Bioresour. Technol.* 370, 128535. https://doi.org/10.1016/j.biortech.2022.128535

Huang, K.-X., Vadiveloo, A., Zhou, J.-L., Yang, L., Chen, D.-Z., Gao, F., 2023. Integrated culture and harvest systems for improved microalgal biomass production and wastewater treatment. *Bioresour. Technol.* 376, 128941. https://doi.org/10.1016/j.biortech.2023.128941

Jia, H., Yuan, Q., 2016. Removal of nitrogen from wastewater using microalgae and microalgae–bacteria consortia. *Cogent Environ. Sci.* 2, 1275089. https://doi.org/10.1080/23311843.2016.1275089

Johnson, I., Girijan, S., Tripathy, B.K., Ali, M.A.S., Kumar, M., 2020. Algal–bacterial symbiosis and its application in wastewater treatment. In M.P. Shah, S. Rodriguez-Couto, S.S. Şengör (eds.), *Emerging Technologies in Environmental Bioremediation*, pp. 341–372. Elsevier. https://doi.org/10.1016/b978-0-12-819860-5.00015-8

Khan, M.I., Shin, J.H., Kim, J.D., 2018. The promising future of microalgae: Current status, challenges, and optimization of a sustainable and renewable industry for biofuels, feed, and other products. *Microb. Cell Fact.* 17, 36. https://doi.org/10.1186/s12934-018-0879-x

Khan, S., Naushad, M., Iqbal, J., Bathula, C., Sharma, G., 2022. Production and harvesting of microalgae and an efficient operational approach to biofuel production for a sustainable environment. *Fuel* 311, 122543. https://doi.org/10.1016/j.fuel.2021.122543

Kim, G.-Y., Roh, K., Han, J.-I., 2019. The use of bicarbonate for microalgae cultivation and its carbon footprint analysis. *Green Chem.* 21, 5053–5062. https://doi.org/10.1039/c9gc01107b

Kumari, S., Das, S., 2023. Bacterial enzymatic degradation of recalcitrant organic pollutants: Catabolic pathways and genetic regulations. *Environ. Sci. Pollut. Res.* 30, 79676–79705. https://doi.org/10.1007/s11356-023-28130-7

Legrand, J., Artu, A., Pruvost, J., 2021. A review on photobioreactor design and modelling for microalgae production. *React. Chem. Eng.* 6, 1134–1151. https://doi.org/10.1039/d0re00450b

Leng, L., Yang, L., Chen, J., Hu, Y., Li, H., Li, H., Jiang, S., Peng, H., Yuan, X., Huang, H., 2021. Valorization of the aqueous phase produced from wet and dry thermochemical processing biomass: A review. *J. Clean. Prod.* 294, 126238. https://doi.org/10.1016/j.jclepro.2021.126238

Li, H., Li, X., Cui, L., Asraful Alam, M., Lu, W., 2023. Coupling nutrient removal and biodiesel production by cultivation of Chlorella sp. in cafeteria wastewater: Assessment of the effect of wastewater disinfection. *Desalin. Water Treat.* 291, 101–107. https://doi.org/10.5004/dwt.2023.29477

Li, S., Hu, T., Xu, Y., Wang, J., Chu, R., Yin, Z., Mo, F., Zhu, L., 2020. A review on flocculation as an efficient method to harvest energy microalgae: Mechanisms, performances, influencing factors and perspectives. *Renew. Sustain. Energy Rev.* 131, 110005. https://doi.org/10.1016/j.rser.2020.110005

Li, S., Li, X., Ho, S.-H., 2022. How to enhance carbon capture by evolution of microalgal photosynthesis? *Sep. Purif. Technol.* 291, 120951. https://doi.org/10.1016/j.seppur.2022.120951

Liyanaarachchi, V.C., Premaratne, M., Ariyadasa, T.U., Nimarshana, P.H. V, Malik, A., 2021. Two-stage cultivation of microalgae for production of high-value compounds and biofuels: A review. *Algal Res.* 57, 102353. https://doi.org/10.1016/j.algal.2021.102353

Luo, Y., Henderson, R.K., Le-Clech, P., 2019. Characterisation of organic matter in membrane photobioreactors (MPBRs) and its impact on membrane performance. *Algal Res.* 44, 101682. https://doi.org/10.1016/j.algal.2019.101682

Lutzu, G.A., Ciurli, A., Chiellini, C., Di Caprio, F., Concas, A., Dunford, N.T., 2021. Latest developments in wastewater treatment and biopolymer production by microalgae. *J. Environ. Chem. Eng.* 9, 104926. https://doi.org/10.1016/j.jece.2020.104926

Magwaza, S.T., Magwaza, L.S., Odindo, A.O., Mditshwa, A., 2020. Hydroponic technology as decentralised system for domestic wastewater treatment and vegetable production in urban agriculture: A review. *Sci. Total Environ.* 698, 134154. https://doi.org/10.1016/j.scitotenv.2019.134154

Maltsev, Y., Maltseva, K., Kulikovskiy, M., Maltseva, S., 2021. Influence of light conditions on microalgae growth and content of lipids, carotenoids, and fatty acid composition. *Biology (Basel)* 10, 1060. https://doi.org/10.3390/biology10101060

Manikandan, A., Suresh Babu, P., Shyamalagowri, S., Kamaraj, M., Muthukumaran, P., Aravind, J., 2021. Emerging role of microalgae in heavy metal bioremediation. *J. Basic Microbiol.* 62, 330–347. https://doi.org/10.1002/jobm.202100363

Mao, G.H., Zhang, Z.H., Fei, F., Ding, Y.Y., Zhang, W.J., Chen, H., Ali, S.S., Zhao, T., Feng, W.W., Wu, X.Y. and Yang, L.Q., 2019. Effect of Grifola frondosa polysaccharide on anti-tumor activity in combination with 5-Fu in Heps-bearing mice. *Int. J. Biol. M*acromol. 121, 930–935.

Metsoviti, M.N., Papapolymerou, G., Karapanagiotidis, I.T., Katsoulas, N., 2019. Comparison of growth rate and nutrient content of five microalgae species cultivated in greenhouses. *Plants (Basel, Switzerland)* 8, 279. https://doi.org/10.3390/plants8080279

Min, K.H., Kim, D.H., Ki, M.-R., Pack, S.P., 2022. Recent progress in flocculation, dewatering, and drying technologies for microalgae utilization: Scalable and low-cost harvesting process development. *Bioresour. Technol.* 344, 126404. https://doi.org/10.1016/j.biortech.2021.126404

Molina, D., de Carvalho, J.C., Júnior, A.I.M., Faulds, C., Bertrand, E., Soccol, C.R., 2019. Biological contamination and its chemical control in microalgal mass cultures. *Appl. Microbiol. Biotechnol.* 103, 9345–9358. https://doi.org/10.1007/s00253-019-10193-7

Molinuevo-Salces, B., Riaño, B., Hernández, D., Cruz García-González, M., 2019. Microalgae and wastewater treatment: Advantages and disadvantages. In M. Asraful Alam, J.-L. Xu, Z. Wang (eds.), *Microalgae Biotechnology for Food, Health and High Value Products*, 415–432. Springer. https://doi.org/10.1007/978-981-13-2264-8_20

Moondra, N., Jariwala, N.D., Christian, R.A., 2020. Microalgal-bacterial consortia: An alluring and novel approach for domestic wastewater treatment. *Water Conserv. Manag.* 4, 51–56. https://doi.org/10.26480/wcm.01.2020.51.56

Morán-Valencia, M., Nishi, K., Akizuki, S., Ida, J., Cuevas-Rodríguez, G., Cervantes-Avilés, P., 2023. Nitrogen removal from wastewater by an immobilized consortium of microalgae–bacteria in hybrid hydrogels. *Water Sci. Technol.* 87, 527–538. https://doi.org/10.2166/wst.2023.001

Moreno Osorio, J.H., Pollio, A., Frunzo, L., Lens, P.N.L., Esposito, G., 2021. A review of microalgal biofilm technologies: Definition, applications, settings and analysis. *Front. Chem. Eng.* 3. https://doi.org/10.3389/fceng.2021.737710

Muhammad, G., Alam, M.A., Xiong, W., Lv, Y., Xu, J.-L., 2020. Microalgae biomass production: An overview of dynamic operational methods. *Microalgae Biotechnol. Food, Heal. High Value Prod.* https://doi.org/10.1007/978-981-15-0169-2_13

Mujtaba, G., Lee, K., 2016. Advanced treatment of wastewater using symbiotic co-culture of microalgae and bacteria. *Appl. Chem. Eng.* 27, 1–9. https://doi.org/10.14478/ace.2016.1002

Mustafa, S., Bhatti, H.N., Maqbool, M., Iqbal, M., 2021. Microalgae biosorption, bioaccumulation and biodegradation efficiency for the remediation of wastewater and carbon dioxide mitigation: Prospects, challenges and opportunities. *J. Water Process Eng.* 41, 102009. https://doi.org/10.1016/j.jwpe.2021.102009

Nagarajan, D., Chen, C.-Y., Lee, D.-J., Chang, J.-S., 2023. Photobioreactors for microalgae-based wastewater treatment. In R. Sirohi, A. Pandey, S. Sim, J.-S. Chang D.-J. Lee (eds.), *Current Developments in Biotechnology and Bioengineering*, pp. 121–152. Elsevier. https://doi.org/10.1016/b978-0-323-99911-3.00002-6

Nagarajan, D., Lee, D.-J., Varjani, S., Lam, S.S., Allakhverdiev, S.I., Chang, J.-S., 2022. Microalgae-based wastewater treatment: Microalgae-bacteria consortia, multi-omics approaches and algal stress response. *Sci. Total Environ.* 845, 157110. https://doi.org/10.1016/j.scitotenv.2022.157110

Najjar, Y.S.H., Abu-Shamleh, A., 2020. Harvesting of microalgae by centrifugation for biodiesel production: A review. *Algal Res.* 51, 102046. https://doi.org/10.1016/j.algal.2020.102046

Neethu, K.C., Dhinesh, G., Dharani, G., 2020. Analysis of paddlewheel mixing system for algal biomass production in raceway ponds using simulation techniques. *J. Exp. Zool. India* 23, 475–483.

Nguyen, L.N., Vu, M.T., Vu, H.P., Johir, M.A.H., Labeeuw, L., Ralph, P.J., Mahlia, T.M.I., Pandey, A., Sirohi, R., Nghiem, L.D., 2022. Microalgae-based carbon capture and utilization: A critical review on current system developments and biomass utilization. *Crit. Rev. Environ. Sci. Technol.* 53, 216–238. https://doi.org/10.1080/10643389.2022.2047141

Nwoba, E.G., Chuka-Ogwude, D., Vadiveloo, A., Ogbonna, J.C., 2022. Process control strategies applied to microalgae-based biofuel production. In E. Jacob-Lopes, L.Q. Zepka, M.M. Maroneze (eds.), *3rd Generation Biofuels*, pp. 105–134. Woodhead Publishing. https://doi.org/10.1016/b978-0-323-90971-6.00031-0

Nwoba, E.G., Parlevliet, D.A., Laird, D.W., Alameh, K., Moheimani, N.R., 2019. Light management technologies for increasing algal photobioreactor efficiency. *Algal Res.* 39, 101433. https://doi.org/10.1016/j.algal.2019.101433

Oruganti, R.K., Katam, K., Show, P.L., Gadhamshetty, V., Upadhyayula, V.K.K., Bhattacharyya, D., 2022. A comprehensive review on the use of algal-bacterial systems for wastewater treatment with emphasis on nutrient and micropollutant removal. *Bioengineered* 13, 10412–10453. https://doi.org/10.1080/21655979.2022.2056823

Ozcelik, D., Koray Sakarya, F., Tezel, U., Haznedaroglu, B.Z., 2022. Recent advancements in algae–bacteria consortia for the treatment of domestic and industrial wastewater. In G.N. Demirer, S. Uludag-Demirer (eds.), *Integrated Wastewater Management and Valorization Using Algal Cultures*, pp. 13–50. Elsevier. https://doi.org/10.1016/b978-0-323-85859-5.00002-6

Pechsiri, J.S., Thomas, J.-B.E., El Bahraoui, N., Fernandez, F.G.A., Chaouki, J., Chidami, S., Tinoco, R.R., Martin, J.P., Gomez, C., Combe, M., Gröndahl, F., 2023. Comparative life cycle assessment of conventional and novel microalgae production systems and environmental impact mitigation in urban-industrial symbiosis. *Sci. Total Environ.* 854, 158445. https://doi.org/10.1016/j.scitotenv.2022.158445

Penesyan, A., Paulsen, I.T., Kjelleberg, S., Gillings, M.R., 2021. Three faces of biofilms: A microbial lifestyle, a nascent multicellular organism, and an incubator for diversity. *NPJ Biofilms Microbiomes* 7, 80. https://doi.org/10.1038/s41522-021-00251-2

Peter, A.P., Koyande, A.K., Chew, K.W., Ho, S.-H., Chen, W.-H., Chang, J.-S., Krishnamoorthy, R., Banat, F., Show, P.L., 2022. Continuous cultivation of microalgae in photobioreactors as a source of renewable energy: Current status and future challenges. *Renew. Sustain. Energy Rev.* 154, 111852. https://doi.org/10.1016/j.rser.2021.111852

Qin, L., Alam, M.A., Wang, Z., 2019. Open pond culture systems and photobioreactors for microalgal biofuel production. In L. Qin, M. Asraful Alam, Z. Wang (eds.), *Microalgae Biotechnology for Development of Biofuel and Wastewater Treatment*, pp. 45–74. Springer. https://doi.org/10.1007/978-981-13-2264-8_3

Qu, W., Zhang, C., Chen, X., Ho, S.-H., 2021. New concept in swine wastewater treatment: Development of a self-sustaining synergetic microalgae-bacteria symbiosis (ABS) system to achieve environmental sustainability. *J. Hazard. Mater.* 418, 126264. https://doi.org/10.1016/j.jhazmat.2021.126264

Qv, M., Dai, D., Liu, D., Wu, Q., Tang, C., Li, S., Zhu, L., 2023. Towards advanced nutrient removal by microalgae-bacteria symbiosis system for wastewater treatment. *Bioresour. Technol.* 370, 128574. https://doi.org/10.1016/j.biortech.2022.128574

Ranjbar, S., Malcata, F.X., 2022. Is genetic engineering a route to enhance microalgae-mediated bioremediation of heavy metal-containing effluents? *Molecules* 27, 1473. https://doi.org/10.3390/molecules27051473

Raper, E., Stephenson, T., Anderson, D.R., Fisher, R., Soares, A., 2018. Industrial wastewater treatment through bioaugmentation. *Process Saf. Environ. Prot.* 118, 178–187. https://doi.org/10.1016/j.psep.2018.06.035

Rayen, F., Behnam, T., Dominique, P., 2019. Optimization of a raceway pond system for wastewater treatment: A review. *Crit. Rev. Biotechnol.* 39, 422–435. https://doi.org/10.1080/07388551.2019.1571007

Raza, N., Rizwan, M., Mujtaba, G., 2022. Bioremediation of real textile wastewater with a microalgal-bacterial consortium: An eco-friendly strategy. *Biomass Convers. Biorefinery.* 14, 1–13. https://doi.org/10.1007/s13399-022-03214-5

Reetu, Clifford, M., Prakash, R., Rai, M.P., 2023. Latest advances and status analysis of nanomaterials for microalgae photosystem, lipids and biodiesel: A state of art. *J. Environ. Chem. Eng.* 11, 109111. https://doi.org/10.1016/j.jece.2022.109111

Rezvani, F., Sarrafzadeh, M.-H., Oh, H.-M., 2020. Hydrogen producer microalgae in interaction with hydrogen consumer denitrifiers as a novel strategy for nitrate removal from groundwater and biomass production. *Algal Res.* 45, 101747. https://doi.org/10.1016/j.algal.2019.101747

Rohers, F., Dalsasso, R.L., Nadaleti, W.C., Matias, M.S., de Castilhos Júnior, A.B., 2021. Physical–chemical pre-treatment of sanitary landfill raw leachate by direct ascending filtration. *Chemosphere* 285, 131362. https://doi.org/10.1016/j.chemosphere.2021.131362

Sahu, S., Kaur, A., Singh, G., Kumar Arya, S., 2023. Harnessing the potential of microalgae-bacteria interaction for eco-friendly wastewater treatment: A review on new strategies involving machine learning and artificial intelligence. *J. Environ. Manag.* 346, 119004. https://doi.org/10.1016/j.jenvman.2023.119004

Saifullah, A.Z.A., Karim, M.A., Ahmad-Yazid, A., 2014. Microalgae: An alternative source of renewable energy. *Am. J. Eng. Res.* 3, 330–338.

Saleh, H.M., Hassan, A.I., 2023. Green conversion of carbon dioxide and sustainable fuel synthesis. *Fire* 6, 128. https://doi.org/10.3390/fire6030128

Sales, R., Abreu, P.C., 2014. Use of natural pH variation to increase the flocculation of the marine microalgae Nannochloropsis oculata. *Appl. Biochem. Biotechnol.* 175, 2012–2019. https://doi.org/10.1007/s12010-014-1412-2

Sánchez Zurano, A., Gómez Serrano, C., Acién-Fernández, F.G., Fernández-Sevilla, J.M., Molina-Grima, E., 2020. Modeling of photosynthesis and respiration rate for microalgae–bacteria consortia. *Biotechnol. Bioeng.* 118, 952–962. https://doi.org/10.1002/bit.27625

Sathinathan, P., Parab, H.M., Yusoff, R., Ibrahim, S., Vello, V., Ngoh, G.C., 2023. Photobioreactor design and parameters essential for algal cultivation using industrial wastewater: A review. *Renew. Sustain. Energy Rev.* 173, 113096. https://doi.org/10.1016/j.rser.2022.113096

Sauer, K., Stoodley, P., Goeres, D.M., Hall-Stoodley, L., Burmølle, M., Stewart, P.S., Bjarnsholt, T., 2022. The biofilm life cycle: Expanding the conceptual model of biofilm formation. *Nat. Rev. Microbiol.* 20, 608–620. https://doi.org/10.1038/s41579-022-00767-0

Saxena, R., Rodríguez-Jasso, R.M., Chávez-González, M.L., Aguilar, C.N., González-Fernández, C., Quijano, G., Ruiz, H.A., 2023. A plausible scenario for the third generation of biofuels from microalgae. In M.P. Shah (ed.), *Green Approach to Alternative Fuel for a Sustainable Future*, pp. 107–121. Elsevier. https://doi.org/10.1016/b978-0-12-824318-3.00013-8

Sayara, T, Khayat, S., Saleh, J., Van Der Steen, P., 2021. Bioremediation of nutrients from municipal wastewater using algal-bacterial photobioreactor: Real field scale demonstration for improving effluent quality. *J. Water Chem. Technol.* 43, 261–268. https://doi.org/10.3103/s1063455x21030103

Sayara, Tahseen, Khayat, S., Saleh, J., Abu-Khalaf, N., van der Steen, P., 2021. Algal–bacterial symbiosis for nutrients removal from wastewater: The application of multivariate data analysis for process monitoring and control. *Environ. Technol. In*nov. 23, 101548. https://doi.org/10.1016/j.eti.2021.101548

Singh, V., Mishra, V., 2021. Exploring the effects of different combinations of predictor variables for the treatment of wastewater by microalgae and biomass production. *Biochem. Eng. J.* 174, 108129. https://doi.org/10.1016/j.bej.2021.108129

Sirohi, R., Kumar Pandey, A., Ranganathan, P., Singh, S., Udayan, A., Kumar Awasthi, M., Hoang, A.T., Chilakamarry, C.R., Kim, S.H., Sim, S.J., 2022. Design and applications of photobioreactors: A review. *Bioresour. Technol.* 349, 126858. https://doi.org/10.1016/j.biortech.2022.126858

Solmaz, A., Işık, M., 2020. Optimization of membrane photobioreactor; the effect of hydraulic retention time on biomass production and nutrient removal by mixed microalgae culture. *Biomass Bioenergy* 142, 105809. https://doi.org/10.1016/j.biombioe.2020.105809

Stirk, W.A., van Staden, J., 2022. Bioprospecting for bioactive compounds in microalgae: Antimicrobial compounds. *Biotechnol. Adv.* 59, 107977. https://doi.org/10.1016/j.biotechadv.2022.107977

Su, Y., 2021. Revisiting carbon, nitrogen, and phosphorus metabolisms in microalgae for wastewater treatment. *Sci. Total Environ.* 762, 144590. https://doi.org/10.1016/j.scitotenv.2020.144590

Sun, X.-M., Ren, L.-J., Zhao, Q.-Y., Ji, X.-J., Huang, H., 2019. Enhancement of lipid accumulation in microalgae by metabolic engineering. *Biochim. Biophys. Acta Mol. Cell Biol. Lipids* 1864, 552–566. https://doi.org/10.1016/j.bbalip.2018.10.004

Sur, S., Mahmoud, A., Ebrahimi Khabbazi, A., Pavlov, E., Bilton, A.M., 2016. Computational modeling and field evaluation of an innovative solar updraft aeration system for aquaculture in the developing world. Vol. 2A, *42nd Design Automation Conference*. https://doi.org/10.1115/detc2016-59572

Talapatra, N., Gautam, R., Mittal, V., Ghosh, U.K., 2023. A comparative study of the growth of microalgae-bacteria symbiotic consortium with the axenic culture of microalgae in dairy wastewater through extraction and quantification of chlorophyll. *Mater. Today Proc.* 80, 2268–2273. https://doi.org/10.1016/j.matpr.2021.06.227

Talapatra, N., Mittal, V., Ghosh, U.K., Gautam, R., 2021. Evaluation of the potential of four microalgal-bacterial symbiotic association on wastewater treatment and lipid accumulation. *The 17th International Conference on Environmental Science and Technology*, Athens, Greece. https://doi.org/10.30955/gnc2021.00513

Tavares, J., Silva, T.P., Paixão, S.M., Alves, L., 2023. Development of a bench-scale photobioreactor with a novel recirculation system for continuous cultivation of microalgae. *J. Environ. Manag.* 332, 117418. https://doi.org/10.1016/j.jenvman.2023.117418

Tiwari, A.K., Pal, D.B., 2022. Nutrients contamination and eutrophication in the river ecosystem. In S. Madhav, S. Kanhaiya, A. Srivastav, V. Singh, P. Singh (eds.), *Ecological Significance of River Ecosystems*, pp. 203–216. Elsevier. https://doi.org/10.1016/b978-0-323-85045-2.00001-7

Tong, C.Y., Honda, K., Derek, C.J.C., 2023. A review on microalgal-bacterial co-culture: The multifaceted role of beneficial bacteria towards enhancement of microalgal metabolite production. *Environ. Res.* 228, 115872. https://doi.org/10.1016/j.envres.2023.115872

Touliabah, H.E.-S., El-Sheekh, M.M., Ismail, M.M., El-Kassas, H., 2022. A review of microalgae- and cyanobacteria-based biodegradation of organic pollutants. *Molecules* 27, 1141. https://doi.org/10.3390/molecules27031141

Vargas-Estrada, L., Longoria, A., Arenas, E., Moreira, J., Okoye, P.U., Bustos-Terrones, Y., Sebastian, P.J., 2021. A review on current trends in biogas production from microalgae biomass and microalgae waste by anaerobic digestion and co-digestion. *BioEnergy Res.* 15, 77–92. https://doi.org/10.1007/s12155-021-10276-2

Vergara, C., Muñoz, R., Campos, J.L., Seeger, M., Jeison, D., 2016. Influence of light intensity on bacterial nitrifying activity in algal-bacterial photobioreactors and its implications for microalgae-based wastewater treatment. *Int. Biodeterior. Biodegrad.* 114, 116–121. https://doi.org/10.1016/j.ibiod.2016.06.006

Wang, X., Hong, Y., 2022. Microalgae biofilm and bacteria symbiosis in nutrient removal and carbon fixation from wastewater: A review. *Curr. Pollut. Rep.* 8, 128–146. https://doi.org/10.1007/s40726-022-00214-x

Whitton, R., Le Mével, A., Pidou, M., Ometto, F., Villa, R., Jefferson, B., 2016. Influence of microalgal N and P composition on wastewater nutrient remediation. *Water Res.* 91, 371–378. https://doi.org/10.1016/j.watres.2015.12.054

Wicker, R.J., Kwon, E., Khan, E., Kumar, V., Bhatnagar, A., 2023. The potential of mixed-species biofilms to address remaining challenges for economically-feasible microalgal biorefineries: A review. *Chem. Eng. J.* 451, 138481. https://doi.org/10.1016/j.cej.2022.138481

Wolf, J., Chapman, R., Deepika, C., Pietri, M., Bensalem, S., Hankamer, B., 2023. High-throughput screening to accelerate microalgae-based phycochemical production. In A. Abomohra, S. Ende (eds.), *Value-Added Products from Algae*. Springer. https://doi.org/10.1007/978-3-031-42026-9_10

Xu, K., Zou, X., Xue, Y., Qu, Y., Li, Y., 2021. The impact of seasonal variations about temperature and photoperiod on the treatment of municipal wastewater by algae-bacteria system in lab-scale. *Algal Res.* 54, 102175. https://doi.org/10.1016/j.algal.2020.102175

Yan, H., Lu, R., Liu, Y., Cui, X., Wang, Y., Yu, Z., Ruan, R., Zhang, Q., 2022. Development of microalgae-bacteria symbiosis system for enhanced treatment of biogas slurry. *Bioresour. Technol.* 354, 127187. https://doi.org/10.1016/j.biortech.2022.127187

Yang, H., Xin, X., 2023. CO_2 capture and lipid production performance of microalgae in the S-shaped photobioreactor under different culture modes. *Enzyme Microb. Technol.* 165, 110194. https://doi.org/10.1016/j.enzmictec.2023.110194

Yang, J., Shi, W., Fang, F., Guo, J., Lu, L., Xiao, Y., Jiang, X., 2020. Exploring the feasibility of sewage treatment by algal–bacterial consortia. *Crit. Rev. Biotechnol.* 40, 169–179. https://doi.org/10.1080/07388551.2019.1709796

Yu, J., Hu, H., Wu, X., Zhou, T., Liu, Y., Ruan, R., Zheng, H., 2020. Coupling of biochar-mediated absorption and algal-bacterial system to enhance nutrients recovery from swine wastewater. *Sci. Total Environ.* 701, 134935. https://doi.org/10.1016/j.scitotenv.2019.134935

Yusoff, F.M., Nagao, N., Imaizumi, Y., Toda, T., 2019. Bioreactor for microalgal cultivation systems: Strategy and development. *Biofuel Biorefinery Technol.* 2019, 117–159. https://doi.org/10.1007/978-3-030-14463-0_4

Zdarta, J., Jesionowski, T., Pinelo, M., Meyer, A.S., Iqbal, H.M.N., Bilal, M., Nguyen, L.N., Nghiem, L.D., 2022. Free and immobilized biocatalysts for removing micropollutants from water and wastewater: Recent progress and challenges. *Bioresour. Technol.* 344, 126201. https://doi.org/10.1016/j.biortech.2021.126201

Zetterlund, S., Schwartz, O., Sandberg, M., Venkatesh, G., 2023. Computational modelling to advise and inform optimization for aeration and nutrient-dosing in wastewater treatment: Case study from pulp and paper mill in south-central Sweden. *J. Water Process Eng.* 56, 104508. https://doi.org/10.1016/j.jwpe.2023.104508

Zhang, B., Li, W., Guo, Y., Zhang, Z., Shi, W., Cui, F., Lens, P.N.L., Tay, J.H., 2020. Microalgal-bacterial consortia: From interspecies interactions to biotechnological applications. *Renew. Sustain. Energy Rev.* 118, 109563. https://doi.org/10.1016/j.rser.2019.109563

Zhao, D., Cheah, W.Y., Lai, S.H., Ng, E.-P., Khoo, K.S., Show, P.L., Ling, T.C., 2023. Symbiosis of microalgae and bacteria consortium for heavy metal remediation in wastewater. *J. Environ. Chem. Eng.* 11, 109943. https://doi.org/10.1016/j.jece.2023.109943

Zhou, X.-R., Wang, R., Tang, C.-C., Varrone, C., He, Z.-W., Li, Z.-H., Wang, X.C., 2023. Advances, challenges, and prospects in microalgal-bacterial symbiosis system treating heavy metal wastewater. *Chemosphere* 345, 140448. https://doi.org/10.1016/j.chemosphere.2023.140448

7 Potential of Macroalgae (Seaweeds) in the Treatment of Wastewaters

Mukul Machhindra Barwant, Vaneeta Chandrashekhar Karande, Shalagha Sharma, Gholamareza Abdi, Varsha Sadashiv Patil, and Avinash Shivaji Jondhale

7.1 INTRODUCTION

Environmental preservation has recently rose to the top of the societal agenda. Human activities play a major role in environmental deterioration. Environmental regulations are growing stricter with relation to emissions brought on by human activities as a result. These technologies have all had negative consequences on the environment, including emissions into the atmosphere, energy usage, the use of natural resources, whether renewable or not, and the production of byproducts, despite the fact that they have all greatly improved environmental quality. The five environmental consequence categories of abiotic depletion, eutrophication, global warming, acidification, and human toxicity were applied to all resource usage and emissions. Operating system effects on the environment are not now a top priority for decision-makers. The inclusion of this type of environmental evaluation tool in the decision-making process, however, may be motivated by the growing social sensitivity to environmental issues. The five environmental consequence categories of abiotic depletion, eutrophication, global warming, acidification, and pollution were applied to all resource usage and emissions. Each category's potential resources and emissions were all taken into consideration. Algae serve as the primary producers in various aquatic environments, and their activities can have both positive and negative effects on water quality and pollution. On the one hand, organic effluent enrichment can provide nutrients to certain algal species, leading to substantial growth or 'blooms' that can degrade water quality and affect its utility. On the other hand, specific algae adapted to organic waste-polluted water play a vital role in the natural purification of aquatic ecosystems. Therefore, while algae can contribute to water pollution, they also have a role in the self-cleansing processes of water bodies. In recent decades, river pollution

DOI: 10.1201/9781003459286-7

has witnessed a significant increase, primarily due to the runoff of pollutants from agricultural regions and the discharge of industrial, mining, and urban wastewater. Historically, pesticides were predominantly detected in surface runoff from agricultural zones. However, in urban areas, wastewater has emerged as a major pathway for pesticide contamination in the environment. The prevalence of pesticides in urban wastewater treatment plants (WWTPs) can be attributed to non-agricultural activities. These activities encompass a wide range of sources, including but not limited to industrial, residential, and commercial sectors, contributing to the presence of pesticides in urban WWTPs (Köck-Schulmeyer et al., 2013; Cahill et al., 2011). A diverse range of sources contributes to pesticide contamination in urban areas. These sources encompass golf courses, educational institutions, parks, cemeteries, industrial sites, electric utilities, road networks, railways, pipelines, public health initiatives (such as mosquito control and rodent management programs), and aquatic environments. Additionally, non-agricultural activities like commercial forestry, horticulture, and plant nurseries are significant contributors to pesticide residues in urban settings. To assess the environmental impact of pesticides originating from wastewater, a novel index has been introduced. This index represents a groundbreaking approach as it considers both the concentration and toxicity of pesticides present in wastewater. It offers a comprehensive framework for evaluating the environmental significance of pesticides derived from wastewaters. Furthermore, techniques for assessing the risk of pesticides in natural water bodies have been developed and are currently available for implementation (Köck-Schulmeyer et al., 2013). As a result of increased pesticide production and use over the past 40 years that has been emphasized, a worry over the possible adverse effects of these pesticides in water bodies and their influence on human health and ecosystems has emerged (Burrows et al., 2002).

Wastewaters are one of the main entry points for pesticide contamination into the ecosystem through intensive agriculture and surface runoff (Dagnac et al., 2009). In today's context, it is imperative that every strategy consider its global environmental impact. Failing to do so could lead to environmental changes and negative consequences. Even seemingly eco-friendly systems like WWTPs have environmental repercussions due to factors such as energy consumption, chemical usage, emissions into the atmosphere, and the generation of sludge. These impacts are further magnified during post-treatment processes. Over the past few decades, there has been a notable increase in environmental consciousness, with the environment now recognized as a pivotal aspect of any endeavor. Consequently, contemporary strategies must be designed with a keen awareness of the global environmental landscape. Neglecting this aspect could result in shifts in environmental burdens and their consequences (Vryzas et al., 2009; Kuster et al., 2009). The majority of home and industrial processes produce wastewater that contains dangerous contaminants. In this situation, water supplies must be safeguarded at all costs. Technologies used today to treat wastewater integrate physical, chemical, and biological processes with steps to remove both soluble and insoluble contaminants from effluents (Crini et al., 2018). Exactly what is water pollution? Different criteria can be used to determine water contamination. When a substance or substances that might harm the water are poured into it, water pollution occurs. Numerous things may pollute water, such as energy consumption, radioactive waste, urban growth, sewage and wastewater, pesticides and chemical fertilizers, industrial waste, and mining operations. Water will get contaminated because it is utilized; any activity, whether domestic, agricultural, or industrial in nature, creates effluent that contains undesirable pollutants that may also be hazardous. Consistent efforts must be undertaken in this circumstance to protect water supplies (Morin-Crini et al., 2017). In most cases, pollutants are removed physically, chemically, and biologically. Due to the complexity of industrial effluents, there is presently no single method capable of providing effective treatment. In reality, a combination of many technologies is typically used to achieve the desired water quality in the most economical way. Water constitutes the most crucial element for life on Earth, characterized by its pure form that lacks odor, color, and taste. Unfortunately, the introduction of industrial waste into water bodies has elevated pollution levels in aquatic ecosystems, necessitating greater water resources for both domestic and industrial purposes. In municipal WWTP effluents, organic micropollutants have become a pervasive presence, typically measured in the ng-g/L range due to multiple human-induced discharges (Reddy & Narsaiah, 2021).

Pharmaceutical compounds, generated primarily through batch processes, introduce a diverse array of chemicals into wastewater during activities involving extensive water usage, such as equipment cleaning, solid cake washing, and extraction processes. The prevalence of pharmaceutical compounds in drinking water can be attributed to two primary sources: the production processes within the pharmaceutical industry itself and the widespread use of pharmaceutical products. As a result, these compounds find their way into both urban and agricultural wastewater streams. Consequently, wastewater generated during various stages of pharmaceutical and drug manufacturing contains a broad spectrum of contaminants (Crini et al., 2018). In recent times, algae have gained significant recognition for their role in biological WWT, primarily because of their remarkable capacity to absorb various substances such as plant nutrients, heavy metals, pesticides, organic and inorganic pollutants, as well as radioactive particles, effectively sequestering them within their cell structures. Over the past five decades, the utilization of microalgae in biological WWT systems has gained widespread acceptance, with the growing consensus that these algal-based systems are just as effective as conventional treatment methods. The distinct advantages offered by algal WWT systems have positioned them as cost-effective alternatives to more complex and expensive treatment solutions, particularly in the context of municipal wastewater purification. Additionally, algae harvested from treatment ponds often serves as valuable supplements of nitrogen (N) and phosphorus (P) for agricultural applications. Furthermore, these algae can be subjected to fermentation processes to harness energy in the form of methane. Notably, algae have the unique capability to accumulate and store highly hazardous compounds such as selenium, zinc (Zn), and arsenic (As), thereby aiding in the removal of these contaminants from aquatic environments. Radiation pollution represents another significant concern, as some water sources contain natural radioactive elements, while others become radioactive due to contamination. Many species of algae exhibit the capacity to absorb and accumulate various radioactive minerals, even when present in high concentrations within the water, serving as a valuable mechanism for reducing radioactive contamination in aquatic ecosystems. The amount of pollutants coming from hospitals shows that some substances, notably organohalogens and medications that have partially been metabolized, mostly escape WWTP unmodified. Hospital wastewater demonstrates the presence of heavy metals, including silver and mercury (Hg), as well as a high concentration of chlorinated compounds. Hospital effluent discharge into the municipal sewage network without previous treatment, as is the situation with urban classic effluents, is one of the main environmental problems hospital effluents present. The flow and polluting load of the hospital effluents have already generated saturation effects in the municipal WWTP, allowing pollutants to be discharged into the environment after exceeding the unitary process' maximum efficiency. In fact, the interaction between hospital contaminants and components of aquatic ecosystems illustrates a danger connected with the existence of hazardous chemicals, or substances having the potential to harm the environment and living creatures. Long-term development depends critically on the management of wastewater and the monitoring of its quality. A current approach to environmental protection includes the ecotoxicological assessment of effluents as a complement to the classic chemical tests. This study looked at the impact of effluent quality in WWTPs. Despite the fact that dilution may greatly lessen the effects, the data showed that the whole effluent from the WWTP was detrimental to the test species. Municipal WWTP effluent may benefit from using ecotoxicological evaluation techniques, which might help current water management practices (Pereao et al., 2021).

The aquatic environments, WWTPs are thought to be the main sources of chemicals of emerging concern (CECs), which have the potential to disrupt aquatic ecosystems. In aquatic environments, WWTPs are thought to be the main sources of CECs, which have the potential to disrupt aquatic ecosystems. The remaining sludge may be burned, disposed of at landfills where it may seep into groundwater, or used as fertilizer in agricultural areas where it may flow off into bodies of water while the final effluent is released into surface waterways (Golovko et al., 2021).

Surface water pollution stands as one of the most critical environmental challenges facing us today. Globally, there are two predominant and persistent forms of water pollution: organic pollution, characterized by the excessive presence of organic matter in aquatic ecosystems, which ultimately leads to eutrophication. It is a widely acknowledged fact that contaminated water can significantly degrade water quality, restricting its utility for various purposes. Organic pollution transpires when substantial quantities of organic chemicals from diverse origins find their way into rivers, lakes, and oceans through sources like untreated or treated domestic sewage, urban runoff, industrial discharges, and agricultural runoff. The consequences of organic pollution on water quality are multifaceted. As organic materials decompose in water, they can consume dissolved oxygen at a rate faster than it can be naturally replenished, resulting in oxygen depletion, which poses severe threats to aquatic life. Additionally, organic effluents typically contain substantial amounts of suspended particles, reducing the available light for photosynthetic organisms, primarily algae. Furthermore, organic waste originating from human and animal sources may carry a high load of disease-causing pathogens, further exacerbating the hazards associated with organic pollution in aquatic environments (Sen et al., 2013). In general, a high detergent concentration in the aquatic environment results in the formation of a layer of foam on the water's surface, which means that these substances slow down the rate at which oxygen from the air permeates the water, impairing aquatic organisms' ability to absorb dissolved oxygen, allowing contaminated water to seep into the soil, and having a negative impact on the local soil flora. Detergents with enzymatic activity have been shown to be effective in the photosynthetic process of plants, but they harm plant germination (Golovko et al., 2021).

7.2 THE POTENTIAL OF SEAWEEDS IN WASTEWATER TREATMENT

To remove dangerous metals from wastewater, algae can either be utilized for biosorption, which uses non-living organisms, or for bioaccumulation, which uses live cells. As a result, bioaccumulation rather than the form of adsorption known as biosorption is more usually used to remove contaminants from wastewater. Biosorption is one of the techniques utilized in environmental biotechnology. Seaweeds have particularly high biosorption capabilities due to the presence of macromolecules in the algal cell wall, such as polysaccharides and proteins, which offer diverse functional groups, including amino, hydroxyl, carboxyl, and sulfate, capable of binding metal ions (Michalak, 2020). Industrial effluents and aquaculture farms typically cause severe environmental issues like ecological degradation and excessive pollution. Eutrophication, which causes hypoxia and increases the frequency of harmful microalgal blooms, is frequently caused by the presence of key nutrients like N and P in water bodies (Arumugam et al., 2018). The other use is for the removal of dangerous elements, including copper (Cu), nickel (Ni), lead (Pb), Zn, and cadmium (Cd), from industrial effluent. Metals can be found naturally, mined, or recycled from commercial waste. Brown seaweeds like *Sargassum*, *Laminaria*, and *Ecklonia* are better at collecting hazardous metals than the green seaweeds *Ulva* and *Enteromorpha*. Consequently, it acts as a biological indicator of heavy metal contamination (Pati et al., 2016). Although treating wastewater on-site is the most efficient way to prevent pollution, few companies and aquacultures have these facilities. N and P may be naturally removed from wastewater using seaweed. Agricultural, industrial, and sewage water discharges commonly contain the N molecules, nitrate and ammonia-N, which seaweeds may utilize (Wang et al., 2020a). Seaweeds make good bioremediation agents since they develop autotrophically. Several polymers, such as cellulose, pectin, hemicellulose, and arabinogalactan proteins, make up the cell walls of seaweed. Seaweed possesses negatively charged cell walls as a result of its dominating functional groups, which include carboxyl, amines, and phosphoryl. By attracting contaminants containing cationic groups to the seaweed's surface, this starts the sorption process (Farghali et al., 2023; Wang et al., 2020b). Seaweeds are now being employed to reduce the levels of N- and P-containing compounds in treated wastewater before it is discharged into rivers or oceans. Eutrophication, a process that can occur naturally, can also be accelerated by the introduction of water rich in dissolved fertilizers into nearby lakes and streams or by the release of sewage effluent into rivers and coastal waters. This can lead to the excessive and unintended growth of aquatic or marine plants. A significant feature of many seaweed species is their ability to absorb more P than required for optimal growth. The most resilient species in this regard are typically intertidal and estuary species, particularly green seaweeds like *Enteromorpha* and *Monostroma* (Pati et al., 2016). The ability of several seaweed species and their derivatives to remove a range of heavy metals from solutions has been documented by certain authors. Three different kinds of brown seaweeds were *Durvillea potatorum*, *Ecklonia maxima*, and *Lessonia flavicans*. The two derivatives of seaweed produced during industrial seaweed processing were alginate fibers and dealgined seaweed waste. The ability of different biomasses to remove the heavy metal ions Cu, Ni, Pb, Zn, and Cd from solution was compared while being continuously agitated. The effect of having numerous metal ions in solution was investigated to determine whether any competitive effects may have affected metal-ion absorption. It has been noted that these seaweeds' capacity to remove metals from the environment has ramifications for both the economy and the environment (Aderhold et al., 1996). Several researchers have published their findings using biomass-based adsorbents. For example, the brown algae *Cystoseira barbata* and *Cystoseira crinite* were shown to be effective chromium (Cr) adsorbents. Three other species of algae with distinctive natural colors—*Ulva lactuca* (green), *Gracilaria gracilis* (red), and *Fucus vesiculosus* (brown)—were examined for Hg absorption. The capacity to bind metals and our understanding of intrinsic adsorption mechanisms have recently been greatly enhanced by efforts to develop novel algae- and seaweed-based adsorbents employing extraction, nanoparticles, molecular, and chemical modes. The amount of porosity, the existence of functional groups such as amine, carbonyl, hydroxyl, thiol, and carboxyl, and additional naturally occurring materials that promote metal binding such as lignins, cellulose, lipids, and others all have an effect on the removal of metal (Znad et al., 2022).

Macroalgae have proven to be effective in removing various pollutants from wastewater, including dyes from industries such as paper, textiles, and printing, as well as N, P, and phenolic compounds often found in sources like aquaculture. Additionally, they can efficiently eliminate heavy metals from different origins. These heavy metals include common ones like As, Cd, Hg, and Pb, along with essential microelements such as Zn, Cu, manganese (Mn), and cobalt (Co), which are required in smaller quantities for proper plant and animal growth. In a unique application, macroalgae, particularly the species Oedogonium, were cultivated in wastewater at a coal-fired power plant to serve as feedstock for biochar production. Seaweeds offer versatile potential in the fields of bioremediation and biomonitoring for pollutants due to their capability to extract nutrients and heavy metal ions from water. They have been recognized as a valuable resource for cleaning polluted soils, treating eutrophicated water, managing aquaculture effluents (particularly in marine integrated multitrophic aquaculture), and addressing industrial effluent concerns. Seaweeds can be used both as dead biomass (biosorption) and live biomass (bioaccumulation) in environmental biotechnology applications. For instance, the red algae species *Porphyra leucostica* has been identified as an effective, cost-efficient, and biodegradable biomaterial for reducing heavy metal contamination in the environment and wastewater. When it comes to bioremediation of heavy metals like boron, Pb, Cd, Cu, and solid and liquid Cr, crucial species include *Caulerpa racemosa* and Ulvalactuca. These macroalgae also make significant contributions to removing and absorbing substances such as ammonia, phosphate, nitrate, and nitrite from wastewater and industrial effluents, fostering plant development and growth.

TABLE 7.1
Seaweed Used in Removal Heavy Metals (Farghali et al., 2023)

Type of Seaweed	Name of Seaweed	Wastewater Treatment Source	Contaminant
Red Seaweed	*Gracilaria lemaneiformis*	Aquaculture	Phosphate and N
	Gracilaria tikvahiae	Shrimp wastewater	N
	Agarophyton tenuistipitatum	Brackish water	N and P
	Hydropuntia edulis		
	Gracilaria corticata	Aqueous solution	Co
Brown Seaweed	*Turbinaria ornata*	Municipal wastewater	Pb
	Sargassum sp.	Simulated wastewater	Ni (ii) ion and Cu (ii) ion
	Sargassum sp.	Synthetic wastewater	Cd (ii) ion and Zn (ii) ion
	Sargassum filipendula	Simulated wastewater	Silver, Cd, Cr, Cu, Ni, Pb, and Zn ions
	Cystoseira crinite and Cystoseira barbata	Aqueous and wastewater solutions	Cr (III) and Cr (VI)
	Sargassum muticum	Mining-influenced water	As
	Caulerpa scalpelliformis	industrial wastewater	Zn
	Sargassum polycystum	Simulated water	Multi-metals
Green seaweed	*Ulva rigida*	Simulated wastewater	As cation (3+), As cation (5+), antimony (3+), selenium (4+) ions, and selenium (6+) ions

7.2.1 Traditional Methods of Wastewater Treatment and Their Limitations

Water makes up 65% of our bodies and covers 71% of the Earth's surface. According to estimates, between 70% and 80% of the entire amount of water delivered is wasted. Numerous pathogenic bacteria, germs, suspended particles, nutrients, minerals, poisonous metals, etc. are present in the wastewater. Sonune and Ghate (2004) state that the main goal of WWT is typically to enable the disposal of industrial and human effluents without endangering public health or causing unacceptable harm to the environment. WWT and reuse are crucial since they are related to public health. Wherever contact, inhalation, or ingestion of a drug or microbiological element of health concern occurs, the presence of pathogenic organisms and contaminated compounds in wastewater raises the prospect of adverse health impacts (Crook and Surampalli, 1996). Reducing the amount of suspended sediments, oxygen-demanding substances, dangerous microorganisms, and dissolved inorganic compounds has been the main objective of WWT for lots of years. But in recent years, there has been a greater emphasis on enhancing municipal treatment procedures for the disposal of solid waste. Organic matter, nutrients, and sediments are removed from wastewater via physical, chemical, and biological processes in typical WWT (Krantz and Kifferstein, 2005).

7.2.2 Conventional Wastewater Treatment Methods

Solids, organic debris, and nutrients are removed from wastewater using physical, chemical, and biological techniques in traditional WWT. Preliminary, primary, secondary, and tertiary are the many phases.

Initial Processing: The primary goal of the initial processing stage is to segregate materials that float on the surface of wastewater, such as deceased animals, loose branches, paper debris, rags, and also heavy inorganic solids that settle at the bottom. This phase is instrumental in eliminating substances like oils and grease from the sewage. Through this treatment, the biochemical oxygen demand (BOD) of the wastewater is typically reduced by approximately 15%–30%. Several units are involved in preliminary treatment, including screening, a detritus tank (also known as a grit chamber), comminutors, floatation units, and skimming tanks.

Screening: Employed for the removal of suspended debris and matter floating on the surface. Detritus tank (grit chamber): used to eliminate sand and grit from the wastewater. Comminutors: these devices are responsible for the grinding and fragmentation of larger suspended solid (SS) particles. Floatation units and skimming tanks: utilized to effectively extract oils and greases from the wastewater.

Primary Treatment: In the primary treatment phase, physical procedures involving sedimentation and flotation are employed to eliminate both organic and inorganic solids from wastewater. This primary treatment step results in the removal of approximately 5%–50% of the initial BOD (BOD5), 50%–70% of the overall total SS, and approximately 65% of the oil and grease present in the wastewater. While primary sedimentation effectively removes organic N, organic P, and heavy metals that are associated with solid particles, it does not impact the colloidal and dissolved constituents. In many highly industrialized nations, primary treatment is the minimal prerequisite for wastewater irrigation. This type of treatment is generally considered adequate for irrigating crops that are not intended for human consumption, as well as for orchards, vineyards, and certain processed food crops (Ahuja et al., 2014). Primary sedimentation tanks typically come in either round or rectangular basins, with depths ranging from 3 to 5 m, and they maintain a hydraulic retention time of 2–3 hours. The solids that settle at the bottom of these tanks, referred to as primary sludge, are typically removed from a central well and subsequently pumped to sludge processing facilities. Additionally, any scum present on the surface of the tank is eliminated, either through the use of water jets or mechanical means, and directed to sludge processing units.

Secondary Treatment: The effluent emerging from the primary sedimentation tank still contains a significant portion of the initially present unstable elementary organic matter, ranging from 60% to 80%. To further enhance the treatment, it is essential to address the colloidal organic matter that bypasses the primary clarifiers. In the secondary or biological treatment phase of sewage, the focus is on modifying the nature of this organic matter, transforming it into more stable forms primarily through oxidation or nitrification (Ebrahimi and Najafpour, 2016). Secondary sewage treatment encompasses a range of methods, broadly categorized into two groups: filtration and the activated sludge process. Within the domain of secondary treatment, various filtration methods are employed, including contact beds, intermittent sand filters, and trickling filters (Fujie et al., 1983). Tertiary treatment: tertiary treatment steps in when specific constituents within the wastewater cannot be effectively separated through secondary treatment alone. This final purification stage is dedicated to enhancing wastewater quality before it is either reused, recycled, or released into the environment. Tertiary treatment is particularly adept at removing inorganic compounds and substances like N and P (Table 7.2).

TABLE 7.2
Summery of Traditionally Used Different Techniques for Wastewater Treatment

Treatment Method	Objectives	Efficiency	References
Preliminary treatment	Removing grit, oil, grease, big floating, and SS matter, as well as other large elements that are frequently present in raw wastewater	The BOD of wastewater is reduced by 15%–30% thanks to this treatment	Amina (2017)
Filtration	It is used to get rid of pollutants, including bacteria, oils, fats, and suspended particles	Filtration can reduce it by up to 99% by removing particles smaller than 100 mg/L and oil smaller than 25 mg/L	Tchobanoglous et al. (1991) and Ebrahimi and Najafpour (2016)
Sedimentation and gravity separation	This method involves leaving water undisturbed or semi-undisturbed in various types of tanks for varying lengths of time in order to remove suspended particles, grits, and silts. The settling of the SSs is influenced by gravity	Gravity separation alone can remove up to 60% of suspended particles	Gupta et al. (2012), Tchobanoglous et al. (1991), and Ahuja et al. (2014)
Flotation	By securing them to air or gas, flotation eliminates pollutants such as greases, oils, biological compounds, and suspended particles	Up to 75% of suspended particles and up to 95% of grease and oil may be removed with this technique	Tchobanoglous et al. (1991) and Ahuja et al. (2014)

7.2.3 Limitations

Current WWT methods face numerous challenges, including heightened chemical reliance, sludge disposal dilemmas, escalated energy consumption, and expanded spatial requirements. Moreover, conventional systems struggle with effectively removing stubborn organic constituents, accommodating higher wastewater volumes beyond their design limits, and coping with a shortage of skilled labor. These operational and technological limitations have spurred researchers to explore innovative categories of advanced WWT approaches, aiming to address these pressing issues. Advanced wastewater techniques encompass a spectrum of enhancements, such as membrane technology adoption, Advanced Oxidation Processes (AOPs) integration, minimizing sludge generation, and exploring ways to repurpose generated sludge rather than resorting to disposal at landfills. Additionally, there is a growing emphasis on cost-effective adsorption materials, reduced reliance on chemicals or bioflocculants, and the utilization of novel nanoparticles in WWT. Despite the substantial body of research in these areas, there remain avenues for further advancements in WWT methods as the literature continues to grapple with evolving concerns. The synergy of contemporary wastewater technologies in conjunction with traditional methods holds promise for achieving more efficient WWT, alongside heightened opportunities for water reuse and recycling (Veress et al., 2020).

7.3 BENEFITS OF USING SEAWEEDS FOR WASTEWATER TREATMENT

The problems associated with global warming are now developing alarmingly and are getting more intertwined with one another. As a result of climate change, extreme weather phenomena, including heat waves, droughts, and floods, are becoming more often and intense, endangering human security. Meanwhile, problems like soil degradation, biodiversity loss, disease spread, and water shortages are made worse by global warming. People's well-being, political stability, and patterns of economic development have all been predicted to be impacted by these shifts (Global Carbon Project, 2020). As a result of various human activities that cause global warming, carbon dioxide (CO_2), a substantial greenhouse gas (GHG) found in the Earth's atmosphere and responsible for up to 60% of all GHGs, has been rising. In the 1950s, CO_2 emissions started to rise quickly and peaked in 2000 at 25.23 billion tons. The CO_2 emission grew by 32% between 2000 and 2010, reaching 34.81 billion metric tons in 2020 (Global Carbon Project, 2020). Utilizing alternative energy sources and improving CO_2 capture and storage methods are two practical strategies to minimize CO_2 emissions over the long run. To capture and store CO_2, several biological, chemical, or physical procedures can be applied. However, there are a number of technological and budgetary limitations with current technology. Therefore, it is essential that new technologies are created and existing ones are improved. In contrast to physical or chemical approaches, biological CO_2 fixation appears to be a more economical and ecologically benign solution. The process through which photosynthetic organisms absorb CO_2 and help regulate atmospheric CO_2 levels is called biological CO_2 fixation (Daneshvar et al., 2022). Additionally, during the past few decades, carbon sequestration utilizing microalgae has been acclaimed as one of the most significant and successful technologies in the world (Alami et al., 2021). Microalgae can also endure a broad range of environmental conditions without competing with people or other animals for sustenance. Peptides, carbohydrates, lipids, pigments, vitamins, and minerals are just a few of the functional components found in microalgae, all of which have been demonstrated to have a range of positive health effects. Due to its potential applications in a variety of disciplines, including medicine, aquaculture, animal and human nutritional supplements, agricultural production, and bioenergy, microalgae have attracted a lot of attention. Recent developments have improved the usage of microalgae for environmental problems, including CO_2 absorption or coupling with WWT. The benefits and drawbacks of microalgal production for use in the green economy have only been briefly discussed in a few publications.

7.4 CHALLENGES AND CONSIDERATIONS IN IMPLEMENTING SEAWEED-BASED WASTEWATER TREATMENT

Water is the most crucial natural resource for all living things and is necessary for life to survive on this planet. Problems with water contamination now have a significant influence on our society. Water resources appear to be under serious quantitative and qualitative threats today, despite the fact that the strategic importance of fresh water is more widely acknowledged than ever before and that issues relating to appropriate water management can be found on almost every scientific, social, or political agenda worldwide. As a result, there are grave hazards to the availability and quality of water resources globally as a result of industrialization and rapid economic expansion (Abdel-Raouf et al., 2012). Nutrients are discharged into the environment as a result of wastewater release, and aquatic bodies experience dystrophic and hypertrophic conditions as a result. In order to lessen and control the pollution produced by wastewater, several governmental entities have the ability to embrace the circular economy idea. However, recycling or reusing wastewater would necessitate a fit-for-purpose strategy that could lower energy and production costs related to WWT and improve resource recovery from wastewater (Capodaglio, 2020). To reuse and discharge conventional wastewater into the environment, a number of significant issues must be overcome. The typical WWT procedures have detrimental effects on the environment, including high energy use, resource waste of recyclable materials, and excessive solid waste disposal. The traditional WWT methods for raising water quality typically include physical and chemical treatment techniques. Chemical treatment techniques are the most successful or efficient WWTs for reducing turbidity, adjusting pH levels, and removing dissolved materials from water. This treatment includes ozonation, UV radiation, coagulation, chlorination, and chloramination (Tohamy et al., 2022; Basu et al., 2021; Aswara-El et al., 2020). But standard procedures employ general principles for primary and subsequent therapy for WWT. Therefore, initial treatment aims to remove solid particles, while secondary treatment uses microbial activities to break down organic materials. Despite the fact that these techniques are more efficient, these treatments require a lot of work and money to maintain. (Crini & Lichtfouse, 2019). In order to degrade and convert WWT contaminants into biomass and related gases, Dalvi et al. (2021) demonstrate how biological techniques depend on microorganisms' metabolic activity. This lowers the levels of BOD and chemical oxygen demand (COD) in the effluents and improves their quality. Under aerobic or anaerobic circumstances, all biological processes use extremely affordable therapies. There are some significant limitations to investigating biological approaches. According to Abdelfattah et al. (2023), biological WWTs only provide a limited level of long-term design and operating flexibility. Inadequate elimination of pollutants, pollutants with complex chemical structures that are resistant to degradation using traditional techniques, and the necessity for a wide land area and a lengthy treatment process were also highlighted. Additionally, Roy and Saha (2021) revealed these significant limitations in biological techniques. Additionally, biological treatments comprised mixing biodegradation processes employing a variety of microorganisms, such as bacteria, fungus, yeast, and microalgae. In recent years, the production of microalgae biomass has proven to be an efficient method for treating tertiary wastewater and has gained recognition as a viable and sustainable solution (Aigars & Talis, 2017). Contrary to bacteria-based WWT, which emits a considerable quantity of CO_2, algae are a photosynthetic organism that produces oxygen and assimilates CO_2. According to Li et al. (2019), the development of algae in wastewater can provide useful biomass while also serving as a WWT service. Therefore, using wastewater directly minimizes the need for freshwater while simultaneously providing an affordable and efficient supply of nutrients (Yang et al., 2011). Nowadays, low-cost microalgae biomass farming mostly uses WWT to extract high-value products and provide electricity. In order to achieve water treatment and biomass production performance, investigations on algae-based WWT have expanded to numerous sizes and approaches, spanning a variety of contaminant removals (Gentili & Fick, 2016; Yadav et al., 2017). Currently, microalgae-based bioremediation is being used to upgrade the classic western water treatment system. It is efficiently used to remove any liquid or solid contaminants contained in municipal wastewater, agricultural effluent, and industrial wastewater utilizing traditional methods. Then, using this WWT, they may be made into goods with value added (Abdelfattah et al., 2023). In summary, the application of microalgae in WWT exhibits significant potential for effectively purifying agricultural effluent, municipal wastewater, and industrial wastewater. However, there are several challenging factors that hinder the widespread adoption of microalgae-based WWT systems. Li et al. (2023) have illustrated these challenges and potential solutions diagrammatically in Figure 7.1. Consequently, four primary challenges are associated with WWT using microalgae. The first challenge revolves around the complex and costly pretreatment of wastewater. Wastewater characteristics vary significantly due to differences in source materials and manufacturing processes. For instance, research conducted by Deng et al. (2017) and (2020) revealed that N in silk industry wastewater primarily exists in the form of organic compounds, whereas in swine manure, it is predominantly present as NH_4^+-N. Additionally, different processing units within the silk industry yield wastewater with distinct physicochemical properties. Moreover, agricultural wastewater contains a substantial quantity of toxic compounds, including insecticides, pesticides, heavy metals, and antibiotics (Wang et al., 2018). To prepare most wastewater for microalgae utilization, dilution and pretreatment are necessary, incurring additional treatment costs. Sharmila et al. (2020) have reported that common methods for costly wastewater pretreatment include UV radiation, chemical oxidation, anaerobic digestion (AD), filtration, and autoclaving. In light of these challenges, it is imperative to develop more accessible,

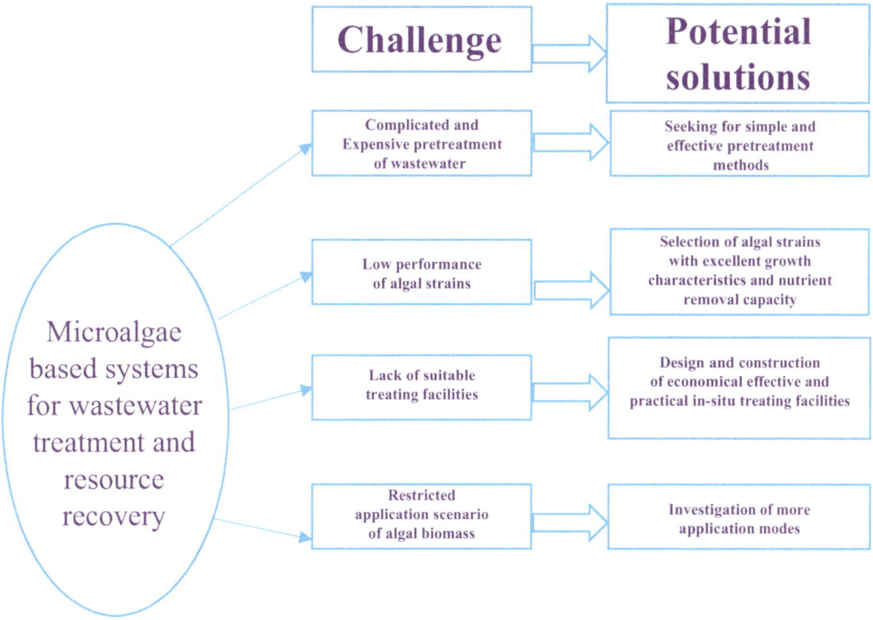

FIGURE 7.1 Challenges and potential solutions of microalgae-based systems for WWT and resource recovery (Li et al., 2023).

convenient, efficient, and cost-effective wastewater pretreatment methods to facilitate the wider application of microalgae-based WWT systems. Another significant challenge in WWT utilizing microalgae is the limited performance of algal strains. Lu et al. (2018) have highlighted the difficulty in controlling microalgal populations in microbial WWT, primarily due to the potential contamination from other microbial communities. Microalgal species exhibit substantial variations in their characteristics, including nutrient removal efficiency, tolerance to toxic compounds, adaptability, and cell composition (Lu et al., 2020). Certain microalgal strains are particularly inhibited by various compounds present in livestock wastewater. However, Franchino et al. (2016) have observed that this inhibition can result in low pollutant removal efficiencies and limited biomass productivity. The selection of a high-performance algal strain for WWT and resource recovery is crucial, as pointed out by López-Sánchez et al. (2022). Furthermore, there is a lack of suitable treatment facilities designed specifically for microalgae-based WWT. Frequently, microalgae cultivation facilities are directly adapted or substituted into the field of microalgae WWT. Open and closed systems are the common choices for WWT using microalgae. Open systems encompass natural lagoons, ponds, artificial ponds, and reservoirs. Leong et al. (2021) have mentioned that open systems like raceway reactors, waste stabilization ponds, and advanced high-rate algal ponds have been employed because of their cost-effectiveness and scalability. However, open systems come with several disadvantages, such as suboptimal light utilization, CO_2 diffusion to the atmosphere, susceptibility to contamination by predators and heterotrophs, evaporative losses, and the need for extensive land areas (Cheng et al., 2021). Consequently, there is a pressing need to enhance the cost-effectiveness and construction of microalgae-based

WWT systems. Li et al. (2023) have comprehensively outlined the challenges and potential solutions in WWT using microalgae. One significant challenge is the limited application scope of algal biomass. While many microalgae have found applications in various industrial sectors, including feed, food, pharmaceuticals, and cosmetics, their cultivation on wastewater introduces heavy metals, pollutants, and pathogens into the biomass, rendering it unsuitable for specific products. Therefore, exploring alternative application modes for microalgal biomass is essential for the future. The Food and Agriculture Organization of the United Nations (FAO) reported 87.0 tons of microalgae cultivation globally, with China accounting for the biggest proportion with 86.6 tons (Ferreira de Oliveira & Bragotto, 2022). China, in particular, has experienced significant microalgae production. Due to their abundance in fatty acids, proteins, carbs, lipids, and different micronutrients, microalgae have also been used in the market as healthy food items in the form of capsules, powders, and tablets (Tang et al., 2020; Okeke et al., 2022; Kusmayadi et al., 2021). According to Ganesan et al. (2019), the CSIR-CSMCRI successfully developed the tube-net technique for the cultivation of *Kappaphycus alvarezii* and *Gracilari dura* on the Gujarat coast. Tube-net systems are resilient to choppy sea conditions and offer high biomass yields. Through active R & D efforts, a number of different culture systems that can tolerate high water dynamics in open ocean locations need to be created. An important obstacle that must be addressed is seasonal dependence on Indian seaweed culture. In India, the Southwest and Northeast monsoon seasons produce strong seawater turbulence and large tidal fluctuations and are occasionally accompanied by cyclones and typhoons. Only seed bank preservation is allowed to be grown at this time. The majority of macroalgae (seaweeds) and microalgae are found in or near marine water, freshwater, and saline water. Compared to microalgae grown in photobioreactors, macroalgae require a significantly larger area and more work to cultivate. In terms of food production and medicine, both have about equal potential. Microalgae can be grown in high salinity and effluent, which has helped to address some environmental issues. The choice between microalgae and macroalgae, then species selection, and the appropriate culture conditions should all be considered in an overview of the many processes involved in the production of algae (Osama et al., 2021). See Table 7.1 (Farghali et al., 2023).

7.5 CASE STUDIES AND SUCCESS STORIES OF USING SEAWEEDS FOR WASTEWATER TREATMENT

Nearly 30% of global aquaculture production is made up mostly of algae, which includes microalgae and seaweeds. The foundation of life on Earth was microalgae and seaweeds, but only lately have some of them been used commercially. The potential for seaweed and microalgae production has gained more attention due to the diverse contributions seaweeds and microalgae provide to human health, the environment, and ecosystem services. Geographically disproportionate output and consumption suggest that seaweed and microalgae agriculture have a lot of room to grow. Consequently, their application in the manufacture of foods, feed, cosmetics, medicines, and other products is now a commercial reality that is growing every year. As a result, it has been suggested that microalgae and seaweeds be employed to improve the sustainability of current industrial activities, such as WWT and biofuel generation. Because of this, the widespread usage of microalgae and seaweeds is regarded as a "green revolution" in human sustainability. The use of microalgae and seaweeds in WWT has proven to be versatile and effective, as demonstrated by various case studies. However, it's essential to recognize that the success of these approaches can be influenced by factors like local conditions, species selection, and operational considerations. The utilization of microalgae and seaweeds in WWT has garnered attention due to their capacity to efficiently remove nutrients and contaminants from wastewater while also generating biomass for various applications. Here are some case studies and success stories highlighting the use of microalgae and seaweeds in WWT. As Olugin (2012) noted, microalgae exhibit the remarkable ability to thrive in diverse environments, from warm tropical regions to cold mountainous areas and polar regions. Their advantages include not requiring fertile land, usable water, and the ability to grow in contaminated waters. Historical examples include the harvesting of *Spirulina* by ancient Mexican populations in the vicinity of Lake Texcoco, a practice also observed in parts of Africa near Lake Chad (Abdulqader et al., 2000). Microalgae have historically played a role in agriculture by fixing atmospheric N to enhance crop production, particularly rice (Muller-Feuga, 2013). Another significant but often overlooked contribution of microalgae is in aquaculture, where natural phytoplankton forms the basis of aquatic feed systems, supporting the production of millions of tons of fish and mollusks in oceans and lakes (Muller-Feuga, 2013). The production of microalgae for various purposes began in the 1950s, with early focus on *Chlorella* cultivation in tubular photobioreactors (Burlew, 1953). Oswald introduced raceway reactors for microalgae production for WWT (Oswald & Golueke, 1968). *Spirulina* and *Dunaliella* are examples of microalgae cultivated at a commercial scale for various applications today. The production capacity for microalgae has been steadily increasing, with an annual growth rate exceeding 10% (Benemann, 2013). Enzing et al. (2014) reported that microalgae-based products in the food and feed sector in Europe contributed significantly to a global marine biotechnology market valued at 2.4 billion in 2011, with an expected annual growth rate of 10%. Large food ingredient companies like BASF, Unilever, and Dow Chemical have invested in microalgae production projects due to their potential in the health food market. Seaweeds have the capability to utilize common N compounds, ammonia-N, and nitrate, often present in agricultural, industrial, and sewage wastewater discharges (Wang et al., 2020a). Large-scale seaweed farms in China have shown remarkable success in nutrient removal and mitigating coastal water eutrophication. For example, seaweed farming in China removed approximately 75 megatonnes of N and 9.5 megatonnes of P (Xiao et al., 2017). The potential for global seaweed harvest by 2050 is estimated to be 500 million tonnes, utilizing a significant portion of water N and P (Jagtap & Meena, 2022). Seaweeds like *Gracilaria lemaneiformis* have been shown to segregate substantial amounts of P, carbon, and N from seawater (Duan et al., 2019). Due to species variety and seasonal change in the chemical properties of the biomass, AD of seaweeds has also shown excellent biogas outputs (Milledge et al., 2019). For example, *Spirogyra neglecta* and *Cladophora glomerata* produced biogas that was high in biomethane (Baltrenas & Misevicius, 2015). *Fucus vesiculosus* and other seaweeds have been successfully used as bioadsorbents to remove cationic dyes from aqueous solutions (Monika et al., 2022).

The development and recycling of biomass production offer solutions to food and feed shortages, energy challenges, climate change mitigation, population growth, and environmental pollution. Seaweeds, as feedstocks, can reduce reliance on fossil fuel resources and support the synthesis of cost-effective and eco-friendly products, biofuels, and sustainable biorefinery processes. While seaweed-based biorefineries are still emerging compared to lignocellulosic biomass from terrestrial plants, they hold great potential (Mohamed et al., 2023). Globally, seaweeds have the potential to sequester substantial amounts of carbon, with the majority being exported to deep water and some being buried in coastal sediments. The future of seaweed utilization includes applications in various sectors, such as antimicrobial, antioxidant, antiviral, anticancer, and more, for products ranging from cosmetics to skincare (Mohamed et al., 2023). In conclusion, the implementation of microalgae and seaweeds in WWT has proven highly promising and effective in addressing diverse environmental challenges. Several case studies and success stories showcase their capabilities in removing pollutants and nutrients from wastewater while offering additional benefits like biomass production and carbon sequestration. Though challenges remain, such as optimizing cultivation techniques and addressing water quality concerns, these success stories underscore the importance of continued research collaboration between academia and industry to achieve more sustainable and efficient WWT processes.

7.6 HOW SEAWEEDS CAN REMOVE POLLUTANTS FROM WASTEWATER

There are substantial environmental stresses brought on by the expanding human population, such as soil, water, and air pollution. Industrialization and urbanization exacerbate pollution, which has a detrimental impact on climate change and ecological stability. Pesticides, antibiotics, steroids, hydrocarbons, medicines, heavy metals, and their byproducts make up the majority of the chemical structure of these pollutants, which eventually cause considerable environmental impact (Ahmad et al., 2021). Water pollution is one of the most serious environmental issues as a result of these pressures, and several reliable procedures are employed to purify the water, but they are highly expensive and not practical. To address the issues connected to conventional approaches, new organic WWT techniques are now being used. Microalgae-based biological water treatment systems have gained popularity, and it is now widely accepted that they are just as effective as traditional water treatment systems. In order to remediate wastewater, limit pollution, and generate energy from microbial biomass, microalgae culture in waste water is a popular practice today. There has been a lot of research done by different biologists on the use of microalgae for WWT, but there is less information accessible on the usage of seaweed. However, due to the presence of macromolecules in the algal cell wall, such as polysaccharides and proteins, which offer diverse functional groups including amino, hydroxyl, carboxyl, and sulfate capable of binding metal ions, seaweeds exhibit high biosorption characteristics (Michalak, 2020). Microalgae possess the remarkable capability to eliminate a wide array of pollutants, including nutrients, heavy metals, and organic and inorganic toxic substances, from wastewater by harnessing sunlight, CO_2, and various nutrients. The key advantage of employing algae for WWT lies in their ability to capture solar energy through chloroplast cells, absorbing CO_2 and nutrients from the wastewater to synthesize biomass and release oxygen. This oxygen is adequate to meet the needs of most aerobic bacteria while they break down any remaining organic matter in the treated wastewater. Furthermore, microalgae release simpler organic compounds into the water, fostering an environment conducive to bacterial growth. Bacteria, in turn, provide the necessary CO_2 for algae growth, stimulate the release of vitamins and organic growth factors, and influence the environmental pH to support algae growth (Sekaran et al., 2013). According to Mantzorou et al. (2018), some microalgal species are obligate heterotrophs and rely on organic carbon from industrial waste products like ethanol, glucose, acetate, and glycerol. Others are mixotrophs, utilizing both facultative CO_2 and organic carbon. They assimilate P and N and convert them into valuable products like proteins, carbohydrates, and lipids. Given that wastewater is typically rich in nutrients, the integration of microalgae into WWT reduces production costs and environmental impact. Combining microalgae with different types of wastewater leads to cost-effective treatment with reduced energy consumption and enhanced environmental sustainability compared to traditional WWT methods. Seaweeds have garnered considerable attention due to their diverse applications, particularly in WWT. Biosorption using seaweed is an attractive method, given its reliance on naturally occurring raw materials and high efficiency with low-cost investment in WWT. Wastewater, a byproduct of various industrial processes and activities, including domestic, textiles, petroleum, and agriculture, is invariably contaminated with organic and inorganic substances. Seaweeds have been extensively studied as adsorbents in WWT, offering an alternative to costly activated carbon (Nithiya et al., 2018). Heavy metals, such as Cr, Ni, Cu, As, Cd, Hg, and Pb, pose significant threats due to their toxic and carcinogenic properties. Various methods, including reverse osmosis, ion exchange, ultrafiltration, and nanofiltration, have been employed to remove these metals, but they often have limitations. In contrast, adsorption, an affordable, efficient, versatile, and easy-to-use technique, has become the preferred choice for heavy metal removal (Rezania et al., 2016). Seaweed biosorption, with its high efficiency, low cost, and minimal environmental impact, holds great promise. Phenolic pollutants, commonly found in wastewater from chemical factories and oil refineries, are toxic, carcinogenic, and mutagenic. They are challenging to treat using conventional methods alone. However, seaweeds have demonstrated effectiveness in the biological adsorption of phenolic compounds (Rubin et al., 2006). Seaweed-based biosorption relies on hydrophobic interactions and donor-acceptor interactions. The process is akin to chlorophenols' adsorption onto activated carbon through donor-acceptor complexation (Aksu et al., 2001). Dyes, another class of organic pollutants in wastewater, originate from various industries, including textiles, paper and pulp, food, cosmetics, and carpet manufacturing. These dyes are often toxic and pose threats to aquatic life. Seaweed has been found to adsorb dyes effectively due to its active functional groups, such as hydroxyl, carboxyl, carbonyl, amine, and sulfate, which facilitate biosorption. Kinetic studies have shown that the process aligns with the pseudo-second-order kinetic model, indicating a chemical reaction involved in removing dyes from wastewater (Mokhtar et al., 2017). In conclusion, seaweed's environmental friendliness, coupled with the affordability and accessibility of raw materials, makes seaweed biosorption a viable alternative to current WWT technologies for efficiently removing pollutants. Additionally, the integrated biorefinery approach involving microalgae and seaweeds not only addresses environmental issues but also serves as a producer of high-value biocompounds such as biofuel, biodiesel, and other valuable compounds.

7.7 THE ROLE OF SEAWEEDS IN NUTRIENT REMOVAL AND RESOURCE RECOVERY

As previously highlighted, algae encompass a diverse group of aquatic photosynthetic organisms, ranging from single-celled microalgae to larger macroalgae. Microalgae, typically a few to several hundred micrometers in size, are rapid reproducers, efficiently absorbing nutrients into their cells under favorable conditions (Nguyen et al., 2022). Similarly, macroalgae or seaweeds, exhibit rapid growth, sometimes reaching lengths of up to 10 m. They possess a thallus structure with gas-filled compartments and an anchoring structure known as a holdfast for attachment (Suganya et al., 2016). Due to their broad range of physiological and biochemical characteristics, both microalgae and macroalgae are proficient at accumulating nutrients, such as N and P, from various aquatic environments. They produce diverse bioactive compounds like pigments, carbohydrates, proteins, and lipids, which find applications in various industries (Suganya et al., 2016; Khoo et al., 2019; Liu et al., 2020). Algae effectively extract nutrients from their surroundings by assimilating N and phosphorylating them. Wastewater contains a mix of organic and inorganic components, including proteins, amino acids, carbohydrates, and lipids, as well as inorganic substances like calcium, phosphate, sodium, potassium, N, chlorine, and sulfate, all of which promote microalgae growth (Li et al., 2019). Previous research has demonstrated that a variety of functional groups within the algal cell wall facilitate pollutant binding, a phenomenon known as biosorption. This rapid and reversible process is independent of microalgae metabolism, making it applicable to both living and non-living biomass (Michalak et al., 2013; Spain et al., 2021). Additionally, certain compounds are absorbed by living algal cells and biodegraded to produce valuable substances. To fully grasp the mechanisms behind pollutant removal, it is essential to understand microalgae's cultivation and growth conditions in wastewater (Ubando, 2021). By integrating WWT, industrial CO_2 sequestration, and the production of valuable microalgal biomass, microalgae-driven processes have the potential to transform WWT from a linear economy to a circular economy (Vaz et al., 2023). Nutrients, organic contaminants, heavy metals, COD, and biological oxygen demand may all be effectively removed or reduced from wastewater by microalgae (Song et al., 2022). For sectors including biofuels, cosmetics, medicines, and nutraceuticals, algae biomass grown in wastewater has the potential to be a lucrative resource (Li et al., 2019). This article conducts a literature review and discussion on WWT and nutrient recovery using microalgae and seaweeds. Different algae species exhibit varying capacities for nutrient removal. Researchers have extensively reported the use of microalgae like *Chlorella, Scenedesmus*, and *Phormidium* for domestic and industrial WWT (Liu et al., 2021, 2022; Morillas-España et al., 2021; Shayesteh et al., 2021). Notably, *Chlorella vulgaris* has been a popular choice due to its adaptability to diverse environments, high growth rates, and tolerance to varying pH, temperature, and salinity levels. Additionally, Valderrama et al. (2002) achieved substantial removal efficiencies, with *Chlorella vulgaris* removing 61% of COD, 76.6% of N, and 28% of P from wastewater produced during ethanol and citric acid production. Seaweeds have garnered attention as well, with reports suggesting that seaweed aquaculture can be an efficient method for remediating nutrients from eutrophic waters like urbanized estuaries, a concept referred to as nutrient bioextraction (Kim et al., 2014). Studies in China have indicated the significant role of seaweed cultivation in nutrient removal, with large-scale seaweed farms removing substantial amounts of N and P (Xiao et al., 2017; Fei, 2004; Shi et al., 2013). Research by Ji-Sook Park et al. (2021) highlighted the nutrient bioextractability of algae and shellfish aquaculture as a cost-effective solution to eutrophication in coastal waters. Additionally, studies have shown that seaweeds, such as *Ulva fasciata*, can effectively reduce ammonia and nitrate levels, enhancing water quality in aquaculture (Ratih Ida Adharini et al., 2021). In summary, the utilization of extractive seaweed aquaculture technologies for nutrient recovery holds the potential to improve water quality, create jobs, and enhance natural resources at a relatively low cost.

7.8 CONCLUSION

One of the most often utilized biotic systems to lessen the strain on nutrient eutrophication brought on by wastewater is the use of seaweeds in WWT. The seaweeds employed in the WWT are physicochemically unique, which restricts their commercialization. As a result, the majority of this biomass is discarded or used to produce energy (where the danger is thought to be low). Such biomass does, however, have the potential to be employed in a number of businesses with greater value in an effort to create a circular economy. However, in order to overcome the current biomass recovery bottleneck, technologies or procedures must be developed to ensure that seaweeds used in WWT can be used in general applications. Numerous scholars have been interested in seaweeds as a result of the large number of documented uses, which has highlighted the significance of seaweeds during this time. The various roles and applications of seaweeds outlined above will necessitate ongoing production of high-quality raw seaweed resources. Seaweed-based industries are predicted to have a chance to improve socioeconomic conditions. As one of the most fascinating and complicated living resources, seaweeds unquestionably provide a variety of potential that we still need to explore. This essay sought to provide a comprehensive overview of seaweeds used in WWT. Using seaweed for biosorption is undoubtedly a promising technique because it uses naturally occurring raw materials, has better efficiency, and requires little upfront investment. It is notable that only a small number of research treated actual wastewater; the majority employed simulation. Therefore, it is advised that future research concentrate more on WWT utilizing actual wastewater. Because they don't closely mirror the real wastewater properties, which are somewhat complicated in nature, synthetic or simulated WWT using seaweed may have significant drawbacks. Furthermore, it's possible that the outcomes of simulating wastewater won't be appropriate to the actual use of WWTPs.

REFERENCES

Abdelfattah, A., Sameh, S. A., Hassan, R., Eslam, I. A., Reham, E., Shih-Hsin, H., Tamer, E., Shengnan, L., Mostafa, M., El-Sheekh., Michael, S., Michael, K., & Jianzhong, S. (2023). Microalgae-based wastewater treatment: Mechanisms, challenges recent advances and future prospects. *Environmental Science and Ecotechnology*, 13:100205.

Abdel-Raouf, N., Al-Homaidan, A. S., & Ibaheem, B. M. (2012). Microalgae and wastewater treatment. Saudi Journal of Biological Sciences, 19:257–275.

Abdulqader, G., Baranti, L., & Tredici, M. R. (2000). Harvest of Arthrospira plantensis from Lake Kossorom (Chad) and its household usage among the Kanembu. Journal of Applied Phycology, 12:493–498.

Aderhold, D., Williams, C. J., & Edyvean, R. G. J. (1996). The removal of heavy-metal ions by seaweeds and their derivatives. Bioresource Technology, 58(1):1–6.

Adharini, R. I., Murwantoko, M., Probosunu, N., Setiawan, R. Y., & Satriyo, T. B. (2021). The effectiveness of seaweeds as biofilter for reducing wastewater nutrient and preventing water pollution from hybrid grouper culture. Jurnal Ilmiah Perikanan dan Kelautan, 13:133–143. https://doi.org/10.20473/jipk.v13i2.28105

Ahmad, I., Abdullah, N., Koji, I., Yuzir, A., & Mohamad, S. E. (2021). Potential of microalgae in bioremediation of wastewater. Bulletin of Chemical Reaction Engineering & Catalysis, 16:413–429.

Ahuja, S., Larsen, M. C., Eimers, J. L., Patterson, C. L., Sengupta, S., & Schnoor, J. L. (eds.). (2014). Comprehensive Water Quality and Purification. Elsevier, Amsterdam.

Aigars, L., & Talis, J. (2017). Reivew on challenges and limitations for algae-based wastewater treatment. Construction Science, 20:17–25.

Aksu, Z., & Yener, J. (2001). A comparative adsorption/biosorption study of mono-chlorinated phenols onto various sorbents. Waste Management, 21:695–702. https://doi.org/10.1016/S0956-053X(01)00006-X

Alami, A. H., Alasad, S., Ali, M., and Alshamsi, M. (2021). Investigating algae for CO_2 capture and accumulation and simultaneous production of biomass for biodiesel production. Science of the Total Environment, 759, 143529. https://doi.org/10.1016/j.scitotenv.2020.143529

Amina, T. (2017). Wastewater screening & classification of screens (complete list) wastewater treatment. Environmental Engineering 17:(2017).

Arumugam, N., Chelliapan, S., Kamyab, H., Thirugnana, S., Othman, N., & Nasri, N. S. (2018). Treatment of wastewater using seaweed: A review. International Journal of Environmental Research and Public Health, 15(12):2851.

Aswara-El, E. I., Gaber, S. E., & Zahran, M. M. (2020). Characterization of biosynthesized silver nanoparticles by Haplophyllum tuberculatum plant extract under microwave irradiation and detecting their antibacterial activity against some wastewater microbes. Desalination and Water Treatment, 195:275–285.

Baltrenas, P., & Misevicius, A. (2015). Biogas production experimental research using algae. Journal of Environmental Health Science and Engineering, 13:18.

Basu, S. A., Dutta, S., Mukherjee, K., & Hossain, S. T. (2021). Exploration of green technology for arsenic removal from groundwater by oxidation and adsorption using arsenic-oxidizing bacteria and metal nanoparticles. In: M. P. Shah, S. R. Couto, & V. Kumar (eds.), New Trends in Removal of Heavy Metals from Industrial Wastewater, Elsevier, pp. 177–211.

Benemann, J. (2013). Microalgae for biofuels and animal feeds. Energies, 6:5869–5886.

Burlew, J. S. (1953). Algal Culture from Laboratory to Pilot Plant. Carnegie Institution, Washington, pp. 235–281.

Burrows, H. D., Santaballa, J. A., & Steenken, S. (2002). Reaction pathways and mechanisms of photodegradation of pesticides. Journal of Photochemistry and Photobiology B: Biology, 67(2):71–108.

Cahill, M. G., Caprioli, G., Stack, M., Vittori, S., & James, K. J. (2011). Semi-automated liquid chromatography–mass spectrometry (LC–MS/MS) method for basic pesticides in wastewater effluents. Analytical and Bioanalytical Chemistry, 400:587–594.

Capodaglio, A. G. (2020). Fit-for-purpose urban wastewater reuse: Analysis of issues and available technologies for sustainable multiple barrier approaches. Critical Reviews in Environmental Science and Technology, 51:1619–1666.

Cheng, Y. W., Lim, J. S. M., Chong, C. C., Lam, M. K., Lim, J. W., & Tan, I. S. (2021). Unravelling CO_2 capture performance of microalgae cultivation and other technologies via comparative carbon balance analysis. Journal of Environmental Chemical Engineering, 9:106519.

Crini, G., & Lichtfouse, E. (2019). Advantages and disadvantages of techniques used for wastewater treatment. Environmental Chemistry Letters, 17:145–155.

Crini, G., & Lichtfouse, E. (2019). Advantages and disadvantages of techniques used for wastewater treatment. Environmental Chemistry Letters, 17:145–155.

Crini, G., Lichtfouse, E., Wilson, L. D., & Morin-Crini, N. (2018). Adsorption-oriented processes using conventional and non-conventional adsorbents for wastewater treatment. In: G. Crini, & E. Lichtfouse (eds.), Green Adsorbents for Pollutant Removal: Fundamentals and Design, Springer, pp. 23–71.

Crook, J., & Surampalli, R. Y. (1996). Water reclamation and reuse criteria in the US. Water Science and Technology, 33(10–11):451–462.

Dagnac, T., Garcia-Chao, M., Pulleiro, P., Garcia-Jares, C., & Llompart, M. (2009). Dispersive solid-phase extraction followed by liquid chromatography–tandem mass spectrometry for the multi-residue analysis of pesticides in raw bovine milk. Journal of Chromatography A, 1216(18), 3702–3709.

Dalvi, V., Naaz, F., Nigam, H., Jain, R., Samuchiwal, S., Kalia, S., Kumar, R., Mathur, M., Bano, F., Malik, A., & Singh, A. (2021). Removal of pollutants from wastewater via biological methods and shifts in microbial community profile during treatment process. In: M. P. Shah, & S. Rodriguez-Couto (eds.), Wastewater Treatment Reactors, Elsevier, pp. 19–38.

Deng, X. Y., Gao, K., Zhang, R. C., Addy, M., Lu, Q., & Ren, H. Y. (2017). Growing Chlorella vulgaris on thermophilic anaerobic digestion swine manure for nutrient removal and biomass production. Bioresource Technology, 243:417–425.

Deng, X., Li, D., Xue, C., Chen, B., Dong, J., & Tetteh, P. A. (2020). Cultivation of Chlorella sorokiniana using wastewaters from different processing units of the silk industry for enhancing biomass production and nutrient removal. Journal of Chemical Technology & Biotechnology, 95:264–273.

Duan, Y., Yang, N., Hu, M., Wei, Z., Bi, H., Huo, Y., & He, P. (2019). Growth and nutrient uptake of Gracilaria lemaneiformis under different nutrient conditions with implications for ecosystem services: A case study in the laboratory and in an enclosed mariculture area in the East China Sea. Aquatic Botany, 153, pp. 11–12.

Ebrahimi, A, & Najafpour, G. D. (2016). Biological treatment processes: Suspended growth vs. attached growth. Iranica Journal of Energy & Environment 7(2):114–123.

Enzing, C., Ploeg, M., Barbosa, M., & Sijtsma, L. (2014). Microalgae-based products for the food and feed sector: An outlook for Europe. EUR 26255. Luxembourg (Luxembourg): Publications Office of the European Union; 2014. JRC85709.

Farghali, M., Mohamed, I. M., Osman, A. I., & Rooney, D. W. (2023). Seaweed for climate mitigation, wastewater treatment, bioenergy, bioplastic, biochar, food, pharmaceuticals, and cosmetics: A review. Environmental Chemistry Letters, 21(1):97–152.

Fei, X. (2004). Solving the coastal eutrophication problem by large scale seaweed cultivation. Hydrobiologia, 512:145–151.

Ferreira de Oliveira, A. P., & Bragotto, A. P. A. (2022). Microalgae-based products: Food and public health. Future Foods, 6:100157.

Franchino, M., Tigini, V., Varese, G. C., Mussat, S. R., & Bona, F. (2016). Microalgae treatment removes nutrients and reduces ecotoxicity of diluted piggery digestate. Science of the Total Environment, 569(570):40–45.

Fujie, K., Bravo, H. E., & Kubota, H. (1983). Operational design and power economy of a rotating biological contactor. Water Research, 17(9):1153–1162.

Ganesan, M., Nitin, T., Vishal, G., Venu, M. S., Radhakrishna, C. R., & Levine, I. A. (2019). Seaweed resources in India – Current status of diversity and cultivation: Prospects and challenges. Botanica Marina, 62(5):463–482.

Gentili, F. G., & Fick, J. (2016). Algal cultivation in urban wastewater: An efficient way to reduce pharmaceutical pollutants. Journal of Applied Phycology, 29:1–8.

Global Carbon Project 2020. (2021). Global carbon atlas. Available at: https://www.globalcarbonatlas.org/en/CO2-emissions (Accessed November 15, 2021).

Golovko, O., Örn, S., Sörengård, M., Frieberg, K., Nassazzi, W., Lai, F. Y., & Ahrens, L. (2021). Occurrence and removal of chemicals of emerging concern in wastewater treatment plants and their impact on receiving water systems. Science of the Total Environment, 754:142122.

Gupta, V. K., Ali, I., Saleh. T. A., Nayak, A., & Agarwal, S. (2012). Chemical treatment technologies for waste-water recycling: An overview. RSC Advances, 2(16):6380–6388.

Jagtap, A. S., & Meena, S. N. (2022). Seaweed farming: A perspective of sustainable agriculture and socio-economic development. In: M. K. Jhariya, R. S. Meena, A. Banerjee, & S. N. Meena (eds.), Natural Resources Conservation and Advances for Sustainability, Elsevier, Amsterdam, pp. 493–501.

Khoo, C. G., Dasan, Y. K., Lam, M. K., & Lee, K. T. (2019). Algae biorefinery: Review on a broad spectrum of downstream processes and products. Bioresource Technology, 292:121964.

Kim, J. K., Kraemer, G. P., & Yarish, C. (2014). Field scale evaluation of seaweed aquaculture as a nutrient bioextraction strategy in Long Island sound and the Bronx River estuary. Aquaculture, 433, 148–156.

Köck-Schulmeyer, M., Villagrasa, M., de Alda, M. L., Céspedes-Sánchez, R., Ventura, F., & Barceló, D. (2013). Occurrence and behavior of pesticides in wastewater treatment plants and their environmental impact. Science of the Total Environment, 458:466–476.

Krantz, D., & Kifferstein, B. (2005). Water pollution and society. Journal of Environmental Law Practice, 4(1).

Kusmayadi, A., Leong, Y. K., Yen, H. W., Huang, C. Y., & Chang, J. S. (2021). Microalgae as sustainable food and feed sources for animals and humans - biotechnological and environmental aspects. Chemosphere, 271:129800

Kuster, M., de Alda, M. L., & Barceló, D. (2009). Liquid chromatography–tandem mass spectrometric analysis and regulatory issues of polar pesticides in natural and treated waters. Journal of Chromatography A, 1216(3):520–529.

Leong, Y. K., Huang, C. Y., & Chang, J. S. (2021). Pollution prevention and waste phycoremediation by algal-based wastewater treatment technologies: The applications of high-rate algal ponds (HRAPs) and algal turf scrubber (ATS). Journal of Environmental Management, 296:113193.

Li, K., Liu, Q., Fang, F., Luo, R., Lu, Q., Zhou, W., Huo, S., Cheng, P., Liu, J., Addy, M., Chen, P., Chen, D., & Ruan, R. (2019). Microalgae-based wastewater treatment for nutrients recovery: A review. Bioresource Technology, 291:121934.

Li, L., Kun, G., Mengting, Y., Qilin, Z., Meng, Z., & Deng, X. (2023). Challenges and potential solutions of microalgae–based systems for wastewater treatment and resource recovery. Frontiers in Bioengineering and Biotechnology, 11:1210228.

Liu, J., Pemberton, B., Lewis, J., Scales, P. J., & Martin, G. J. O. (2020). Wastewater treatment using filamentous algae: A review. Bioresource Technology, 298:122556.

Liu, X. Y., Hong, Y., Zhai, Q.-Y., Zhao, G.-P., Zhang, H.-K., & Wang, Q. (2022). Performance and mechanism of Chlorella in swine wastewater treatment: Roles of nitrogen-phosphorus ratio adjustment and indigenous bacteria. Bioresource Technology, 358:127402. https://doi.org/10.1016/j.biortech.2022.127402

Liu, X. Y., Hong, Y., Zhao, G. P., Zhang, H. K., Zhai, Q. Y., & Wang, Q. (2021). Microalgae-based swine wastewater treatment: Strain screening, conditions optimization, physiological activity and biomass potential. Science of the Total Environment, 807:151008. https://doi.org/10.1016/j.scitotenv.2021.151008

López-Sánchez, A., Silva-Gálvez, A. L., Aguilar-Juárez, Ó., Senés-Guerrero, C., Orozco-Nunnelly, D. A., & Carrillo-Nieves, D. (2022). Microalgae-based livestock wastewater treatment (MbWT) as a circular bioeconomy approach: Enhancement of biomass productivity, pollutant removal and high-value compound production. Journal of Environmental Management, 308:114612.

Lu, L., Guest, J. S., Peters, C. A., Zhu, X., Rau, G. H., & Ren, Z. J. (2018). Wastewater treatment for carbon capture and utilization. Nature Sustainability, 1(12), 750–758.

Lu, W., Asraful Alam, M., Liu, S., Xu, J., & Parra Saldivar, R. (2020). Critical processes and variables in microalgae biomass production coupled with bioremediation of nutrients and CO_2 from livestock farms: A review. Science of the Total Environment, 716:135247.

Mantzorou, A., Navakoudis, E., Paschalidis, K., & Ververidis, F. (2018). Microalgae: A potential tool for remediating aquatic environments from toxic metals. International Journal of Environmental Science and Technology, 15:1815–1830.Michalak, I. (2020). The application of seaweeds in environmental biotechnology. Advances in Botanical Research, 95:85–111. Academic Press.

Michalak, I., Chojnacka, K., & Witek-Krowiak, A. (2013). State of the art for the biosorption process—A review. Applied Biochemistry and Biotechnology, 170:1389–1416. https://doi.org/10.1007/s12010-013-0269-0

Milledge, J. J., Birthe, V. N., Supattara, M., & Patricia, J. H. (2019). A brief review of anaerobic digestion of algae for bioenergy. Energies, 12(6):1166.

Mohamed, F., Israa, M., Mohamed, A., Ahmed, I., Osman, D., & Rooney, W. (2023). Seaweed for climate mitigation, wastewater treatment, bioenergy, bioplastic, biochar, food, pharmaceuticals, and cosmetics: A review. Environmental Chemistry Letters, 21:97–152.

Mokhtar, N., Aziz, E. A., Aris, A., Ishak, W. F. W., & Ali, N. S. M. (2017). Biosorption of azo-dye using marine macro-alga of Euchema spinosum. Journal of Environmental Chemical Engineering, 5:5721–5731. https://doi.org/10.1016/j.jece.2017.10.043

Monika, Y., Sonal, T., & Rajendrasinh, J. (2022). Removal of organic dyes using Fucus vesiculosus seaweed bioadsorbent an ecofriendly approach: Equilibrium, kinetics and thermodynamic studies. Environmental Chemistry and Ecotoxicology, 4:67–77.

Morillas-España, A., Sánchez-Zurano, A., Lafarga, T., Morales-Amaral, M. D. M., Gómez-Serrano, C., Acién-Fernández, F. G., & González-López, C. V. (2021). Improvement of wastewater treatment capacity using the microalga Scenedesmus sp. and membrane bioreactors. Algal Research, 60:102516. http://dx.doi.org/10.1016/j.algal.2021.102516

Morin-Crini, N., Crini, G., & Roy, L. (2017). Eaux Industrielles Contaminées. Presses universitaires de Franche-Comté.

Muller-Feuga, A. (2013). Microalgae for aquaculture: The current global situation and future trends. In: A. Richmond, & Q. Hu (eds.), Handbook of Microalgal Culture: Applied Phycology and Biotechnology, John Wiley and Sons, Ltd. Oxford, UK, 613–627.

Nguyen, L. N., Vu, M. T., Vu, H. P., Johir, M. A., Labeeuw, L., Ralph, P. J., Mahlia, T. M., Pandey, A., Sirohi, R., & Nghiem, L. D. (2022). Microalgae-based carbon capture and utilization: A critical review on current system developments and biomass utilization. Critical Reviews in Environmental Science and Technology, 53:1–23.

Nithiya, A., Shreeshivadasan, C., Hesam K., Sathiabama, T., Norazli O., & Noor S. N. (2018). Treatment of wastewater using seaweed: A review. International Journal of Environmental Research and Public Health, 15(12):2851.

Okeke, E. S., Ejeromedoghene, O., Okoye, C. O., Ezeorba, T. P. C., Nyaruaba, R., & Ikechukwu, C. K. (2022). Microalgae biorefinery: An integrated route for the sustainable production of high-value-added products. Energy Conversion and Management: X, 16:100323.

Olugin, E. J. (2012). Dual purpose microalgae-bacteria-based systems that treat wastewater and produce biodiesel and chemical products within a biorefinery. Biotechnology Advances, 30:1031–1046.

Osama, A., Hadeel, H., & Moussa, M. S. (2021). Potential of household photobioreactor for algae cultivation. Journal of Water and Climate Change, 12:2147–2181.

Oswald, W. J., & Golueke, C. G. (1968). Large scale production of microalgae. In: R. I. Mateless, & S. R. Tannenbaum (eds.), Single Cell Protein. MIT Press, Cambridge, MA, pp. 271–305.

Park, J.-S., Shin, S. K., Wu, H., Yarish, C., Yoo, H. I., & Kim, J. K. (2021). Evaluation of nutrient bioextraction by seaweed and shellfish aquaculture in Korea. Journal of World Aquaculture Society, 52:1118–1134. https://doi.org/10.1111/jwas.12786

Pati, M. P., Sharma, S. D., Nayak, L. A. K. S. H. M. A. N., & Panda, C. R. (2016). Uses of seaweed and its application to human welfare: A review. International Journal of Pharmacy and Pharmaceutical Sciences, 8(10):12–20.

Pereao, O., Akharame, M. O., & Opeolu, B. (2021). Effects of municipal wastewater treatment plant effluent quality on aquatic ecosystem organisms. Journal of Environmental Science and Health, Part A, 56(14):1480–1489.

3053. https://doi.org/10.1016/j.rser.2012.02.055

Reddy, G., & Narsaiah, T. B. (2021). Evaluation of activated carbon efficacy for the treatment of pharmaceutical wastewater Yeddla. International Journal of Engineering Research and Applications 11(5), (Series-VII):31–35.

Rezania, S., Taib, S. M., Din, M. F. M., Dahalan, F. A., & Kamyab, H. (2016). Comprehensive review on phytotechnology: Heavy metals removal by diverse aquatic plants species from wastewater. Journal of Hazardous Materials, 318:587–599. https://doi.org/10.1016/j.jhazmat.2016.07.053

Roy, M., & Saha, R. (2021). Dyes and their removal technologies from wastewater: A critical review. In: S. Bhattacharyya, N. K. Mondal, J. Platos, V. Snášel, & P. Krömer (eds.), Intelligent Environmental Data Monitoring for Pollution Management, Academic Press, pp. 127–160.

Rubin, E., Rodriguez, P., Herrero, R., & de Vicente, M. E. S. (2006). Biosorption of phenolic compounds by the brown alga Sargassum muticum. Journal of Chemical Technology & Biotechnology, 81:1093–1099. https://doi.org/10.1002/jctb.1430

Sekaran, G., Karthikeyan, S., Nagalakshmi, C., & Mandal, A. B. (2013). Integrated Bacillus sp. immobilized cell reactor and Synechocystis sp. algal reactor for the treatment of tannery wastewater. Environmental Science and Pollution Research, 20:281–291.

Sen, B., Alp, M. T., Sonmez, F., Kocer, M. A. T., & Canpolat, O. (2013). Relationship of algae to water pollution and waste water treatment. Water Treatment, 14:335–354.

Sharmila, V. G., Banu, J. R., Kim, S. H., & Kumar, G. (2020). A review on evaluation of applied pretreatment methods of wastewater towards sustainable H_2 generation: Energy efficiency analysis. International Journal of Hydrogen Energy, 45:8329–8345.

Shayesteh, H., Vadiveloo, A., Bahri, P. A., & Moheimani, N. R. (2021). Can CO_2 addition improve the tertiary treatment of anaerobically digested abattoir effluent (ADAE) by Scenedesmus sp. (Chlorophyta)? Algal Research, 58:102379. https://doi.org/10.1016/j.algal.2021.102379

Shi, H., Zheng, W., Zhang, X., Zhu, M., & Ding, D. (2013). Ecological–economic assessment of monoculture and integrated multi-trophic aquaculture in Sanggou Bay of China. Aquaculture, 410:172–178.

Song, Y., Wang, L., Qiang, X., Gu, W., Ma, Z., & Wang, G. (2022). The promising way to treat wastewater by microalgae: Approaches, mechanisms, applications and challenges. Journal of Water Process Engineering, 49:103012. https://doi.org/10.1016/j.jwpe.2022.103012

Sonune, A., & Ghate, R. (2004). Developments in wastewater treatment methods. Desalination, 167: 55–63.

Spain, O., Plöhn, M., & Funk, C. (2021). The cell wall of green microalgae and its role in heavy metal removal. Physiologia Plantarum, 173(2):526–535.

Suganya, T., Varman, M., Masjuki, H. H., & Renganathan, S. (2016). Macroalgae and microalgae as a potential source for commercial applications along with biofuels production: A biorefinery approach. Renewable and Sustainable Energy Reviews, 55:909–941.

Tang, D. Y. Y., Khoo, K. S., Chew, K. W., Tao, Y., Ho, S. H., & Show, P. L. (2020). Potential utilization of bioproducts from microalgae for the quality enhancement of natural products. Bioresource Technology, 304:122997.

Tchobanoglous, G., Burton, F. L., & Stensel, H. (1991). Wastewater engineering. Management, 7:1–4.

Tohamy, R., Al, S. S., Ali, F., Li, K. M., Okasha, Y. A., Mahmoud, T., Elsamahy, H., Jiao, Y., & Fu, J. (2022). A critical review on the treatment of dye-containing wastewater: Ecotoxicological and health concerns of textile dyes and possible remediation approaches for environmental safety. Ecotoxicology and Environmental Safety, 231:113160.

Ubando, A. T. (2021). Microalgal biosorption of heavy metals: A comprehensive bibliometric review. Journal of Hazardous Materials, 402:123431.

Valderrama, L. T., Del Campo, C. M., Rodriguez, C. M., de Bashan, L. E., & Bashan, Y. (2002). Treatment of recalcitrant wastewater from ethanol and citric acid production using the microalga Chlorella vulgaris and the macrophyte Lemna minuscula. Water Research, 36:4185–4192.

Vaz, S. A., Badenes, S. M., Pinheiro, H. M., and Martins, R. C. (2023). Recent reports on domestic wastewater treatment using microalgae cultivation: Towards a circular economy. Environmental Technology & Innovation, 30:103107. https://doi.org/10.1016/j.eti.2023.103107

Veress, M., Bartik, A., Benedikt, F., Hammerschmid, M., Fuchs, J., Müller, S., & Hofbauer, H. (2020). Development and techno-economic evaluation of an optimized concept for industrial bio-SNG production from sewage sludge. In: Proceedings of the 28th European Biomass Conference, Florenz, Italy.

Vryzas, Z., Vassiliou, G., Alexoudis, C., & Papadopoulou-Mourkidou, E. (2009). Spatial and temporal distribution of pesticide residues in surface waters in northeastern Greece. Water Research, 43(1):1–10.

Wang, L., Jayawardena, T. U., Yang, H. W., Lee, H. G., & Jeon, Y. J. (2020). The potential of sulfated polysaccharides isolated from the brown seaweed Ecklonia maxima in cosmetics: Antioxidant, anti-melanogenesis, and photoprotective activities. Antioxidants, 9(8):724.

Wang, S., Zhao, S., Benjamin, Uzoejinwa, B., Zheng, A., Wang, Q., Huang, J., & Abomohra, A. E.-F. (2020b). A state-of-the-art review on dual purpose seaweeds utilization for wastewater treatment and crude bio-oil production. Energy Conversion and Management, 222:113253. https://doi.org/10.1016/j.enconman.2020.113253

Wang, T., Ai, S., Zhou, Y., Luo, Z., Dai, C., & Yang, Y. (2018). Adsorption of agricultural wastewater contaminated with antibiotics, pesticides and toxic metals by functionalized magnetic nanoparticles. Journal of Environmental Chemical Engineering, 6:6468–6478.

Wang, Y., Yang, Q., Chen, J., Yang, J., Zhang, Y., Chen, Y., Xiaoqiang, L., Wei, D., Aiping, L., Shih-Hsin, H., & Jo-Shu, C. (2020c). Adsorption behavior of Cr(VI) by magnetically modified Enteromorpha prolifera based biochar and the toxicity analysis. Journal of Hazardous Materials, 395:122658.

Yadav, K. K., Gupta, N., Kumar, V., & Singh, J. K. (2017). Bioremediation of heavy metals from contaminated sites using potential species: A review. Indian Journal of Environmental Protection, 37(1):65–84.

Yang, J., Xu, M., Zhang, X., Hu, Q., Sommerfeld, M., & Chen, Y. (2011). Life-cycle analysis on biodiesel production from microalgae: Water footprint and nutrients balance. Bioresource Technology, 102(1):159–165.

Znad, H., Awual, M. R., & Martini, S. (2022). The utilization of algae and seaweed biomass for bioremediation of heavy metal-contaminated wastewater. Molecules, 27(4):1275.

8 Open and Closed Photobioreactors for Algal Wastewater Treatment

R. Raksasat, U. Suparmaniam, H. Rawindran, and J.W. Lim

8.1 INTRODUCTION: BACKGROUND AND DRIVING FORCES

A staggering 80% of worldwide sewage remains untreated, indiscriminately released into natural water bodies such as seas, oceans, and rivers [1]. This reckless discharge not only contaminates drinking water sources but also wreaks havoc on delicate ecosystems. Exposure to wastewater pollutants poses severe threats, leading to coral bleaching, diseases, and mortality in aquatic life, as well as increasing the risk of human ailments such as diarrhea, cholera, dysentery, typhoid, hepatitis A, and polio [2,3]. Consuming chemically contaminated water exacerbates conditions and can potentially causing cancer, hormone disruption, altered brain function, and harm to the immune, reproductive, and cardiovascular systems, including kidney failures [2,3]. Consequently, wastewater undergoes rigorous treatment utilizing various techniques, with traditional methods like disinfection, especially chlorination, ozone treatment, ultraviolet light exposure, and chloramines being commonly employed to purify its quality [4]. The volatility of chlorine compounds renders incomplete oxidization detrimental to effluent toxicity, generating byproducts in the process [4]. Hence, in the quest for sustainable solutions to address the escalating environmental challenges associated with wastewater treatment, the potential of algae has emerged as a beacon of promise [5].

Algae, a diverse group of photosynthetic organisms, encompass a vast array of single-celled and multicellular species that thrive in aquatic environments. Ranging from microscopic phytoplankton to macroscopic seaweeds, algae play a pivotal role in global ecosystems by converting sunlight into energy through photosynthesis [6]. Moreover, algae exhibit immense potential for applications in various industries, such as biofuel production, pharmaceuticals, cosmeceuticals, nutraceuticals, aquaculture, the food sector, and even for the generation of bioplastics and biodegradable materials, marking them as versatile and pivotal contributors to both environmental sustainability and industrial innovation [7]. In addition, their unparalleled adaptability and capacity to produce O_2, sequester CO_2, and serve as a fundamental link

 DOI: 10.1201/9781003459286-8

in aquatic food chains underscore their ecological significance [8]. Noteworthily, algae can serve as an excellent candidate for wastewater bioremediation due to their inherent ability to absorb and assimilate nutrients like nitrogen (N) and phosphorus (P) from wastewater sources for their cellular growth [9]. The O_2 produced through photosynthesis in turn helps in creating conditions that support aerobic bacteria, which are essential for the decomposition of organic matter [5,10,11]. Harnessing the biomass generated by algae in wastewater treatment not only purifies water but also acts as a potent source for biofuel production, embodying a dual-purpose strategy that simultaneously tackles environmental concerns and advances the production of renewable energy resources [7]. Each algal species exerts a unique influence on the efficacy of treatment methodologies. Notably, the integration of photobioreactors (PBRs) amplifies the potential for harnessing the photosynthetic prowess of algae. In this chapter, we explore the synergies between algal diversity and PBR technologies for the sustainable purification of water resources, which also unveils the collective potential of these biological and technological systems to propel environmentally friendly wastewater treatment practices forward.

8.2 WASTEWATER SOURCES AND TREATMENT TECHNOLOGIES

Every year, a staggering 380 billion cubic meters of municipal wastewater, stemming from diverse sources including petrochemical industries, pulp and paper manufacturing, food processing, farming, households, and natural runoff effluents like stormwater, inundate the global landscape [12]. Projections indicate a concerning 24% increase by the year 2030 [12]. It is estimated that approximately 359.4 billion cubic meters of global wastewater will be generated annually, with a concerning 48% of its production, equivalent to around 171.3 billion cubic meters per year, expected to remain untreated and discharged into water bodies [13]. Wastewater exhibits a diverse array of characteristics contingent upon its origin. Yet, regardless of the source, common impurities persist, comprising organic substances, vital nutrients like N and P, suspended solids, and hazardous elements such as heavy metals and toxic chemicals. Additionally, harmful microorganisms, encompassing viruses, bacteria, fungi, and parasites, further contribute to the intricate composition of wastewater. The exposure to elevated concentrations of these pollutants poses a grave threat to ecosystems and jeopardizes animal life, including human health. Adverse effects range from severe stomach cramps and diarrhea to vomiting and infections [14,15]. A pertinent illustration is drawn from the findings of Affam et al. [16], who reported that a typical wastewater sample collected from a local downstream area contained 2,030 mg/L of total suspended solids (TSSs), 548 mg/L of chemical oxygen demand (COD), and 9.6 mg/L of biochemical oxygen demand (BOD). These heightened TSS and COD values signify increased bacterial pollution, indicative of elevated levels of nutrients, metals, and organic matter in the effluent. The consequences of such contaminated wastewater underscore the urgency and importance of effective treatment measures to mitigate environmental impact and safeguard public health.

The regulation of wastewater discharge limits varies based on its designated purpose and the governing organization's specifications. According to the Central Pollution Control Board located in India, the acceptable limit of wastewater discharged on inland surface water is defined at 30 mg/L of BOD, 250 mg/L of COD, 10 mg/L of oil and grease concentration, 100 mg/L of TSSs, 20 mg/L of Kjeldahl N content, and pH at 7.5 [17]. The World Health Organization (WHO) imposes the maximum permissible limits of heavy metals presented in drinking water as follows: 0.2 mg/L of Al, 0.1 mg/L of Mn, 0.3 mg/L of Fe, 0.07 mg/L of Ni, and 3 mg/L of Zn [18]. Similarly, Malaysia's environmental law sets the effluent standard limit upon discharging into other inland waters or Malaysian waters at 50 mg/L of BOD concentration, 100 mg/L of COD level, 10 mg/L of oil and grease, 100 mg/L of TSS, 5 mg/L of Fe, and 1 mg/L of each Mn, Ni, and Zn.

The common conventional process for wastewater treatment is known as chlorination disinfection due to its economics and simple procedure by adding chlorine for a chemical process to oxidize harmful organisms such as pathogens, bacteria, and viruses presented in effluent [19]. Chlorine will weaken the microbe's cell membrane and inhibit its respiratory system and DNA activity accordingly. On the downside, protozoan cyst species are tolerant toward chlorine activity, and the byproduct is formed due to the interaction between chlorine and organic substances [19,20]. Another uncomplicated wastewater treatment, typically used, is the boiling technique, in which the sewage is heated to a high temperature to deactivate living pathogenic organisms [21]. However, the high energy input to elevate the temperature, as well as the time-consuming are required and the removal of turbidity, including heavy metals, taste, and odor are incapable by application of this method [22]. Apart from the traditional wastewater treatments, membrane-involved methods, like membrane bioreactors could occur membrane fouling, resulting in frequent maintenance is needed and high energy consumption for the cleaning process [23,24]. Thus, an innovative promising method to purify wastewater by employing algae-assisted has been investigated.

8.3 ALGAE AND ITS BENEFITS

Algae is a large-diverse group of eukaryotic photosynthetic organisms present in aquatic ecosystems, including salt or marine water, fresh water, and brackish water bodies, while the photosynthetic bacteria, like cyanobacteria, will be called blue-green algae instead of algae [25–27]. The indispensable pigment chlorophyll assumes a pivotal role in converting sunlight into a nutritive substrate for algal growth through the process of photosynthesis. Remarkably, over 50% of primary photosynthesis is attributed to algae, making them key contributors to this vital ecological process [26,27]. In essence, algae emerge as primary producers of atmospheric O_2, highlighting their profound impact on global O_2 levels. Beyond sunlight, the prerequisites for algal photosynthetic activity encompass essential factors such as water sources, optimal environmental conditions (including pH and temperature), nutrients, ample space, and the presence of CO_2 in the air [28,29]. Belonging to the phyla Rhodophyta (recognized for red algae containing phycoerythrin pigment), *Chlorophyta* (characterized by green algae with chlorophyll pigment), and *Phaeophyta* (comprising brown algae with carotenoid pigment), exhibit a diverse taxonomy of algae. They can be broadly categorized into two major types: microalgae and macroalgae. Microalgae, represented by unicellular entities like phytoplankton, contribute to this classification. On the other hand, macroalgae encompasses multicellular plants, exemplified by seaweed [26,30]. Prominent among microalgal species are dinoflagellates, diatoms, and chlorella, each harboring specific types such as *Alexandrium catenella*, *Aureococcus anophagefferens*, *Prorocentrum donghaiense*, *Phaeocystis*, *Karenia mikimotoi*, *Noctiluca scintillans*, and *Skeletonema costatum*; while *Enteromorpha* spp., *Codium isthmocladum*, *Ulva*, and *Sargassum* are the dominant macroalgal species [31,32].

The growth mechanism of microalgae can be divided into five phases as reported by Price and Farag [29]. Initially, algae slightly grow up in the lag phase prior to entering the rapid growth in the exponential phase, where the cell will be usually cleaved. Once any of the essential resources required for algal development are limited, the production rate of algal cell concentration over cultivation time is then reduced, which is called the declining relative growth phase. Algae is subsequently stepped into a stationary phase, in which the growth ceases but the cell division remains active at a gradual rate, upon which the energy sources are completely deficient. Finally, algal cell starts to perish in the death or lysis phase due to the resource's scarcity [29]. Precautionary measures against overgrown algae, particularly in warmer waters, are essential to counter the occurrence of eutrophication, manifested as algal blooms. This phenomenon arises from excessive nutrient levels, predominantly N and P, present in wastewater or stormwater runoff that ultimately finds its way into waterways [33]. The surge in algal population, characterized by water discoloration, often turning green or reddish-brown due to pigments, can lead to detrimental effects. Notably, the infamous "red tides", attributed to microscopic algae in seawater with peridinin pigment, result in a vivid red coloration [31,32]. Brown tides, caused by marine algae like pelagophytes (e.g., *Aureococcus anophagefferens*), turn water brown [31,32]. The environmental repercussions include sunlight blockage, O_2 depletion, and the production of gaseous toxins due to the dense accumulation of microalgae [26]. While less harmful occurrences, such as green spectacles (possibly caused by unicellular *Phaeocystis* algae but mainly contributed to by macroalgae) and golden tides of *Ulva* and *Sargassum* macroalgal blooms, may not excrete toxins, their massive accumulation, especially under anoxic conditions, poses a potential threat to coastal ecosystems and biota. This threat stems from either nuisance or the production of toxic hydrogen sulfide (H_2S) upon the degradation of biomass compounds [31,32,34,35].

While uncontrolled algae growth can pose environmental challenges, adopting careful management practices opens the door to a multitude of benefits across diverse sectors. Notably, the simplicity of the algal cultivation process and its high efficiency in carbon fixation stand out as key advantages. Beyond these, the utilization of algal biomass holds significant potential for various applications. The rich nutrient content, encompassing organic substances and bioactive metabolites, particularly prevalent in red and brown algae, offers substantial benefits to agriculture and ecosystems. When employed as a biofertilizer, algal biomass not only accelerates the decomposition process but also strengthens soil bonding, reducing soil loss during stormwater runoff. Furthermore, this application enhances the water retention capacity of the soil, contributing to a more resilient and sustainable agricultural and environmental landscape [7,36]. The deposition of algae into the soil could assist the aeration as well as oxidation in soil, which facilitates the nutritional uptake, particularly potassium constituted in algae to nourish its root part and enhance the drought tolerance; thereby, stunted plant growth and unproductive harvesting yield are prevented [36]. The diversity of algal species, including its pigments, is beneficial to color utilization-associated industries such as artificial chemical dyes, coloring agents, and food coloring due to their economical and sustainable resources [7]. Cultivated as aquaculture in the food industry, certain edible algae, such as seaweed rich in essential micronutrients, and extracted algal compositions like carrageenan, a crucial thickener additive stabilizing various products like icecream, toothpaste, and cosmetics, are poised to play a significant role in providing food additives and algal-based meals. The global market value for these algal products is projected by Choudhury [37] to reach an estimated US$ 349.1 million by 2033 [7].

An indispensable product derived from algal biomass is the well-known agar, a gelatinous substance composed of polysaccharides extracted from the cell walls of red algae, specifically ogonori (*Gracilaria verrucosa*) and tengusa. Agar serves a dual purpose, acting as both an additive to stabilize, texturize, and thicken various foods and beverages and as a medium for bacterial cultivation or the gel electrophoresis process [7,38]. Additionally, the remarkable potential of microscopic diatom skeletons, formed through the biomineralization process using silica, has been uncovered. These fossilized diatoms can be harnessed to create diatomaceous earth, offering versatile applications in both the food industry (as a pesticide, insecticide, or anti-caking agent) and non-edible industrial sectors (notably water filtration due to its exceptional water-absorption capabilities or for use in dynamite production) [39,40]. Microalgae, particularly, exhibit a remarkable capability to biodegrade plastic polymers through their consumption of organic polymers as sources of feeding energy for growth. Subsequently, they metabolize these polymers under either aerobic or anaerobic mechanisms, resulting in the production of final algal biomass [41]. The enzymes synthesized by algae, inclusive of a toxin system, play a crucial role in weakening the chemical bonds within the polymer structure. This, in turn, reduces the energy required to reach the activation energy for polymer degradation, thereby enhancing the efficiency of biodegradation [41,42]. Additionally, polymers derived from algae, based on proteins and carbohydrates, present a promising avenue for environmentally friendly materials. These algae-derived polymers have the potential to be utilized in the production of biodegradable plastics, commonly referred to as bioplastics. In essence, the symbiotic relationship between microalgae and plastic degradation not only underscores their ecological significance but also positions them as key contributors to sustainable and eco-friendly material solutions [41].

The exploration of algal biomass as a sustainable raw material for biofuel production, specifically bioethanol and biodiesel, has gained significant attention in addressing the limitations of traditional fossil fuels and previous feedstocks. Currently, fossil fuels, comprising 35% crude oil, 29% coal, and 24% natural gas, stand as the world's primary non-renewable energy source [43]. However, their combustion releases a plethora of harmful gases, including greenhouse gases and pollutants, contributing to global warming, climate change, and environmental degradation [44]. To mitigate these issues, the International Energy Agency projects a 25% increase in global biofuel production by 2024 [45]. Initially, first-generation biofuels derived from edible crops raised concerns over large agricultural areas and potential food shortages [43]. Consequently, non-edible biomass, such as lignocellulosic materials from agricultural byproducts and waste cooking oil, replaced them in second-generation biofuel production [43,46]. Despite the challenges of complex transformation processes and high operating costs associated with second-generation feedstock, algae emerge as a promising third-generation alternative [47]. Algae offer benefits such as reduced environmental impact, high performance in terms of yield and biofuel quality, widespread availability, and economic viability. Notably, biodiesel produced from algal lipids display significant fatty acid methyl esters (FAME) compositions. For instance, lipid extracted from *Chlorella sorokiniana* microalgae for biodiesel production exhibits a predominant myristic acid (14:0) composition at 19.5%, along with palmitic acid (16:0) and stearic acid (18:0) at 16.4% [48]. Additionally, the cultivation of algae for biofuel production contributes to carbon capture, with approximately 1.83 tons of atmospheric CO_2 captured by 1 ton of algal biomass [43]. Liu et al. [43] propose that macroalgae offer a simpler and more cost-effective harvesting process compared to microalgae. Essentially, harnessing algal biomass for diverse applications holds the promise of a sustainable solution that not only mitigates environmental impacts but also extends to potential applications in wastewater treatment, a topic we will delve into in subsequent discussions.

8.4 THE POTENTIAL OF ALGAE FOR WASTEWATER TREATMENT

In the ongoing quest for sustainable solutions to environmental challenges, the utilization of algae for wastewater treatment stands out as a pioneering and impactful approach. Algae, with their remarkable ability to thrive in diverse aquatic environments, offer a natural and efficient means of wastewater remediation. In algal-based wastewater treatment plants, nutrient-rich wastewater, specifically N and P, crucial for promoting algal growth, is treated under aerobic conditions facilitated by the O_2 produced by algae. During cultivation phases, algal cells uptake the essential nutrients, concurrently stripping ammonia N. This synergistic process leads to the subsequent decomposition of organic contaminants by aerobic bacteria, underscoring the efficiency of algal-driven purification systems [5,10,11]. In the initial stages, algae cultivation within wastewater mediums involves exposure to sufficient natural or artificial illumination, harnessed and collected by PBRs. It is crucial to acknowledge that surplus light, not utilized for photosynthesis, transforms into thermal energy, thereby elevating system temperatures and potentially compromising algal biomass integrity [49]. To optimize the photosynthesis reaction, CO_2 is then introduced into the PBRs, ensuring a consistent and even supply to algal cells through continuous circulation facilitated by centrifugal or airlift pumps, thereby enhancing gas-liquid transport efficiency [50,51]. Under optimal conditions, CO_2 will be photosynthesized into O_2 and glucose by algae, as shown in Eq. 8.1:

$$6CO_2 + 6H_2O + \text{sunlight} \rightarrow C_6H_{12}O_6 + 6O_2 \tag{8.1}$$

To maintain optimal algal growth, excess O_2 is efficiently degassed through a de-passing port, preventing inhibition [50]. Upon achieving desired productivity, algae separation employs two primary methods: (i) the flocculation process, aggregating biomass for subsequent filtration, flotation, or gravity sedimentation, and (ii) thickening post-separation before entering the dewatering stage [52]. On the other hand, cultivation in open raceway ponds, characterized by natural sunlight exposure, offers a cost-effective approach with lower energy requirements compared to PBRs. These ponds are equipped with paddle wheels for circulation and baffles for turbulence, optimizing the algae cultivation system [53].

Optimizing various operational parameters, such as nutrient concentration, pH, temperature, initial contaminant concentration, light intensity, and aeration duration, is crucial for enhancing both the wastewater treatment efficiency and biomass productivity of algae [10,54]. The choice of algal strains further influences treatment performance, with considerations like photoinhibition thresholds and strain-specific responses to light intensity variations [10]. Bohutskyi et al. [55] observed a noteworthy phenomenon in the context of photoinhibition within *Chlorella sorokiniana* microalgae. Surprisingly, despite exposing the microalgae to a high range of light intensity (140–200 μmol photon/m^2/s), no evidence of photoinhibition was detected. This discovery accentuates the resilience of *Chlorella sorokiniana* to elevated light levels. Conversely, the study revealed that different microalgal strains responded diversely to light stress, with some species exhibiting significant growth inhibition once the light intensity surpassed the saturation point of 150 μmol photon/m^2/s [55]. The optimization of irradiance intensity and prolonged exposure to light has been found to significantly enhance nutrient removal in wastewater treatment. In a study, *Chlorella vulgaris* demonstrated remarkable efficiency, achieving a 99% reduction in N substances and a substantial 67.6% reduction in P under optimal conditions. This optimal performance was achieved with a specific light intensity set at 180 μE/m^2/s and a continuous light exposure cycle of 24 hours [56]. However, a reduction in light intensity to 36 μE/m^2/s under the same cycle led to a decline in removal efficiency to 97.1% for N and 29.3% for P. Likewise, a shortened light period to 10:14 hours of light:dark ratio, maintaining the optimal light intensity, resulted in 86.2% N removal and 18.3% P reduction [56]. Parallel findings in *Pseudokirchneriella subcapitata* microalgae reiterated this sensitivity to operational parameters. A 14-hour reduction in the light period led to a decrease of 31.1% in N and 47.3% in P removal, while a lower light intensity reduced performance by 12% for N and 33% for P removal [56]. These insights underscore the intricate relationship between irradiance conditions and treatment efficacy, offering valuable guidance for optimizing nutrient removal in algal-based wastewater treatment systems.

Furthermore, a group of researchers explored the impact of light quality, comparing white LEDs to blue and red LEDs. The findings unveiled that white LEDs significantly enhanced the efficiency of *Chlorella pyrenoidosa* microalgae's growth, contrasting with the less favorable performance observed under blue light [57]. Notably, the red LEDs exhibited the poorest performance among the three color variations, shedding light on the nuanced relationship between light characteristics and microalgal responses [57]. The performance and biomass quality of algal wastewater treatment are also intricately influenced by the concentration of nutrients present in the wastewater. Insufficient N in the feeding medium is linked to diminished protein accumulation in algal biomass, with a consequential increase in lipid or carbohydrate production [58]. This observation aligns with the findings of Zhu et al. [59], who asserted that elevating the initial nutrient concentration leads to a decrease in lipid accumulation in *Chlorella zofingiensis* biomass. The nexus between nutrient availability and the compositional dynamics of algal biomass underscores the pivotal role of nutrient concentration in shaping the outcomes of wastewater treatment processes, shedding light on a critical parameter for optimizing both performance and product quality in algal-based remediation strategies [49]. Maintaining an optimal surrounding temperature is imperative for maximizing cultivation performance and preventing the production of unsaturated lipids in algal biomass [60]. Algae, categorized as psychrophilic (17°C), mesophilic (20°C–25°C), and thermophilic (40°C), exhibit varying temperature preferences, exemplified by strains like *Asterionella formosa*, *Chaetoceros* sp., and *Anacystis nidulans* [10]. Additionally, the color and salinity conditions of wastewater play a pivotal role, impacting photosynthetic efficiency. Darker hues and higher salinity hinder photosynthesis by obstructing light permeation and limiting algae accessibility due to reduced CO_2 availability, often converted into carbonate (CO_3^{2-}) in higher pH conditions [10,61]. Conversely, severe acidity should be cautiously addressed, as it can impede algal growth, diminishing wastewater treatment efficiency [62].

Alazaiza et al. [10] demonstrated the remarkable efficacy of microalgae in removing pollutants from domestic wastewater through the implementation of air-bubbling reactors and fluorescent lamps. Impressively, this approach achieved removal efficiencies of 84% for COD, 95% for ammoniacal N, and 97% for P. However, the study revealed a critical limitation: algal growth inhibition occurred as light intensity decreased, attributed to elevated concentrations of suspended particles and turbidity in the sewage [55]. The exacerbation of COD concentration was linked to prolonged aeration periods, resulting in the decay of accumulated algal biomass and a subsequent rise in organic concentrations [10]. Notably, the synergistic action of microalgae generating oxygen, coupled with the biological oxidation of carbon-rich compounds by heterotrophic bacteria, contributed to the removal of over 90% of pathogenic bacteria from wastewater. This dual mechanism proved instrumental in achieving substantial reductions in both COD and BOD, underscoring the potential of microalgae-based systems in fostering efficient and comprehensive wastewater treatment [10,63]. In an insightful study conducted by Xin et al. [62], the influence of various N sources present in wastewater on *Scenedesmus sp.* microalgal cultivation was systematically explored. Notably, the choice of N source exerted a profound impact on the different phases of algal growth. Ammonium-derived N demonstrated the fastest initial algal growth, albeit leading to the lowest growth at the stationary phase [62]. This phenomenon was attributed to the acidity induced by the presence of ammonium in the cultivation environment. In contrast, the nitrate-N source emerged as the most efficient, achieving an impressive 90.4% total N removal and over 99% P removal [62]. This underscored its superior performance compared to urea and ammonium sources. Ammonium, characterized by the least dense algal population at the stable phase, exhibited a limited nutrient purification capacity, achieving only 31.1% total N and 76.4% P reduction [62]. These findings elucidate the pivotal role of N source selection in governing the dynamics of microalgal cultivation and its consequential impact on nutrient removal efficiency. All in all, optimizing proper

TABLE 8.1
The Efficiency of Algal Wastewater Treatment Systems under Various Operating Conditions

Algae	Optimal Conditions	Removal Efficiency	Biomass Production	References
Chlorella Vulgaris	• 50% wastewater concentration • Wastewater: algae (125:25 mL)	• 84% of COD • 95% of total N • 97% of total P	0.033 g/L/day	[10]
	• 180 μE/m²/s of light intensity	• 99% of total N • 67.6% of total P	0.133 g/L/day	[56]
Pseudokirchneriella subcapitata	• Light (L): Dark (D) at 24:0 hours	• 100% of total N • 1.2% of total P	0.022 g/L/day	
	Except for *P. subcapitata* productivity obtained at 10:14 hours of L:D and 36 μE/m²/s of light intensity			
Chlorella sorokiniana	• L:D at 24:0 hours	• 82.2% of total N • 83.8% of total P	0.488 g/L/day	[54]
Chlorella pyrenoidosa	• L:D at 18:6 hours • 24 hours of illumination time • White light	—	0.0851 g/L/day	[57]
Chlorella zofingiensis	• 230 μE/m²/s of light intensity • L:D 24:0 hours • High initial N medium (148 mg/L)	• 80% of total N	—	[59]
Scenedesmus sp.	• L:D at 14:10 hours • Employing nitrate-N source	• 90.4% of total N • >99% of total P	—	[62]

operating parameters is essential for efficiently removing impurities, a crucial consideration highlighted by various algal wastewater treatment performances detailed in Table 8.1.

8.5 PHOTOBIOREACTORS (PBRS) SYSTEM FOR ALGAL WASTEWATER TREATMENT

The application of PBRs in wastewater treatment represents a cutting-edge approach, transforming contaminants into valuable biochemical products, particularly algal biomass through cell-associated biotransformation processes. Distinguishing themselves from conventional bioreactors, PBRs stand out for their eco-friendly and efficient attributes, leveraging light as the primary energy source instead of relying on chemical reactions [64]. Functioning as enclosed-illuminated systems, PBRs cultivates photoautotrophic organisms, predominantly algae, on a large scale. Operating under controlled conditions, this closed environment minimizes the risk of contamination, prevents CO_2 and water loss, and ensures desired biomass production through the photosynthesis process [65,66]. Notably, approximately 71% of microalgae production is achieved through PBRs, highlighting their pivotal role in sustainable biomass cultivation, while 68% of macroalgal biomass is sourced from harvesting wild stock natural resources [67]. This paradigm shifts toward PBRs underscores a transformative stride in wastewater purification, aligning with the principles of environmental sustainability and resource optimization.

8.5.1 Agitated PBRs

Ensuring optimal agitation is a critical factor in the effective operation of PBR systems. The primary objective is to prevent the sedimentation of algal cells at the reactor bottom and to achieve a homogeneous blend of solid organisms, liquid culture medium, and air phases [51]. Striking the right balance is essential, as excessive agitation leading to turbulent shear stress can result in cell collapse. Therefore, a careful approach is needed to avoid overmixing [68]. The methods employed for bubble distribution are diverse, with four main categories: airlift, bubble, stirrer or impeller, and the use of a pump [65]. Each of these approaches play a crucial role in maintaining an optimal environment within the PBR, facilitating efficient nutrient distribution and gas exchange without compromising the structural integrity of the algal cells. This nuanced interplay between agitation techniques and their impact on cellular well-being underscores the significance of a well-calibrated and precisely controlled agitation strategy in PBR systems.

Airlift PBRs represent an innovative class of bubble vertical columns propelled pneumatically, distinguishing themselves by the incorporation of an internal baffle or concentric draft tube [64,68,69]. This design, depicted in Figure 8.1, offers enhanced mass transfer capabilities compared to bubble columns but introduces a higher degree of complexity. Unlike bubble columns, airlift PBRs lack a draft tube. The choice between an internal or external connection of the draft tube, as illustrated in Figure 8.1, adds another layer of design intricacy. The absence of a draft tube in bubble columns, while simpler, comes with limitations, as the use of stirrers or impellers can potentially damage organism cells [64,68,69]. The internal-loop configuration, exemplified in Figure 8.1a, utilizes a draft tube to facilitate the upward flow of liquid in the riser current, with CO_2 aeration occurring through a sparger. The gas is released at the column's top, while the denser, bubble-free liquid descends in the downcomer current for recirculation in the airlift reactor [69]. Another internal airlift configuration involves employing an internal baffle, as shown in Figure 8.1b, where air dispersal and downcomer liquid flow are managed on opposite sides of the baffle. In the external-loop system illustrated in Figure 8.1c, the draft tube connects the top and bottom of the column externally, creating a more efficient degassing and circulation process. This external configuration, characterized by a greater density differential after gas exhaustion, further accelerates the circulating process compared to internal arrangements [69].

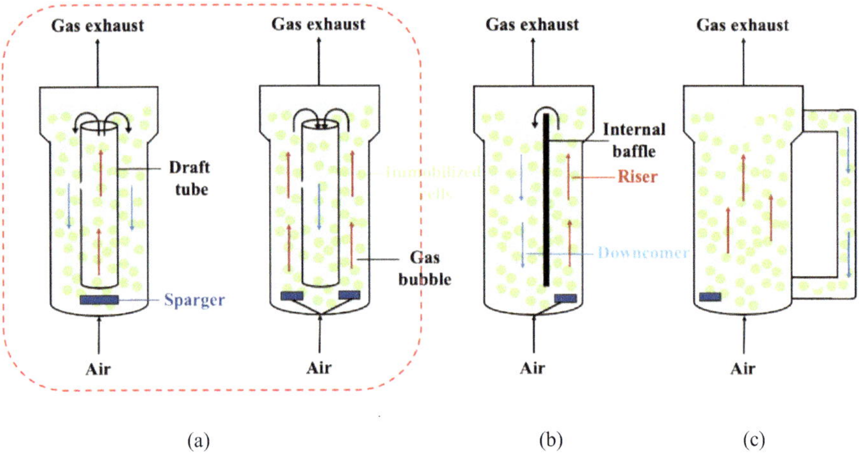

FIGURE 8.1 Different types of airlift PBR configurations: (a) draft-tube internal loop; (b) internal baffle system; and (c) external loop.

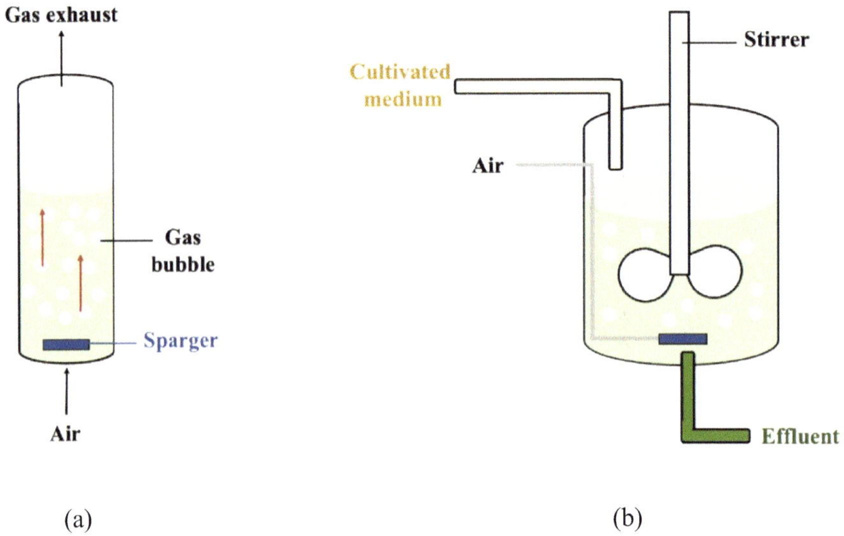

FIGURE 8.2 (a) Bubble column PBRs and (b) stirred-tank PBRs.

The configuration of bubble column PBR involves the generation and aeration of bubbles from the bottom, as depicted in the straightforward setup illustrated in Figure 8.2a. The critical consideration of an optimal gas flow rate becomes paramount, given its direct impact on the ability of algal cells to absorb light [70]. It is noteworthy that bubble size plays a pivotal role, as larger bubbles can accelerate liquid velocity in the downcomer. However, this acceleration may not guarantee sufficient contact time between cells and light absorption, potentially leading to shear stress on

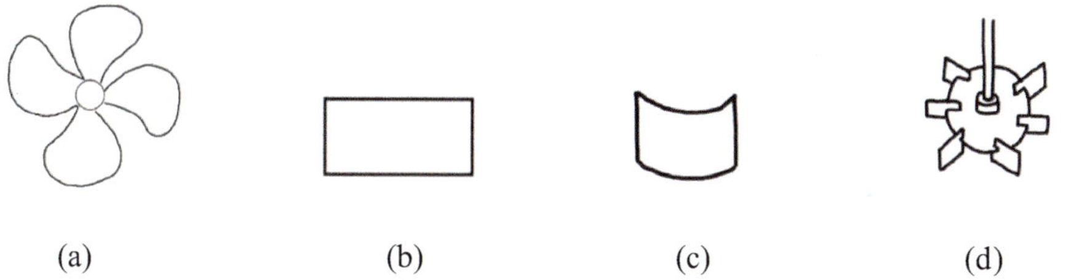

FIGURE 8.3 The significant impellers: (a) propellers; (b) flat blade; (c) curved blade; and (d) rushton turbine used for algal wastewater treatment in close PBRs.

immobilized cells [70]. Demonstrating the efficacy of bubble column PBRs, Mubarak et al. [71] conducted a study on *Chlorella pyrenoidosa* cultivation. In this system, a biomass concentration of 1.18 g/L was achieved, accompanied by a remarkable 19%wt lipid accumulation. Notably, this outperformed cultivation in traditional glass tanks, boasting a 28.6% higher biomass concentration and a 12.8% increase in lipid production [71]. Moreover, the versatility of bubble column PBR was underscored by their ability to effectively remove 84.1% of ammonium and 90.7% of phosphate pollutants present in human urine [72]. This successful treatment employed a mix of microalgae species, including *Chlorella* sp., *Spirulina* sp., *Scenedesmus* sp., and *Synechocystis* sp., all under artificial white fluorescent lights [72]. The results highlight the potential of bubble column PBRs not only in enhancing algal biomass and lipid production but also in addressing wastewater challenges with significant efficiency.

Conventional stirred-tank PBRs, illustrated in Figure 8.2b, traditionally rely on mechanical impellers for aeration, accompanied by baffles aimed at mitigating the vortex, a mixing blind spot induced by excessive agitation. However, this design compromise leads to a reduced surface area to volume ratio compared to bubble columns, resulting in suboptimal light absorption efficiency [73]. The impact of impeller types, depicted in Figure 8.3, has been examined in medium solutions, revealing distinct characteristics: (i) propellers prove effective for low-viscosity fluids, generating axial current flow parallel to the impeller axis; (ii) blades induce a radial turbulent flow, giving rise to vortex formation; and (iii) turbines produce a combination of radial and tangential currents, creating a robust fluid flow pattern [51,74]. This exploration underscores the critical role of PBR design elements in shaping fluid dynamics and, consequently, the overall efficiency of light absorption in algal cultivation systems.

8.5.2 Solid Particulate Packed PBRs

The primary bed reactors employed in various applications can be categorized into two distinctive configurations: fixed bed (packed bed) and fluidized bed (moving bed) reactors, as depicted in Figure 8.4. Packed bed PBRs take on a tubular design suitable for gas-solid phases, fostering the growth of immobilized microbial cells on the surface of solid beds, whether porous or rigid particles, packed as photo-biocatalyst in a column. This design, activated by ample light, demonstrates enhanced capabilities for sustained algal cultivation over extended periods with minimal shear

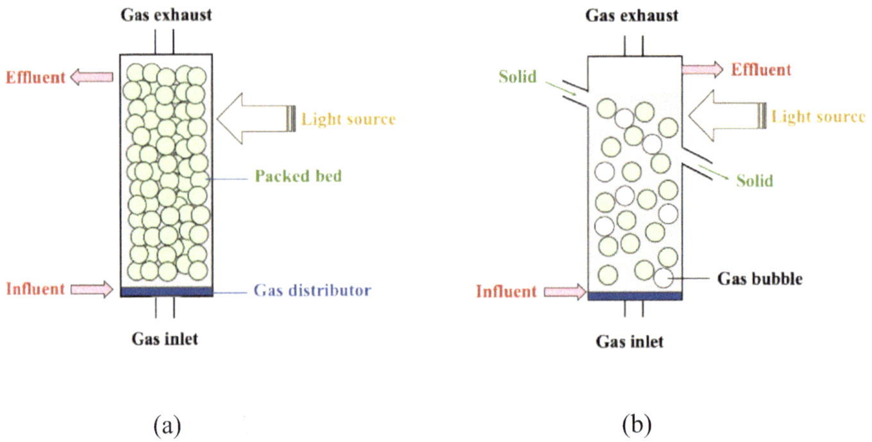

FIGURE 8.4 (a) Packed bed PBRs and (b) fluidized bed PBRs.

stress [75–77]. The surrounding medium or nutrient-laden wastewater flows through the packed bed column, with controlled fluid velocity and downward flow driven by gravity [64]. Research by Borges et al. [78] showcased the efficacy of packed bed PBRs, employing TiO_2 supported on glass spheres as a semiconductor catalyst, achieving over 99% removal of paracetamol contaminants in wastewater. In contrast, fluidized bed PBRs, akin to bubble columns, manage fluid velocity through column expansion and constriction, facilitating efficient heat and mass transfer. Despite its effectiveness, considerations for weight, including appropriate density, become crucial to prevent floating or sedimentation issues during the process [64,79,80]. Rosli et al. [81] highlighted the success of attached *Chlorella vulgaris* growth on support material, achieving peak productivity at 0.692 g/L biomass concentration under optimal conditions in a fluidized bed bioreactor. Meanwhile, Kube et al. [82] demonstrated improved municipal secondary effluent treatment through the cultivation of *Chlorella vulgaris* and *Scenedesmus abundans* microalgae in fluidized bed PBRs. *Scenedesmus abundans* exhibited prolonged operating times and remarkable efficiency in wastewater purification, with recorded removal rates of 46.4% for total N and 89.8% for total P over 22 days.

8.5.3 PBR Configurations with Solar Receivers

PBR configurations with solar receivers encompass diverse reactor types designed to harness sunlight for photosynthesis. Tubular PBRs, exemplified in Figure 8.5, offer flexibility in installation, with options for vertical columns or horizontal setups comprising straight, coiled, or looped transparent tubes. Despite providing a large illumination surface, tubular PBRs face challenges such as low gas exchange, high shear stress, and energy requirements. Issues like toxic O_2 fouling hindering algae growth and biomass accumulation within the tubes further impede efficiency [83,84]. In an illustrative case, *Neochloris oleoabundans* microalgae achieved a

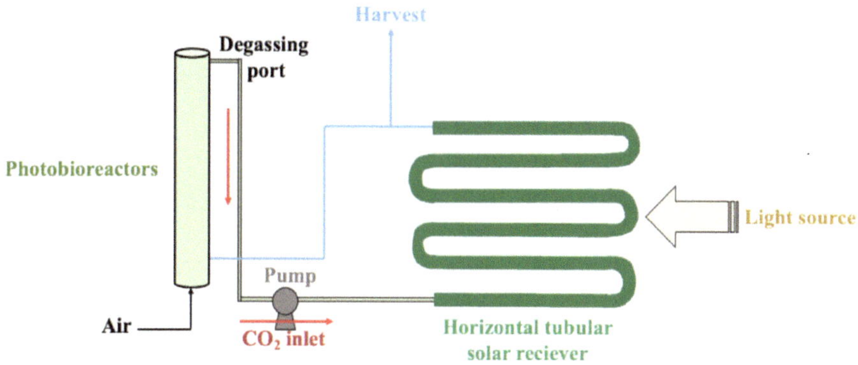

FIGURE 8.5 Horizontal tubular PBR system.

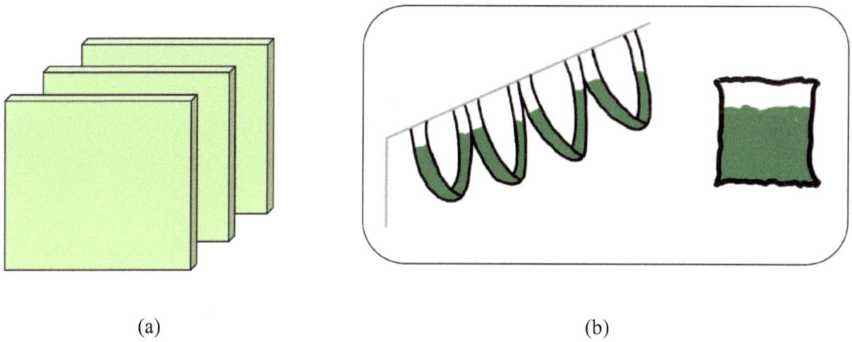

FIGURE 8.6 Common types of PBR configurations: (a) flat panel and (b) hanging bag apart of tubular system.

biomass concentration of 0.43 g/L in tubular PBRs equipped with airlift circulation, demonstrating municipal wastewater purification capabilities with impressive removal rates of 75% for NH_4^+ and 99.6% for PO_4^{3-} [85]. Flat panel PBRs (Figure 8.6a) present an alternative, plate-assembled system with lower power consumption and reduced O_2 accumulation. Despite its efficiency in biomass production, challenges lie in controlling temperature and limited light penetration depth [83]. *Chlorella sorokiniana* showcased its adaptability in waste medium cultivation within flat panel PBRs, achieving a specific growth rate of 0.11 h^{-1} and a lipid content of 19% w/w [48]. Notably, hanging bag PBRs (Figure 8.6b) offer a low-cost, simple design using sterile polythene tubing. However, the single-use nature raises environmental concerns through incineration, and the cleaning process demands significant water, detergent, and energy resources [86]. In transparent dialysis bag PBRs, *Chlorella vulgaris* exhibited efficient nutrient purification, removing 95.12% of NH_4^+-N and 76.87% of total P from wastewater [87]. Beyond conventional reactors, PBR systems also explore non-conventional designs like pyramid, nature-inspired, and hybrid configurations [65].

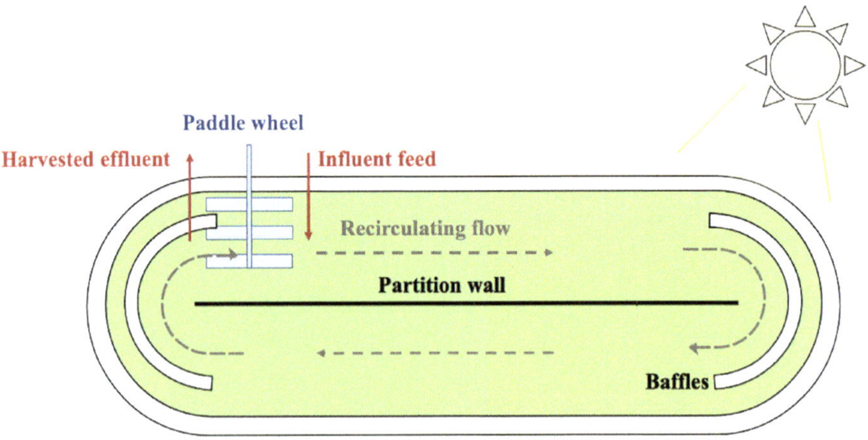

FIGURE 8.7 A simple model of a raceway pond.

8.5.4 Open System for Algal Wastewater Treatment

In contrast to closed systems, open systems, often referred to as "raceway ponds," present a distinct approach in the cultivation of photoautotrophs. These U-shaped shallow artificial ponds, illustrated in Figure 8.7, operate as closed-loop circulation systems, leveraging natural sunlight for the cultivation of organisms such as *Spirulina* spp. and cyanobacteria. Despite accounting for a significant 83% of *Spirulina* spp. production, open systems have limitations that render the biomass unsuitable for manufacturing applications involving chemicals, pharmaceuticals, and food ingredients [65,67,88]. The uncontrolled growth conditions and exposure to external contaminants, including parasites, rotifers, and ciliates preying on algae, result in severe damage to biomass productivity [65,67,88]. Paddle wheels, strategically positioned perpendicular to the liquid flow in open systems, replace impellers, promoting effective mass transfer between the medium and algae while ensuring uniform contact with the growth energy source [51]. The configuration of paddle wheels, be it flat, zigzagged, back-curved, or forward-curved, influences power consumption for agitation in the open system [51]. Notably, a combination of algal strains, *Scenedesmus obliquus*, *Chlorella vulgaris*, and *Oscillatoria tenuis*, cultivated in a raceway pond system, illuminated under 425 μmol/m^2/s of light intensity with a light:dark cycle ratio of 14:10 hours, demonstrated impressive wastewater treatment capabilities, removing over 86.61% of COD, 73.68% of total N, and 89.85% of total P [89]. The comparison of advantages and disadvantages of various PBR configurations is succinctly summarized in Table 8.2.

8.5.5 An Innovative and Hybrid Photobioreactors

The clean technology has become more attractive nowadays for future prospective of greener wastewater treatment processes. Membrane is the most used method coupled with conventional PBRs due to its high treatment efficiency and less space as well as

TABLE 8.2
Advantages and Disadvantages of Different PBR Configurations for Algal Cultivation

PBRs	Advantages	Disadvantages	References
Airlift	• Better degassing • Higher mixing efficiency due to continuously circulating pattern	• High operating cost	[69,71,90]
Bubble column	• Low cost • High surface area to volume ratio	• Lower mixing efficiency	
Stirred tank	• Good temperature control • Easy to clean	• Immobilized cells could be damaged by impeller • Low surface area to volume ratio	[64,68,73]
Packed bed	• Easy scaling up • Able to support the long operating period	• High exothermicity	[64]
Fluidized bed	• Efficient heat and mass transfer • Constant temperature	• Poor efficiency at high gas velocity	[79,80]
Tubular PBRs	• Large surface area	• Low gas exchange • High energy input • O_2 is accumulated	[83,84]
Flat panel PBRs	• Efficient productivity • Less energy input and O_2 accumulated	• Difficult to control temperature • Short dept of light absorption	[83]
Hanging bag PBRs	• Low cost and simple installation	• Environmental polluted during disposal process • Wastewater is produced in cleaning process	[86]
Raceway pond	• Less energy consumption • Low cost • Large production capacity	• Higher risk of contamination • Uncontrolled conditions	[73]

low operating time required [91]. Furthermore, the application of membranes could lower the footprint and enhance the concentration of attained algal biomass recovery, which would subsequently facilitate the harvesting process [92]. Marbelia et al. [92] had investigated membrane PBR illuminated by two fluorescence lamps system to cultivate *Chlorella vulgaris* microalgae for wastewater treatment. The results had shown that the optimal biomass concentration achieved at 0.59 ± 0.1 g/L upon which the membrane was applied, while a conventional PBR obtained only slightly over 0.2 ± 0.17 g/L of biomass concentration [92]. In a similar way, the efficiency of membrane PBR to remove nutrients, particularly N and P, in wastewater has been reported to reach up to more than 80% at low dilution rates [92]. However, the limitation of membrane PBR could occur at high dilution rates, i.e., the P removal was reduced to 50% at 0.4 and 0.5 day^{-1} of dilution rate due to the algae washout [92]. The performance of membrane PBRs was corresponding with Senatore et al. [93], who had assessed the environmental sustainability of this system and revealed that 80% of CO_2 can be eliminated and approximately 1.01 g/L of biomass concentration was recorded in novel membrane PBRs equipped with an innovative self-forming dynamic membrane. Besides the coupling with the membrane, the potential of combination between two or more system configurations called hybrid PBRs using hydrodynamics have also been explored [94]. The cultivation of *Scenedesmus obliquus* CPCC05 in promising hybrid PBR efficiently performed CO_2 conversion and O_2 release rates at 45.32 and 33.98 kgO_2/m^3/day, respectively, with the maximum specific growth rates of 0.96 day^{-1} [94]. The investigation of both membrane coupling and hybrid PBRs has been assessed in order to design novel promising PBRs to enhance the quality of the wastewater treatment process in scaling up plants while reducing environmental footprints.

8.6 CONCLUSION

The compelling potential of algae in wastewater treatment lies in its demonstrated ability to efficiently eliminate hazardous contaminants in an environmentally friendly way. The effectiveness of this process is intricately linked to several variables, including the type of illumination employed, initial pH levels, and temperature. Optimizing these operating conditions becomes imperative to maximize the removal efficiency of pollutants. Harnessing an ample supply of sunlight and CO_2 induces algal photosynthesis, culminating in the production of O_2 as the final output. This dual functionality not only enhances wastewater treatment but also contributes positively to the environment by serving to capture greenhouse gases. Furthermore, the metabolic processes of algae play a pivotal role in mitigating the hazards posed by nutrient-rich effluents. Algal metabolism transforms these nutrients into valuable biomass, opening avenues for applications such as additives in manufacturing or serving as a promising alternative feedstock for biofuel production. With the flexibility of operating under various PBRs, each tailored to specific conditions, there remains a pressing need for future studies to pinpoint the most suitable systems. This research is crucial to attaining optimal biomass characteristics according to diverse usage purposes, further solidifying algae's position as a versatile and impactful solution in the realm of wastewater treatment and environmental sustainability.

REFERENCES

1. Carr S. More than 80% of the world's sewage is discharged into the environment untreated. We can fix this. 2020. Available from: https://octogroup.org/news/more-80-worlds-sewage-discharged-environment-untreated/#:~:text=More%20than%2080%25%20of%20the%20world%27s%20sewage%20is%20discharged%20untreated,(known%20as%20ocean%20defecation). Accessed on 25 December 2023.
2. Haghighi AS. What effects does water pollution have on human health? 2020. Available from: https://www.medicalnewstoday.com/articles/water-pollution-and-human-health. Accessed on 25 December 2023.

3. Reef Resilience Network. Impacts on marine life. Available from: https://reefresilience.org/management-strategies/wastewater-pollution/impacts-on-marine-life/#:~:text=Impacts%20from%20wastewater%20pollution%20have,coral%2C%20fish%2C%20and%20shellfish. Accessed on 25 December 2023.
4. Ishaq MS, Afsheen Z, Khan A, et al. Disinfection methods. *Photocatal Appl Attributes*. 2018;3–19.
5. Olson G. Cleaning up wastewater through algae. 2015. Available from: https://sustainability-innovation.asu.edu/news/archive/11515-2/#:~:text=Algae%20benefit%20wastewater%20treatment%20by,to%20current%20wastewater%20treatment%20practices. Accessed on 17 December 2023.
6. Pawlita-Posmyk M, Wzorek M, Płaczek M. The influence of temperature on algal biomass growth for biogas production. *MATEC Web Conf. EDP Sci.* 2018;240:04008.
7. Wilson A. The incredible uses of algae. 2019. Available from: https://medium.com/@amykwilson/the-incredible-uses-of-algae-798a8f676fe. Accessed on 17 December 2023.
8. Bose P. The use of algae to reduce CO_2 emissions. 2022. Available from: https://www.azocleantech.com/article.aspx?ArticleID=1499#:~:text=Strategies%20to%20Reduce%20Carbon%20Dioxide%20Emission&text=These%20studies%20have%20shown%20that,and%20biofuels%2Dbased%20power%20plants. Accessed on 25 December 2023.
9. Qadir M, Drechsel P, Jiménez Cisneros B, et al. February. Global and regional potential of wastewater as a water, nutrient and energy source. *Nat Resour Forum*. 2020;44(1):40–51. Oxford, UK: Blackwell Publishing Ltd.
10. Alazaiza MY, He S, Su D, et al. Sewage water treatment using *Chlorella Vulgaris* microalgae for simultaneous nutrient separation and biomass production. *Separations*. 2023;10(4):229.
11. Haider MN, Liu CG, Tabish TA, et al. Resource recovery of the wastewater-derived nutrients into algal biomass followed by its cascading processing to multiple products in a circular bioeconomy paradigm. *Fermentation*. 2022;8(11):650.
12. ReliefWeb. Wastewater as a resource - May 2022. *OCHA Services*. 2022. Available from: https://reliefweb.int/report/world/wastewater-resource-may-2022. Accessed on 25 December 2023.
13. Jones ER, Van Vliet MT, Qadir M, et al. Country-level and gridded estimates of wastewater production, collection, treatment and reuse. *Earth Syst Sci Data*. 2021;13(2):237–254.
14. Healthy WA. Household wastewater and health risks. Department of Health. Government of Western Australia. Available from: https://www.healthywa.wa.gov.au/Articles/F_I/Household-wastewater-and-health-risks#:~:text=If%20you%20come%20into%20contact,stomach%20cramps%2C%20diarrhoea%20or%20vomiting). Accessed on 22 December 2023.
15. Jamal H. What is the composition of wastewater? 2017. Available from: https://www.aboutcivil.org/waste-water-composition. Accessed on 22 December 2023.
16. Affam AC, Wong CC, Seyam MA, et al. Preparation, characterization and adsorption study of granular activated carbon/iron oxide composite for the removal of boron and organics from wastewater. *E3S Web Conf. EDP Sci.* 2018;34:02006.
17. Saxena S, Choudhary MP. Performance evaluation of dairy wastewater treatment plant. *Int Res J Eng Technol*. 2017;4:1287–1291.
18. Jonah S, Obadaki YY, Koki IB. Metal Analysis of Dissolved Sediments Discharged from Samaru Stream into Kubanni Reservoir of the Ahmadu Bello University Zaria, Nigeria. 2016.
19. Safe Drinking Water Foundation. What is chlorination? Available from: https://www.safewater.org/fact-sheets-1/2017/1/23/what-is-chlorination#:~:text=This%20method%20was%20first%20used,be%20the%20potable%20water%20source. Accessed on 27 December 2023.

20. Centers for Disease Control and Prevention. Disinfection by-products (DBPs) factsheet. 2022. Available from: http://medbox.iiab.me/modules/en-cdc/www.cdc.gov/biomonitoring/THM-DBP_FactSheet.html. Accessed on 27 December 2023.
21. Dyer-Howe M, Payero J. 2.4 Disinfection by boiling and chlorination. Available from: https://www.oas.org/dsd/publications/unit/oea59e/ch23.htm. Accessed on 27 December 2023.
22. Shrestha LG, Shrestha R, Spuhler D. Boiling. Available from: https://sswm.info/sswm-university-course/module-6-disaster-situations-planning-and-preparedness/further-resources-0/boiling#:~:text=Boiling%20only%20kills%20pathogens%20and,by%20settling%20or%20filtration%20method. Accessed on 27 December 2023.
23. Barrera-Díaz CE, Balderas-Hernández P, Bilyeu B. Electrocoagulation: Fundamentals and prospectives. In CA Martínez-Huitle, MA Rodrigo, O Scialdone (eds.), Electrochemical Water and Wastewater Treatment. Butterworth-Heinemann. 2018;61–76.
24. Monera Technologies Corporation. Wastewater treatment: The membrane bioreactor (MBR). Available from: https://www.moneratec.com/blog/membrane-bioreactor-mbr/#:~:text=Membrane%20Bioreactor%20Disadvantages&text=This%20typically%20requires%20continuous%20air,time%20and%20eventually%20need%20replacing. Accessed on 23 December 2023.
25. Andersen RA, Lewin RA. Algae. *Ency Britannica*. 2023. Available from: https://www.britannica.com/science/algae. Accessed on 17 December 2023.
26. Centers for Disease Control and Prevention. Causes and ecosystem impacts. 2022. Available from: https://www.cdc.gov/harmful-algal-blooms/about/harmful-algal-blooms-contributing-factors-and-impacts.html?CDC_AAref_Val=https://www.cdc.gov/habs/environment.html. Accessed on 17 December 2023.
27. Suganya T, Varman M, Masjuki HH, et al. Macroalgae and microalgae as a potential source for commercial applications along with biofuels production: A biorefinery approach. *Renew Sust Energ Rev*. 2016;55:909–941.
28. Oak Ridge Institute for Science and Education. ORISE lesson plan: Growing algae. Available from: https://orise.orau.gov/resources/k12/documents/lesson-plans/biology-growing-algae.pdf. Accessed on 17 December 2023.
29. Price K, Farag IH. Resources conservation in microalgae biodiesel production. *Int J Eng Tech Res*. 2013;1(8):49–56.
30. Gallimore W. Marine metabolites: Oceans of opportunity. In SB McCreath, YN Clement (eds.), Pharmacognosy. Academic Press. 2017;377–400.
31. Central and Northern California Ocean Observing System. Types of algal blooms. Available from: https://www.cencoos.org/focus-areas/habs/types-of-algal-blooms/. Accessed on 17 December 2023.
32. Feng Y, Xiong Y, Hall-Spencer JM, et al. Shift in algal blooms from micro-to macroalgae around China with increasing eutrophication and climate change. *Glob Chang Biol*. 2023;30:e17018.
33. St. Johns River Water Management District. Understanding algal blooms. Available from: https://www.sjrwmd.com/education/algae/#:~:text=Algal%20blooms%20can%20be%20dramatic,shallow%2C%20slow%2Dflowing%20water. Accessed on 17 December 2023.
34. Bermejo R, Green-Gavrielidis L, Gao G. Editorial: Macroalgal blooms in a global change context. *Front Mar Sci*. 2023;10:1204117.
35. McLachlan A, Defeo O (eds.). Human impacts. In *The Ecology of Sandy Shores* (3rd Ed.). Academic Press. 2018;375–420.
36. Ainsworth J. The benefits of algae as a fertilizer. 2021. Available from: https://algenair.com/blogs/news/the-benefits-of-algae-as-a-fertilizer#:~:text=In%20fact%2C%20marine%20algae%20has,improve%20the%20plants%27%20drought%20resistance. Accessed on 17 December 2023.

37. Choudhury NR. Algae-based food additive market outlook from 2023 to 2033. *Future Market Insights*. 2023. Available from: https://www.futuremarketinsights.com/reports/algae-based-food-additive-market. Accessed on 17 December 2023.
38. Marcus JB (ed.). Food science basics: Healthy cooking and baking demystified: The science behind healthy foods, cooking and baking. In *Culinary Nutrition*. Academic Press. 2013;51–97.
39. Bunch TR, Bond C, Buhl K, et al. Diatomaceous earth general fact sheet; National Pesticide Information Center, Oregon State University Extension Services. 2013. Available from: https://npic.orst.edu/factsheets/degen.html. Accessed on 20 December 2023.
40. Leech J. What are the benefits of diatomaceous earth? 2023. Available from: https://www.healthline.com/nutrition/what-is-diatomaceous-earth. Accessed on 20 December 2023.
41. Chia WY, Tang DYY, Khoo KS, et al. Nature's fight against plastic pollution: Algae for plastic biodegradation and bioplastics production. *Environ Sci Ecotechnol*. 2020;4:100065.
42. Bhuyar P, Muniyasamy S, Govindan N. Biodegradation of plastic waste by using microalgae and their toxins. *J Biotechnol Biomater*. 2019;8(55):10–3389.
43. Liu J, Zhou F, Abed AM, et al. Macroalgae as a potential source of biomass for generation of biofuel: Artificial intelligence, challenges, and future insights towards a sustainable environment. *Fuel*. 2023;336:126826.
44. U.S. Environmental Protection Agency. Sources and solutions: Fossil fuels. 2023. Available from: https://www.epa.gov/nutrientpollution/sources-and-solutions-fossil-fuels#:~:text=When%20fossil%20fuels%20are%20burned,referred%20to%20as%20nitrogen%20oxides. Accessed on 20 December 2023.
45. Sugar Asia Magazine. China to become a leader of global biofuel output up to 25% by 2024. 2020. Available from: https://sugar-asia.com/china-to-become-a-leader-of-global-biofuel-output-up-to-25-by-2024/. Accessed on 20 December 2023.
46. Dahman Y, Dignan C, Fiayaz A, et al. An introduction to biofuels, foods, livestock, and the environment. In D Verma, E Fortunati, S Jain, X Zhang (eds.), *Biomass, Biopolymer-Based Materials, and Bioenergy*. Woodhead Publishing. 2019;241–276.
47. Pagani D. The importance of green mobility. In NE Bassam (ed.), *Distributed Renewable Energies for Off-Grid Communities*. Elsevier. 2021;345–391.
48. Banerjee S, Dasgupta S, Atta A, et al. Flow rate optimization in a flat-panel photobioreactor for the cultivation of microalgae for mitigating waste gas. *Water*. 2023;15(15):2824.
49. Wasanasathian A, Peng CA. Algal photobioreactor for production of lutein and zeaxanthin. In S-T Yang (ed.), *Bioprocessing for Value-Added Products from Renewable Resources*. Elsevier Science. 2007;491–505.
50. Delavar MA, Wang J. Bioreactor concepts, types, and modeling. In MA Delavar, J Wang (eds.), *Advanced Methods and Mathematical Modeling of Biofilms*. Academic Press. 2022;195–245.
51. de Souza Kirnev PC, de Souza Vandenberghe LP, Soccol CR, et al. Mixing and agitation in photobioreactors. In H Thatoi, S Mohapatra, SK Das (eds.), *Innovations in Fermentation and Phytopharmaceutical Technologies*. Elsevier. 2022;13–35.
52. Energy Education. Harvesting algae. Available from: https://energyeducation.ca/encyclopedia/Harvesting_algae#:~:text=This%20can%20be%20done%20by,algae%20clumps%20to%20the%20bottom. Accessed on 23 December 2023.
53. Lu Q, Xiao Y, Chen F. Application of microalgae biotechnology for the sustainable development of aquaculture. *Adv Bioenergy*. 2021;6(1):117–163.
54. Rani S, Chowdhury R, Tao W, et al. Microalga-mediated tertiary treatment of municipal wastewater: Removal of nutrients and pathogens. *Sustainability*. 2021;13(17):9554.
55. Bohutskyi P, Kligerman DC, Byers N, et al. Effects of inoculum size, light intensity, and dose of anaerobic digestion centrate on growth and productivity of *Chlorella* and *Scenedesmus* microalgae and their poly-culture in primary and secondary wastewater. *Algal Res*. 2016;19:278–290.

56. Gonçalves AL, Simões M, Pires JCM. The effect of light supply on microalgal growth, CO_2 uptake and nutrient removal from wastewater. *Energy Convers Manag.* 2014;85:530–536.
57. Li J, Bin H, Lin J, et al. Effects of light-emitting diodes under capped daily energy consumption with combinations of electric power and photoperiod on cultivation of *Chlorella pyrenoidosa*. *Bioresour Technol.* 2016;205:126–132.
58. Gojkovic Z, Lu Y, Ferro L, et al. Modeling biomass production during progressive nitrogen starvation by North Swedish green microalgae. *Algal Res.* 2020;47:101835.
59. Zhu L, Wang Z, Shu Q, et al. Nutrient removal and biodiesel production by integration of freshwater algae cultivation with piggery wastewater treatment. *Water Res.* 2013;47(13):4294–4302.
60. Wang Y, He B, Sun Z, et al. Chemically enhanced lipid production from microalgae under low sub-optimal temperature. *Algal Res.* 2016;16:20–27.
61. Huber M, Blaha-Robinson K. Algae culture and pH. Available from: https://algaeresearchsupply.com/pages/algae-culture-and-ph#:~:text=Carbon%20dioxide%20(CO2)%20is,doing%20photosynthesis%2C%20hurting%20culture%20growth. Accessed on 24 December 2023.
62. Xin L, Hong-Ying H, Ke G, et al. Growth and nutrient removal properties of a freshwater microalga *Scenedesmus* sp. LX1 under different kinds of nitrogen sources. *Ecol Eng.* 2010;36(4):379–381.
63. Abdel-Raouf N, Al-Homaidan AA, Ibraheem I. Microalgae and wastewater treatment. *Saudi J Biol Sci.* 2012;19(3):257–275.
64. Atlas Scientific. 6 Types of bioreactors. 2023. Available from: https://atlas-scientific.com/blog/types-of-bioreactors/. Accessed on 22 December 2023.
65. Chanquia SN, Vernet G, Kara S. Photobioreactors for cultivation and synthesis: Specifications, challenges, and perspectives. *Eng Life Sci.* 2022;22(12):712–724.
66. Znad H. Microalgae culture technology for carbon dioxide biomitigation. In O Konur (ed.), *Handbook of Algal Science, Technology and Medicine.* Academic Press. 2020;303–316.
67. Araújo R, Vázquez CF, Sánchez LJ, et al. Current status of the algae production industry in Europe: An emerging sector of the blue bioeconomy. *Front Mar Sci.* 2021;7:626389.
68. Kumar L, Mohan L, Anand R, et al. A review on unit operations, challenges, opportunities, and strategies to improve algal based biodiesel and biorefinery. *Front Chem Eng.* 2022;4:998289.
69. Doran PM (ed.). Reactor engineering. In *Bioprocess Engineering Principles* (2nd Ed.). Academic Press. 2013;761–852.
70. Zarei Z, Malekshahi P, Trzcinski AP, et al. Investigation of hydrodynamic parameters in an airlift photobioreactor on CO_2 biofixation by *Spirulina* sp. *Sustainability.* 2022;14(12):7503.
71. Mubarak M, Shaija A, Prashanth P. Bubble column photobioreactor for *Chlorella pyrenoidosa* cultivation and validating gas hold up and volumetric mass transfer coefficient. *Energ Source Part A Recov Util Environ Eff.* 2019;45:1–15.
72. Patil SS, Behera B, Sen S, et al. Performance evaluation of bubble column photobioreactor along with CFD simulations for microalgal cultivation using human urine. *J Environ Chem Eng.* 2021;9(1):104615.
73. Singh RN, Sharma S. Development of suitable photobioreactor for algae production: A review. *Renew Sust Energ Rev.* 2012;16(4):2347–2353.
74. McCabe WL, Smith JC, Harriott P. *Unit Operations of Chemical Engineering.* McGraw-Hill. 1993.
75. Mitchell DA, de Lima Luz LF, Krieger N, et al. 2.25 - Bioreactors for solid-state fermentation. In M. Moo-Young (ed.), *Comprehensive Biotechnology* (2nd Ed.). Academic Press. 2011;347–360.

76. Sen P, Nath A, Bhattacharjee C. Packed-bed bioreactor and its application in dairy, food, and beverage industry. In C Larroche, MÁ Sanromán, G Du, A Pandey (eds.), *Current Developments in Biotechnology and Bioengineering*. Elsevier. 2017;235–277.
77. Warnock JN, Bratch K, Al-Rubeai M. Packed bed bioreactors. In J Chaudhuri, M Al-Rubeai (eds.), *Bioreactors for Tissue Engineering: Principles, Design and Operation*. Springer Netherlands. 2005;87–113.
78. Borges ME, García DM, Hernández T, et al. Supported photocatalyst for removal of emerging contaminants from wastewater in a continuous packed-bed photoreactor configuration. *Catalysts*. 2015;5(1):77–87.
79. Cocco R, Karri SBR, Knowlton T. Introduction to fluidization. *American Institute of Chemical Engineers*. 2014. Available from: https://www.aiche.org/sites/default/files/cep/20141121.pdf. Accessed on 27 December 2023.
80. Nakamura H, Watano S. Numerical simulation of particle fluidization behaviors in a rotating fluidized bed. In Asia-Pacific Chemical Reaction Engineering Symposium (Ed.), *Studies in Surface Science and Catalysis*. 2006;505–508. Elsevier.
81. Rosli SS, Lim JW, Lam MK, et al. Cultivation of microalgae in fluidized bed bioreactor: Impacts of light intensity and CO_2 concentration. *IOP Conf Series Mater Sci Eng*. 2020;736(2):022018.
82. Kube M, Mohseni A, Fan L, et al. Energy and nutrient recovery by treating wastewater with fluidised-beds of immobilised algae. *J Water Process Eng*. 2020;38:101585.
83. Jeevanandam J, Harun MR, Lau SY, et al. Microalgal biomass generation via electroflotation: A cost-effective dewatering technology. *Appl Sci*. 2020;10(24):9053.
84. Skjånes K, Andersen U, Heidorn T, et al. Design and construction of a photobioreactor for hydrogen production, including status in the field. *J Appl Phycol*. 2016;28:2205–2223.
85. Valev D, Silva Santos H, Tyystjärvi E. Stable wastewater treatment with *Neochloris oleoabundans* in a tubular photobioreactor. *J Appl Phycol*. 2020;32:399–410.
86. Cui J, Purton S, Baganz F. Characterisation of a simple ‘hanging bag’ photobioreactor for low-cost cultivation of microalgae. *J Chem Technol Biotechnol*. 2022;97(3):608–619.
87. Shao H, Sun Y, Jiang X, et al. Towards biomass production and wastewater treatment by enhancing the microalgae-based nutrients recovery from liquid digestate in an innovative photobioreactor integrated with dialysis bag. *J Environ Manage*. 2022;317:115337.
88. Rahimpour MR, Biniaz P, Makarem MA. Integration of microalgae into an existing biofuel industry. In F Dalena, A Basile, C Rossi (eds.), *Bioenergy Systems for the Future*. Woodhead Publishing. 2017;481–519.
89. Zhang Q, Yu Z, Zhu L, et al. Vertical-algal-biofilm enhanced raceway pond for cost-effective wastewater treatment and value-added products production. *Water Res*. 2018;139:144–157.
90. Lifeasible. Large-scale algae culture service by vertical tubular PBRs. Available from: https://www.lifeasible.com/algae/large-scale-algae-culture-service-by-vertical-tubular-pbrs.html. Accessed on 27 December 2023.
91. U.S. Environmental Protection Agency. Wastewater management fact sheet. *Membrane Bioreactors*. 2007. Available from: https://www.epa.gov/sites/default/files/2019-08/documents/membrane_bioreactor_fact_sheet_p100il7g.pdf. Accessed on 29 February 2024.
92. Marbelia L, Bilad MR, Passaris I, et al. Membrane photobioreactors for integrated microalgae cultivation and nutrient remediation of membrane bioreactors effluent. *Bioresour Technol*. 2014;163:228–235.
93. Senatore V, Buonerba A, Zarra T, et al. Innovative membrane photobioreactor for sustainable CO_2 capture and utilization. *Chemosphere*. 2021;273:129682.
94. Deprá MC, Mérida LG, de Menezes CR, et al. A new hybrid photobioreactor design for microalgae culture. *Chem Eng Res Des*. 2019;144:1–10.

9 Microbial Nanomaterials for the Treatment of Wastewaters

Sanam Iram Soomro, Hatice Kübra Kara, Ebru Guller, Raziye Nur Özçiçek, Rasit Bakir, and Hüseyin Kara

9.1 INTRODUCTION

Water is a natural resource on Earth, and its availability in a pure form is crucial for human beings and other living organisms, as life without water is unimaginable. Due to its remarkable properties, such as its ability to dissolve many substances, water is often referred to as the universal solvent. Currently, water contamination is a major global issue caused by factors such as inadequate sewage treatment, industrial waste, marine dumping, radioactive materials, and certain agricultural practices (Reddy, Venugopal, and Subrahmanyam 2007). Water pollution not only harms the environment but also contributes to air pollution, posing significant health risks to humans. Additionally, it negatively impacts economic growth and the social fabric of affected societies and countries. A recent UN report highlighted that access to clean, fresh water is a global challenge in the 21st century, as the survival of living organisms is threatened by contaminated water (Sawai 2003). Water contamination occurs when unwanted substances enter water bodies or reservoirs, rendering the water unsuitable for drinking and other uses. To combat this growing problem, various chemical, physical, and mechanical methods have been developed, and researchers continue to explore new technologies to enhance water purification processes at low costs (Sayes et al. 2005). One such promising field is nanotechnology, which offers an efficient, low-cost solution for water purification, boasting advantages such as high pollutant removal efficiency and reusability (Sclafani, Mozzanega, and Herrmann 1997). In the past, nanomaterials have been successfully applied in fields like medical science and catalysis. Today, as the world grapples with serious water quality issues, experts recognize that nanomaterials are an effective option for wastewater treatment due to their unique properties, including nanoscale size, large surface area, high reactivity, excellent solution mobility, mechanical strength, porosity, hydrophilicity, dispersibility, and hydrophobicity (Sclafani, Mozzanega, and Herrmann 1997). These materials have been shown to successfully remove heavy metals such as lead (Pb) and molybdenum (Mo), as well as various organic and inorganic pollutants and harmful microbes (Hou et al. 2020).

DOI: 10.1201/9781003459286-9

TABLE 9.1
Outlines the Different types of Water Pollutants, their Sources, and their Harmful Effects

Water Pollutants	Source of Pollutants	Effect of Pollutants	References
Pathogens	Viruses and bacteria	Waterborne diseases	Umar et al. (2019)
Agricultural pollutants	Agricultural chemicals	Directly affect freshwater resources	Tang, Gong, and Wang (2016)
Sediments and suspended solids	Land cultivation, demolition, and mining operations	Damaging fish spawning, affecting aquatic environment of insects and fishes	Richter and Ayers (2018)
Inorganic pollutants	Metals compounds, trace elements, inorganic salts, heavy metals, and mineral acids	Aquatic flora and fauna, public health problems	Sizmur et al. (2017)
Organic pollutants	Detergents, insecticides, and herbicides	Aquatic life problems, carcinogenic	Wang et al. (2019)
Industrial pollutants	Municipal pollutant water	Caused water and air pollution	Liu et al. (2018)
Radioactive pollutants	Different isotopes	Bones, teeth, and skin and can cause severe damage	Bayoumi and Saleh (2018)
Nutrient pollutants	Plant debris and fertilizer	Eutrophication process	Ma et al. (2018)
Macroscopic pollutants	Marine debris	Plastic pollution	Hlongwane et al. (2019)
Sewage and contaminated water	Domestic wastewater	Waterborne diseases	Rajasulochana and Preethy (2016)

According to the World Health Organization (WHO), approximately 1.7 million people die annually from water pollution, and four billion cases of waterborne diseases are reported each year (Alijanzadeh Maliji et al. 2023) (Table 9.1).

In recent years, numerous natural and engineered nanomaterials have demonstrated strong antimicrobial properties, such as chitosan, silver nanoparticles (nAgs), photocatalytic TiO_2, fullerol aqueous fullerene nanoparticles (nC60), and carbon nanotubes (CNTs). Unlike traditional chemical disinfectants, these antimicrobial nanomaterials are not potent oxidants and remain relatively inert in water, reducing the risk of producing harmful disinfection byproducts (DBPs). When effectively integrated into treatment processes, they offer the potential to replace or enhance conventional disinfection methods (Qi et al. 2004).

Another promising application of antimicrobial nanomaterials is in decentralized or point-of-use water treatment and reuse systems. Interest in decentralized water treatment has grown due to concerns over water loss, quality degradation linked to aging distribution networks, rising energy costs for water transport, and the increasing need for alternative water sources and wastewater reuse in regions facing water shortages or national security concerns. Functional nanomaterials with antimicrobial properties could play a key role in developing high-performance, small-scale, or point-of-use systems, improving the resilience of water supply networks, supporting non-centralized systems, and providing emergency solutions in the aftermath of catastrophic events (Bermeo et al. 2022).

9.2 RISK FACTORS OF MICROBIALS ON HUMAN HEALTH

Water contamination is a global concern, primarily caused by both human activities and geological factors. Elevated levels of pollutants in drinking water can lead to serious health risks, depending on individual vulnerability and how these pollutants interact with the body. Groundwater passing through sedimentary soils and rocks can also carry a wide range of substances, including harmful microorganisms with high pathogenic potential. The effects of these natural contaminants depend largely on their types and concentration (Clasen et al. 2004). Changes in environmental factors can lead to water contamination. Microbial pollutants, including bacteria, viruses, protozoa, and parasitic worms, often spread through water contaminated by human or animal waste, either intentionally or accidentally. Bacteria responsible for diseases such as typhoid, cholera, gastroenteritis, diarrhea, and dysentery are commonly found in water (Brown et al. 2008). Human and animal waste, along with protozoa, contaminate rivers, streams, and lakes, causing waterborne illnesses that result in nausea, vomiting, dehydration, and stomach cramps. Viruses such as polio, adenovirus, and hepatitis are frequently detected in contaminated water (Ali et al. 2021).

Coliform bacteria, a primary indicator of water contamination, are found in higher concentrations compared to other pathogens (Pandey et al. 2014). Cyanobacteria, although present in smaller amounts, can produce toxins that pose serious health risks, even at low concentrations, potentially leading to death. Consuming water contaminated with these microbes can result in diseases such as amoebiasis, cholera, shigellosis, gastroenteritis, and hepatitis B (Van Apeldoorn et al. 2007). In recent decades, at least 50 emerging infectious agents have been identified, with 10% of them being bacterial (Terreni, Taccani, and Pregnolato 2021). Globally, approximately, 200 million people die annually due to waterborne illnesses, both fatal and non-fatal. Opportunistic (Alvarez 2021) pathogens like *Pseudomonas aeruginosa*, *Mycobacterium avium*, and *Legionella pneumophila*, along with other non-tuberculous mycobacteria, are increasingly responsible for waterborne diseases with significant economic impact (Alvarez 2021). Contaminated drinking water can cause severe health problems, including dysentery, jaundice, meningitis, encephalopathy, and typhoid. According to a WHO report, water-related diseases are responsible for approximately 502,000 deaths globally each year (Pal et al. 2018). In India alone, a study revealed that from 2013 to 2017, diarrhea caused 6,514 deaths, with 10% of those being children under the age of 5, highlighting the urgent need for action. Viral hepatitis claimed 2,143 lives during the same period, making it the second deadliest waterborne disease (Goswami and Pugazhenthi 2020).

Even at low levels, microbial contamination can have a significant impact on aquatic ecosystems and human health, disrupting the natural processes of waste breakdown. Consequently, there is an increased focus on the removal of microbial contaminants from various water sources to safeguard public health and the environment. The permissible limits for various microorganisms in water as declared by different agencies, primarily the WHO and ISI. For coliform bacteria, the limit is set at less than 1 μg/100 mL (Maheux et al. 2014), emphasizing the importance of keeping coliform levels minimal to prevent waterborne diseases. For more pathogenic bacteria like *Salmonella* and *Shigella*, the permissible limit is slightly higher at 10 μg/250 mL (Mishra 2005). *P. aeruginosa*, which is known for causing respiratory and other infections, has a limit of 14 μg/250 mL. Sulfite-reducing bacteria, which indicate anaerobic conditions, are allowed at a concentration of 1 g/50 mL. *Vibrocholera*, the bacterium responsible for cholera, is permitted up to 1 μg/250 mL (Hardy 1981).Lastly, for yeast and molds, the permissible limit is much lower at 0.1 μg/250 mL, reflecting the need to control fungal contaminants (Chithra et al. 2022). These limits are set to ensure water safety and protect public health.

9.3 MICROBIAL NANOMATERIALS

Naturally occurring chitin and certain peptides have long been recognized for their antimicrobial properties. However, it is only in recent years that these materials have been engineered into nanoparticles (Qi et al. 2004). These nanoparticles show promise for low-cost and simple disinfection solutions, particularly in developing countries. The antibacterial mechanism of natural peptides, such as silk moth cecropins, involves the formation of nanoscale channels in bacterial cell membranes, leading to osmotic collapse (Gazit 2007). Antimicrobial peptides were initially synthesized based on known protein structures (Ghadiri et al. 1993), and their synthesis enables the development of customized antimicrobial nanostructures (Gazit 2007). Researchers are optimizing engineered peptides in terms of size, morphology, coatings, derivatization, and other properties for specific antimicrobial uses. Chitosan, derived from chitin in the shells of arthropods, has long been noted for its antibacterial properties, and it is only more recently that this polymer has been transformed into nanoparticles. Nanoscale chitosan and its derivatives exhibit antimicrobial activity against bacteria, viruses, and fungi. It has been found to be more effective against fungi and viruses than bacteria, and within bacteria, its activity is stronger against Gram-positive than Gram-negative strains. Additionally, chitosan can inhibit the replication of bacteriophages in bacteria and induce resistance to viral diseases in plants (Li et al. 2008; Hossain et al. 2014). Minimum inhibitory concentrations (MICs) for chitosan range from 18 to 5,000 ppm, depending on the organism, pH, molecular weight, degree of polymerization, and the presence of lipids and proteins. Several mechanisms have been proposed to explain chitosan's antimicrobial activity. One theory suggests that positively charged chitosan particles interact with negatively charged cell membranes, increasing membrane permeability and eventually causing rupture and leakage of intracellular components. This aligns with the observation that both natural chitosan macromolecules and engineered nanoparticles are ineffective at pH values above 6 due to the lack of protonated

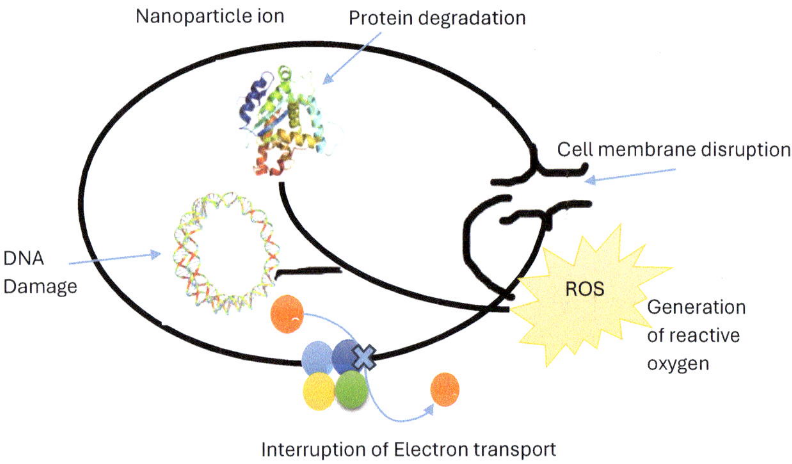

FIGURE 9.1 The general mechanism of microbial nanomaterial action in water treatment.

amino groups. For this reason, chitosan derivatives containing quaternary ammonium groups, such as *N*,*N*,*N*-trimethyl chitosan, *N*-propyl-*N*,*N*-dimethyl chitosan, and *N*-furfuryl-*N*,*N*-dimethyl chitosan, exhibit stronger antibacterial activity, which increases as pH decreases. Another proposed mechanism suggests that chitosan chelates trace metals, inhibiting enzyme activities. In fungal cells, chitosan is believed to penetrate the cell wall and nucleus, where it binds to DNA and inhibits RNA synthesis (Li et al. 2008; Qi et al. 2004) (Figure 9.1).

9.3.1 Silver Nanomaterial

The antimicrobial properties of Ag compounds and Ag ions have long been recognized and utilized in a variety of applications, ranging from disinfecting medical devices and household items to water treatment. However, the precise mechanisms behind their toxicity are not yet fully understood. Ag ions interact with thiol groups in proteins, leading to the inactivation of respiratory enzymes and the generation of reactive oxygen species (ROS). Additionally, $Ag+$ ions have been shown to inhibit DNA replication and alter the structure and permeability of cell membranes. Moreover, Ag ions are photoactive when exposed to UVA and UVC irradiation, which enhances UV-induced bacterial and viral inactivation. It is suggested that the complexation of $Ag+$ with cysteine accelerates the photodimerization of viral DNA, contributing to the synergistic inactivation of *Haemophilus influenzae* phage and MS2 phage. Several mechanisms have been proposed to explain the antimicrobial action of nAgs: (i) attachment of nanoparticles to the bacterial surface, altering membrane properties. nAg has been shown to degrade lipopolysaccharide molecules, form "pits" in the membrane, and significantly increase membrane permeability; (ii) penetration of nAg into bacterial cells, causing DNA damage; and (iii) dissolution of nAg, releasing antimicrobial $Ag+$ ions. The physicochemical properties of nAg play a crucial role in its antimicrobial activity. In general, particles smaller than 10 nm are more toxic to bacteria like *Escherichia coli* and *P. aeruginosa*. nAgs between 1 and

10 nm have been found to inhibit certain viruses by binding preferentially to their gp120 glycoproteins, preventing the virus from attaching to host cells. Furthermore, triangular nAg nanoplates with highly reactive <111> planes exhibit greater toxicity than nAg rods, nAg spheres, or Ag+ ions (Li et al. 2008; Morsi et al. 2017; Zahoor et al. 2021).

9.3.2 Titanium Dioxide Nanomaterial

Titanium dioxide (TiO_2) stands out as one of the most commonly utilized semiconductor photocatalysts, with its extensive application and research outpacing other nanomaterials. Activated by UV-A light, TiO_2's photocatalytic abilities have been harnessed in various environmental solutions, particularly in eliminating pollutants from water and air. Over the years, significant research has explored the use of TiO_2 for its potential in inactivating bacteria. It has shown effectiveness against both Gram-negative and Gram-positive bacteria, although the latter tend to be more resistant due to their spore-forming capabilities. Furthermore, the scope of TiO_2's efficacy extends to viruses, including poliovirus, hepatitis B, herpes simplex virus, and bacteriophages, with studies demonstrating its potential in viral inactivation. The concentration of TiO_2 needed for bacterial inactivation generally ranges from 100 to 1,000 ppm, varying based on factors such as particle size, light intensity, and wavelength. Its antibacterial properties stem from its ability to generate ROS, like hydroxyl radicals and peroxide, through UV-A exposure. This UV-A absorption capacity makes TiO_2 highly efficient in solar disinfection applications, and experiments have shown rapid bacterial inactivation when exposed to sunlight. Moreover, TiO_2's photocatalytic disinfection potential is bolstered by advancements that allow it to be activated by visible light, such as through metal doping, particularly with Ag. Ag-doped TiO_2 has been shown to significantly improve its efficiency, not only enhancing UV-driven photocatalysis but also enabling activation under visible light. Ag's surface plasmons absorb visible light, which can induce electron transfer to TiO_2, resulting in charge separation and subsequent activation by visible light. This mechanism enhances TiO_2's photocatalytic activity, promoting electron-hole separation and increasing the available surface area for adsorption. Therefore, Ag/TiO_2 is considered a highly promising photocatalytic material due to its enhanced photoactivity and responsiveness to visible light, making it a cutting-edge solution for environmental purification and antimicrobial applications under natural light conditions (Li et al. 2008; Vargas and Rodríguez-Páez 2017).

9.3.3 Zinc Oxide Nanoparticles

Zinc oxide (ZnO) nanoparticles are widely recognized for their effectiveness in removing harmful coliform bacteria from water, making them a popular choice in wastewater treatment applications. Several studies have highlighted the efficiency of ZnO nanoparticles in eliminating coliforms from urban wastewater treatment plants, where they act as powerful antibacterial agents against a broad spectrum of bacteria. While their precise antibacterial mechanism is still being explored, one key factor believed to contribute to their effectiveness is the photocatalytic production of hydrogen peroxide, which plays a significant role in bacterial inactivation.

In addition to this, ZnO nanoparticles are thought to exert their antibacterial effects by disrupting bacterial cell membranes and penetrating the cell envelope, which leads to the inhibition of bacterial growth. This mechanism makes ZnO a versatile material in combating various types of bacteria. The role of Zn^{2+} ions, which are released as ZnO dissolves, is another aspect under investigation. These ions may potentially bind to microbial membranes, prolonging the lag phase of bacterial growth and further enhancing the antibacterial properties of ZnO nanoparticles. Despite the promising antibacterial activity of ZnO, there is ongoing debate regarding the impact of nanoparticle size on their toxicity and effectiveness. Some studies suggest that smaller ZnO particles exhibit greater antibacterial potency compared to larger particles, while others have found no significant size-dependent effects. This area remains an active field of research as scientists seek to better understand the mechanisms that govern ZnO's interaction with bacteria and optimize its use in water purification and antibacterial applications. Additionally, ZnO's biocompatibility and its potential for use in various environmental and biomedical fields further underscore its value. Its ability to operate under different conditions, including exposure to sunlight for photocatalytic activation, enhances its versatility for sustainable water treatment and microbial control (Li et al. 2008; Xin et al. 2022; Baruah, Pal, and Dutta 2012).

9.3.4 Cobalt Ferrite Oxide ($CoFe_2O_4$) Nanomaterials

Cobalt ferrite oxide ($CoFe_2O_4$) nanoparticles and copper surfaces, known for their superior anisotropic properties, have attracted attention in antibacterial applications. While these materials are not yet widely implemented in large-scale water treatment systems, they hold significant potential for future scalability. $CoFe_2O_4$ nanoparticles, in particular, exhibit strong magnetic properties, which enable them to effectively capture pathogens, making them promising candidates for innovative antimicrobial strategies. Copper surfaces are also known for their potent antibacterial characteristics, primarily due to their ability to disrupt bacterial cell membranes and generate ROS that lead to cell death. Both copper and $CoFe_2O_4$ have shown considerable promise in small-scale experiments, with copper surfaces being especially effective in environments requiring continuous antimicrobial protection. Despite their advantages, the full-scale application of these materials in water treatment systems remains a challenge. The cost, potential toxicity, and the need for optimization of the materials for different environmental conditions are factors that need to be addressed. Nonetheless, metal oxides like cobalt ferrite, combined with the antimicrobial properties of copper, represent a promising avenue for future research and development in microbial removal technologies for water treatment. Additionally, the unique magnetic properties of $CoFe_2O_4$ allow for the potential integration of external magnetic fields, which could enhance pathogen removal efficiency by facilitating better control and recovery of the nanoparticles after treatment. This feature could make $CoFe_2O_4$ a valuable material in designing more efficient, environmentally friendly water purification systems in the future (Mmelesi et al. 2021).

9.3.5 Fullerenes Nanomaterial

C_{60} have shown great promise as antibacterial nanomaterials in water treatment due to their unique structural and chemical properties. These carbon-based nanospheres have a large surface area that facilitates the physical adsorption of bacterial cells through van der Waals forces and hydrophobic interactions, leading to the destabilization of bacterial membranes. This physical interaction can cause leakage of essential intracellular contents and eventual cell death. One of the most powerful antibacterial mechanisms of C_{60} is their ability to generate ROS such as singlet oxygen ($^1O_2$), superoxide anions (O_2^-), and hydroxyl radicals (OH) when exposed to UV or visible light. These ROS induce oxidative stress, which damages critical bacterial components, including proteins, lipids, and nucleic acids, overwhelming the bacteria's defense mechanisms and leading to their destruction. Furthermore, C_{60} can act as electron acceptors, disrupting the bacterial electron transport chain, which is essential for energy production and overall cellular function, further contributing to bacterial inactivation. Additionally, C_{60} prevent the formation of biofilms—a protective layer that bacteria often use to resist antimicrobial agents—by inhibiting bacterial aggregation and penetrating existing biofilms, allowing for deeper antibacterial action. These nanomaterials are not only highly effective but are also considered environmentally safe and non-toxic to humans at concentrations used for water treatment. Their stability and potential for recyclability make them cost-effective, sustainable, and efficient for long-term use in water purification systems, ensuring cleaner and safer water supplies (Li et al. 2008; Chae, Hotze, and Wiesner 2014).

9.3.6 Carbon Nanomaterials

Various types of carbon nanomaterials are combined with other categories of nanomaterials and utilized as effective adsorbents. The literature indicates that carbon-based materials are extensively employed as nanomaterials. This study focuses on recent advancements in nanomaterials used alongside carbon materials for microbial removal applications. However, there is a limited selection of carbon-based materials specifically targeted for this purpose (Chithra et al. 2022; Morsi et al. 2017; Ali and Ahmad 2020).

Carbon Nanotubes (CNTs): CNTs have emerged as highly effective agents for removing pathogenic bacteria from water due to their strong antibacterial properties. They are capable of targeting a broad range of pathogens, including harmful bacteria such as *E. coli* and *Salmonella*, and even certain viruses. CNTs are particularly useful for water purification because they can not only eliminate pathogens but also adsorb various pollutants from water. Their unique structure, which is formed by rolling graphene sheets into cylindrical tubes, gives them these exceptional properties. There are two primary types of CNTs: single-walled NTs (SWNTs), which consist of a single graphene tube with diameters ranging from 1 to 5 nm, and multi-walled NTs (MWNTs), which comprise multiple nested tubes. These tubes can vary in length, from as short as 100 nm to several micrometers. While both types of CNTs exhibit antibacterial activity, SWNTs tend to be more toxic to cells, including mammalian cells, compared to MWNTs (Chithra et al. 2022). CNTs can cause bacterial cell death through mechanisms such as physical interactions with cell membranes, leading to structural damage, or through oxidative stress that compromises cell integrity. However, the use of CNTs in practical applications is often limited by their tendency to aggregate in water, making it difficult to fully disperse them. To overcome this, surfactants or polymers like sodium dodecyl benzenesulfate (SDBS), polyvinylpyrrolidone (PVP), or Triton-X are frequently used to enhance their dispersion in water. Moreover, CNTs have shown promise in preventing microbial attachment and biofouling on surfaces, which could be beneficial in maintaining cleaner, safer water systems. Despite their potential, several challenges remain, such as managing CNT aggregation and understanding how natural organic matter (NOM) in water affects their stabilization and bioavailability. These factors must be addressed to fully exploit the antimicrobial properties of CNTs in water treatment technologies. In addition to their antibacterial properties, further research into their antiviral effects could unlock new applications for CNTs in broader antimicrobial and disinfection systems, offering a multi-faceted solution for water purification and pathogen control (Chithra et al. 2022; Morsi et al. 2017).

Graphene is a single layer of carbon atoms arranged in a hexagonal lattice, making it the thinnest and one of the strongest materials known. Its unique structure gives it exceptional flexibility and transparency while also making it an outstanding conductor of both heat and electricity. These properties make graphene highly desirable for a range of applications, particularly in areas requiring advanced material performance. Additionally, SWCNTs, which are closely related to graphene, demonstrate strong antimicrobial properties. MWCNTs share similar activity, with the presence of functional groups like -OH and -COOH further enhancing their ability to remove microbes. The effectiveness of these NTs is especially pronounced in longer varieties, which have been shown to perform better in microbial removal (Chithra et al. 2022; Morsi et al. 2017).

Activated carbon, often regarded as a highly effective adsorbent, is widely used due to its porous structure, which significantly increases its surface area and adsorption capacity compared to regular charcoal. It is commonly applied in water purification, particularly for removing bacteria. There are two primary types of activated carbon used for this purpose: granular activated carbon (GAC) and powdered activated carbon (PAC). Both types are effective in water treatment, with GAC being used for larger-scale applications and PAC for more refined treatments. The extensive porosity of activated carbon makes it ideal for trapping and eliminating impurities, making it a preferred choice for bacterial removal (Chithra et al. 2022).

Biochar is gaining attention for its ability to function as a surface adsorbent, much like activated carbon. It is a carbon-rich byproduct formed through the thermal decomposition of organic biomass. Recently, biochar has been explored as a low-cost alternative for removing pathogenic organisms from water. Its production from agricultural byproducts makes it an appealing option due to its affordability and abundance. Biochar's surface characteristics are effective in water treatment, and its potential for large-scale implementation, especially in developing regions, offers a sustainable solution for microbial removal (Chithra et al. 2022; Morsi et al. 2017).

Carbon nanofibers (NFs) are widely used in composite materials, known for their high conductivity and ability to interact with microbial surfaces, making them effective for removing bacteria and viruses. These NFs provide significant mechanical strength to membranes, which enhances their durability in water treatment applications. Their high environmental stability ensures long-lasting antimicrobial effectiveness, making carbon NFs a reliable choice for advanced filtration systems. Additionally, the versatility of carbon-based materials allows them to be combined with other substances, creating strong covalent bonds that further enhance their properties.

Carbon-based materials like graphene, activated carbon, biochar, and carbon NFs all demonstrate immense potential in various applications, particularly in water purification and antimicrobial treatments. Their unique properties, such as high conductivity, mechanical strength, and environmental stability, make them essential tools for developing efficient, sustainable, and cost-effective solutions to global water challenges (Chithra et al. 2022; Morsi et al. 2017).

9.3.7 Polymer/Metal Oxide Nanocomposites in Water Treatment

Polymer/metal oxide nanocomposites, such as TiO_2 combined with thiolene polymer, have demonstrated a higher effective modulus, making them highly efficient in filtration processes. These nanocomposites enhance the mechanical strength of the material, contributing to the removal of biological contaminants from water. Typically fabricated on glass or silicon substrates using repeated surfactant modifications, these films effectively eliminate various pollutants and microbial contaminants. For removing waterborne pathogens, metals such as Ag, Zn, zeolites, and titanium are often used. In the preparation of nanocomposites, metals like Ag, titanium, or Zn are incorporated into the polymer matrix to enhance antimicrobial properties.

TiO_2-graphene with polyaniline nanocomposites has been found particularly effective for water disinfection. These nanocomposites are capable of eliminating most microbial pollutants due to the large surface area of graphene, which allows for uniform polymerization of polyaniline on its surface and between intercalated graphene sheets. The resulting graphene-TiO_2/polyaniline films, often prepared via electrodeposition, offer advantages such as high conductivity, biocompatibility, and rapid redox reactions. Their photocatalytic properties make them highly antimicrobial, particularly against pathogens like *Enterobacter ludwigii* and *E. coli*. In the field of viral removal, Ag-doped titanium oxide nanomaterials have shown great potential, providing effective virus removal from water. Additionally, magnesium nanoparticle-infused biochar has demonstrated efficacy in removing microorganisms through adsorption and matrix pore size mechanisms, further contributing to its role in water purification. Polysulfone nanocomposites are employed as support membranes in reverse osmosis technology, which is highly effective in removing organisms from water. Polyvinylidene fluoride (PVDF) nanocomposites with TiO_2 have also addressed challenges in removing cyanobacteria toxins and are capable of degrading hydrocarbons in the process. These TiO_2/PVDF layers, produced using the sol–gel method, exhibit exceptional hydrophilicity, increased porosity, and superior permeability compared to traditional polymer films. This simple and cost-effective doping method makes large-scale production of nanocomposite films feasible.

nAgs embedded in polyurethane have been particularly effective in removing bacteria such as *Bacillus subtilis* and *E. coli* from water. The optimal structural arrangement of TiO_2 nanoparticles and thiolene monomers enhances their adhesive properties, while catecholic anchors derived from eugenol bind TiO_2 nanoparticles to the polymer matrix, promoting strong adherence of microorganisms to surfaces.

Polypropylene-TiO_2 nanocomposites also demonstrate strong antimicrobial activity due to their photocatalytic properties. The inclusion of TiO_2 nanoparticles significantly increases the dielectric constant of the nanocomposite, improving conductivity over pristine polypropylene (Chithra et al. 2022; Morsi et al. 2017; Opoku et al. 2017).

Cellulose and Carbon-Based Materials: Natural polymers, particularly cellulose, play a crucial role in water purification. Cellulose NFs (CNFs), derived from sources such as agricultural residues and wood, have proven to be effective in removing waterborne pathogens due to their porous structure. These NFs, when combined with metals such as Ag, form highly effective nanocomposites for microbial removal. Nanocellulose-Ag composites are particularly useful in targeting biological contaminants while also removing other pollutants from water. The rigid surface of cellulose allows microbial loads to adhere to the membrane during water treatment, enhancing its filtration capability. CNFs are popular due to their ease of formation, structural modifiability, and inherent strength, making them efficient in removing bacteria like *E. coli* from water. Chitosan-based composites, formed by combining chitosan with metals like TiO_2, MgO, Ag, and ZnO, are another promising material for microbial water treatment. Chitosan, a biodegradable polysaccharide derived from chitin, has NFs with unique properties that make it effective in various water treatment applications (Chithra et al. 2022; Morsi et al. 2017).

Graphene Nanomesh and Other Nanofibers: Graphene nanomesh is another material gaining popularity in water filtration due to its high porosity and ability to adsorb bacteria. When combined with NFs like polyacrylonitrile (PAN) and boehmite, which are synthesized through electrospinning technology, it achieves highly effective microbial removal. Higher concentrations of boehmite with PAN have been shown to reach 99.99% removal of *E. coli* from drinking water (Chithra et al. 2022; Morsi et al. 2017).

Clay-Polymer Composites: Clay-polymer composites, often made with starch-based polymers, are used for water disinfection through filtration techniques. The polycation on the clay surface plays a crucial role in bacterial removal. However, the ratio of clay to polymer is critical for achieving maximum bactericidal effects. Although these composites have some limitations, such as cross-reactivity and conformational instability, recent advances in alternative polymer molecules have improved their efficiency and broadened their applicability in water treatment systems.

Polymer-metal oxide nanocomposites, cellulose-based materials, and graphene nanomesh exhibit considerable potential for enhancing water purification processes.

Their ability to effectively remove microbial contaminants and pollutants, coupled with their scalability and cost-effectiveness, makes them essential materials for future advancements in clean water technologies (Chithra et al. 2022; Morsi et al. 2017; Opoku et al. 2017) (Table 9.2).

TABLE 9.2

Overview of Nanomaterials and their Efficiency in Microbial Removal for Water Treatment Applications

No	Nanomaterials Used	Types of Removal Applications Used	Targeted Microorganisms	Removal Efficiency
1	Iron oxide	Adsorption	*E. coli*	97%
2	ZnO	ROS	Coliforms	>4 log reduction
3	Titanium oxide	Filtration	Bacteria	99.28%
4	Ag	Membrane treatment	*Methylobacterium* species, *Gordonia* sp., *E. coli*, *Salmonella* sp.	80% for *E. coli*, 9% for *Salmonella sp.*
5	Cobalt ferric oxide	ROS	Bacteria	50%–70%
6	SWCNT	Adsorption	*Candida albicans*, *Salmonella enterica*, *Bacillus cereus*	64%–79.4%
7	Magnesium oxide	ROS and shape distortion	*E. coli*, *Staphylococcus epidermidis*, *P. aeruginosa*	72%
8	Calcium oxide	Membrane filtration	*Enterococcus faecalis*	45%
9	Chitosan/Graphene Oxide (GO)	Adsorption mechanism	*Staphylococcus aureus*	67%
10	Aluminum-based nanoparticles	Electrolysis	*Pseudocohnilemsbus pussillu*	88%
11	CuFeO	Ion exchange	*E. coli*, Total coliform	94.40%
12	Chitosan nanoparticles	Membrane filtration	*Klebsiella pneumoniae*, *E. coli*	25%
13	Fe@MgO	Microfiltration technique	*E. coli*, *Staphylococcus aureus*	90%
14	Fe_3O_4@SiO_2	Ion-exchange technique	*Cryptosporidium parvum* oocysts, *E. coli*, *Clostridium perfringens*	96%
15	Polyethylenimine/TiO_2	Adsorption and separation	Fecal Coliform	40%–50%
16	Polyrhodanine/MnO_2	Electrokinetic coagulation	*P. aeruginosa*	72%
17	Polyurethane–clay	Adsorption	*E. coli*, Enterococci	83%
18	TiO_2/Ag nanoparticle	Photochemical technique	*Candida glabrata*	94%
19	Cellulose fibers embedded nAgs	Filtration technique	*E. coli* ATCC 11229, *Enterococcus faecalis* ATCC 47077	Over log 6
20	nAg alginate composite beads	ROS technique	*E. coli* K-12	Disinfection efficiency over 5-log

(*Continued*)

TABLE 9.2 (*Continued*)
Overview of Nanomaterials and their Efficiency in Microbial Removal for Water Treatment Applications

No	Nanomaterials Used	Types of Removal Applications Used	Targeted Microorganisms	Removal Efficiency
21	Ag nanoparticles loaded kaolin clay	Reverse osmosis purification technique	*Salmonella* spp., *E. coli*	70%–80%
22	Magnetic Fe_2O_3–AgBr	Adsorption	*E. coli* K-12	9%–12%
23	Gold nanoparticles	ROS technique	*E. coli*	0.90%
24	PAN/boehmite	Filtration technique	*E. coli*	74%

Source: Table 9.2 is obtained from one review Chithra et al. (2022).

Various environmental factors significantly impact the removal of biological contaminants such as bacteria. Key elements like ionic strength and pH play a vital role in determining the efficiency of bacterial elimination. Similar observations have been made in microbial inactivation, another mechanism for removing microorganisms. For instance, studies using nAgs embedded in chitosan cryogels and Zn phosphide nanowires have shown bacterial reduction rates of approximately 3-log and 4-log, respectively. The process involves the penetration and destruction of microbial cells through the generation of hydroxyl radicals, which damage bacterial cell walls, thereby increasing removal efficiency. The binding of positive ions to bacterial surfaces further enhances the effectiveness of bacterial removal. Research on nAg alginate composite beads and cellulose fibers embedded with nAgs has demonstrated significant bacterial reductions, achieving approximately 5-log and 6-log removal, respectively. Microbial tolerance, especially regarding pH changes, is a critical factor that influences microbial response and accumulation at both physiological and molecular levels. Maintaining a regulated pH is essential for microbial metabolic activity and other processes that impact their removal. Certain microorganisms, such as bacteria and yeast, can maintain a stable pH gradient between their intracellular and extracellular environments during these removal processes, which helps them survive under acidic conditions. Interestingly, viral adsorption appears to be more affected by NOM than bacterial adsorption. While current studies have primarily focused on bacteriophages, future research should examine the interaction between nanomaterials and viruses, as their behavior might differ. Understanding how these nanomaterials can be optimized for virus treatment is critical for advancing water purification technologies. Additionally, certain bacterial traits, such as spore production and mechanisms for repairing cell damage, contribute to the varying tolerance of different bacteria to nanomaterials. Further research is necessary to better understand these bacterial responses, which would help improve the application of nanomaterials in microbial removal (Chithra et al. 2022; Opoku et al. 2017).

9.4 ADVANTAGES AND DISADVANTAGES OF MICROBIAL NANOMATERIAL TREATMENT

Microbial nanomaterial treatment offers several notable advantages, particularly in the removal of harmful microorganisms from water and other environments. Nanomaterials, due to their small size and high surface-area-to-volume ratio, have enhanced reactivity and antimicrobial properties. This allows for efficient microbial inactivation through mechanisms such as cell wall penetration, the release of ROS, and ion exchange, leading to the destruction of pathogens. Moreover, nanomaterials like Ag, ZnO, and TiO_2 exhibit potent antimicrobial effects even at low concentrations, making them highly effective in treating a wide range of microorganisms, including bacteria, fungi, and viruses. Their flexibility to be used in various forms—such as coatings, filtration systems, and membranes—adds to their versatility in microbial treatment applications. Additionally, nanomaterials can be tailored for specific microbial contaminants, enabling more targeted treatments. This adaptability allows for the development of hybrid nanomaterials, such as chitosan-graphene oxide composites, which have been proven to enhance bacterial removal. In comparison to conventional treatment methods, microbial nanomaterials also have the potential for longer-lasting antimicrobial action, which reduces the need for frequent reapplication and maintenance. Furthermore, they can provide a sustainable solution with fewer harmful byproducts when designed with eco-friendly materials.

Despite these advantages, there are also significant challenges associated with microbial nanomaterials treatment. One of the primary concerns is the potential environmental and health risks posed by the release of nanomaterials into ecosystems. Nanoparticles, particularly metals like Ag and Zn, can accumulate in the environment and may be toxic to aquatic life, potentially disrupting ecological balance. Additionally, the long-term impacts of exposure to nanomaterials on human health are not yet fully understood, with concerns about their ability to enter the bloodstream and cause adverse effects at the cellular level. As a result, stringent regulations and assessments are needed to ensure the safe use of nanomaterials. Another drawback is the cost of producing nanomaterials at scale. Synthesis processes can be expensive, and incorporating these materials into treatment systems may not always be economically viable, especially for large-scale water treatment plants. The lack of widespread infrastructure for the recovery and recycling of nanomaterials further adds to operational costs. Additionally, microbial resistance to certain nanomaterials may develop over time, similar to antibiotic resistance, which could reduce the long-term effectiveness of nanomaterials in microbial treatment. Therefore, while microbial nanomaterials treatment holds great promise, these limitations must be addressed through continued research and innovation (Umar et al. 2022; Jamkhande et al. 2019).

9.5 ADVANCES IN WATER PURIFICATION: MICROBIAL CONTAMINATION AND INNOVATIVE TREATMENT METHODS

As the issue of microbial contamination in water becomes more prevalent, innovative approaches are continuously evolving to address it. Several techniques are employed to combat microbial contamination, including filtration, adsorption, disinfection, and remediation. Among these, adsorption has proven to be particularly effective in removing microbial pollutants, as it combines chemical and physical processes to eliminate compounds responsible for water's color, taste, and odor. It is favored for its high removal efficiency, cost-effectiveness, low energy consumption, and minimal waste production. Additionally, adsorbents can be created from renewable sources, though some challenges like cost and the need for specialized distillation processes persist.

Nanomaterials have also gained attention for their potential to effectively remove pathogenic microorganisms from water. The use of microwave and ultrasonic modifications enables the creation of multifunctional nanomaterials that interact with bacterial cell walls to tackle contamination. However, these methods can be expensive. Remediation techniques, such as phytoremediation, chemical, microbial, and mycoremediation, are eco-friendly alternatives. They don't require pretreatment and present lower risks of metal poisoning, though further research is needed to optimize their effectiveness. These techniques, known for their flexibility and cost-efficiency, are often applied in small-scale industries.

Electrochemical water disinfection technologies are another innovative approach, producing reactive chlorine species that eliminate harmful bacteria without introducing chlorine. While highly effective, these methods require intensive labor, and the lifespan of electrodes can be affected by factors like current density. Similarly, slow sand filtration physically removes particles and biologically eliminates pathogens via a biofilm layer within the sand bed. This method improves water quality in terms of physical, chemical, and microbiological aspects without the use of chemicals, although it is limited to small-scale applications.

Membrane filtration, including nanofiltration and ultrafiltration, has gained popularity due to its high flux and low energy requirements. It is effective in separating pathogens with minimal pretreatment, although one drawback is that viruses and mycoplasmas can sometimes pass through the membrane's pores. Nanotechnology has contributed significantly to the field of water disinfection in recent years. Nanomaterials are particularly advantageous due to their large surface area, which enhances the removal of microorganisms without producing harmful metabolic byproducts. However, challenges such as catalyst recovery and a reduction in reaction rates still need to be addressed. Overall, while a variety of techniques have proven successful in microbial removal, each has its advantages and limitations. Continued research and development are essential to improving the efficiency and scalability of these methods, particularly in tackling the growing challenge of waterborne pathogens and ensuring access to clean, safe water (Gehrke, Geiser, and Somborn-Schulz 2015; Yusuf et al. 2023; Lustenberger and Castro-Muñoz 2022).

9.6 FUTURE SUGGESTIONS

In recent decades, the demand for clean water has become increasingly challenging to meet. The primary motivations for ensuring the delivery of clean water are the cost and efficiency of the processes involved. Given their high capabilities, reduced manufacturing costs, and improved efficiency, a wide range of nanomaterials has emerged as promising candidates for enhancing microbial removal from water sources. The main challenge now lies in applying these lab-scale concepts in real-world industrial settings, as the readiness for commercialization of nanomaterials varies significantly.

To effectively gauge the efficiency and applicability of these nanoparticles in removing contaminants from water, it is essential to test them in real natural and industrial wastewater under realistic and worst-case scenarios. Assessing the long-term performance of nanomaterials in water and wastewater treatment applications is also crucial. Therefore, we have compared the efficiencies of various nanomaterials in microbial removal, providing valuable insights for their implementation in water purification efforts. Another limitation of nano-engineered water treatment techniques is their inflexibility for mass development, making them less compatible with conventional treatment methods in many situations. However, nano-engineered materials hold significant potential for transforming water treatment, particularly in decentralized systems, point-of-use applications, and the degradation of serious contaminants. Additionally, there is considerable research focus on developing more efficient nanomaterials, including metal/metal nanocomposites, metal oxide/polymer nanocomposites, and innovative metal oxide/metal oxide nanocomposites. Addressing pricing issues and the commercialization of wastewater treatment systems is also vital. Overall, these diverse applications of nanomaterials could play a crucial role in providing access to clean drinking water worldwide (Mishra et al. 2021; Oliveira et al. 2011).

9.7 CONCLUSION

In conclusion, nanomaterials hold great promise for revolutionizing microbial removal in water treatment, offering a safer and more efficient alternative to traditional chemical disinfectants. Their ability to target and eliminate harmful microorganisms through mechanisms such as ROS production and membrane disruption makes them invaluable in addressing water contamination issues. However, while their potential is clear, more research is needed to fully assess their effectiveness in large-scale systems and understand their broader environmental and societal impacts. By advancing our knowledge of these materials and their interactions with microbial communities, we can unlock new possibilities for cleaner, safer water treatment technologies.

ACKNOWLEDGMENT

We would like to express our sincere gratitude to the Scientific and Technological Research Council of Turkey (TUBITAK) for their generous support through the 2221-TUBITAK Scholarship program under the number 1059B212200690.

REFERENCES

Ali, Sadaf, Shiekh Amir, Shafat Ali, Muneeb U Rehman, Sabhiya Majid, and Ali Mohd Yatoo. 2021. Water pollution: Diseases and health impacts. In Gowhar Hamid Dar, Khalid Rehman Hakeem, Mohammad Aneesul Mehmood, and Humaira Qadri (eds.), *Freshwater Pollution and Aquatic Ecosystems*, pp. 1–23. Apple Academic Press.

Ali, Zarshad, and Rashid Ahmad. 2020. Nanotechnology for water treatment. *Environmental Nanotechnology* 3:143–163.

Alijanzadeh Maliji, Behzad, Ali Babayeemehr, Karen Rohani, Sanaz Mehrabani, and Faeze Aghajanpour. 2023. Role of the World Health Organization in management of gastrointestinal diseases caused by contaminated water in children in the Middle East: A review article. *Journal of Pediatrics Review* 11(1):59–66.

Alvarez, Marisol. 2021. Health risk assessment of local populations ingesting water with naturally occurring arsenic and fecal related contaminants in Lake Atitlan, Guatemala. University of South Florida.

Andrée F. Maheux, Dominique K. Boudreau, Marc-Antoine Bisson, Vanessa Dion-Dupont, Sébastien Bouchard, Martine Nkuranga, Michel G. Bergeron, Manuel J. Rodriguez. 2014. Molecular method for detection of total coliforms in drinking water samples. *Applied and Environmental Microbiology* 80(14):4074–4084.

Baruah, Sunandan, Samir K Pal, and Joydeep Dutta. 2012. Nanostructured zinc oxide for water treatment. *Nanoscience & Nanotechnology-Asia* 2(2):90–102.

Bayoumi, Talat A, and Saleh, Hosam M. 2018. Characterization of biological waste stabilized by cement during immersion in aqueous media to develop disposal strategies for phytomediated radioactive waste. *Progress in Nuclear Energy* 107:83–89.

Bermeo, Lizeth, Kristina Ivanova, Leonardo Martín Pérez, Eva Forés, Sílvia Pérez-Rafael, Juan C Casas-Zapata, Jordi Morató, and Tzanko Tzanov. 2022. Sono-enzymatically embedded antibacterial silver-lignin nanoparticles on cork filter material for water disinfection. *International Journal of Molecular Sciences* 23(19):11679.

Brown, C Hendricks, Wei Wang, Sheppard G Kellam, Bengt O Muthén, Hanno Petras, Peter Toyinbo, Jeanne Poduska, Nicholas Ialongo, Peter A Wyman, and Patricia Chamberlain. 2008. Methods for testing theory and evaluating impact in randomized field trials: Intent-to-treat analyses for integrating the perspectives of person, place, and time. *Drug and Alcohol Dependence* 95:S74–S104.

Chae, So-Ryong, Ernest M Hotze, and Mark R Wiesner. 2014. Possible applications of fullerene nanomaterials in water treatment and reuse. In Anita Street (ed.), *Nanotechnology Applications for Clean Water*, pp. 329–338. Elsevier.

Chithra, A, Rajaseetharama Sekar, P Senthil Kumar, and G Padmalaya. 2022. A review on removal strategies of microorganisms from water environment using nanomaterials and their behavioural characteristics. *Chemosphere* 295:133915.

Clasen, Thomas F, Joseph Brown, Simon Collin, Oscar Suntura, and Sandy Cairncross. 2004. Reducing diarrhea through the use of household-based ceramic water filters: A randomized, controlled trial in rural Bolivia. *The American Journal of Tropical Medicine and Hygiene* 70(6):651–657.

Gazit, Ehud. 2007. Self-assembled peptide nanostructures: The design of molecular building blocks and their technological utilization. *Chemical Society Reviews* 36(8):1263–1269.

Gehrke, Ilka, Andreas Geiser, and Annette Somborn-Schulz. 2015. Innovations in nanotechnology for water treatment. *Nanotechnology, Science and* Applications 8:1–17.

Ghadiri, M Reza, Juan R Granja, Ronald A Milligan, Duncan E McRee, and Nina Khazanovich. 1993. Self-assembling organic nanotubes based on a cyclic peptide architecture. *Nature* 366(6453):324–327.

Goswami, Kakali Priyam, and G Pugazhenthi. 2020. Credibility of polymeric and ceramic membrane filtration in the removal of bacteria and virus from water: A review. *Journal of Environmental Management* 268:110583.

Hardy, J. (1981). The enumeration, isolation and characterization of sulphate-reducing bacteria from North Sea waters. *Journal of Applied Microbiology*, 51:505–516. https://doi.org/10.1111/J.1365-2672.1981.TB01269.X

Hlongwane, Gloria Ntombenhle, Patrick Thabang Sekoai, Meyya Meyyappan, and Kapil Moothi. 2019. Simultaneous removal of pollutants from water using nanoparticles: A shift from single pollutant control to multiple pollutant control. *Science of the Total Environment* 656:808–833.

Hossain, Fahim, Oscar J Perales-Perez, Sangchul Hwang, and Félix Román. 2014. Antimicrobial nanomaterials as water disinfectant: Applications, limitations and future perspectives. *Science of the Total Environment* 466:1047–1059.

Hou, Ting, Shanshan Yu, Minfeng Zhou, Min Wu, Jie Liu, Xiaoli Zheng, Jinxing Li, Joseph Wang, and Xiaolei Wang. 2020. Effective removal of inorganic and organic heavy metal pollutants with poly (amino acid)-based micromotors. *Nanoscale* 12(8):5227–5232.

Jamkhande, Prasad Govindrao, Namrata W Ghule, Abdul Haque Bamer, and Mohan G Kalaskar. 2019. Metal nanoparticles synthesis: An overview on methods of preparation, advantages and disadvantages, and applications. *Journal of Drug Delivery Science and Technology* 53:101174.

Li, Qilin, Shaily Mahendra, Delina Y Lyon, Lena Brunet, Michael V Liga, Dong Li, and Pedro JJ Alvarez. 2008. Antimicrobial nanomaterials for water disinfection and microbial control: Potential applications and implications. *Water Research* 42(18):4591–4602.

Liu, Chang, Tao Hong, Huaifeng Li, and Lili Wang. 2018. From club convergence of per capita industrial pollutant emissions to industrial transfer effects: An empirical study across 285 cities in China. *Energy Policy* 121:300–313.

Lustenberger, Sharon, and Roberto Castro-Muñoz. 2022. Advanced biomaterials and alternatives tailored as membranes for water treatment and the latest innovative European water remediation projects: A review. *Case Studies in Chemical and Environmental Engineering* 5:100205.

Ma, Haiyuan, Yan Guo, Yu Qin, and Yu-You Li. 2018. Nutrient recovery technologies integrated with energy recovery by waste biomass anaerobic digestion. *Bioresource Technology* 269:520–531.

Mahendra Pal, Yodit Ayele, Angesom Hadush, Sumitra Panigrahi and Vijay J Jadhav 2018. Public health hazards due to unsafe drinking water. *Air & Water Borne Diseases* 7(1000138):2.

Mishra, Bishwambhar, Sunita Varjani, Gopalakrishnan Kumar, Mukesh Kumar Awasthi, Sanjeev Kumar Awasthi, Raveendran Sindhu, Parameswaran Binod, Eldon R Rene, and Zengqiang Zhang. 2021. Microbial approaches for remediation of pollutants: Innovations, future outlook, and challenges. *Energy & Environment* 32(6):1029–1058.

Mishra, Prakash Chandra. 2005. Some aspects of the quality of water in and around Rourkela.

Mmelesi, Olga Kelebogile, Ngonidzashe Masunga, Alex Kuvarega, Thabo TI Nkambule, Bhekie B Mamba, and Kebede K Kefeni. 2021. Cobalt ferrite nanoparticles and nanocomposites: Photocatalytic, antimicrobial activity and toxicity in water treatment. *Materials Science in Semiconductor Processing* 123:105523.

Morsi, Rania E, Ahmed M Alsabagh, Shimaa A Nasr, and Manal M Zaki. 2017. Multifunctional nanocomposites of chitosan, silver nanoparticles, copper nanoparticles and carbon nanotubes for water treatment: Antimicrobial characteristics. *International Journal of Biological Macromolecules* 97:264–269.

Oliveira, J, A Cunha, F Castilho, JL Romalde, and MJ Pereira. 2011. Microbial contamination and purification of bivalve shellfish: Crucial aspects in monitoring and future perspectives–A mini-review. *Food Control* 22(6):805–816.

Opoku, Francis, Ephraim M Kiarii, Penny P Govender, and Messai Adenew Mamo. 2017. *Metal Oxide Polymer Nanocomposites in Water Treatments*. Vol. 8: IntechOpen, London, UK.

Pandey, Pramod K, Philip H Kass, Michelle L Soupir, Sagor Biswas, and Vijay P Singh. 2014. Contamination of water resources by pathogenic bacteria. *AMB Express* 4:1–16.

Qi, Lifeng, Zirong Xu, Xia Jiang, Caihong Hu, and Xiangfei Zou. 2004. Preparation and antibacterial activity of chitosan nanoparticles. *Carbohydrate Research* 339(16):2693–2700.

Rajasulochana, Perepi, and V Preethy. 2016. Comparison on efficiency of various techniques in treatment of waste and sewage water: A comprehensive review. *Resource-Efficient Technologies* 2(4):175–184.

Reddy, M Pratap, A Venugopal, and M Subrahmanyam. 2007. Hydroxyapatite-supported Ag–TiO_2 as Escherichia coli disinfection photocatalyst. *Water Research* 41(2):379–386.

Richter, Kenneth E, and Jennifer M Ayers. 2018. An approach to predicting sediment microbial fuel cell performance in shallow and deep water. *Applied Sciences* 8(12):2628.

Sawai, J. 2003. Quantitative evaluation of antibacterial activities of metallic oxide powders (ZnO, MgO and CaO) by conductimetric assay. *Journal of Microbiological Methods* 54(2):177–182.

Sayes, Christie M, Andre M Gobin, Kevin D Ausman, Joe Mendez, Jennifer L West, and Vicki L Colvin. 2005. Nano-C60 cytotoxicity is due to lipid peroxidation. *Biomaterials* 26(36):7587–7595.

Sclafani, Antonino, Marie-Noëlle Mozzanega, and Jean-Marie Herrmann. 1997. Influence of silver deposits on the photocatalytic activity of Titania. *Journal of Catalysis* 168(1):117–120.

Sizmur, Tom, Teresa Fresno, Gökçen Akgül, Harrison Frost, and Eduardo Moreno-Jiménez. 2017. Biochar modification to enhance sorption of inorganics from water. *Bioresource Technology* 246:34–47.

Tang, Kai, Chengzhu Gong, and Dong Wang. 2016. Reduction potential, shadow prices, and pollution costs of agricultural pollutants in China. *Science of the Total Environment* 541:42–50.

Terreni, Marco, Marina Taccani, and Massimo Pregnolato. 2021. New antibiotics for multidrug-resistant bacterial strains: Latest research developments and future perspectives. *Molecules* 26(9):2671.

Umar, Khalid, Tabassum Parveen, Meraj Alam Khan, Mohamad Nasir Mohamad Ibrahim, Akil Ahmad, and Mohd Rafatullah. 2019. Degradation of organic pollutants using metal-doped TiO_2 photocatalysts under visible light: A comparative study. *Desalination and Water Treatment* 161:275–282.

Umar, Wajid, Muhammad Zia ur Rehman, Muhammad Umair, Muhammad Ashar Ayub, Asif Naeem, Muhammad Rizwan, Husnain Zia, and Rama Rao Karri. 2022. Use of nanotechnology for wastewater treatment: Potential applications, advantages, and limitations. In Janardhan Reddy Koduru, Rama Rao Karri, Nabisab Mujawar Mubarak, and Erick R. Bandala (eds.), *Sustainable Nanotechnology for Environmental Remediation*, 223–272. Elsevier.

Van Apeldoorn, Marian E, Hans P Van Egmond, Gerrit JA Speijers, and Guido JI Bakker. 2007. Toxins of cyanobacteria. *Molecular Nutrition & Food Research* 51(1):7–60.

Vargas, Mónica Andrea, and Jorge E Rodríguez-Páez. 2017. Amorphous TiO_2 nanoparticles: Synthesis and antibacterial capacity. *Journal of Non-Crystalline Solids* 459:192–205.

Wang, Jing, Zhenjun Wang, Carolina LZ Vieira, Jack M Wolfson, Guiyou Pingtian, and Shaodan Huang. 2019. Review on the treatment of organic pollutants in water by ultrasonic technology. *Ultrasonics Sonochemistry* 55:273–278.

Xin, Ziming, Qianqian He, Shuangao Wang, Xiaoyu Han, Zhongtian Fu, Xinxin Xu, and Xin Zhao. 2022. Recent progress in ZnO-based nanostructures for photocatalytic antimicrobial in water treatment: A review. *Applied Sciences* 12(15):7910.

Yusuf, Ahmed, Hussein Kehinde Amusa, Jamiu Olanrewaju Eniola, Adewale Giwa, Oluwadamilola Pikuda, Abdallah Dindi, and Muhammad Roil Bilad. 2023. Hazardous and emerging contaminants removal from water by plasma-based treatment: A review of recent advances. *Chemical Engineering Journal Advances* 14:100443.

Zahoor, Muhammad, Nausheen Nazir, Muhammad Iftikhar, Sumaira Naz, Ivar Zekker, Juris Burlakovs, Faheem Uddin, Abdul Waheed Kamran, Anna Kallistova, and Nikolai Pimenov. 2021. A review on silver nanoparticles: Classification, various methods of synthesis, and their potential roles in biomedical applications and water treatment. *Water* 13(16):2216.

10 Treatment of Wastewater Using Nanoparticles from Seaweeds (Marine Algae)

J. Immanuel Suresh and A. Evangelina

10.1 INTRODUCTION

"Seaweed" is a term that encompasses numerous marine plants and algae that thrive in the ocean as well as in other aquatic environments like rivers, lakes, and water bodies. Compared to terrestrial plants, aquatic plants (microalgae and macroalgae) play a more significant role in several bioindustries due to their abundant sources of bioactive compounds and their potential source for serving as biomaterials. The vast coastline of India, spanning over 7,500 km, offers a promising environment for seaweed growth, particularly in regions like Tamil Nadu and Gujarat coasts and Lakshadweep and the Andaman Nicobar Islands. The eastern coast around Visakhapatnam, as well as the southern coast encompassing Nagapattinam, Gulf of Mannar, Tiruchendur, Tuticorin, and Kerala, also boast thriving seaweed beds. Additionally, Veraval and the Gulf of Kutch along the western coast exhibit favorable conditions. Despite the presence of approximately 844 seaweed species suitable and represented for commercial cultivation, their widespread implementation is yet to occur in India. Considering the 844 seaweed species, India possesses around 434 species of red seaweeds, 194 species of brown seaweeds, and 216 species of green seaweeds.

Seaweeds are abundant along the Tamil Nadu and Gujarat coasts and around the Lakshadweep and Andaman and Nicobar Islands. Rich seaweed beds occur around Mumbai, Ratnagiri, Goa, Karwar, Varkala, and Vizhinjam, and Pulicat in Tamil Nadu, Andhra Pradesh, and Chilka in Orissa (Department of Fisheries, 2020). Nearly 60 species are commercially important, out of approximately 700 species of marine algae found in both intertidal and deep-water regions of the Indian coast. In this chapter, we will be discussing the nanoparticles (NPs) extracted from different marine seaweeds and their potential in treating the wastewater. Meanwhile, the quick growth of nanotechnology has added a lot of intrigue in the use of nanomaterials from various substances, including seaweeds, and has enhanced systems for controlling and cleaning contaminants in the water, soil, and air segments (Mohamed, 2017; Taghipour et al., 2019) by eco-friendly methods as well as for a sustainable environment.

DOI: 10.1201/9781003459286-10

10.2 WHAT ARE NANOPARTICLES?

A NP or ultrafine particle, is usually defined as a particle of matter that is between 1 and 100 nanometers (nm) in diameter. NPs have been extensively utilized in bioremediation (a process of detoxifying or degrading contaminants present in soil, wastewater, and industrial sludge by biological means) due to their high specific surface area that improves the removal reaction. The integration of nanomaterials and bioremediation has great potential to be an effective, efficient, and sustainable solution for a clean environment. Algae-based green synthesis of NPs has shown the diverse capability to be employed in various fields, including bioremediation of wastewater, mitigation of environmental pollutants, antibacterial, antifungal, anticancer activity, electronics, optics, cosmetics, coating applications, sensing, space technology, food packaging, light emitters, mechanics, chemical industries, and nonlinear optical devices (Hussein et al., 2019; Khan et al., 2019). Green synthesis employs a clean, safe, cost-effective, and environmentally friendly process of constructing nanomaterials. Microorganisms such as bacteria, yeast, fungi, algal species, and certain plants act as substrates for the green synthesis of nanomaterials (Alprol et al., 2023).

10.3 HOW ARE NPS EXTRACTED FROM SEAWEEDS?

Broadly, the mixing of the plant or algal extract of interest with an aqueous solution of the metal salt at room temperature is carried out. Even though different temperatures have been reported, many silver (Ag) and gold (Au) NPs have been prepared at room temperature (Li et al., 2011). The most notable AgNPs have been extensively studied over the recent years on account of their fascinating, unique physical and chemical properties, leading to diverse applications in many fields (Zhang et al., 2016).

In the present, various interdisciplinary branches help to develop and fabricate nanomaterials via two approaches:

- **Top-down approach** (reducing the size of bulk materials)

  The reduction of the bulk materials to attain nanoscale size, the top-down approach involves exploiting the base material in large quantities. This reduction can be accomplished through both physical and chemical techniques (Pathak et al., 2019), which include photochemical and chemical reduction as well as electrochemical changes using various techniques such as laser and thermal ablation and mechanical milling. These techniques result in stabilized NPs, whereas the most common technique used for this approach is the lithographic interferometric (Agarwal et al., 2017). However, it is important to note that, despite its easier implementation, this approach may induce changes in the properties and surface chemistry of the prepared NPs (Jamkhande et al., 2019).
- **Bottom-up approach** (fabricating nanostructures in a controlled manner)

  On the other hand, the bottom-up approach involves the controlled assembly of atoms or molecules to form nanostructured building units that eventually form the desired NPs. Both chemical and biological techniques can be employed to achieve this.

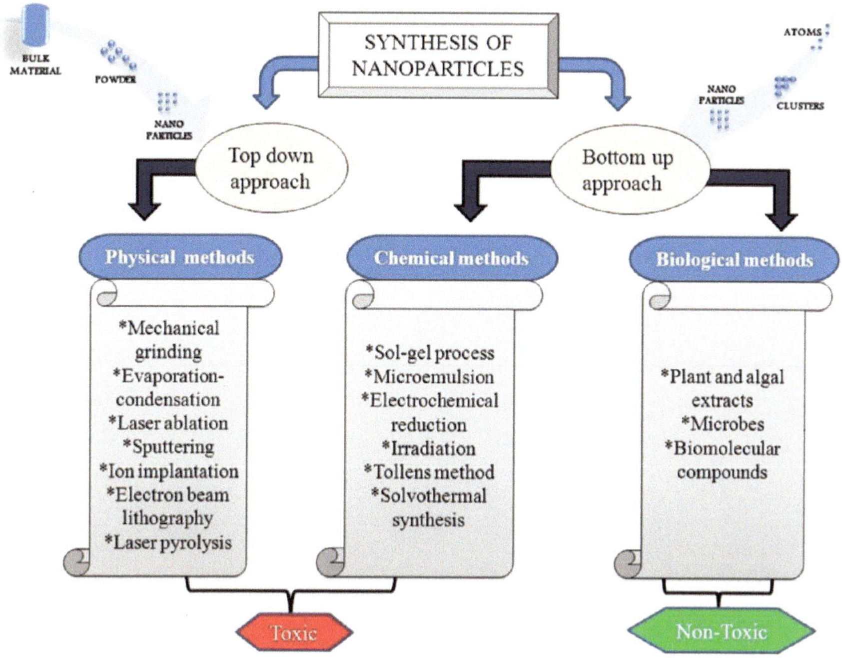

FIGURE 10.1 Methods of synthesizing NPs.

Several studies have clearly shown that biological methods for NP synthesis have the fewest drawbacks compared to other methods. For instance, physical methods have high expenditures in time, energy, and cost, as well as a low production rate, whereas chemical synthesis is attained by the involvement of toxic chemicals and solvents, resulting in noxious derivatives and toxic byproducts (Pathak et al., 2019) (Figure 10.1).

10.4 GREEN SYNTHESIS OF NPs

Currently, the formation of NPs using green synthesis is gaining momentum as an emerging approach. This technique involves the production of NPs utilizing biological resources, effectively addressing and overcoming the limitations imposed by physical and chemical methods (Grammatikopoulos et al., 2016). A notable example involves AgNPs from agricultural wastes, algal extracts, etc. (Wolny-Koładka et al., 2022).

The environmentally sound nature of this approach eliminates the need for harmful (noxious) and toxic chemicals, as well as extreme conditions for obtaining NPs, making it both cost-effective and suitable for large-scale production (Patel et al., 2022). The advantages of the green synthesis approach encompass simplicity, reproducibility, biocompatibility, eco-friendliness, and economic viability. Different biological agents have their own mechanisms to acknowledge metal ions for the

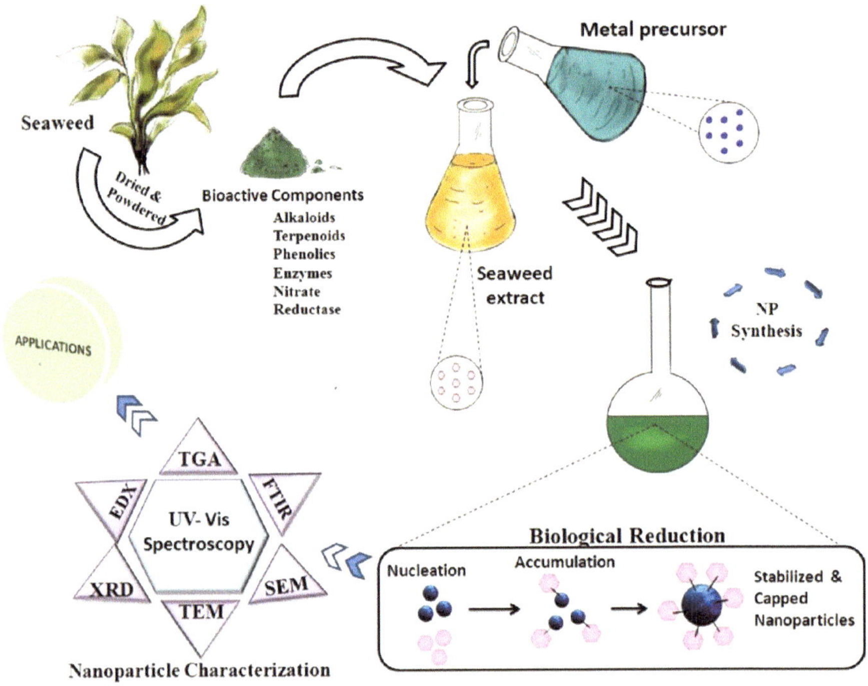

FIGURE 10.2 Syntheses of NPs.

synthesis of respective NPs. Major advantages of using biological resources over non-biological resources as it is obtained via biomolecules present in the biological system, which maintains the properties of NPs, hereby toxic chemical agents are not required (Ovais et al., 2018).

To carry out the process of biosynthesizing metal NPs from desired algae for the experiment, the steps are as follows:

i. Heating or boiling a desired algal extract (dried and powdered) in water or an organic solution for a specific duration.
ii. Additionally, molar solutions of ionic metallic compounds are prepared.
iii. Both the algae and the solutions are incubated under controlled conditions for a definite period of time, either with regular stirring or without stirring (Figure 10.2).

For the time being, the algal-based nanosynthesis has developed into a separate branch known as "phyco-nanotechnology". Several studies on the biosynthesis of NPs have been conducted using seaweed extracts. However, the use of microalgae for NP synthesis is very limited.

10.5 SOURCES AND CONSEQUENCES OF WASTEWATER POLLUTION

Wastewater from several textile industries is a complex mixture of many polluting substances ranging from organochloride-based pesticides to heavy metals associated with dyes and the dyeing process (Correia et al., 1994). It is estimated that around 10,000 different dyes are commercially used extensively in textile processing, paper, printing, pharmaceuticals, food, tanning, and in other industries (McMullan et al., 2001). A wide range of organic and inorganic compounds are sources of contamination. Heavy metals are considered one of the most common and hazardous pollutants (Mitra et al., 2022). The chief sources of water contamination are sewage water, industrial wastewater, improper use of pesticides, and fertilizers. Heavy metals are major inorganic contaminants present in soil, wastewater, and sediments that cannot be degraded, making it a major concern of environmental pollution. Metals, like copper (Cu), iron (Fe), manganese (Mn), and zinc (Zn) are essential for life processes, whereas others, like cadmium (Cd), nickel (Ni), and mercury (Hg), have no biological function and are considered very toxic even at lower concentrations of 0.001–0.1 mg/L, often resulting in harmful disorders at a higher concentration through anthropogenic sources. The dyes or their degradation products are toxic, mutagenic, and carcinogenic in nature. The toxic compounds from the untreated dye effluents and the heavy metals in industrial wastewater effluents are known to alter the quality of the water and have been proved to be hazardous to both aquatic and human life and the environment.

Example: Azo dyes are capable of producing intestinal cancer.

Treatment of the wastewater containing the toxic chemicals, heavy metals, and dyes is very difficult due to the wide range of pH levels, salt concentration, and high range of colorants in effluents. Several conventional technologies and remediation methods, such as excavation and landfill, thermal treatment, acid leaching, and electroreclamation, have already been used for their removal, but most of them are not suitable and have been limited due to their high installation as well as operating costs, low efficiency, sensitivity to a variable wastewater input, and excessive use of chemicals, whereas no optimum results were obtained.

The major causes are anthropogenic sources such as mining (Kpan et al., 2014; Modoi et al., 2014; Cobbina et al., 2013; Dhakate and Singh, 2008; Mishra et al., 2008; Bidone et al., 2001), smelting procedures (Kapusta and Sobczyk, 2015; Asami, 1988; Whitby and Hutchinson, 1974), industrial activities, like chemical and metallurgical industries (Nakayama et al., 2010; Cortes et al., 2003), cement industry (Al-Khashman and Shawabkeh, 2006; Isikli et al., 2003; Schuhmacher et al., 2002), electroplating, plastics manufacturing, transports, pesticides, fertilizers (Bolan et al., 2005), and agriculture (Vaalgamaa and Conley, 2008). The natural environment's metal concentration is further increased by long-range atmospheric pollution transport.

Pollution of the natural environment by heavy metals and dyes is a worldwide problem because these metals are indestructible, and most of them have toxic effects on living organisms when they exceed a certain concentration (Ghrefat and Yusuf, 2006). They can disturb important biochemical processes, constituting an important threat for the health of plants and animals as they consume water by contact with their external surfaces, through ingestion, and also from inhalation of airborne particles and vaporized metals (Mudgal et al., 2010a). Furthermore, if the food chain is polluted with toxic metals and metalloids, it would be an important route of human exposure and cause several dangerous effects on humans (Mudgal et al., 2010b).

10.5.1 Remediation for Reduction of Heavy Metals

However, using microorganisms and plants for remediation purposes is thus a possible and remarkable solution for heavy metal pollution, since it includes sustainable remediation technologies to rectify and re-establish the natural condition of soil (Doelman et al., 1994). Currently, a vast array of biological materials, especially bacteria, algae, yeasts, and fungi, have received increasing attention for heavy metal removal and recovery due to their good performance, higher productivity, low cost, and large available quantities. Bioremediation techniques are more beneficial than traditional methods because they can be implemented on site, thus reducing risks for personnel. Effluent volume generated by bioremediation is much smaller, reducing the problem of sludge disposal. Additionally, since the technology is based on natural processes, the public considers it more acceptable and preferable than other conventional methods.

Especially, seaweeds were identified as potential indicators for biomonitoring heavy metals in coastal water (Akcali and Kucuksezgin, 2011). Seaweeds had been regarded as hinge species to monitor the environmental conditions in coastal ecosystems for decades in Europe (Reis et al., 2014). There are few (brown, green, and red) seaweeds that have been designated as biomonitors for Cd, Cu, Zn, and lead (Pb) in Asia (Chakraborty and Owens, 2014; Zhang et al., 2009). Selecting suitable native seaweed species having high potential for biological purification of NPs is essential in the future. Therefore, seaweeds may reflect the ecological status of coastal environments and can be used to improve the coastal environments (Paraskevi et al., 2015).

10.5.1.1 Potential Bioremediation Effects of Seaweed *Gracilaria lemaneiformis* on Heavy Metals in Coastal Sediment (Luo et al., 2020)

In the present study, the bioremediation potential of cultivated *Gracilaria lemaneiformis* on heavy metals in **Nan'ao** (Island) coastal sediment from a mariculture zone in South China was assessed. The collection of the sediment samples occurred from five different zones (*Gracilaria* cultivation zone, G; fish culture zone, F; shellfish culture zone, S; transition zone, T; and control zone, C) from December 2014 to July 2015. The concentrations of Cd, Pb, Cu, and Zn in the sediments varied significantly across different mariculture areas. Upon the pollution load index, principal components and cluster analysis indicated that heavy metal concentrations were the highest at the fish culture zone, while the concentrations were the lowest at the *Gracilaria* cultivation zone. The bioremediation effects of *Gracilaria* cultivation on heavy metals in the sediments were also demonstrated. *Gracilaria* had strong adsorption capacities for heavy metals from seawater, showing the highest heavy metal bio-concentration factors in May (higher seaweed biomass period). Consequently, the results suggested that *Gracilaria* cultivation influences the heavy metal concentrations in the sediments from the typical coastal mariculture zone. *Gracilaria* cultivation has the potential to bioremediate heavy metals in the coastal sediments. Therefore, *Gracilaria* cultivation can add environmental advantages and ecological values to coastal mariculture zones.

This study exhibited that the cultivated *Gracilaria* could enrich heavy metals, including Cd, Pb, Cu, and Zn, in coastal seawater, thereby decreasing the heavy metal concentrations of surface sediments in the mariculture zone. This seaweed is commendable co-cultured with marine economically animals for bioremediation.

As an efficient nutrient pumper, *Gracilaria lemaneiformis* (red seaweeds) may be used in integrated aquaculture systems to reduce eutrophication, control red tides, remediate contaminants, and sequestrate CO_2. All those will contribute to the improvement of coastal water environments. Seaweeds also create habitats, providing shelters and serving as feeding/nursery grounds for many organisms, which helps protect the diversity of marine organisms.

10.5.2 Do Red Seaweed NPs Enhance Bioremediation Capacity of Toxic Dyes from Aqueous Solution?

Dyes are one of the most dangerous pollutants found in industrial effluents. Their presence in water bodies prevents light penetration, prohibits aquatic flora from photosynthesis, and causes harm to all ecosystem components. Approximately 50% of all dye output in the world is wasted during the dying process in tanning, pharmaceutical, plastic, paints, etc., it ends up in textile effluents. Furthermore, in natural aquatic habitats, the presence of very low quantities of dyes in water (less than 1 ppm of some dyes) is highly visible, difficult to biodegrade, and difficult to eliminate. Dyes are also more stable and difficult to biodegrade due to their complex aromatic molecular structures. Dyes can also induce allergic reactions, dermatitis, and skin irritation in humans, as well as cancer and mutation. As a result, removing them from industrial effluents before they are discharged into the environment is most essential. Over the past years, a variety of technologies have been used to treat dye-containing wastewaters, such as flocculation, membrane filtration, precipitation, electrochemical, and adsorption (physical and chemical methods), which have all been successful in removing different dyes and pigments from the aqueous phase have been limited due to their cause of harmfulness to the environment. Each method has its package of advantages and cons. The adsorption approach is the most effective method for removing synthetic dyes from aqueous effluents, where different species of seaweeds can play a major role since they are highly efficient and low-cost dye uptake mediums. The red seaweed (*Pterocladia capillacea)* species are transferred from the water effluent to a solid phase, leading to a low effluent volume. After that, the adsorbent can be regenerated or stored in a dry location away from the environment.

In general, *P. capillaceous*, a marine red alga, is multicellular and usually small to moderate in size, with hollow leaves that have a cartilaginous texture. They are found in the Mediterranean Sea on rocks near the coast and in shallow water. The liquid extract of this species previously shows significant application in the field of agricultural growth enhancers and aquaculture feed additives. On the other hand, the dried form of *P. capillacea* shows a positive impact in wastewater treatments. NPs of this seaweed have been extensively utilized in bioremediation due to their high specific surface area that improves the removal reaction. Algal-cell NPs are increasingly becoming essential in many biological applications. Many studies have reported the NPs of the red seaweed species *P. capillacea* showed antimicrobial activities, anticancer activities, biomedical applications, and bioremediation capacity of toxic dyes from an aqueous solution. However, only few information are specified of the NPs of this algae having the potential for bioremediation of toxic dyes from aqueous solution.

Mansour et al. (2022a) performed batch investigations to evaluate the possibilities of using *P. capillacea* NPs for the elimination of harmful Ismate Violet 2R (IV2R) from an aqueous solution that was carried out. Contact time, adsorbent dose, pH, starting concentration, temperature, and the amount of *P. capillacea* NPs doped were studied. The adsorption isotherm, kinetic, and thermodynamic data approaches were used to investigate the adsorbents' equilibrium sorption behavior. The optimal isotherm to correlate the experimental data was determined by fitting experimental data to various isotherm equations. Moreover, the adsorption process of the dye onto these adsorbents was examined using BET, SEM, FTIR, and UV analysis. Before and after IV2R removal, the NPs were characterized using BET, SEM, FTIR, and UV. The maximum percentage of methylene blue (MB) removal (87.2%) was achieved at a NP biomass dosage of 0.2 g, pH of 2.60 mg/L, and 30°C with a 30 minutes equilibrium time, which were shown to be optimum conditions.

The adsorption ability of *P. capillacea* NPs for the IV2R dye increased with an increasing initial IV2R concentration and decreasing temperature degree. The interaction of IV2R ions with the surface of the adsorbent resulted in the development of distinct aggregates. The dye ions were attracted to the active sites of the adsorbents by electrostatic attraction or by complexation. Low cost, environmental friendliness, high uptake capacity, and non-toxicity are only a few of the benefits of this natural adsorbent. As a result, they concluded that this red seaweed can be used to remove IV2R dye from aqueous solutions as an effective adsorbent. This biomass can be used to provide a solution for the removal of different pollutants.

10.5.3 Biosynthesis of AgNPs from Seaweeds

AgNPs are an efficient NP for the treatment of dye effluent that have been explored in many studies in recent years. The room temperature-mediated biosynthesis of AgNPs by *Ulva lactuca* has been observed to have high efficiency for the photocatalytic degradation of methyl orange dye (Kumar et al., 2013).

AgNPs produced by *G. amansii* (brown algae) may serve as potential anti-fouling coatings for a variety of biomedical and environmental applications (Pugazhendhi et al., 2018).

Therefore, the possible reducing agents, such as polysaccharides, polyphenols, proteins/enzymes, and other chelating agents, play a vital role in the formation of

NPs. However, the methods for achieving desirable size, shape, stability, and controlled crystal growth structure also require further investigations (Sharma, 2019).

The aqueous extract obtained from the green macroalgae *Caulerpa serrulata* was used in the green synthesis of AgNPs capable of dealing with the organic dye congo red, promoting its reduction (Aboelfetoh et al., 2017). *Sargassum myriocystum* (brown macroalgae) was used in the synthesis of AgNPs exhibiting photocatalytic activity to deal with the organic pollutant MB (Balaraman et al., 2020). Another brown macroalgae species, *Sargassum serratifolium,* could also be used to synthesize AgNPs to deal with pollution caused by organic dyes methyl orange, MB, and rhodamine B (Kim et al., 2021).

This finding supports the use of AgNPs for degrading organic compounds and dyes under ambient temperature and visible light illumination in an eco-friendly manner. Moreover, the biological NPs from seaweeds can be scaled up to satisfy the increasing demands of eradicating toxic wastewater effluents from textile industries in the future.

10.6 BIOSYNTHESIS OF IRON OXIDE NPS FROM SEAWEEDS

Iron oxide (Fe_2O_3/Fe_3O_4) and cerium oxide (CeO_2) exhibit high reactivity and photolytic properties against wastewater in comparison to other metal oxide NPs such as titanium dioxide (TiO_2) and zinc oxide (ZnO). Whereas Fe_2O_3/Fe_3O_4 and CeO_2 are effective adsorbents for water purification due to their unique characteristics and properties, including a large surface area, their affinity toward diverse functionalized groups (Chaturvedi et al., 2020), high reactivity, selectivity, and a high degree of functionalization (Trivedi and Bergi, 2021; Khan et al., 2019). The current use of iron NPs in wastewater treatment relies on adsorptive and photocatalytic mechanisms. Among metal-based NPs, iron-based NPs received greater attention for their ability to oxidize, precipitate, reduce, and adsorb the contaminants such as nitroaromatic compounds, inorganic anions, phosphates, radioelements, nitrates, phenols, organic dyes, and chlorinated and halogenated organic compounds, making them applicable for water and wastewater purification (Prasad et al., 2019; Araújo et al., 2015). Further, these are cheaper and the second most abundant metal on Earth (Bhateria and Singh, 2019). Water, being the most crucial and essential resource on Earth, has become contaminated due to rapid industrialization and increasing population growth, resulting in the release of heavy metals, pathogens, insecticides, and various organic and inorganic constituents (Chaturvedi et al., 2020). These environmental and human health issues are of significant concern presently and have intensified the application challenges of conventional treatment technologies (Ibrahim et al., 2016; Song et al., 2015).

Examples:

- *Padina pavonica* (Brown algae) – Reduction of compounds containing nitrogen and phosphorous
- *Spirulina platensis* (Blue–green algae) – Reduction of crystal violet and methylene orange

TABLE 10.1
Industrial Application of Seaweeds for Wastewater Treatment

S.No.	Industrial Application	Type of Seaweed Used for Treatment
1.	The biological indicators of heavy metal pollution i.e., (Cu, Ni, Pb, Zn, and Cd)	**Brown seaweeds** (*Sargassum*, *Laminaria*, and *Ecklonia*) **Green seaweeds** (*Ulva* and *Enteromorpha)*
2.	High removal efficiency of crystal violet and congo red dye from aqueous solution	**Green seaweed** *Codium decorticatum* alga (CDA)
3.	For elimination of MB dye (MBD) from an aqueous solution	**Brown seaweed** *S. latifolium*
4.	The removal of MB and rhodamine B (RB)	*F. vesiculosus* seaweed bioadsorbent (FVSB)
5.	Removal of nutrient and heavy metal from aqueous solution from industries	*C. taxifolia* and *K. alvarezii*
6.	Effective in degrading methyl orange dye	*U. lactuca* and *Hypnea musciformis*

10.7 BIOSYNTHESIS OF ZINC NPS FROM SEAWEEDS

Mansour et al. (2022a) investigated green ZnO NPs synthesized from *P. capillacea*, and their ability to remove an organic hazardous dye (IV2R) from an aqueous solution was assessed. Hassan et al. (2015) reported the use of Fe_2O_3/Fe_3O_4 NPs made by green synthesis for the elimination of pyrene and benzo (a) pyrene (PAHs) from water. The adsorption factors were observed. These variables include the dose of NPs, pH, temperature, and initial PAH concentration. According to this study, the results showed that green FeO_3 NPs have a maximal adsorption capacity of 2.8 and 0.029 mg/g for both pyrene and benzo (a) pyrene, respectively. Fouda et al. (2022) examined the photocatalysis, tanning wastewater treatment, and characterization of psycho-synthesized ZnO NPs using *Ulva fasciata*. The study found that the catalytic activity to remove the MB dye increased in the presence of visible light compared to the dark conditions, the maximum amount of dye removed was 84.9% as compared to 53.4%. MB dye was efficiently corrupted under sunlight in the presence of *Ul*-ZnO NPs (Table 10.1).

10.8 APPLIANCES OF ALGAE-MEDIATED NP

Algae belonging to the *Cyanophyceae, Chlorophyceae, Phaeophyceae*, and *Rhodophyceae* families have been used as nanomachines by intracellular and extracellular synthesis of Au, Ag, and other metallic NPs. Algae are an attractive medium for the processing of diverse nanomaterials, owing to the inclusion of bioactive compounds in their cell extracts, such as pigments and antioxidants that serve as biocompatible reductants. AgNPs synthesized in an environmentally friendly manner effectively. As a result, NPs produced by algae can compete with standard drugs and have been shown to have antibacterial, anticancer, and antifungal effects. Aside from medical uses, metal NPs have a broad variety of applications in computing, optics, cosmetics, and other areas.

FIGURE 10.3 Appliances of seaweed NPs.

Seaweeds are the sole source of phytochemicals such as agar, carrageenan, mannitol, iodine, laminarin, furcellarin, and sodium alginate (Holdt and Kraan, 2011). They also exhibit antimicrobial, antifungal, antiviral, and anticancer properties (Pereira and Costa-Lotufo, 2012). In recent days, approximately 32,000 tonnes of wet seaweed have been harvested to meet a wide range of industrial and domestic requirements worldwide. Metal NPs (especially Ag and Au NPs) derived from seaweeds can be an appropriate alternative for NPs obtained from plants (Desai and Shankar, 2008) (Figure 10.3).

10.8.1 Recent Studies Carried Out in Various Seaweeds

Among all the classes of seaweeds, brown ones have been the most widely studied (Romera et al., 2007; Saldarriaga-Hernandez et al., 2010). In the past 5 years, the most commonly studied species being *Ascophyllum* sp., *Colpomenia* sp., *Cystoseira* sp., *Fucus* sp., *Laminaria* sp., *Pelvetia* sp., and *Sargassum* sp., have a great capacity to recover metallic pollutants, such as Cd, Pb, Ni, Ag, chromium (Cr), Cu, arsenic (As), and Zn, from different sources with high adsorption capacity among other seaweeds. Nevertheless, *Sargassum dentifolium* stood out by removing 99.68% of total hexavalent Cr from water. Additionally, *Sargassum filipendula* showed a high capacity to absorb Pb, with a maximum uptake of 367.94 mg/g (approximately 96% of total Pb) (Verma et al., 2018, 2016).

However, *Sargassum* species are recognized as highly invasive, impacting non-native ecosystems and being considered a threat to marine biodiversity (Saldarriaga-Hernandez et al., 2020; Pinteus et al., 2018). Thus, developing bioremediation strategies based on *Sargassum* biomass would simultaneously contribute to mitigating the negative impacts of these species.

Seaweed that stands out, not only for the number of published studies but also due to its high ability to capture different metallic ions from water, is the green seaweed *Ulva lactuca*, also called "sea lettuce"—is widely abundant in most shorelines around the world, being found in high amounts in sheltered areas (Singh et al., 2023). Studies conducted with the dried biomass of these seaweeds recorded 60%–99% uptake capacity for Cr, Cu, Hg, and Pb (Pandya et al., 2017; Patel et al., 2016; Figueira et al., 2016). On the other hand, studies conducted with live *U. lactuca* also presented high absorption rates of different metallic pollutants, particularly Hg (98%), Pb (87%), and Cu (86%).

Since *U. lactuca* showed high tolerance to changes in pH, salinity, and pollutant concentration, the use of live seaweed may bring additional advantages, particularly in multi-integrated systems (Pinteus, 2022).

10.9 CONCLUSION

The utilization of seaweeds and their NPs has garnered significant attention among researchers, highlighting the significance of seaweeds in contemporary times. The various applications and benefits of seaweeds, especially in the treatment of wastewater effluents previously discussed, will foster the continuous production and provision of high-quality raw seaweed materials in diverse sectors. It is anticipated that industries reliant on seaweeds possess the potential to enhance socioeconomic conditions and can sustain the environment at a substantial rate. The objective of this chapter was to present a comprehensive evaluation of the utilization of seaweed NPs for the cost-effective and highly efficient treatment of the textile industrial effluents.

REFERENCES

Aboelfetoh, E.F., El-Shenody, R.A., and Ghobara, M.M. Eco-friendly synthesis of silver nanoparticles using green algae (Caulerpa serrulata): Reaction optimization, catalytic and antibacterial activities. *Env. Monit. Assess.* 2017;189:349. https://doi.org/10.1007/s10661-017-6033-0

Agarwal, H., Venkat Kumar, S., and Rajeshkumar, S. A review on green synthesis of zinc oxide nanoparticles: An eco-friendly approach. *Resour. Technol.* 2017;3:406–413.

Akcali, I., and Kucuksezgin, F. A biomonitoring study: Heavy metals in macroalgae from eastern Aegean coastal areas. *Marine Pollut. Bull.* 2011;62(3):637–645. ISSN-0025-326X. https://doi.org/10.1016/j.marpolbul.2010.12.021. https://www.sciencedirect.com/science/article/pii/S0025326X1100004X

Al-Khashman, O.A., and Shawabkeh, R.A. Metal distribution in soils around the cement factory in Southern Jordan. *Env. Poll.* 2006;140:387–394.

Alprol, A.E., Mansour, A.T., Abdelwahab, A.M., and Ashour, M. Advances in green synthesis of metal oxide nanoparticles by marine algae for wastewater treatment by adsorption and photocatalysis techniques. *Catalysts* 2023;13:888. https://doi.org/10.3390/catal13050888

Araújo, R., Castro, A.C.M., and Fiúza, A. The use of nanoparticles in soil and water remediation processes. *Mater. Today Proc.* 2015;2:315–320. https://doi.org/10.1016/j.matpr.2015.04.055

Asami, T. Soil pollution by metals from mining and smelting activities. In *Chemistry and Biology of Solid Waste: Dredged Material and Mine Tailings*. Eds W. Salomons and U. Forstner, pp. 143–149. Springer, Berlin, 1988.

Balaraman, P., Balasubramanian, B., Kaliannan, D., Durai, M., Kamyab, H., Park, S., et al. Phyco-synthesis of silver nanoparticles mediated from marine algae Sargassum myriocystum and its potential biological and environmental applications. *Waste Biomass Valoriz.* 2020;11(10):5255–5271. https://doi.org/10.1007/s12649-020-01083-5

Bhateria, R., and Singh, R. A review on nanotechnological application of magnetic iron oxides for heavy metal removal. *J. Water Process Eng.* 2019; 31:100845. https://doi.org/10.1016/j.jwpe.2019.100845

Bidone, E.D., Laybauer, L., Castilhos, Z.C., and Maddock, J.L. Environmental risk increase due to heavy metal contamination caused by a copper mining activity in Southern Brazil. *An. Acad. Bras. Cienc.* 2001;73(2):277–286.

Bolan, N.S., Adriano, D.C., and Naidu, R. Phosphorus-trace element interactions in soil plant system. In *Phosphorus: Agriculture and the Environment.* Eds A. Sharpley and J.T. Sims, pp. 317–352. American Society of Agronomy, Crop Science Society of America, Soil Science Society of America, USA, 2005.

Chakraborty, S., and Owens, G. Metal distributions in seawater, sediment and marine benthic macroalgae from the South Australian coastline. *Int. J. Environ. Sci. Technol.* 2014;12. https://doi.org/10.1007/s13762-013-0310-4

Chaturvedi, V.K., Kushwaha, A., Maurya, S., Tabassum, N., Chaurasia, H., and Singh, M.P. Wastewater treatment through nanotechnology: Role and prospects. In *Restoration of Wetland Ecosystem: A Trajectory towards a Sustainable Environment.* Eds A.K. Upadhyay, R. Singh, and D.P. Singh, pp. 227–247. Springer, 2020. https://doi.org/10.1007/978-981-13-7665-8_14

Cobbina, S.J., Myilla, M., and Michael, K. Small scale gold mining and heavy metal pollution assessement of drinking water sources in Datuku in the Talensi-Nabdam district. *Int. J. Scientific Tech. Res.* 2013;2(1):96–100.

Correia, V.M., Stephenson, T., and Judd, S.J. Characterisation of textile wastewaters: A review. *Env. Technol.* 1994;15:917–929.

Cortes, O.E.J., Barbosa, L., and Kiperstok, A. Biological treatment of industrial liquid effluent in copper production industry. *58th Congresso Anual da ABM (ASSociacao Brasileira de Metalurgia e Materiais).* Rio De Janeiro, Brazil, 2003;18:1135–1144.

Desai, S., & Shankar, M. R. (2008). Polymers, composites and nano biomaterials: Current and future developments. In Bio-Materials and Prototyping Applications in Medicine (pp. 15–26). Boston, MA: Springer US.

Department of Fisheries. Seaweed cultivation. 2020. Available online: https://dof.gov.in/sites/default/files/2020-07/Seaweed_Cultivation.pdf. Accessed on November 18, 2023.

Dhakate, R., and Singh, V.S. Heavy metal contamination in groundwater due to mining activities in Sukinda valley, Orissa: A case study. *J. Geogr. Reg. Plann.* 2008;1(4):58–67.

Doelman, P., Jansen, E., Michels, M., and Van Til, M. Effects of heavy metals in soil on microbial diversity and activity as shown by the sensitivity resistance index, an ecologically relevant parameter. *Biol. Fertil. Soils* 1994;17:171–184.

Figueira, P., Henriques, B., Teixeira, A., Lopes, C.B., Reis, A.T., Monteiro, R.J.R., Duarte, A.C., Pardal, M.A., and Pereira, E. Comparative study on metal biosorption by two macroalgae in saline waters: Single and ternary systems. *Environ. Sci. Pollut. Res.* 2016;23:11985–11997. https://doi.org/10.1007/s11356-016-6398-6

Fouda, A., Eid, A.M., Abdelkareem, A., Said, H.A., El-Belely, E.F., Alkhalifah, D.H.M., Alshallash, K.S., and Hassan, S.E.-D. Phyco-synthesized zinc oxide nanoparticles using marine macroalgae, *Ulva fasciata* Delile, characterization, antibacterial activity, photocatalysis, and tanning wastewater treatment. *Catalysts* 2022;12:756.

Ghrefat, H., and Yusuf, N. Assessing Mn, Fe, Cu, Zn and Cd pollution in bottom sediments of Wadi Al-Arab Dam, Jordan. *Chemosphere* 2006;65(11):2114–2121.

Grammatikopoulos, P., Stephan, S., Jerome, V., Singh, V., and Sowwan, M. Nanoparticle design by gas-phase synthesis. *Adv. Phys.* X. 2016;1:81–100.

Hassan, S.S., El Azab, W.I., Ali, H.R., and Mansour, M.S. Green synthesis and characterization of ZnO nanoparticles for photocatalytic degradation of anthracene. *Adv. Nat. Sci. Nanosci.* Nanotechnol. 2015;6:045012.

Holdt, S.L., and Kraan, S. Bioactive compounds in seaweed: Functional food applications and legislation. *J. Appl. Phycol.* 2011;23:543–597. https://doi.org/10.1007/s10811-010-9632-5.

Hussein, E.A.M., Mohammad, A.A.-H., Harraz, F.A., and Ahsan, M.F. Biologically synthesized silver nanoparticles for enhancing tetracycline activity against Staphylococcus aureus and Klebsiella pneumoniae. *Braz. Arch. Biol.* Technol. 2019. https://doi.org/10.1590/1678-4324-2019180266

Ibrahim, R.K., Hayyan, M., AlSaadi, M.A., Hayyan, A., and Ibrahim, S. Environmental application of nanotechnology: Air, soil, and water. *Env. Sci. Pollut. Res.* 2016;23:13754–13788. https://doi.org/10.1007/s11356-016-6457-z

Isikli, B., Demir, T.A., Ürer, S.M., Berber, A., Akar, T., and Kalyoncu, C. Effects of chromium expo-sure from a cement factory. *Env. Res.* 2003;91:113–118.

Jamkhande, P.G., Ghule, N.W., Bamer, A.H., and Kalaskar, M.G. Metal nanoparticles synthesis: An overview on methods of preparation, advantages and disadvantages, and applications. *J. Drug Deliv. Sci.* Technol. 2019;53:101174.

Kapusta, P., and Sobczyk, L. Effects of heavy metal pollution from mining and smelting on enchytraeid communities under different land management and soil conditions. *Sci. Total Env.* 2015;536:517–526.

Khan, N.A., Khan, S.U., Ahmed, S., Farooqi, I.H., Dhingra, A., Hussain, A., and Changani, F. Applications of nanotechnology in water and wastewater treatment: A review. *Asian J. Water Environ. Pollut.* 2019;16:81–86. https://doi.org/10.3233/AJW190051

Kim, B., Song, W.C., Park, S.Y., and Park, G. Green synthesis of silver and gold nanoparticles via Sargassum serratifolium extract for catalytic reduction of organic dyes. *Catalysts.* 2021;11(3):1–13. https://doi.org/10.3390/catal11030347

Kpan, J.D.A., Opuku, B.K., and Gloria, A. Heavy metal pollution in soil and water in some selected towns in Dunkwa-on-offin district in the central region of Ghana as a result of small scale gold mining. *J. Agric. Chem. Env.* 2014;3:40–47.

Kumar, P., Govindaraju, M., Senthamilselvi, S., and Premkumar, K. Photocatalytic degradation of methyl orange dye using silver (Ag) nanoparticles synthesized from Ulva lactuca. *Colloids Surf. B. Biointerfaces.* 2013;103:658–661. https://doi.org/10.1016/j.colsurfb.2012.11.022

Li, X., Xu, H., Chen, Z.-S., and Chen, G. Biosynthesis of nanoparticles by microorganisms and their applications. *J. Nanomater.* 2011;2011:270974. https://doi.org/10.1155/2011/270974

Luo, H., Wang, Q., Liu, Z., Wang, S., Long, A., and Yang, Y. Potential bioremediation effects of seaweed Gracilaria lemaneiformis on heavy metals in coastal sediment from a typical mariculture zone. *Chemosphere.* 2020;245:125636.

Mansour, A.T., Alprol, A.E., Ashour, M., Ramadan, K.M.A., Alhajji, A.H.M., Abualnaja, K.M. Do red seaweed nanoparticles enhance bioremediation capacity of toxic dyes from aqueous solution? *Gels*. 2022a;8:310. https://doi.org/10.3390/gels8050310

Mansour, A.T., Alprol, A.E., Abualnaja, K.M., El-Beltagi, H.S., Ramadan, K.M.A., and Ashour, M. Dried Brown Seaweed's phytoremediation potential for methylene blue dye removal from aquatic environments. *Polymers*. 2022b;14:1375. https://doi.org/10.3390/polym14071375

Mansour, A.T., Alprol, A.E., Khedawy, M., Abualnaja, K.M., Shalaby, T.A., Rayan, G., Ramadan, K.M., and Ashour, M. Green synthesis of zinc oxide nanoparticles using red seaweed for the elimination of organic toxic dye from an aqueous solution. *Materials*. 2022c;15:5169.

McMullan, G., Meehan, C., Conneely, A., Kirby, N., Robinson, T., Nigam, P., Banat, I., Marchant, R., and Smyth, W.F. Microbial decolourisation and degradation of textile dyes. *Appl. Microbiol. Biotechnol.* 2001;56:81–87. https://doi.org/10.1007/s002530000587

Mishra, V.K., Upadhyaya, A.R., Pandey, S.K., and Tripathi, B.D. Heavy metal pollution induced due to coal mining effluent on surrounding aquatic ecosystem and its management through naturally occurring aquatic macrophytes. *Bioresour. Tech.* 2008:99(5):930–936.

Mitra, S., Chakraborty, A.J., Tareq, A.M., Emran, T.B., Nainu, F., Khusro, A., ... Simal-Gandara, J. Impact of heavy metals on the environment and human health: Novel therapeutic insights to counter the toxicity. *Journal of King Saud University – Science*, 2022;34(3):101865. https://doi.org/10.1016/j.jksus.2022.101865

Modoi, O., Roba, C., Török, Z., and Ozunu, A. Environmental risks due to heavy metal pollution of water resulted from mining wastes in NW Romania. *Env. Eng. Manage. J.* 2014:13(9):2325–2336.

Mohamed, E.F. Nanotechnology: Future of environmental air pollution control. *Environ. Manag. Sustain. Dev.* 2017;6:429. https://doi.org/10.5296/emsd.v6i2.12047

Mudgal, V., Madaan, N., and Mudgal, A. Heavy metals in plants: Phytoremediation: Plants used to remediate heavy metal pollution. *Agric. Bio. J. North America.* 2010a;1(1):40–46.

Mudgal, V., Madaan, N., Mudgal, A., Singh, R.B., and Mishra, S. Effect of toxic metals on human health. *Open Nutraceuticals J.* 2010b;3:94–99.

Nakayama, M.M.S., Ikenaka, Y., Muzandu, K., Choongo, K., Oroszlany, B., Teraoka, H., Mizuno, N., and Ishizuka, M. Heavy metal accumulation in lake sediments, fish (Oreochromis niloticus and Serranochromis thumbergi) and crayfish (Cherax quadricarinatus) in lake Itezhitezhi and lake Kariba, Zambia. *Archives Env. Contam. Toxicology.* 2010;59(2):291–300.

Ovais, M., Khalil, A., Ayaz, M., Ahmad, I., Nethi, S., and Mukherjee, S. Biosynthesis of metal nanoparticles via microbial enzymes: A mechanistic approach. *Int. J. Mol.* Sci. 2018;19:4100.

Pandya, K.Y., Patel, R.V., Jasrai, R.T., and Hatt, N.B. Optimization of Cr and Cu biosorption by green marine algae *Caulerpa racemosa* Var. Cylindracea & *Ulva lactuca*. *Int. J. Adv. Res.* 2017;5:923–939.

Paraskevi, M., and Anastasia, C. Theodoros K trace element seasonality in marine macroalgae of 446 different functional-form groups. *Marine Env. Res.* 2015;103:18–26.

Patel, G.G., Doshi, H.V., and Thakur, M.C. Biosorption and equilibrium study of copper by marine seaweeds from North West Coast of India. *J. Environ. Sci. Toxicol. Food Technol.* 2016;10:54–64.

Patel, M. Green synthesis of nanoparticles: A solution to environmental pollution. In *Handbook of Solid Waste Management*. Eds C. Baskar, S. Ramakrishna, S. Baskar, R. Sharma, A. Chinnappan, and R. Sehrawat, pp. 1965–1993. Springer, Berlin/Heidelberg, Germany, 2022.

Pathak, J., Ahmed, H., Singh, D.K., Pandey, A., Singh, S.P., and Sinha, R.P. Recent developments in green synthesis of metal nanoparticles utilizing cyanobacterial cell factories. In *Nanomaterials in Plants, Algae, and Microorganisms*. Eds D.K. Tripathi, P. Ahmad, S. Sharma, D.K. Chauhan, and N.K. Dubey, pp. 237–265. Academic Press, Cambridge, MA, 2019.

Pereira, R.C., and Costa-Lotufo, L.V. Bioprospecting for bioactives from seaweeds: Potential, obstacles and alternatives. *Revista Brasileira de Farmacognosia* 2012;22:894–905. https://doi.org/10.1590/S0102-695X2012005000077

Pinteus, S., Lemos, M., Alves, C., Neugebauer, A., Silva, J., Thomas, O., Botana, L., Gaspar, H., and Pedrosa, R. Marine invasive macroalgae: Turning a real threat into a major opportunity: The biotechnological potential of *Sargassum muticum* and *Asparagopsis armata*. *Algal Res.* 2018;34:217–234. https://doi.org/10.1016/j.algal.2018.06.018

Pinteus, S., Susano, P., Alves, C., Silva, J., Martins, A., and Pedrosa, R. Seaweed's role in energetic transition-from environmental pollution challenges to enhanced electrochemical devices. *Biology (Basel)*. 2022;11(3):458. https://doi.org/10.3390/biology11030458. PMID: 35336831; PMCID: PMC8945715.

Prasad, R., Kumar, V., Kumar, M., and Choudhary, D. (eds.). *Nanobiotechnology in Bioformulations. Nanotechnology in Life Sciences*, pp. 187–224. Springer., 2019. https://doi.org/10.1007/978-3-030-17061-5

Pugazhendhi, A., Prabakar, D., Jacob, J.M., Karuppusamy, I., and Saratale, R.G. Synthesis and characterization of silver nanoparticles using *Gelidium amansii* and its antimicrobial property against various pathogenic bacteria. *Microb. Pathog.* 2018;114:41–45. https://doi.org/10.1016/j.micpath.2017.11.013

Reis, P.A., Cassiano, J., Veiga, P., Rubal, M., and Sousa-Pinto, I. Fucus spiralis as monitoring tool of metal 456 contamination in the northwest coast of Portugal under the European Water framework 457 directives. *Env. Monit. Assess.* 2014;186:5447–5460.

Romera, E., González, F., Ballester, A., Blázquez, M.L., and Muñoz, J.A. Comparative study of biosorption of heavy metals using different types of algae. *Bioresour. Technol.* 2007;98:3344–3353. https://doi.org/10.1016/j.biortech.2006.09.026

Saldarriaga-Hernandez, S., Hernandez-Vargas, G., Iqbal, H.M.N., Barceló, D., and Parra-Saldívar, R. Bioremediation potential of Sargassum sp. biomass to tackle pollution in coastal ecosystems: Circular economy approach. *Sci. Total Env.* 2020;715:136978. https://doi.org/10.1016/j.scitotenv.2020.136978

Schuhmacher, M., Bocio, A., Agramunt, M.C., Domingo, J.L., and de Kok, H.A.M. PDCC/F and metal concentration in soil and herbage samples collected in the vicinity of a cement plant. *Chemosphere*. 2002;48:209–217.

Sharma, D. Kanchi, S., and Bisetty, K. Biogenic synthesis of nanoparticles: A review. *Arabian Journal of Chemistry*. 2019;12(8):3576–3600. https://doi.org/10.1016/j.arabjc.2015.11.002

Singh, A., Tyagi, P., Ranjan, R., Sushkova, S.N., Minkina, T., Burachevskaya, M., and Rajput, V.D. Bioremediation of hazardous wastes using green synthesis of nanoparticles. *Processes*. 2023;11:141. https://doi.org/10.3390/pr11010141

Song, H.J., You, S., Jia, X.H., and Yang, J. MoS_2 nanosheets decorated with magnetic Fe_3O_4 nanoparticles and their ultrafast adsorption for wastewater treatment. *Ceram. Int.* 2015;41:13896–13902. https://doi.org/10.1016/j.ceramint.2015.08.023

Sundararaj, V. In *Seaweed Culture*. Ed M.C.J. Milton. Loyola College Publication, Chennai, India, 2008.

Taghipour, S., Hosseini, S.M., and Ataie-Ashtiani, B. Engineering nanomaterials for water and wastewater treatment: Review of classifications, properties, and applications. *New J. Chem.* 2019;43:7902–7927. https://doi.org/10.1039/c9nj00157c

Trivedi, R., and Bergi, J. Application of bio nanoparticles in wastewater treatment. In *Advanced Oxidation Processes for Effluent Treatment Plants*. Eds M.P. Shah, pp.177–197. Elsevier, 2021. https://doi.org/10.1016/b978-0-12-821011-6.00010-4

Vaalgamaa, S., and Conley, D.J. Detecting environmental change in estuaries: Nutrient and heavy metal distributions in sediment cores in estuaries from Gulf of Finland, Baltic Sea. *Estuarine Coastal Shelf Sci.* 2008;76(1):45–56.

Verma, A., Kumar, S., and Kumar, S. Biosorption of lead ions from the aqueous solution by *Sargassum filipendula*: Equilibrium and kinetic studies. *J. Environ. Chem. Eng.* 2016;4:4587–4599. https://doi.org/10.1016/j.jece.2016.10.026

Verma, A., Kumar, S., Balomajumder, C., and Kumar, S. Efficacy of Sargassum filipendula for the removal of Pb^{2+}, Cd^{2+} and Ni^{2+} ions from aqueous solution: A comparative study. *Desalin. Water Treat.* 2018;129:216–226. https://doi.org/10.5004/dwt.2018.22788

Whitby, L.M., and Hutchinson, T.C. Heavy- metal pollution in the Sudbury mining and smelting region of Canada. II. Soil toxicity tests. *Env. Conser.* 1974;1(3):191–200.

Wolny-Koładka, K., Malina, D., Suder, A., Pluta, K., and Wzorek, Z. Bio-based synthesis of silver nanoparticles from waste agricultural biomass and its antimicrobial activity. *Processes.* 2022;10:389.

Zhang, W., Feng, H., Chang, J., Qu, J., Xie, H., and Yu, L. Heavy metal contamination in surface sediments of Yangtze River intertidal zone: An assessment from different indexes, *Env. Pollut.* 2009;157(5):1533–1543. https://doi.org/10.1016/j.envpol.2009.01.007. https://www.sciencedirect.com/science/article/pii/S0269749109000219

Zhang, X.-F., Liu, Z.-G., Shen, W., and Gurunathan, S. Silver nanoparticles: Synthesis, characterization, properties, applications, and therapeutic approaches. *Int. J. Mol. Sci.* 2016;17:1534. https://doi.org/10.3390/ijms17091534

11 Metagenomics
Hidden Details of Microbes Involved in Wastewater Treatment

O. Rebai, R. Ben Abdallah, N. Belalkhel, and S. Fattouch

11.1 INTRODUCTION: BACKGROUND AND DRIVING FORCES

Metagenomics represents a pioneering methodology for scrutinizing microorganisms sourced from distinct environments via functional gene screening or sequencing analysis. At its core, metagenomics endeavors to unravel the intricacies of microbial communities, delving into their diversity, genetic relationships, functional capacities, and ecological interactions. As sequencing technologies have advanced from shotgun sequencing to the era of high-throughput next-generation sequencing (NGS) and third-generation sequencing (TGS), the field has witnessed a quantum leap in its capabilities (Handelsman, 2004). Notably, NGS and TGS have emerged as powerful tools for swift pathogen detection, revolutionizing our ability to identify microbial threats. Leveraging novel algorithms further augments our capacity for taxonomic profiling and gene prediction, enhancing our understanding of microbial dynamics. Of particular significance is the realm of functional metagenomics, which serves as a beacon for discovering novel bioactive compounds and functional genes within microbial communities and their metabolites (Zhang et al., 2021). Through this chapter, we embark on a comprehensive exploration of the fundamental processes, classifications, and diverse applications that underscore the profound impact of metagenomics on scientific inquiry and innovation.

Metagenomics plays a pivotal role in understanding and optimizing wastewater treatment processes, offering insights into the complex microbial communities involved and their functional capabilities. In wastewater treatment, metagenomic analysis involves the comprehensive study of microbial DNA extracted directly from environmental samples, providing a holistic view of the microbial diversity and metabolic potential present within wastewater systems (Azli et al., 2022). One key application of metagenomics in wastewater treatment is the identification of microbial species and their functional genes involved in pollutant degradation. Wastewater contains a diverse array of organic contaminants, such as pharmaceuticals, industrial chemicals, and organic matter, which must be efficiently removed

 DOI: 10.1201/9781003459286-11

to meet environmental standards (Kundan et al., 2022). Overall, metagenomics holds great promise for advancing our understanding of microbial communities in wastewater treatment and unlocking novel solutions for improving treatment efficiency, sustainability, and resilience to environmental challenges. By leveraging the power of metagenomic analysis, we can develop more effective and environmentally friendly approaches for managing wastewater and mitigating its impact on the environment. This chapter summarizes the application of metagenomics to assess microbial communities and hidden details of microbes involved in biological wastewater treatment. Innovative technical bottlenecks and future research directions in the application of metagenomics to biological wastewater treatment are discussed.

11.2 METAGENOMICS APPROACHES

Metagenomics is a field of genetics and molecular biology that involves the study of genetic material (DNA or RNA) recovered directly from environmental samples, such as soil, water, air, or biological specimens. Unlike traditional genomics, which focuses on the DNA of individual organisms, metagenomics examines the collective genomes of entire microbial communities within these samples. It provides a comprehensive view of the genetic diversity, composition, and functional potential of microbial ecosystems, including bacteria, archaea, fungi, viruses, and other microorganisms. Metagenomic analysis typically involves high-throughput DNA sequencing, followed by bioinformatics techniques to characterize and interpret the genetic information obtained. This approach enables researchers to explore the structure, dynamics, and ecological roles of microbial communities in various environments, and it has diverse applications in environmental science, biotechnology, agriculture, medicine, and beyond (Sabree et al., 2009).

The basic procedure of metagenomics consists of the following steps. First, all genes in environmental microbial samples are extracted and enriched. Second, the genes are cloned into the vector, which is transformed into host bacteria to establish the metagenomic library. Finally, the metagenomic library is screened and analyzed. The extraction of metagenomic DNA, construction, and screening of metagenomic libraries are essential (Zhang et al., 2021), as shown in Figure 11.1. The process of extracting and enriching metagenomic DNA is pivotal for constructing metagenomic libraries, given that 99% of environmental microorganisms cannot be cultured. Metagenomic DNA extraction requires addressing two key aspects: extracting all genes from microorganisms present in the sample and preserving the integrity and purity of DNA fragments. Direct and indirect extraction methods are differentiated based on whether cells are isolated before DNA extraction (Verma et al., 2017). Sample size and decontamination methods are tailored based on microbial concentration and environmental milieu. Subsequently, enrichment at both the cellular and genetic levels can be performed for DNA samples with low abundance. Library screening methods include function-based screening, sequence-based screening, and substrate-induced gene expression screening (SIGEX). Various sequencing technologies, ranging from high-throughput shotgun sequencing to TGS, are also discussed, along with their respective advantages and disadvantages (Yun and Ryu, 2005).

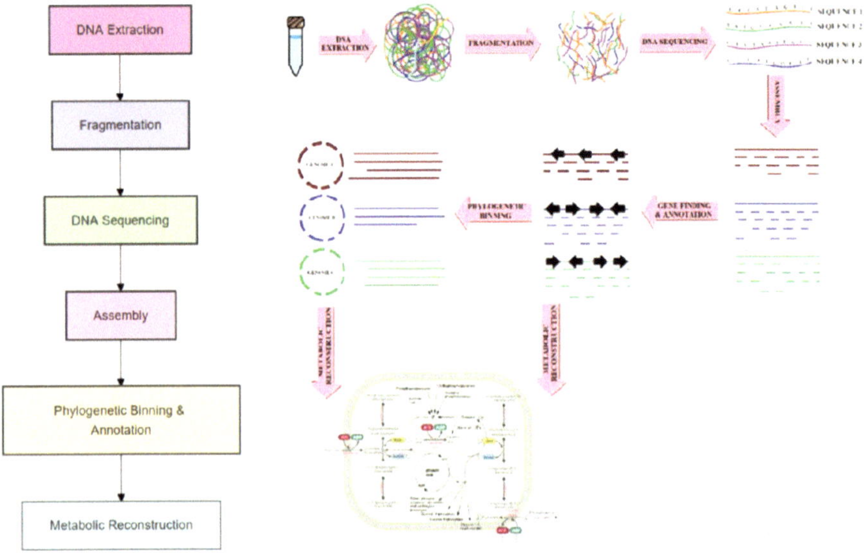

FIGURE 11.1 Key steps of metagenomics study (Kumari et al., 2024).

Finally, an overview of commonly used software for analyzing metagenomic sequencing data is presented, highlighting different data processing steps, from assembly quality to gene functional annotation (Navgire et al., 2022).

11.2.1 Extraction of Metagenomic DNA

Extraction and enrichment of metagenomic DNA pose significant challenges due to the inability to culture 99% of microorganisms in the environment (Nwachukwu and Babalola, 2022). This underscores the necessity of extracting high-concentration, large-fragment environmental microbial DNA, which serves as the foundational step in constructing a metagenomic library. Two primary objectives define the extraction of metagenomic DNA: first, to capture the genetic material of all microorganisms within the sample, and second, to preserve the integrity and purity of DNA fragments. Sample collection mandates meticulous procedures to maximize DNA extraction from environmental microorganisms while retaining large DNA fragments. The extraction process can be categorized into direct and indirect methods based on whether cells are isolated before DNA extraction. Direct extraction employs physical (e.g., freeze-thaw) and chemical (e.g., protease or SDS addition) methods to directly lyse microorganisms and release microbial DNA (Gómez-Acata et al., 2019). While direct extraction efficiently yields total DNA from all samples, it is more suitable for extracting DNA fragments ranging from 1 to 50kb. However, its simplicity is offset by lower purity. Conversely, indirect extraction involves isolating microbial cells before DNA extraction, resulting in higher-purity DNA suitable for extracting larger fragments (20–500kb). Despite its efficacy, the indirect method is more complex and may lead to the loss of some microbial DNA (Gabor et al., 2003). Sample size considerations are contingent upon microbial concentration. High-density microorganism samples like stool may necessitate only anal swab sampling. Conversely, studying low-density microbial communities, such as marine microorganisms, requires capturing and concentrating large volumes of water samples through filtration. Moreover, samples must undergo thorough decontamination procedures. For instance, humic acid in soil often tightly binds to DNA, necessitating its removal during sample preparation. When investigating microorganisms in humans or animals, eliminating DNA contamination from both samples and hosts is imperative.

In cases of DNA samples with low abundance, enrichment at both the cellular and genetic levels can be conducted post-DNA fragment extraction, albeit with potential deviations (Probst et al., 2015). Common gene enrichment methods include stable-isotope probing (SIP), suppression subtractive hybridization (SSH) (Chen and Murrell, 2010; Galbraith et al., 2008; and DNA chip technology (Avarre et al., 2007).

11.2.2 Metagenome Libraries Constructing

In constructing a metagenome library, the initial step involves mechanically cutting or digesting microbial DNA into specific sizes, which are then randomly recombined with suitable clone vectors before being transformed into host cells. The objective is twofold: either to isolate target genes or to preserve the entire genetic makeup of an organism. The approach to building such libraries varies depending on the research objectives, whether they pertain to genes of low or high abundance, single gene segments, or entire gene clusters within metabolic pathways.

Selection of the vector hinges on its compatibility with amplifying the target DNA fragment and its suitability for screening post-transformation and expression. Common vectors encompass plasmids, cosmids, fosmids, bacterial artificial chromosomes (BACs), yeast artificial chromosomes (YACs), and phages. Plasmids are apt for cloning DNA fragments below 10 kb, while cosmids and fosmids suit larger genes or multiple gene fragments (20–40 kb). BACs and YACs facilitate the cloning of DNA fragments spanning 200–450 kb. The versatility of the BAC shuttle vector allows it to be duplicated and expressed in various hosts, thereby broadening the host range (Kakirde et al., 2011). Host cells are typically selected based on their stability with the recombinant vector, transformation efficiency, and ability to express the target gene fragment effectively. Different host strains are chosen according to the research objectives. Although *Escherichia coli* is widely used due to its ease of cultivation, its prokaryotic nature imposes limitations. New host cells, such as *Streptomyces, Pseudomonas*, and *Mycobacterium*, are gradually being adopted to overcome these limitations (Rebets et al., 2017).

Numerous challenges impede the study of foreign gene expression in metagenomics. For instance, target genes may be lost during library construction, or the host cell may fail to recognize the unique promoter, hindering the expression of heterologous gene fragments. To address this issue, multi-host expressions can be employed, as demonstrated by research groups expressing DNA fragments in six different bacteria (Craig et al., 2010). In cases where the target gene exists in cluster form, inserting large DNA segments becomes necessary to obtain complete metabolic enzymes and pathways. Currently, failures in DNA cloning of large fragments may stem from the destructive effects of physical and chemical actions during the extraction process or the limited receptivity of host bacteria.

11.2.3 Screening of Metagenome Library

Screening a metagenome library involves several approaches, including function-driven screening, sequence-driven screening, and SIGEX, alongside techniques like DNA stable-isotope probing (DNA-SIP) and fluorescence *in situ* hybridization.

Function-driven screening methods encompass two main strategies. One approach involves screening for the expression of active products of exogenous genes in host cells, identified by specific characteristics (such as color or plaque) on selective media. Alternatively, host strains and their mutants can be screened based on complementary growth characteristics under selective conditions. While function-driven screening is rapid and straightforward and doesn't rely on known sequence information, it suffers from a low success rate due to dependence on the expression of functional genes in foreign hosts. Therefore, selecting the appropriate host and cloning the full length of a gene or gene cluster becomes crucial, making it suitable for gene screening of large-fragment DNA libraries.

Sequence analysis relies on oligonucleotide primers or probes to screen target genes using molecular biological methods like PCR and gene hybridization. This method is best suited for highly conserved sequence enzyme genes and gene screening of small DNA fragments. While sequence analysis doesn't rely on host cells to express the cloned genes, it cannot screen unknown genes with entirely different sequences from existing ones.

Novel techniques such as microarray, stable-isotope labeling, and fluorescence in situ hybridization can also be employed for sequence screening. DNA microarray technology, for instance, enables the study of key genes in metabolic pathways but lacks the sensitivity and specificity of PCR methods, making it unsuitable for detecting unknown functional genes. SIGEX leverages substrate-induced conditions to express metabolism-related genes or enzyme genes. This high-throughput screening method offers advantages such as inferring the functions of unknown enzymes and genes from the substrate. However, it demands a well-structured and fit target gene in the host and is limited to substrates that can enter the cytoplasm and target gene (Avarre et al., 2007).

Current demands in metagenomics call for high-throughput screening methods. In 2015, a research team explored microfluidic chips integrating sample detection and automatic analysis, known as lab-on-chip (LOC). LOC offers miniaturization, automation, integration, and portability of sample testing but requires advanced technologies like chip sampling, microfluidics, and focusing electrophoresis (Colin et al., 2015).

11.2.4 Metagenomic Sequencing and Analysis Techniques

Rapid progress has been made in sequencing technologies, from shotgun sequencing to high-throughput, NGS, and TGS. Metagenomic DNA can be sequenced from DNA libraries obtained from target gene cloning or directly from samples.

The purity and content of the DNA must be ensured before it is sequenced (Satam et al., 2023).

Second-generation DNA sequencing was established on PCR techniques. The first-generation sequencing is the shotgun sequencing based on the traditional Sanger sequencing technology, and the sequencing using the dideoxy chain termination method is the gold standard of sequencing. The accuracy of shotgun sequencing is higher than that of second and TGS. It has a high accuracy of long reads from 700 to 1,000 bp and can deal with repeated sequences well. The disadvantage is that only one template can be detected at a time, which is slow and time-consuming. NGS uses large-scale parallel sequencing technology, which can simultaneously synthesize millions of complementary chains of sequencing templates and acquire sequence data. High-throughput and low-cost NGS is used worldwide in large-scale genome sequencing and resequencing. The NGS mainly used sequencing by synthesis (SBS) technology and sequence by ligation (SBL). Roche has developed GS FLX and GS Junior sequencing platforms with 454 sequencing technology and SBS, which can read more than 400 bp (Ambardar et al., 2016).

Illumina used SBS and a 3′ blocked reversible terminator for sequencing (Shendure and Ji, 2008; Ansorge, 2009). Based on this technology, the Genome Analyzer platform reads 125 bp. In addition, ion torrent sequencing technology developed by Life Technologies/Thermo Fisher Scientific adopts semiconductor chip technology. Four kinds of dNTPs sequential flow through the chip in turn, if one dNTP is complementary to the DNA molecule in a specific microhole of the chip, the dNTP is synthesized into the DNA strand. Ion torrent has the advantages of low cost, only an electrical signal detection system, and no light/fluorescence detection system (Pennisi, 2010) (Figure 11.2).

Third-generation sequencing, also called de novo sequencing technology, is a milestone in sequencing technology based on single-molecule sequencing technology (SMS) (Xu et al., 2009) and large-scale parallel sequencing technology. There are two methods for the TGS. The first is single-molecule fluorescence sequencing. The representative technology from Pacific Bioscience is the American Helicos SMS (true single molecular sequencing, TSMS) and single molecular real-time sequencing (SMRT) technology. SMRT Principle is SBS. DNA polymerase is anchored to the nanopore. The four fluorescent-labeled dNTPs emit different light in the base-pairing stage. The entering base can be determined according to the light wavelength and peak value. The phosphate group of dNTP is labeled with fluorescence in SMRT. The long reads are determined by DNA polymerase activity. SMRT requires a highly efficient optical detection system.

The second is nanopore sequencing, based on electrical signals and single-molecule long-read detection. The representative company is Oxford Nanopore (Midha et al., 2019). The principle of nanopore sequencing is to use electrophoresis technology to drive individual molecules through nanopores one by one for sequencing. The electrification properties of each ATCG base are different during nanopore sequencing, and the difference in electrical signals detected can correspond to different bases. Compared with NGS, TGS is single-molecule sequencing. TGS has neither PCR amplification nor GC preference and can be sequenced directly without PCR amplification. The TGS has a very long read length, with an average read length

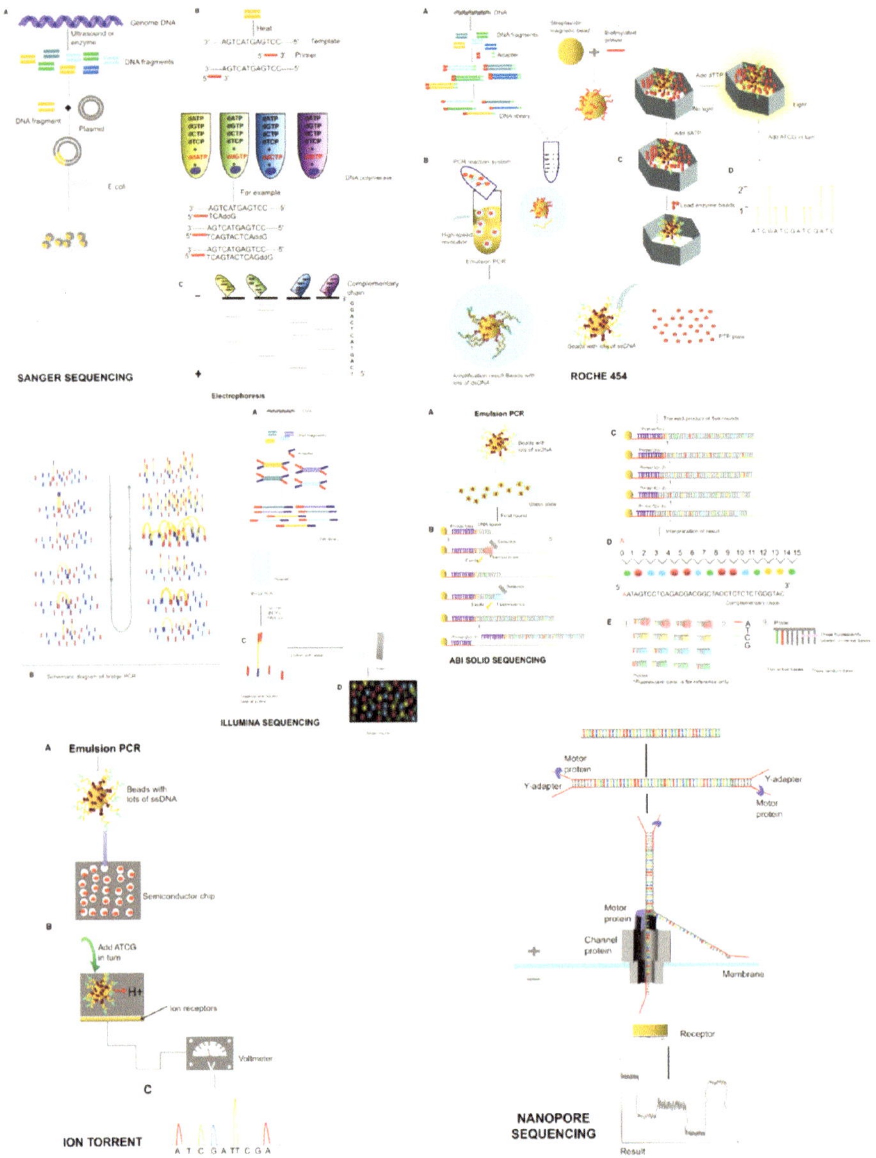

FIGURE 11.2 Innovative metagenomic sequencing and analysis techniques (Zhang et al., 2021).

of 10–15 kb and a maximum read length of up to 40 kb. TGS can directly detect DNA methylation (Timp et al., 2010). The disadvantages of TGS are mainly shown in the following three aspects. First, sequencing errors occur randomly, with error rates of each base as high as 15%, which multiple sequences can overcome and correct. Second, the cost of sequencing is high. Third, SMRT sequencing technology requires enzymes, and currently, better enzymes with much more activity and stability are needed (Athanasopoulou et al., 2021).

Metagenomic gene prediction uses prediction tools to identify potential open reading frames (ORFs) in genomic libraries and to predict the structure and function of unknown sequences of genes. Metagenomic gene prediction includes homology prediction and *de novo* prediction. Homology prediction can be obtained by aligning homologous sequences with the gene database, but the disadvantages depend on the existing gene database, and new genes cannot be annotated. *De novo* prediction is based on the sequence characteristics obtained by sequencing, which can predict known and unknown genes based on different attributes in the coding and non-coding regions (Kim et al., 2013).

Function annotation compares a gene or protein sequence in a specific functional database, associates a gene or protein with a specific function, and helps understand the relevant metabolic pathway. The information on metabolic pathways can be obtained through gene prediction and functional annotation analysis. Bowers RM formulated Genome Sketch Standards of Metagenome-Assembled Genome (MAG) to improve the quality of sequencing analysis in 2017. High-quality drafts require gene integrity >90%, contamination <5%, and inclusion of 16S, 23S, and other highly conserved genes in assembly quantity (Bowers et al., 2017).

11.3 ADVANCES IN METAGENOMICS APPLICATION

Traditionally, microbial research relied heavily on cultured organisms, a method fraught with limitations (Kellenberger, 2001). However, in 1998, a seminal work by Handelsman et al. coined the term "Metagenome," also known as the microbial environmental genome, defining it as "the genome of the total microbiota found in nature." Since then, the term has evolved to encompass the collective genetic material of bacteria and fungi present in environmental samples. Metagenomics emerged as a groundbreaking approach to scrutinizing microbial communities, leveraging functional gene screening or sequencing analysis to unveil their mysteries within specific environments. Metagenomics encompasses a multifaceted exploration of microbial diversity, population structure, genetic and evolutionary relationships, functional activity, cooperative dynamics, and their intricate interplay with the environment. This burgeoning field has swiftly gained momentum across various disciplines, including medicine, agriculture, and environmental protection, catalyzing transformative advancements. The evolution and application of metagenomics detection technology are delineated into three distinct sections (Dweh et al., 2023). First, it delves into the molecular biology techniques underpinning metagenomics, spanning from extracting microbial genes to sequencing target genes, with a particular emphasis on heterologous expression and product screening (Escobar-Zepeda et al., 2015). Second, it navigates through metagenomics' primary classifications or branches, categorized based on research objectives and study subjects. Lastly, it explores the myriad applications of metagenomics across diverse domains, illustrating its versatility and impact in driving innovation and addressing pressing societal challenges (Navgire et al., 2022).

11.4 METAGENOMICS FOR SUSTAINABLE ENVIRONMENTAL MANAGEMENT

There are different fields of microorganism metagenomics applications. Environmental metagenomics applications include the study of microorganisms in oceans, soils, deep oceans, glaciers, craters, and other environments. Metagenomics also has many applications in the medical field, such as identifying pathogenic microorganisms in the intestinal tract, bloodstream infections, lung infections, central nervous system infections, etc. Metagenomic research focuses on antibiotic-resistant bacteria (ARB) and antibiotic-resistant genes (ARGs), biocatalysts, drugs, etc.

Environmental metagenomics has become a powerful tool for exploring microbial life in diverse habitats. It provides a window into the genomic diversity and functional capabilities of microbial communities in their natural environments. This knowledge is essential for advancing our understanding of ecosystem dynamics, microbial ecology, and potential applications of environmental microorganisms.

11.4.1 Impact on Environmental Conservation

Biodiversity assessment	Environmental metagenomics contributes significantly to the assessment of biodiversity in natural ecosystems. By revealing microbial diversity in diverse environments, it helps understand the intricacies of ecosystems and supports conservation efforts.	Tyag et al. (2022); Mendes et al. (2017)
Indicator species and environmental health	Specific microbial species identified through metagenomic analysis can serve as indicators of environmental health. Changes in microbial community composition may reflect environmental disturbances, pollution, or other factors affecting ecosystem integrity.	Song et al. (2022)
Monitoring microbial responses to climate change	Studying microbial communities using metagenomics allows researchers to monitor how microorganisms respond to climate change. This information is essential for predicting and mitigating the impacts of climate change on ecosystems.	Long et al. (2016); Tiedje et al. (2022); Emerso et al. (2015)
Bioremediation strategies	Metagenomic data can guide the development of bioremediation strategies by identifying microbial species that can degrade pollutants. Harnessing the natural capabilities of microbial communities can be a sustainable approach to environmental cleanup.	Malla et al. (2018); Joshi et al. (2024)

11.4.2 Applications in Agriculture and Soil Health

Characterization of the soil microbiome	Environmental metagenomics provides information on the composition and functional capabilities of soil microbiomes. Understanding the soil microbiome is crucial for sustainable agriculture and soil health management.	Nwachukwu and Babalola (2022)
Plant-microbe interactions	Metagenomic studies help elucidate the complex interactions between plants and soil microbes. This knowledge can lead to developing microbial solutions to improve crop productivity, nutrient cycling, and disease resistance.	Vishwakarma et al. (2020); Yadav (2021)

(*Continued*)

(Continued)

Biofertilizers and microbial inoculants	Identifying beneficial microorganisms in soil metagenomes contributes to developing biofertilizers and microbial inoculants. These products can improve nutrient availability, promote plant growth, and reduce the need for chemical fertilizers.	Pirttilä et al. (2021); Yapa et al. (2022)
Suppression of diseases	Understanding the microbial communities of soil ecosystems helps identify microorganisms involved in disease suppression. Exploiting these natural antagonistic interactions can lead to sustainable agricultural disease management strategies.	van Elsas et al. (2008)

11.4.3 Insights into Human Health and Disease

Microbiome-health connections	Environmental metagenomics has illuminated the links between the environment, human microbiomes, and health. Microbial communities in natural environments influence the composition of the human microbiome, which, in turn, plays a crucial role in health and disease.	Aggarwal et al. (2023)
Antibiotics discovery	Metagenomic studies of various environments, such as soil and marine sediments, have led to the discovery of novel antibiotic-producing microorganisms. This has implications for the development of new antibiotics to combat antibiotic-resistant pathogens.	Pillay et al. (2022); Panthee et al. (2022)
Microbial ecology of built environments	Studying the microbial ecology of built environments, including homes and hospitals, through metagenomics provides insights into indoor air quality, potential pathogens, and microbial community dynamics that may impact human health.	Hill and Gilbert (2023)
Disease surveillance and epidemic forecasting	Monitoring microbial diversity in environmental samples, such as water sources, can facilitate disease surveillance and early prediction of outbreaks. Metagenomic approaches make it possible to detect potential pathogens before they significantly affect human health.	Ko et al. (2022)

11.4.4 Metagenomics and Pollution

Metagenomics has emerged as a powerful tool in the arsenal against environmental pollution, offering innovative solutions for bioremediation and controlling contaminants. Bioremediation harnesses the innate purification capabilities of environmental microorganisms to mitigate pollutants or render them harmless within treatment systems. To effectively leverage bioremediation, several conditions must be met: (i) pollutants must be organic and susceptible to degradation by microorganisms. (ii) Microorganisms must possess specific metabolic activities conducive to pollutant breakdown. (iii) Environmental conditions must be conducive to degradation while maintaining microbial activity. Metagenomics plays a pivotal role in this endeavor by uncovering new functional microorganisms and genes, facilitating the construction and screening of microbial strains with heightened degradation efficiency, broad applicability, and stable expression (Offiong et al. 2023).

In wastewater treatment projects, the focus extends beyond the degradation of organic matter to encompass the removal of nitrogen and phosphorus elements. The intricate microbial communities involved in wastewater treatment include anaerobic ammonium oxidation bacteria, phosphorus-accumulating bacteria, and electrochemically active bacteria. Metagenomic investigations of these microorganisms contribute significantly to biological nitrogen removal and enhanced biological phosphorus removal. Functional genes identified within microorganisms demonstrate the capacity to degrade diverse pollutants, including pesticides, plastics, polycyclic aromatic hydrocarbons, and petroleum hydrocarbons, thereby being pivotal in environmental bioremediation efforts. For instance, researchers have unearthed a plethora of novel proteolytic enzymes from yard compost capable of pesticide degradation. Additionally, the screening of benzoate anaerobic degradation genes from Black Sea sediment has facilitated the anaerobic breakdown of aromatic compounds. Moreover, microbial electrochemical systems (MES) utilize electroactive bacteria enriched in anodes to directly harness electrical energy from organic waste, propelling the rapid development of this emerging sewage treatment technology. Through metagenomics, we unlock the potential of microbial communities to combat environmental pollution, paving the way for sustainable and effective strategies in wastewater treatment and beyond (Kumari et al., 2023).

11.4.5 Metagenomics and Wastewater Treatment

Metagenomics is pivotal in understanding and optimizing wastewater treatment processes, offering insights into the complex microbial communities and their functional capabilities. In wastewater treatment, metagenomic analysis involves the comprehensive study of microbial DNA extracted directly from environmental samples, providing a holistic view of the microbial diversity and metabolic potential present within wastewater systems. One critical application of metagenomics in wastewater treatment is identifying microbial species and their functional genes involved in pollutant degradation. Wastewater contains diverse organic contaminants, such as pharmaceuticals, industrial chemicals, and organic matter, which must be efficiently removed to meet environmental standards.

Metagenomic analysis allows researchers to identify microorganisms capable of degrading specific pollutants and elucidate the genetic pathways involved in their degradation. This information can be used to design tailored bioremediation strategies or engineer microbial consortia with enhanced pollutant degradation capabilities.

Furthermore, metagenomics enables monitoring microbial community dynamics in response to changes in environmental conditions or treatment processes. By tracking shifts in microbial populations and gene expression patterns, wastewater treatment plants can optimize operational parameters to enhance treatment efficiency and resilience to environmental fluctuations. For example, metagenomic analysis can reveal microbial indicators of process stability or impending system failures, allowing operators to implement timely interventions to mitigate potential disruptions.

Moreover, metagenomics facilitates the discovery of novel enzymes and metabolic pathways with potential applications in wastewater treatment. Enzymes derived from environmental microorganisms can be harnessed for biotechnological purposes, such as the enzymatic degradation of recalcitrant pollutants or the production of value-added products from wastewater-derived substrates. By mining metagenomic datasets, researchers can identify candidate enzymes with desirable properties for biocatalysis or bioremediation applications.

11.4.6 Identification of Microbial Species and Their Functional Genes Involved in Pollutant Degradation

Identifying microbial species and their functional genes involved in pollutant degradation is critical to metagenomic studies in wastewater treatment. Several strategies and methodologies have been developed to achieve this goal.

11.4.6.1 Shotgun Metagenomics

Shotgun metagenomics involves the sequencing of DNA extracted directly from environmental samples without prior cultivation. By sequencing all DNA in a sample, researchers can obtain comprehensive information about microbial diversity and functional potential (Usyk et al., 2022). Bioinformatic analysis of shotgun metagenomic data allows for identifying microbial species through taxonomic profiling and detecting functional genes associated with pollutant degradation pathways.

11.4.6.2 16S rRNA Gene Sequencing

16S rRNA gene sequencing targets a specific region of the bacterial genome, allowing for the identification and classification of the bacterial taxa present in a sample (Johnson et al., 2019). Researchers can assess microbial community composition and abundance by amplifying and sequencing the 16S rRNA gene from environmental samples. While 16S rRNA gene sequencing provides valuable insights into microbial diversity, it may not capture the total functional potential of microbial communities.

11.4.6.3 Functional Gene Targeting

In addition to taxonomic profiling, metagenomic studies often target specific functional genes associated with pollutant degradation pathways. For example, genes encoding enzymes involved in the degradation of hydrocarbons, aromatic compounds, pesticides, and other pollutants can be amplified and sequenced from environmental samples (Bouhajja et al., 2016). By focusing on functional genes, researchers can directly assess the metabolic potential of microbial communities for pollutant degradation.

11.4.6.4 Metatranscriptomics and Metaproteomics

Metatranscriptomics and metaproteomics involve analyzing microbial gene expression and protein profiles, respectively, in environmental samples (Aguiar-Pulido et al., 2016). These approaches provide insights into which genes and microbial communities actively express proteins under specific conditions, shedding light on pollutant degradation pathways and metabolic activities, as shown in Figure 11.3.

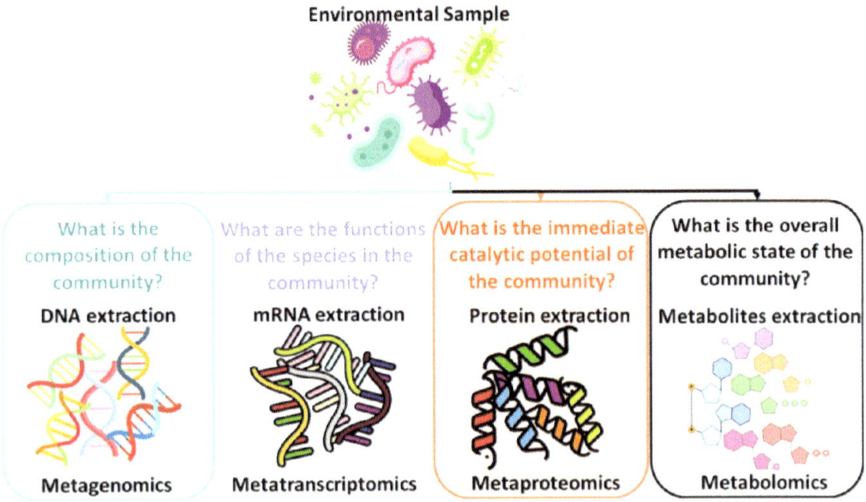

FIGURE 11.3 Subfields of meta-omics: metatranscriptomics and metaproteomics.

11.4.6.5 Bioinformatics Analysis

Bioinformatic analysis is crucial in identifying microbial species and functional genes from metagenomic datasets. Computational tools and databases annotate DNA sequences, predict gene functions, and assign taxonomic classifications to microbial taxa (Hiraoka et al., 2016). Advanced bioinformatics algorithms enable the reconstruction of metabolic pathways and infer microbial interactions based on genomic and transcriptomic data.

11.4.6.6 Validation and Functional Characterization

Once candidate microbial species and functional genes involved in pollutant degradation are identified, validation and functional characterization experiments are conducted to confirm their roles and elucidate their mechanisms of action (Ghosal et al., 2016). This may involve laboratory-based assays, gene expression studies, biochemical analyses, and microbial culture experiments to demonstrate pollutant degradation capabilities and assess environmental relevance.

11.4.7 Metagenomics, Novel Enzymes, and Metabolic Pathways

Metagenomics has emerged as a powerful tool for discovering novel enzymes and metabolic pathways with significant applications in wastewater treatment. Metagenomic studies have identified enzymes produced by environmental microorganisms capable of degrading a wide range of pollutants commonly found in wastewater, including organic compounds, pharmaceuticals, and industrial chemicals (Ufarté et al., 2015). These enzymes, such as oxidases, hydrolases, and dehydrogenases, catalyze the breakdown of pollutants into less harmful or more easily treatable forms, contributing to the overall remediation of wastewater contaminants. Moreover, metagenomic analysis of anaerobic microbial communities in wastewater treatment systems has led to the discovery of novel enzymes and metabolic pathways involved in anaerobic digestion processes. These pathways, including those for methanogenesis and syntrophic acetate oxidation, play crucial roles in converting organic matter into methane and carbon dioxide, facilitating biogas production as a renewable energy source from wastewater sludge.

Metagenomic studies have recently identified microbial enzymes and metabolic pathways involved in biological nitrogen and phosphorus removal from wastewater. For example, genes encoding nitrogenase enzymes responsible for denitrification and ammonium oxidation have been characterized, along with phosphorus-accumulating bacteria harboring polyphosphate kinase genes essential for phosphorus removal via enhanced biological phosphorus removal processes (Liu et al., 2022).

Metagenomics has contributed to the understanding of microbial communities in MES, which harness the electrochemical activity of microorganisms for wastewater treatment and energy recovery. Enzymes involved in extracellular electron transfer, such as cytochromes and hydrogenases, have been identified through metagenomic analysis, elucidating the mechanisms underlying microbial electricity generation and pollutant degradation in MES. Recent metagenomic studies have also uncovered enzymes and pathways involved in the biosynthesis of biopolymers, such as polyhydroxyalkanoates (PHA), from wastewater-derived substrates. PHA-producing microorganisms capable of utilizing various carbon sources present in wastewater have been identified, offering potential applications in the production of biodegradable plastics and other sustainable materials (Godain et al., 2023).

11.5 TRENDS IN METAGENOMIC APPLICATION IN WASTEWATER TREATMENT

In recent years, metagenomic studies in wastewater treatment have undergone significant advancements, leading to several emerging trends that shape research in this field:

11.5.1 Integration of Multi-Omics Approaches

Researchers are increasingly adopting multi-omics approaches, such as metagenomics, metatranscriptomics, metaproteomics, and metabolomics, to obtain comprehensive insights into microbial community structure and function in wastewater treatment systems. By combining data from different omics layers, scientists can elucidate gene expression patterns, protein activities, and metabolite profiles, providing a holistic understanding of microbial activities and interactions.

11.5.2 Longitudinal Monitoring and Time-Series Analysis

There is a growing emphasis on longitudinal monitoring and time-series analysis of microbial communities in wastewater treatment plants to capture temporal dynamics and responses to operational changes (Faust et al., 2015). By conducting regular sampling over extended periods, researchers can track shifts in microbial composition, functional gene abundance, and system performance, facilitating predictive modeling and process optimization.

11.5.3 Microbial Network Analysis and Ecological Modeling

Advanced computational techniques, such as network analysis and ecological modeling, are applied to metagenomic data to decipher microbial interactions and community dynamics in wastewater treatment processes (Zhang et al., 2021). By constructing microbial co-occurrence networks and ecological models, researchers can identify keystone species, predict community stability, and explore ecological principles governing microbial consortia in treatment systems.

11.5.4 Application of Artificial Intelligence and Machine Learning

Artificial intelligence (AI) and machine learning (ML) algorithms are increasingly being utilized to analyze large-scale metagenomic datasets and extract meaningful insights for wastewater treatment optimization (Liu et al., 2023). Researchers can automate data processing, identify microbial biomarkers, and predict system performance by leveraging AI and ML techniques, facilitating data-driven decision-making and real-time monitoring.

11.5.5 Exploration of Microbial Dark Matter

Metagenomic studies are shedding light on the so-called "microbial dark matter," referring to uncultured and poorly characterized microorganisms, constituting a significant portion of microbial communities in wastewater treatment systems (Narayanan and Narayan, 2019). Through metagenomic binning, single-cell genomics, and other genomic approaches, researchers are unraveling the genomes of elusive microorganisms and uncovering novel metabolic pathways and functional capabilities.

11.5.6 Development of Synthetic Biology Tools for Microbial Engineering

With the advent of synthetic biology tools, such as genome editing techniques and genetic circuit design, researchers are exploring the engineering of microbial consortia for enhanced wastewater treatment performance. By redesigning microbial genomes and metabolic pathways, scientists aim to engineer synthetic microbial communities with tailored functions, such as pollutant degradation, nutrient removal, and bioenergy production.

11.6 ROLE OF METAGENOMICS IN WASTEWATER TREATMENT: RECENT ADVANCES AND FUTURE CHALLENGES

Metagenomics is a relatively new but fast-growing field within environmental biology directed at obtaining knowledge on genomes of environmental microbes as well as of entire microbial communities. The recent emergence of new sequencing technologies, resulting in higher sequence output and a dramatic drop in the price of sequencing, will be defining a new era in metagenomics. At this time, the sequencing effort will be taken to a new level to allow addressing new, previously unattainable biological questions as well as accelerating genome-based discovery for medical and biotechnological applications. From the experience of the past 5 years, the power of metagenomics is obvious, while its potential is still waiting to be fully realized.

Metagenomics is advancing steadily, fueled by the continuous evolution of sequencing technologies. This progression has unveiled a crucial realization: the future challenge in metagenomics lies not in generating sequence data but in managing the ensuing computational demands and data storage requirements. Over the past half-decade, the potency of metagenomics has become increasingly apparent, yet its full potential remains untapped. Even ecosystems of seemingly modest complexity present formidable hurdles for genomic exploration, underscoring the imperative for more extensive sampling and the development of specialized assembly and analysis tools. Fortunately, progress in these domains is swift and promising. Nevertheless, the expense of achieving comprehensive metagenomic sequencing, despite leveraging cutting-edge technologies, remains a formidable barrier. While sequencing costs are typically advertised per base, the depth of coverage required varies across different sequencing platforms. For instance, while Sanger technology has achieved remarkable precision over the years, generating reads of up to 1 kb with minimal error rates, forthcoming technologies like 454, though promising similar read lengths, inherently introduce higher error rates, necessitating greater coverage for comparable data quality. Even with cost-effective alternatives like Illumina or SOLiD, achieving adequate sequencing depth—perhaps 40–50 times greater—is imperative to ensure data reliability. Elevating sequencing depth is pivotal for comprehending the intricate dynamics of microbial communities, elucidating their evolutionary pathways, and deciphering their responses to environmental shifts. Moreover, it is indispensable for unlocking novel genes and pathways through metagenomics (Chistoserdova, 2009).

11.7 WASTEWATER CASE STUDIES: SOLUTIONS AND STRATEGIES

To clarify how metagenomics can be applied to wastewater applications and monitoring, we will discuss some case studies that exemplify the application of these techniques. An integrated wastewater management system is the base for sustainable development in urban and peri-urban areas. It is important to quantify and characterize the amount of wastewater for (i) developing effective strategies to treat the wastewater, (ii) applying different technologies, i.e., anaerobic, followed by aerobic and physicochemical, and (iii) using the treated wastewater in a sustainable way taking into account the risks involved, e.g., using treated wastewater for toilet flushing and landscaping (Philip et al., 2019).

11.7.1 A Strategic Framework for Managing Wastewater: A Case Study of Harare, Zimbabwe

Urban water management (UWM) has quantitative and qualitative (hygienic) dimensions, both of which require careful planning if cities are to be managed efficiently in the future. UWM includes the abstraction, treatment, storage, transmission, and distribution of water to consumers, as well as water for city maintenance, rain- and stormwater management, water for ecological functions, and the subsequent collection and processing of the generated wastewater. Water scarcity is a critical problem in some countries.

This strategy focuses on sustainable ways of urban water and nutrient (nitrogen and phosphorus) management in Harare, part of the Lake Chivero catchment. The Dublin Principles on water conservation, environmental protection, reuse, and economic utilization of water were used as a basis for formulating innovative and responsive solutions. External and internal drivers for future evolution of wastewater approaches are identified. Water conservation, treatment, and reuse strategies are developed, including water-saving devices, regulation, leak detection and repair, and including wastewater treatment and reuse. The three levels where wastewater can be handled – on-site, decentralized, and centralized – are discussed and solutions formulated. Based on this, a strategic framework, the so-called "3-Step Strategic Approach," is then developed aimed at the sustainable use of water, nutrients, and other resources in the urban environment. This 3-step strategic approach consists of pollution prevention and minimization (Step 1), reuse after treatment (Step 2), and discharge into the environment with stimulation of self-purification capacity of the receiving environment (Step 3). The 3-step strategic approach was subsequently applied to the Lake Chivero catchment, which demonstrated that these strategies could substantially reduce current water pollution and future water scarcity problems in Harare (Nhapi et al., 2005).

11.7.2 Sustainable Wastewater Management through Decentralized Systems in India

In India, less than 20% of the wastewater generated is getting treated. Recent studies indicate that the supply is roughly equal to the demand for the country, hiding wide regional variations with acute shortages in many parts. Since economic growth implies increased water use, the water situation can be expected to worsen rapidly. Even though water resources are not scarce from the perspective of total available water volumes, the precipitation is highly variable in time and space. In addition, the untreated or partially treated wastewater from human settlements is polluting the existing freshwater bodies, creating a shortage of freshwater for different uses.

The on-site wastewater treatment system was piloted in a single household at the outskirts of Chennai (Medavakkam), Tamil Nadu, India. The house that was chosen for the study had 25 inhabitants, and the per capita water consumption was taken as 135 L/capita/day for design purposes. The single household wastewater was collected and was treated in a battery of treatment units consisting of (i) modified septic tank, (ii) baffled reactor, (iii) aerobic attached growth system, (iv) settler, (v) sand filter, and (vi) disinfection system. This sequence was arrived at based on laboratory scale studies (Philip et al., 2019).

11.7.3 Household Water Treatment and Safe Storage Methods in Acute Emergency Response: Case Study Results from Nepal, Indonesia, Kenya, and Haiti

Household water treatment and safe storage (HWTS) methods, such as boiling or chlorination, have long been recommended in emergencies. A leading provider of safe drinking water to Haiti's underserved communities through water kiosks had several water quality issues to deal with upon implementing their pilot program of 10 water kiosks. There was discussion of using surface water or deep well bore water; however, both were unsuitable for drinking water purposes due to microbiological and mineral contamination. Chlorine-based methods were distributed in all four emergencies and filters in one emergency. Effective use ranged widely, from 0% to 67.5%, with only one pre-existing chlorine program in Haiti and unpromoted boiling use in Indonesia reaching >20%. More successful programs provided an effective HWTS method, with the necessary supplies and training provided, to households with contaminated water who were familiar with the method before the emergency. HWTS can be effective at reducing the risk of unsafe drinking water in the acute emergency context. Genesis Water Technologies was engaged to consult on the best solution for this challenge. An analysis was completed to choose the best water sources available in these communities with lower initial contamination levels. Once this was completed, a system solution was developed with water quality to exceed WHO drinking water standards. The systems were powered by renewable solar energy. The system solution included specialized pre-filtration, disinfection, and a GWT series scale prevention system followed by our GWT series reverse osmosis solution to reduce salts with post-treatment remineralization and disinfection prior to water dispensing. The systems were installed by a local contractor on site in various communities with technical and installation layout assistance provided by Genesis Water Technologies. The systems are performing within their operating parameters, producing, sustainable high-quality drinking water to be sold to the community residents through these kiosks. Plans are to expand this program to eventually include over 300 communities across this country (Daniele et al., 2012).

11.7.4 A Smart Packaged Aspiral™ Plant Provides Stand-Alone Wastewater Treatment to a Highway Service Area in China

The Xiaogan Service Area on the Beijing-HongKong-Macao Highway was being renovated to accommodate an increase in traffic. The service area, which covers 46,289 m^2, includes restaurants, gas stations, mechanics, accommodations, shopping, and parking. Improvements included an upgrade to the area's wastewater treatment capabilities to ensure compliance with new regulations.

The design and equipment of the new plant needed to treat 200 m^3/day on the existing infrastructure. Because the plant is near the service station, which receives visitors throughout the day, it had to be aesthetically pleasing and environmentally friendly, with minimal noise and odor. As with many service stations, the influent includes a high nitrogen concentration, which requires highly efficient treatment. The solution had to be economical, with effluent meeting China's strict Class 1A standards.

An Aspiral™ Smart Packaged wastewater treatment system, based on innovative MABR technology, was selected for the project. Results show that the plant, which was commissioned within 2 months, is treating as much as 200 m^3/day of water that meets or exceeds China's strict Class 1A standard for reuse.

11.8 CONCLUSION

In conclusion, the field of metagenomics has unveiled hidden details of microbial communities thriving in wastewater treatment systems, offering valuable insights and potential solutions for environmental challenges. Through metagenomic studies, researchers have identified novel microbial species, characterized their functional genes, and elucidated metabolic pathways involved in pollutant degradation. One of the key findings of metagenomic research in wastewater treatment is the discovery of diverse microbial populations capable of metabolizing various organic pollutants, including pesticides, plastics, hydrocarbons, and aromatic compounds. These microorganisms possess unique enzymatic machinery and metabolic pathways that enable them to degrade pollutants effectively, contributing to the remediation of contaminated water bodies. Moreover, metagenomic approaches have shed light on the so-called "microbial dark matter," referring to elusive microbial taxa with unknown functions that play crucial roles in wastewater treatment systems. By unraveling the genetic potential of these microbial dark matter communities, researchers can gain a deeper understanding of ecosystem dynamics and identify new targets for bioremediation strategies. Furthermore, metagenomic studies have facilitated the development of innovative wastewater treatment technologies, such as MES, which harness the metabolic activities of electroactive bacteria to recover electrical energy from organic waste. Overall, metagenomics holds great promise for advancing our understanding of microbial ecology in wastewater treatment and accelerating the development of sustainable solutions for environmental pollution. By leveraging the power of genomic analysis, we can harness the natural capabilities of microbial communities to address pressing environmental challenges and safeguard our water resources for future generations.

REFERENCES

Aggarwal, N., Kitano, S., Puah, G.R.Y., Kittelmann, S., Hwang, I.Y., & Chang, M.W. (2023). Microbiome and human health: current understanding, engineering, and enabling technologies. *Chem. Rev.* 123(1), 31–72.

Aguiar-Pulido, V., Huang, W., Suarez-Ulloa, V., Cickovski, T., Mathee, K., & Narasimhan, G. (2016). Metagenomics, metatranscriptomics, and metabolomics approaches for microbiome analysis. *Evol. Bioinform.* 12(Suppl 1), 5–16.

Ambardar, S., Gupta, R., Trakroo, D., Lal, R., & Vakhlu, J. (2016). High throughput sequencing: an overview of sequencing chemistry. *Indian J. Microbiol.* 56(4), 394–404.

Ansorge, W.J. (2009). Next-generation DNA sequencing techniques. *N Biotechnol.* 25(4), 195–203.

Athanasopoulou, K., Boti, M.A., Adamopoulos, P.G., Skourou, P.C., & Scorilas, A. (2021). Third-generation sequencing: the spearhead towards the radical transformation of modern genomics. *Life* 12(1), 30.

Avarre, J.C., de Lajudie, P., & Béna, G. (2007). Hybridization of genomic DNA to microarrays: a challenge for the analysis of environmental samples. *J. Microbiol. Methods* 69(2), 242–248.

Bahiyah, A., Nasharudin, R.M., Rahman, O.A., Azimah, M.Z.N., Fatimah, A.R., & Nurulfiza, I. (2022). Metagenomics insights into the microbial diversity and microbiome network analysis on the heterogeneity of influent to effluent water. *Front. Microbiol.* 13, 779196.

Bouhajja, E., Agathos, S.N., & George, I.F. (2016). Metagenomics: probing pollutant fatein natural and engineered ecosystems. *Biotechnol. Adv.* 34(8), 1413–1426.

Bowers, R.M., Kyrpides, N.C., Stepanauskas, R., Harmon-Smith, M., Doud, D., Reddy, T., et al. (2017). Minimum information about a single amplified genome (MISAG) and a metagenome-assembled genome (MIMAG) of bacteria and archaea. *Nat. Biotechnol.* 35, 725–731.

Chen, Y., & Murrell, J.C. (2010). When metagenomics meets stable-isotope probing: progress and perspectives. *Trends Microbiol.* 18, 157–163.

Chistoserdova, L. (2009). Functional metagenomics: recent advances and future challenges. *Biotechnol. Genetic Eng. Rev.* 26(1), 335–352.

Colin, P.Y., Kintses, B., Gielen, F., Miton, C.M., Fischer, G., Mohamed, M.F., et al. (2015). Ultrahigh-throughput discovery of promiscuous enzymes by picodroplet functional metagenomics. *Nat. Commun.* 6, 10008.

Craig, J.W., Chang, F.Y., Kim, J.H., Obiajulu, S.C., & Brady, S.F. (2010). Expanding small-molecule functional metagenomics through parallel screening of broad-host-range cosmid environmental DNA libraries in diverse proteobacteria. *Appl. Environ. Microbiol.* 76, 1633–1641.

Dweh, T.J., Pattnaik, S., & Sahoo, J.P. (2023). Assessing the impact of meta-genomic tools on current cutting-edge genome engineering and technology. *Int. J. Biochem. Mol. Biol.* 14(4), 62–75.

Escobar-Zepeda, A., Vera-Ponce de León, A., & Sanchez-Flores, A. (2015). The road to metagenomics: from microbiology to DNA sequencing technologies and bioinformatics. *Front. Genet.* 6, 155161.

Faust, K., Lahti, L., Gonze, D., de Vos, W.M., & Raes, J. (2015). Metagenomics meets time series analysis: unraveling microbial community dynamics. *Curr. Opin. Microbiol.* 25, 56–66.

Gabor, E.M., de Vries, E.J., & Janssen, D.B. (2003). Efficient recovery of environmental DNA for expression cloning by indirect extraction methods. *FEMS Microbiol. Ecol.* 44, 153–163.

Galbraith, E.A., Antonopoulos, D.A., & White, B.A. (2008). Application of suppressive subtractive hybridization to uncover the metagenomic diversity of environmental samples. *Methods Mol. Biol.* 410, 295–333.

Ghosal, D., Ghosh, S., Dutta, T.K., & Ahn, Y. (2016). Current state of knowledge in microbial degradation of polycyclic aromatic hydrocarbons (PAHs): a review. *Front. Microbiol.* 7, 1837.

Godain, A., Vogel, T., Monnie, J.-M., Paitier, A., & Haddour, N. (2023). Metaproteomic and metagenomic-coupled approach to investigate microbial response to electrochemical conditions in microbial fuel cells. *Microorganisms* 11, 2695.

Gómez-Acata, E.S., Centeno, C.M., & Falcón, L.I. (2019). Methods for extracting' omes from microbialites. *J. Microbiol. Methods* 160, 1–10.

Handelsman, J. (2004). Metagenomics: application of genomics to uncultured microorganisms. *Microbiol. Mol. Biol. Rev.* 68(4), 669–685.

Handelsman, J., Rondon, M.R., Brady, S.F., Clardy, J., & Goodman, R.M. (1998). Molecularbiological access to the chemistry of unknown soil microbes: a new frontier for naturalproducts. *Chem. Biol.* 5, 245–249.

Hill, M.S., & Gilbert, J.A. (2023). Microbiology of the built environment: harnessing human-associated built environment research to inform the study and design of animal nests and enclosures. *Microbiol. Mol. Biol. Rev.* 87(4), e0012121.

Hiraoka, S., Yang, C.C., & Iwasaki, W. (2016). Metagenomics and bioinformatics in microbial ecology: current status and beyond. *Microbes. Environ.* 31(3), 204–212.

Johnson, J.S., Spakowicz, D.J., Hong, B.Y., Petersen, L.M., Demkowicz, P., Chen, L., et al. (2019). Evaluation of 16S rRNA gene sequencing for species and strain-level microbiome analysis. *Nat. Commun.* 10, 5029.

Joshi, S.S., Jadhav, P.K., Chourasia, N.K., Jadhav, P.J., Pathade, K.N., & Kanimozhi, K. (2024). Bioremediation strategies for soil and water pollution harnessing the power of microorganisms. *Migration Lett.* 21(S8), 1–20.

Kakirde, K.S., Wild, J., Godiska, R., Mead, D.A., Wiggins, A.G., Goodman, R.M., et al. (2011). Gram negative shuttle BAC vector for heterologous expression of metagenomic libraries. *Gene* 475, 57–62.

Kellenberger E. (2001). Exploring the unknown. The silent revolution of microbiology. *EMBO Rep.* 2(1), 5–7.

Kim, M., Lee, K.H., Yoon, S.W., Kim, B.S., Chun, J., & Yi, H. (2013). Analytical tools and databases for metagenomics in the next-generation sequencing era. *Genomics Inform.* 11(3), 102.

Ko, K.K., Chng, K.R., & Nagarajan, N. (2022). Metagenomics-enabled microbial surveillance. *Nat. Microbiol.* 7(4), 486–496.

Kumari, K., Maurye, P., & Sinha, A. (2024). Metagenomics and its application in environmental management. *Agrotech. Today.* 1(11).

Kumari, S., Rajput, V.D., Sushkova, S., & Minkina, T. (2023). Microbial electrochemical system: an emerging technology for remediation of polycyclic aromatic hydrocarbons from soil and sediments. *Environ. Geochem. Health* 45(12), 9451–9467.

Lantagne, D.S., & Clasen, T.F. (2012). Use of household water treatment and safe storage methods in acute emergency response: case study results from Nepal, Indonesia, Kenya, and Haiti. *Environ. Sci. Technol.* 46(20), 11352–11360.

Liu, J., Chu, G., Wang, Q., Zhang, Z., Lu, S., She, Z., Zhao, Y., Jin, C., Guo, L., Ji, J., & Gao, M. (2022). Metagenomic analysis and nitrogen removal performance evaluation of activated sludge from a sequencing batch reactor under different salinities. *J. Environ. Manage.* 323, 116213.

Liu, X., Nie, Y., & Wu, X.L. (2023). Predicting microbial community compositions in wastewater treatment plants using artificial neural networks. *Microbiome* 11, 93.

Long, P.E., Williams, K.H., Hubbard, S.S., & Banfield, J.F. (2016). Microbial metagenomics reveals climate-relevant subsurface biogeochemical processes. *Trends Microbiol.* 24(8), 600–610.

Malla, M.A., Dubey, A., Yadav, S., Kumar, A., Hashem, A., & Abd_Allah, E.F. (2018). Understanding and designing the strategies for the microbe-mediated remediation of environmental contaminants using omics approaches. *Front. Microbiol.* 9, 1132.

Mendes, L.W., Braga, L.P.P., Navarrete, A.A., de Souza, D.G., Silva, G.G., & Tsai, S.M. (2017). Using metagenomics to connect microbial community biodiversity and functions. *Curr. Issues Mol. Biol.* 24(1), 103–118.

Midha, M.K., Wu, M., & Chiu, K.P. (2019). Long-read sequencing in deciphering human geneticsto a greater depth. *Hum. Genet.* 138, 1201–1215.

Narayanan, C.M., & Narayan, V. (2019). Biological wastewater treatment and bioreactor design: a review. *Sustain. Environ. Res.* 29, 33.

Navgire, G.S., Goel, N., Sawhney, G., Sharma, M., Kaushik, P., Mohanta, Y.K., Mohanta, T.K., & Al-Harrasi, A. (2022). Analysis and interpretation of metagenomics data: an approach. *Biol. Proced. Online* 24, 18.

Nhapi, I., Siebel, M. A., & Gijzen, H. J. (2005). A strategic framework for managing wastewater: A case study of Harare, Zimbabwe. Water international, 30(4), 456–467.

Nwachukwu, B.C., & Babalola, O.O. (2022). Metagenomics: a tool for exploring key microbiome with the potentials for improving sustainable agriculture. *Front. Sustain. Food Syst.* 6, 886987.

Offiong, N.A.O., Edet, J.B., Shaibu, S.E., Akan, N.E., Atakpa, E.O., Sanganyado, E., Okop, I.J., Benson, N.U., & Okoh, A. (2023). Metagenomics: an emerging tool for the chemistry of environmental remediation. *Front. Environ. Chem.* 4, 1052697.

Panthee, B., Gyawali, S., Panthee, P., & Techato, K. (2022). Environmental and human microbiome for health. *Life (Basel).* 12(3), 456.

Pennisi, E. (2010). Genomics: semiconductors inspire new sequencing technologies. *Science* 327, 1190.

Philip, L., Ramprasad, C., & Krithika, D. (2019). Sustainable wastewater management through decentralized systems: case studies. In M. Pannirselvam (ed.), *Water Scarcity and Ways to Reduce the Impact: Management Strategies and Technologies for Zero Liquid Discharge and Future Smart Cities*, pp. 15–45. Springer.

Pillay, S., Calderón-Franco, D., Urhan, A., & Abeel, T. (2022). Metagenomic-based surveillance systems for antibiotic resistance in non-clinical settings. *Front. Microbiol.* 13, 1066995.

Pirttilä, A.M., Mohammad Parast Tabas, H., Baruah, N., & Koskimäki, J.J. (2021). Biofertilizers and biocontrol agents for agriculture: how to identify and develop new potent microbial strains and traits. *Microorganisms* 9(4), 817.

Probst, A.J., Weinmaier, T., DeSantis, T.Z., Santo Domingo, J.W., & Ashbolt, N. (2015). New perspectives on microbial community distortion after whole-genome amplification. *PLoS One* 10, e0124158. doi:10.1371/journal.pone.0124158

Rebets, Y., Kormanec, J., Luzhetskyy, A., Bernaerts, K., & Anné, J. (2017). Cloning and expression of metagenomic DNA in *Streptomyces lividans* and subsequent fermentation for optimized production. *Methods Mol. Biol.* 1539, 99–144.

Sabree, Z.L., Rondon, M.R., & Handelsman, J. (2009). Metagenomic. In M. Schaechter (ed.), *Encyclopedia of Microbiology* (Third Edition), pp. 622–632. Academic Press. ISBN 9780123739445.

Samal, K., Mahapatra, S., & Ali, M.H. (2022). Pharmaceutical wastewater as emerging contaminants (EC): treatment technologies, impact on environment and human health. *Energy Nexus.* 6, 100076. ISSN 2772–4271.

Satam, H., Joshi, K., Mangrolia, U., Waghoo, S., Zaidi, G., Rawool, S., Thakare, R.P., Banday, S., Mishra, A.K., Das, G., & Malonia, S.K. (2023). Next-generation sequencing technology: current trends and advancements. *Biology (Basel).* 12(7), 997.

Shendure, J., & Ji, H. (2008). Next-generation DNA sequencing. *Nat. Biotechnol.* 26, 1135–1145.

Song, D., Huo, T., Zhang, Z., Cheng, L., Wang, L., Ming, K., Liu, H., Li, M., & Du, X. (2022). Metagenomic analysis reveals the response of microbial communities and their functions in lake sediment to environmental factors. *Int. J. Environ. Res. Public Health* 19(24), 16870.

Tiedje, J.M., Bruns, M.A., Casadevall, A., Criddle, C.S., Eloe-Fadrosh, E., Karl, D.M., & Zhou, J. (2022). Microbes and climate change: a research prospectus for the future. *mBio* 13(3), e00800–22.

Timp, W., Mirsaidov, U.M., Wang, D., Comer, J., Aksimentiev, A., & Timp, G. (2010). Nanopore sequencing electrical measurements of the code of life. *IEEE Trans. Nanotechnol.* 9(3), 281–294.

Ufarté, L., Laville, É., Duquesne, S., & Potocki-Veronese, G. (2015). Metagenomics for the discovery of pollutant degrading enzymes. *Biotechnol. Adv.* 33(8), 1845–1854.

Usyk, M., Peters, B.A., Karthikeyan, S., McDonald, D., Sollecito, C.C., Vazquez-Baeza, Y., Shaffer, J.P., Gellman, M.D., Talavera, G.A., Daviglus, M.L., Thyagarajan, B., Knight, R., Qi, Q., Kaplan, R., & Burk, R.D. (2022). Comprehensive evaluation of shotgun metagenomics, amplicon sequencing, and harmonization of these platforms for epidemiological studies. *Cell Rep. Methods*. 3(1), 100391.

Van Elsas, J.D., Costa, R., Jansson, J., Sjöling, S., Bailey, M., Nalin, R., Vogel, T.M., & Van Overbeek, L. (2008). The metagenomics of disease-suppressive soils–experiences from the METACONTROL project. *Trends Biotechnol.* 26(11), 591–601.

Verma, S.K., Singh, H., & Sharma, P.C. (2017). An improved method suitable for isolation of high-quality metagenomic DNA from diverse soils. *3 Biotech* 7(3), 171.

Vishwakarma, K., Kumar, N., Shandilya, C., Mohapatra, S., Bhayana, S., & Varma, A. (2020). Revisiting plant–microbe interactions and microbial consortia application for enhancing sustainable agriculture: a review. *Front. Microbiol.* 11, 560406.

Xu, M., Fujita, D., & Hanagata, N. (2009). Perspectives and challenges of emerging single-molecule DNA sequencing technologies. *Small* 5, 2638–2649.

Yadav, A.N. (2021). Beneficial plant-microbe interactions for agricultural sustainability. *J. Appl. Biol. Biotechnol.* 9(1), i–iv.

Yun, J., & Ryu, S. (2005). Screening for novel enzymes from metagenome and SIGEX, as a way to improve it. *Microbiol. Cell Factories* 4, 8.

Zhang, L., Chen, F.X., Zeng, Z., Xu, M., Sun, F., Yang, L., Bi, X., Lin, Y., Gao, Y.J., Hao, H.X., Yi, W., Li, M., & Xie, Y. (2021). Advances in metagenomics and its application in environmental microorganisms. *Front. Microbiol.* 12, 766364.

12 Biofuel Production from Micro/Macroalgae Cultivated with Wastewaters

Ghadir A. El-Chaghaby and Sayed Rashad

12.1 INTRODUCTION

The demand for an eco-friendly replacement for fossil fuels has led to the use of biological resources in the production of biofuels (El-Sheekh et al., 2023a). One of the biological species that serves as useful biomass in the production of biodiesel, an environmentally friendly substitute for petroleum-based transportation fuels, is algae. With advantages such as simple culture, high biomass output, adaptability, etc., algae appear to be the species that offers the most potential for the generation of biodiesel. Cost of production and environmental circumstances are the key issues (Yaashikaa et al., 2022). The quality of the water is profoundly important to the well-being of all life on earth. The rapidly growing human population, industry, urbanization, and chemically enhanced agriculture are all contaminating water supplies. Today, many people struggle to get access to clean water, and poor countries suffer greatly as a result. The above-mentioned characteristics of modern living have resulted in a reduction in the availability of drinkable water, which has caused a sharp increase in the quantity of contaminants in water bodies (Dinka, 2018). Water bodies are being contaminated by a variety of inorganic and organic contaminants, including domestic trash, agricultural runoff, and industrial waste. These contaminants originate from both diffuse and point sources. Such harmful water pollutants frequently accumulate in organisms and other environmental materials, posing significant risks to human health (Siddiqua et al., 2022). There is an urgent need for technologies and materials that are reasonably priced, easy to use, environmentally benign, transportable, economically viable, and chemically resilient in order to meet the increasing demand for clean water around the world. Globally, the economy, the environment, food chains, and biological systems are all negatively impacted by pollutant toxicity (Ramesh et al., 2023). The chapter goes over the many forms of wastewaters (WWs), their primary issues, and their composition. The chapter also discusses recent studies that demonstrated how algae could be successfully grown on WW and how the resulting biomass might be used to produce biofuels. The various

DOI: 10.1201/9781003459286-12

biofuels made from algal biomass grown on WWs are also highlighted. The chapter concludes by discussing the value and benefits of producing biofuels from micro-/macroalgae grown on WW as well as the techno-economics of this method.

12.2 WASTEWATER: TYPES, SOURCES, AND CHARACTERISTICS

One of the most noticeable kinds of pollution is water pollution, which is produced by a variety of sources, including industrial, home, sewage, hazardous chemicals, municipal effluent, medical waste, industrial waste, and more (Soffian et al., 2022). WW comprises a variety of substances, some of which are very harmful to organisms at certain doses. High chemical oxygen demand and biological oxygen demand are indicators of the considerable amounts of inorganic and organic nutrients that are discharged into the environment (Abdelfattah et al., 2023). According to Amaro et al. (2023), three main categories of WW can be distinguished based on their sources, as depicted in Figure 12.1.

Each type of WW is characterized by the presence of specific pollutants. Herbicides, chemical pesticides, animal manure, pathogens, and antibiotics are prevalent in agricultural WWs, which are released from agricultural areas and animal farms (such as runoff water, swine, or poultry) and have high quantities of organic carbon, nitrogen, and phosphorus. Municipal WWs are made up of nitrogen and phosphorus compounds, microplastics, and contaminants of emerging concern (such as pharmaceutical medications and personal care items) that come from domestic sources. On the other hand, industrial WWs can be released from a wide range of sources, including food processing, mining, chemical manufacture, textiles, power

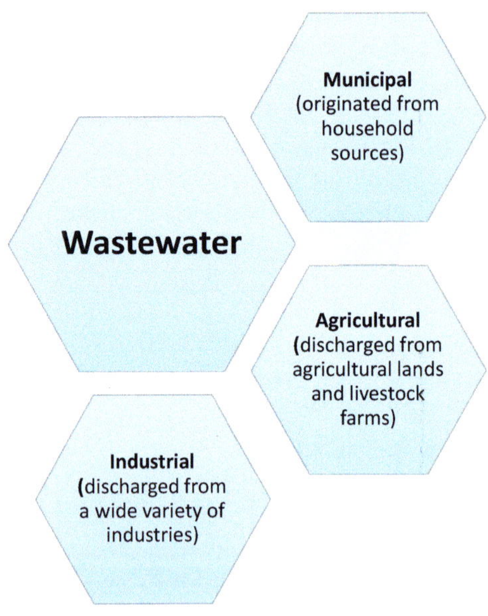

FIGURE 12.1 Main types of WW.

plants, and businesses connected to energy, as well as substances like phenols, dyes, and heavy metals, among others (Wollmann et al., 2019).

The primary components of the WW include organic, inorganic, and artificial materials. In sewage, proteins, carbohydrates, lipids, amino acids, and volatile acids make up the majority of the organic carbon, whereas the inorganic levels included significant amounts of ammonium, bicarbonate, calcium, chlorine, magnesium, phosphate, potassium, sodium, sulfur, and heavy metals (Kaloudas et al., 2021).

12.3 WASTEWATER TREATMENT

In the literature, several methods for WW treatment have been published. They generally consist of physical, chemical, and biological procedures that are thought to be sufficient for treating water in a variety of ways. Their choice is influenced by several variables, including dye concentration, sewage composition, process cost, and other pollutants in WW. Each treatment method's unique characteristics might be advantageous in some ways while being constrained in others (Rashid et al., 2021).

Physical procedures, such as screening, mixing, flocculation, sedimentation, flotation, filtration, and gas transfer, are based on the use of physical forces. Chemical processes can also remove pollutants by introducing chemicals or by accelerating particular chemical reactions. Chemical treatment techniques include disinfection, adsorption, and precipitation. Particularly in industrial treatment settings, physical and chemical approaches are frequently mixed. Contrary to physiochemical processes, biological processes primarily rely on microbial activity to eliminate organic pollutants (such as biodegradable organic waste). Aerobic treatment in ponds, lagoons, trickling filters, and activated sludge plants and anaerobic treatment in equivalent reactor systems are two frequently used biological treatment techniques. Anaerobic and aerobic unit activities are frequently combined in processes (Lakatos, 2018).

According to Crini and Lichtfouse (2019), the WW treatment process involves the following steps, as presented in Figure 12.2.

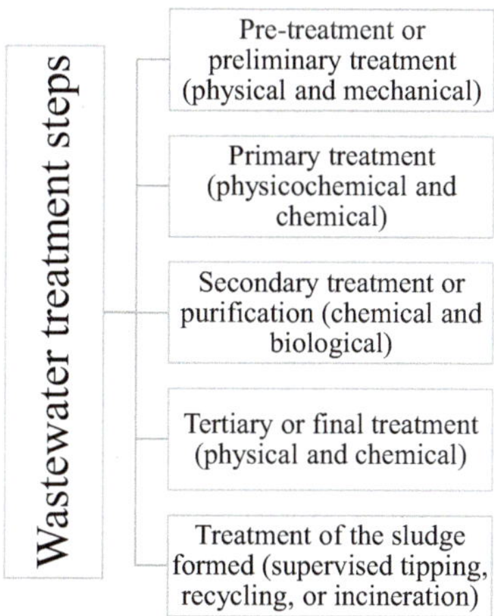

FIGURE 12.2 Steps for WW treatment.

The WW treatment industry may help with the transition to a circular economy since WW contains a large number of resources that may be recovered and utilized again in the economy. WW would therefore be recognized as a source of recovered water, energy, and nutrients like nitrogen and phosphorus rather than as waste. WW treatment plants have become an essential part of the circular economy since resource recovery and clean water production are interwoven (Díaz et al., 2022).

12.4 ALGAE: TYPES, DIVERSITY, AND DISTRIBUTION

Algae are a diverse collection of organisms that engage in photosynthetic processes. They are creatures that live in various types of aquatic environments and can be found in many different and harsh habitats, such as rivers, ponds, oceans, and WW. Macroalgae are multicellular organisms, whereas microalgae are single-celled creatures. Multicellular organisms, macroalgae, or seaweed (which can grow up to 60 m in length), and unicellular organisms, commonly known as microalgae (which range in size from 1 mm to several cm), are included in this division. The coloration of macroalgae can be used as a basis for classification: Brown seaweed (Phaeophyceae), red seaweed (Rhodophyceae), and green seaweed (Chlorophyceae) are the three types of seaweed (Biris-Dorhoi et al., 2020; Rashad and El-Chaghaby, 2020).

Algae cover a large portion of the ocean floor, are a group of unicellular and multicellular organisms categorized according to ecological and morphological features, and are the main producers in aquatic environments. The importance of these microscopic or huge thallus-forming organisms to the human diet is likewise increasing. By analyzing the genomes of algae, which generate about half of the oxygen that humans breathe, we were able to trace the evolution of endosymbiosis and horizontal gene transfer processes. The unique organs and structures found in land plants, including leaves, roots, waxy epidermis, and other organs, are absent from algae. Algae all have green chlorophyll, but this can be hidden by photosynthetic pigments (blue, red, brown, and gold), which give them a characteristic color that is crucial for identification. Due to the lipid content in certain algae, which are minute and can float in surface waters (phytoplankton), other algae are macroscopic and cling to rocks or other structures (seaweed), growing to lengths of more than 50 m. In laboratory settings, several green algae species from the genera *Scenedesmus*, *Chlorella*, *Spirogyra*, *Dunaliella*, and *Botryococcus* can collect lipids (mostly triacylglycerols) in amounts exceeding 70% of dry mass (Jabłońska-Trypuć et al., 2023).

12.5 ALGAE FOR WASTEWATER TREATMENT "PHYCOREMEDIATION"

The development of WW treatment methods has always been studied by researchers. Recently, "photosynthetic wastewater treatment" has been viewed as a novel green method to strongly improve nutrient recovery and industrialize WW reclamation concurrently by utilizing the primary contributor, microalgae, which performs around 40% of photosynthesis on earth (Han et al., 2023).

Phycoremediation or the utilization of algae (micro- or macroalgae), looks to be a sustainable method for removing or biotransforming pollutants from WW, such as heavy metals, fertilizers, and other toxins (Koul et al., 2022).

Freshwater and marine algae can be used to treat WW from a variety of sources, including industrial, municipal, and agricultural operations. In a procedure known as phycoremediation, the algae used in the WW treatment process may remove phosphates and nitrates as well as heavy metals, pesticides, and hydrocarbons (Phang et al., 2015). They are popular species for the bioremediation of WW bodies due to their capacity to use nutrients from the WWs as well as nitrogen and phosphorus throughout their development (Kaloudas et al., 2021).

Algae are desirable candidates for WW treatment, but there are a number of limitations and challenges that must be overcome before using algae for WW treatment. Algae are a challenging option for WW treatment because of their high energy requirements, low biomass productivity, and expense. The method's species and strain are associated with poor levels of biomass productivity. Algae with a high productivity level are the best in treating WW. Not all algae are equally good in removing toxins from WW, which is a disadvantage of using algae for WW treatment (Mohsenpour et al., 2021).

The effectiveness of algae as water purifiers is influenced by the nutrients found in WW. Therefore, it's crucial to choose a species that can endure in the current environment. Processing microalgae and the potential of photobioreactors (PBRs), which demand artificial lighting and chemical sterilizers, raise the cost of manufacture. When choosing the appropriate microalgae for the treatment, the chemical content of the WW should also be taken into account. Other challenges must be overcome in addition to the productivity and economics of biomass. The quantity of suspended particles, the water's clarity, the cost and harvesting methods, the recovery of microalgae, and internal defense are a few of these. Similar to internal observation, turbidity and suspended particles prevent light from penetrating the WW. Light wouldn't be able to reach the bottom due to the rapid increase in cell concentration (Rahman et al., 2020).

The use of algae-based WW treatment is expanding because of its advantages, including decreased emissions and higher-quality manufactured goods. A WW treatment method based on algae produces power, recovers nutrients, and fixes inorganic carbon in the atmosphere. However, the efficiency, durability, and profitability of this treatment method raise questions about its applicability. Genetic engineering and other technological developments give hope that constraints can become possibilities. Investigations are also ongoing on the mechanisms and comprehension of how algae treat WW. Techniques for effective solid/liquid separation, dewatering, and drying still need to be developed. Beyond this, any WW treatment enterprise that uses technology based on algae must be able to make money from it. The required area with the standard treatment system is one of the additional problems. To be a viable alternative, the technology's potential advantages must also be realized and demonstrated. The microalgae used to treat WW must have the required characteristics, including high biomass output, adaptability to different types of WWs, stable biochemical composition, and economic superiority over unwanted species. However, there are a number of significant obstacles, including

(i) the ineffective removal of pollutants, (ii) the resistance of pollutants with complex chemical structures to degradation using conventional processes, and (iii) the limited design and long-term operational flexibility offered by biological WWT (Kandasamy et al., 2023).

12.6 ALGAL METABOLISM AND WASTEWATER TREATMENT MECHANISMS

Algal-based systems show promise as cleaner and more sustainable alternatives to present practices among upcoming WW technologies. Algae's metabolic skills allow them to grow in WW by consuming its organic and nutritional content. According to Nirmalakhandan et al. (2019), a supply of carbon for anabolism, a source of energy for catabolism, and the availability of major and minor nutrients are the three main metabolic needs of all algal systems.

Algal metabolism systems shown in Figure 12.3 are based on the sources of energy and carbon for each of these systems. The carbon and energy sources are inorganics and light for photoautotrophic; organics and light in the case of heterotrophic and organics and/or inorganics and/or light for mixotrophic (Nirmalakhandan et al., 2019).

Microalgae and macroalgae, while growing on WW, were capable of removing WW pollutants, thus remediating WW and producing high biomass in an eco-friendly way. As shown in Table 12.1, different algal species was successfully grown on a variety of WW effluents.

Several mechanisms were proposed for the remediation ability of algae in treating WW. Among these, bio-adsorption, bioaccumulation, and extracellular and intracellular bioremediation are the key mechanisms for eliminating harmful substances from WW by algae. These systems play an important role in the removal of contaminants from WW. Antibiotics, heavy metals, and pesticides were shown to be effectively eliminated by algae via cell membrane, extracellular, and intracellular pathways. Organic pollutants are mostly eliminated by biodegradation and biotransformation, whereas inorganic chemicals are remediated through membrane transporters, biosorption, and bioaccumulation (Bhatt et al., 2022).

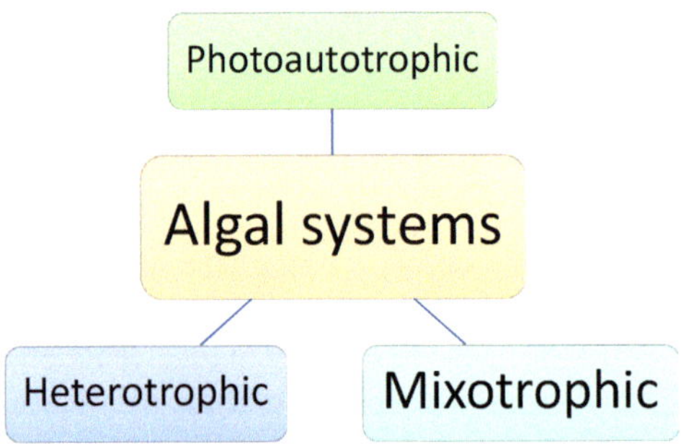

FIGURE 12.3 Classification of algal metabolism.

TABLE 12.1
Algal Species Grown on Different WWs

Wastewater Type	Algae	References
Piggery WW	*Scendesmus obliquus*	Prandini et al. (2016)
Dairy production effluent	*Chlorella pyrenoidosa*	Kothari et al. (2012)
Municipal WW	*Chlorella vulgaris*	Gouveia et al. (2016)
Brewery effluent	*Chlorella vulgaris*	Choi (2016)
Municipal WW	*Chaetomorpha linum*	Ge and Champagne (2017)
Domestic sewage	*Botryococcus braunii* and *Chlorella vulgaris*	Sydney et al. (2011)
Industrial effluent, from copper smelter and refinery	*Spirulina* sp.	Chojnacka et al. (2004)
Pulp and paper industry WW	Green algae (*Chlorella*)	Tarlan et al. (2002)
Downstream WW	*Oedogonium, Tribonema, Spirogyra, and Cladophora genera*	Liu et al. (2023)
Municipal WW	Arthrospira platensis Gomont, 1892	Salman et al. (2023)
Local drainage system WW	*Oedogonium oblangum*	Ijaz et al. (2023)
Industrial WW	*Vaucheria debaryana and Cladophora glomerata*	Khan et al. (2023)
Textile WW	*Shewanella algae B29*	Chaieb et al. (2023)
Domestic WWs	*Desmodesmus sp*	Li et al. (2023)

12.7 FUELS AND BIOFUELS

Due to diminishing oil sources, energy security problems, and rising environmental concerns caused by climate change and greenhouse gas emissions, biofuels have recently gained major relevance (Mahapatra et al., 2021). The use of fossil fuels pollutes the environment and damages ecosystems for both plants and animals. As a result, biofuels have become a viable substitute for fossil fuels. The growing depletion of fossil fuels has made biofuels more appealing (Prabakaran and Karthikeyan, 2023). According to Mahapatra et al. (2021), biofuels are categorized based on the feedstock material into four categories, as depicted in Table 12.2.

12.8 BIOFUELS FROM ALGAL BIOMASS

Algae are already crucial for many industrial applications, including the creation of dietary supplements, sewage treatment, and coloring agents. A sustainable source of raw material for biofuels, algae offer one of the most promising uses. Algae-derived vegetable oil can be refined into a variety of biofuels, including renewable diesel and jet fuel, as well as other chemical components for goods like cosmetics. It can also be utilized directly (as straight vegetable oil that is esterified into biodiesel). Algae's sugars (carbohydrates) can be fermented to produce additional biofuels like ethanol and butanol, as well as other goods like plastics and biochemicals. Algae biomass can be utilized to generate combined heat and power or pyrolyze oil. Renewable

TABLE 12.2
Types of Biofuels

Biofuel Type	Feed Stock	Description
First-generation biofuel	Animal feed crops or other food products	These biofuels are also known as "conventional biofuels" since they are produced using a variety of well-known technologies and processes, such as fermentation, distillation, and transesterification. Since the process is primarily concerned with producing fuel and discarding the remaining non-fuel material as waste, membranes are not necessary in these processes
Second-generation biofuel	Non-food feedstocks such as dedicated energy crops and other lignocellulosic plants, agricultural residues, forest residues, and other waste products	It is a better technique that concentrates on boosting fuel recovery as well as creating secondary raw material. It is an economically viable technology because, in contrast to the first-generation process, it concentrates on creating useful fuels while lowering the cost of energy overall and the quantity of trash produced. For these reasons, researchers frequently support methods like membrane filtering and the integration of several biorefineries to increase biofuel output
Third-generation biofuels	Microalgae	Produced via transesterification or hydrotreatment of the algal oil. These methods can efficiently increase the biofuel yield per year than the first-generation biofuels that use traditional crops
Fourth-generation biofuels	Genetically modified algae and photobiological solar fuels and electro-fuels	The raw materials for fourth-generation solar panels are broadly accessible, economically more advantageous, and never-ending. By causing autolysis of cells and product secretary systems, genetic alteration of microalgal biomass offers a potential use in oil extraction methods. Tools for genome editing are often used in bioinformatics

diesel and jet fuels made from algae can easily replace petroleum-based fuels without requiring engine modifications. They comply with all requirements for the petroleum fuel they swap out (Arora et al., 2021).

Algae-derived bioenergy has long been considered a potential large-scale replacement for fossil fuels. To achieve large-scale algal bioenergy production, a number of restrictions still need to be reduced. These restrictions include high prices, ineffectiveness, insufficient sedimentation of the created biomass, and an inability to regulate the types of microalgae. The use of the entire biomass, cocultivation, the choice of native algal and bacterial strains, the addition of inexpensive organic substrates, optimization of the metabolic pathway, and attachment of the cells to the suspended materials are a few potential solutions. Other options include cultivation in WW (Muthuraman and Kasianantham, 2023).

Algae farming is a sustainable technique that uses 1.83 g of carbon dioxide for every 1 g of biomass produced. As a result, it promotes carbon neutrality. Fresh water and fertile soil are not necessary for the generation of algal biomass.

TABLE 12.3
Biofuels from Algal Biomass

Biofuel	Description	References
Biodiesel	• A blend of monoalkyl esters of long-chain FAs from algal biomass • Has excellent ignition characteristics and reduces pollutants and carbon dioxide emissions by 78%	V et al. (2021)
Bio-oil	• Produced by algal biomass pyrolysis • Utilized as a chemical or may be used directly as fuel for internal combustion engines	V et al. (2021)
Bioethanol	• Preferably produced from macroalgae for their high carbohydrate content • The conversion to occur requires hydrolysis followed by a subsequent fermentation with bacteria or yeast	Alfonsín et al. (2019)
Biomethane	• Organic matter is broken down by a group of microbes during the complicated biological process known as anaerobic digestion without the presence of oxygen • Preferably produced from macroalgae for high polysaccharide and negligible lignin contents	Darko et al. (2022)
Biohydrogen	• Produced by microalgae like cyanobacteria and green algae once they have derived their photosynthetic metabolism • Microalgae biohydrogen production technology primarily consists of photosynthesis biohydrogen production and fermentation biohydrogen production	Wang et al. (2021) and Wang and Yin (2018)

Algae may grow at a rate that is 20–30 times greater than that of other feed crops. It has a high oil content and may produce 10–100 times as much oil as conventional oil sources. Algae can generate a variety of renewable biofuels, including biogas from anaerobic digestion, biodiesel from photobiological production, and hydrogen from photosynthesis. As a result, because algae can produce more fuel than conventional bio-oil sources while using less fresh water and arable land, it is the ideal choice for the manufacture of biodiesel, which can lower greenhouse gas (carbon dioxide) levels (Prabakaran and Karthikeyan, 2023). Biofuels produced from algal biomass include: bioethanol, biodiesel, biohydrogen, bio-oil, and biomethane. Table 12.3 depicts a brief about the forms of biofuels produced from algal biomass.

12.9 INTEGRATING ALGAL GROWTH ON WASTEWATER AND BIOFUEL PRODUCTION (CASE STUDIES)

The algal WW treatment system is regarded as a quick, safe, and low-cost solution for resource recovery and sustainable development. Several algae species have been effectively used for resource recovery from polluted WW and long-term development goals. *Chlorella sp., C. sorokiniana, Nannochloropsis sp., Scenedesmus sp., Tetraselmis sp., and Chlorella pyrenoidosa* were studied in the bioremediation of hazardous contaminants from WW, and the biomass of these algal cultures was used to produce bioenergy (Bhatt et al., 2022). Recently, several studies focused on the dual benefit of treating WW while growing algae for the production of biofuels. In this section, a summary of many relevant studies is discussed.

Rather et al. (2023) investigated how *Chlorella sp.* produced biomass and lipid utilizing WW containing industrial sludge as a cheap supply of nutrients. The characteristics of algal growth, the production of biodiesel, and the rates of nitrogen and phosphate removal were all investigated. *Chlorella sp.* cultured had high cell density and lipid production. Its fatty acids (FAs) also contained a significant amount of heptadecanoic acid, which is beneficial for biodiesel quality.

From particular sewage-fed urban lakes and sewage treatment facilities in India, *Euglena* species, *Spirogyra* species, and *Phormidium* species were collected. These species had a greater proportion of FAs with the desired biofuel qualities (palmitate and stearate, followed by oleic and linoleic acids). The results of the study support the use of algae-based treatment to remove nutrients from WW and produce biofuels or support the long-term development of renewable energy sources (Ramachandra et al., 2013).

A research study into how various cyanobacteria and microalgae can treat WW and generate biodiesel has been conducted in Egypt. Eleven microalgae were identified and were used for WW treatment. WW samples included: agricultural drainage, house drainage, and municipal WWs. The outcomes suggested that the isolated microalgal strains were highly effective in treating WW. A transesterification process was used to extract the oil from the microbial biomass and transform it into biodiesel. The findings showed that the *Oscillatoria Chalybea* examined had a high amount of undecanoic acid (C11:0), making it the best ideal for the manufacture of high-quality biodiesel (Taher et al., 2022).

Two types of microalgae that are commercially used to produce biodiesel: the heterotrophic freshwater microalgae *Chlorella* protothecoides and the marine microalgae *Tetraselmis suecica* were grown on Swine WW by the aid of fungi. The derived biomass from fungal-algal pellets was employed as a feedstock for the pyrolysis-based manufacture of liquid petrochemicals, biogas, and value-added chemicals. Straight-chain C12 to C21 alkanes, which may be employed right away as a glycerine-free component of biodiesel, were produced from the biomass generated by thermochemical conversion, together with biogas, biosolids, and a variety of liquid petrochemicals (Muradov et al., 2015).

In order to assess the viability of producing algal biomass and biodiesel from WW that contained 10%–15% municipal sewage and 85%–90% carpet manufacturing effluents, research was carried out. WW from carpeting was used to extract native algal strains. According to preliminary growth tests, both freshwater and marine algae grew well in WWs. More than 96% of the nutrients in treated WW were removed by a group of 15 native algal strains. This consortium's ability to produce biomass and lipids per hectare per year while using treated WW was 9.2–17.8 tons and 6.82%, respectively. A total of 63.9% of the consortium's algal oil may be transformed into biodiesel (Chinnasamy et al., 2010).

The use of municipal WW in the cultivation of *Amphora sp.* has been proven to be beneficial in terms of their efficient growth rates, biomass, and lipid productivities when combined with traditional nutrient medium. According to the results of this investigation, the combination of municipal WW and De Walne's medium (ratio 80:20) had the greatest lipid content (63.12%). Transesterification was used to transform the acquired lipid into biodiesel (Harini et al., 2023).

The extraction of lipids from the microalgae *Scenedesmus sp.* (SCRE-1) grown in synthetic residential WW was found to be a viable method for producing high-quality liquid biofuels with 80% of high-quality biodiesel and 96% of high-quality green diesel obtained through the transesterification and hydrodeoxygenation reactions, respectively (Chavarría-Hernández et al., 2022) (Table 12.4).

Hamouda et al. (2023) performed a study to investigate the ability of the green alga *Chlorella vulgaris,* the blue-green alga *Synechococcus sp.*, and their consortium to grow and degrade hydrocarbons such as kerosene (k) at various concentrations (0%, 0.5%, 1%, and 1.5%), as well as the use of algal biomasses to produce biofuel. *C. vulgaris* generated the most lipids (32%) after 15 days of algae growth. The results revealed that undecane was detected in significant numbers, *C. vulgaris* (19.9%), *Synechococcus sp.* (82.16%), and algae consortium (79.51%), and moderate amounts of FA methyl ester in *Synechococcus* sp. According to the findings, a consortium of algae can absorb and eliminate kerosene from water while also producing biofuels such as biodiesel and petroleum-based fuels.

Scenedesmus sp., a green alga, was grown in diverse residence WWs. In 20 days, WW yielded 0.68–0.84 (g/L) biomass and 5.6–8.5 (mg/L/day) lipid productivity. Algal culture removed 66% of inorganic phosphorus, 99% of nitrite, 86% of nitrate, 80% of ammoniacal nitrogen, 84% of calcium, and 76% of sulfate from the initial high level of WW nutrients. In these circumstances, the total FA generated by the alga was 43.3% saturated FA, 44.4% monounsaturated FA, and 12.3% polyunsaturated FA. This small-scale investigation might be used to simultaneously remove nutrients and produce algal oil using home WWs (Baldev et al., 2021).

An integrated system was created to treat brewery WW utilizing algal-bacterial aggregates rather than activated sludge. After acid hydrolysis and fermentation using Saccharomyces cerevisiae, the collected biomass was transformed into bioethanol, yielding 0.076 g of bioethanol per gram of biomass (Papadopoulos et al., 2023). A schematic diagram for biofuel production from algae is depicted in Figure 12.4.

12.10 ADVANTAGES AND LIMITATIONS FOR BIOFUEL PRODUCTION FROM ALGAE GROWN ON WASTEWATER

Algae appear to be a viable alternative biological WW cleanup method that is both technically possible and environmentally secure. It is commonly known that algae may thrive in a variety of settings, including autotrophic, mixotrophic, and heterotrophic ones (Alazaiza et al., 2022). The most promising microorganisms for WW treatment and nutrient recovery are algae. They have supplanted other primitive WW treatment techniques due to additional benefits such as CO_2 fixation and pollution reduction, utilization of microalgal biomass to produce biofuels (biodiesel, bioethanol, biomethane, biohydrogen, microalgae-microbial fuel cells, etc.), and other value-added products (Ghaffar et al., 2023). It is imperative to market algae on a wide scale, particularly as an alternative to current WW treatment methods, as well as instruments for environmental management and biological evaluation. Algae-based

TABLE 12.4
Biofuels from Algae Cultivated on WW

Algae	Lipid Content	Product	Source of the WW	References
Planktochlorella nurekis and Chlamydomonas reinhardtii	24.3% and 20.5%	Biodiesel	Cashew nut processing WW	Kumar et al. (2023)
Chaetoceros sp.	21.82%	Biodiesel	Moroccan seawater of Sidi Moussa	Houria et al. (2023)
Microalgal consortia	31.33%	Biodiesel	Municipal WW	Sharma et al. (2020)
Haematococcus pluvialis	12.0 mg/L	Biofuel	Domestic secondary effluent	Wu et al. (2013)
Anabaena ambigua (NCIM No. 2785), the microalga Chlorella pyrenoidosa (NCIM No. 2738), and Scenedesmus abundans	10.36%, 13.13%, and 16.93%	Bioenergy	Dairy WW	Brar et al. (2019)
Chlorella sorokiniana	1.644 mg/mL	Biodiesel	Dairy WW	Hamidian and Zamani (2021)
Desmodesmus sp.	23.51% ± 3.62%	Biodiesel	Municipal WW	Do et al. (2019)
Arthrospira platensis Gomont, 1892		Biodiesel	Municipal WW	Salman et al. (2023)
Chlorella vulgaris	442.9 ± 6.5 mg/L/day	Biodiesel and biogas	Sludge WW	Sakarika and Kornaros (2019)
Nannochloropsis oculata	27.0%	Bioethanol	Municipal WW	Reyimu and Özçimen (2017)
Tetraselmis suecica	6.25%			
Nannochloropsis oceanica	41.97%	Biodiesel	Yeast WW	Senousy et al. (2023)
Chlamydomonas reinhardtii, Monoraphidium braunii, and Scenedesmus obliquus.		Biodiesel	Sewage WW	El-Sheekh, et al. (2023b)
Leptolyngbya sp.	31.34%	Biofuel	Sewage WW	Maity et al. (2014)
Chlorella pyrenoidosa, Scenedesmus obliquus, and Chlorella sorokiniana	25.34%–26.23%	Biohydrogen	Domestic WW	Satheesh et al. (2023)
Scenedesmus obliquus and Chlorella vulgaris	21% and 18%	Biofuels	Dairy WW	de Mendonça et al. (2022)

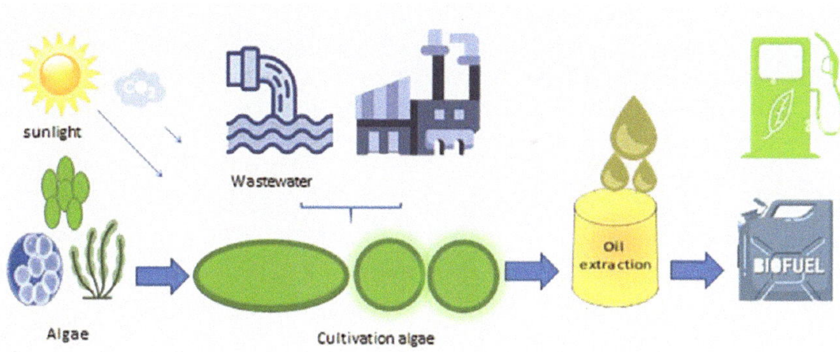

FIGURE 12.4 Production of biofuel from algae cultivated with WW.

eco-technologies can also be used in process integration to create a sustainable final product that could help with energy consumption. The use of commercially available innovative technologies based on algae has shown to be an efficient solution for processes that use fossil fuels and suffer from adverse environmental impacts. Recovered carbohydrates, lipids, pigments, and proteins can be used to make biohydrogen, exopolysaccharides, bioethanol, and biodiesel. In addition to being a component in the synthesis of polyunsaturated FAs and carotenoids, algae may also be employed as a tool for environmental conservation (Gondi et al., 2022).

Despite the several advantages offered by biofuel production from WW-grown algal biomass, there are still some limitations, nevertheless, that must be taken into account. According to Pittman et al. (2011), (i) in all microalgae culture applications, such as algal WW treatment and algal growth for biofuels, efficient and economically sound microalgal harvesting and processing is a critical problem that has to be solved. (ii) Whether the process will produce a positive amount of energy will be a key factor in determining the genuine potential of producing algal biofuel utilizing WW resources. This factor will also impact the method's economic feasibility. Additionally, it will be necessary to assess the process's true sustainability in terms of the use of natural resources and its true carbon neutrality. (iii) The majority of research to date has examined the generation of microalgae lipids, and biomass in lab settings after microalgae were grown in bioreactors or batch culture flasks as opposed to ponds. Additionally, it must be shown that these biomass outputs can be sustained throughout extended culture times.

12.11 A PLAUSIBLE SCENARIO FOR PRODUCING BIODIESEL USING MICROALGAE

A plausible scenario for the third generation of biofuels from microalgae: global mapping of economically viable areas for producing microalgal biofuels with low environmental impact. Microalgae are a widely used source of third-generation biofuels, and their high lipid content has attracted the attention of researchers in the last 10 years. However, using microalgae to produce renewable biofuels is currently regarded as not being economically feasible (Ananthi et al., 2021).

As a response to climate change and global warming, microalgae, which are derived from the global energy crisis, have emerged as an alternative biomass with enormous potential as a renewable and environmentally friendly biofuel feedstock. Furthermore, microalgae have the potential to function as bio-cell factories that produce important industrial commodities like proteins, enzymes, pigments, lipids, bioethanol, and biohydrogen. Under algal biorefinery, these high-value-added bioproducts are a more affordable option to microalgae-based renewable energy (Saxena et al., 2023).

Microalgae-derived third-generation biodiesel is a viable alternative because of its quick biomass weight doubling, year-round harvesting cycles, rapid lipid accumulation, and simplicity in using CO_2, sunlight, and water. However, there are major obstacles to commercialization when it comes to life cycle assessment and economic viability. Microalgae biodiesel production is mostly restricted to laboratory and pilot scales due to high costs. Despite this, several techniques have been developed to reduce the cost of producing microalgal biodiesel; however, the majority of these have not been included in recent research. Instead of cultivating, which can also increase costs, cost-minimization techniques have been applied mainly to dewatering, harvesting, and pretreatment processes to reduce overall production costs (Rafa et al., 2021).

12.12 CHALLENGES AND FUTURE PERSPECTIVES

The demand for affordable, clean, and sustainable fuel sources is rising as a result of industrialization and rising living standards. However, modern societies face serious challenges from the depletion of fossil fuel resources, environmental problems, and growing energy costs. The main raw materials and resource constraints affect biofuels, which are typically made from edible oils leftover from oil plants, animal fats, or leftover edible oils. Because of their high oil capacity and rapid growth, algae resources such as micro- and macroalgae are thought to be suitable for the production of biofuel. In recent years, research has been done on the use of macroalgae as a raw material for biofuel production (Pouresmaieli et al., 2023).

As part of the circular bioeconomy, the biorefinery approach separates biomass and reduces waste components. Due to downstream processing constraints like diluted microalgal culture and the complexity of the algal cell wall, it is still not economically feasible. At least half of the expenses are related to these factors. Thus, to greatly improve biorefineries, unit operations must be integrated and co-optimized. This emphasizes the pressing need for a more effective and economical approach to the biorefinery process. The selection of nutrients, strains, and engineering methods to optimize value-added products is the main focus of research on microalgae biorefinery. Understanding the economics, energy balance, and application of microalgae strains in biorefinery systems is aided by the initial screening of these strains based on metabolites. Lipid productivity and biomass production are also impacted by the selection of PBRs (Ali et al., 2022).

Industrial-scale algal biorefinery requires co-integration and optimization of operation units, including valorizing residual and secondary microalgal products like soluble proteins used as food additives. Integrated processes like WW treatment and nanoparticle (NP) biosynthesis can extract high-value-added compounds, countering low yields. Multiple product synthesis approaches can be successfully applied using these integrated approaches. Future research should incorporate these approaches to make algal biorefinery a profitable business (Abdelfattah et al., 2023).

Because of their advantageous biochemical characteristics, macroalgae are regarded as a viable substitute for the production of renewable energy. Biomass can be used to produce energy through two main types of processes. The first group consists of biochemical reactions, such as fermentation and anaerobic digestion. Methane production from macroalgae is linked to success, according to research on the process of producing biogas from it (Abusweireh et al., 2023).

The exploration of microalgae cultivation on WW for the production of biofuel and bioenergy involves social, economic, and environmental aspects. A few crucial considerations should be made for industrial uses. Potentially hazardous substances like medications, hormones, surfactants, and pharmaceuticals must be taken into account when using WW as a growth medium for microalgae culture. Except for metallic ions, very little is known. The majority of studies on the removal of heavy metals were conducted in lab settings using pure cultures of immobilized microalgae. To confirm their high potential for industrial applications, they must be up-scaled (Udayan et al., 2023).

To avoid inhibiting effects on microalgae growth caused by excess loadings of specific compounds and high turbidity, dilution with freshwater may be required depending on the type of WW. To cut down on drinking water consumption, it is therefore recommended to use WW when none or very little dilution is required. Furthermore, adding undigested manure—which has a significantly higher concentration of organic carbon than digested manure—may be beneficial since it can stimulate the growth of microalgae, which reduces the hydraulic retention time needed to remove nutrients (Ali et al., 2022).

To reduce operating costs and improve treatment quality, future research should concentrate on the organic matter of WW in conjunction with native microalgal-bacterial communities. Furthermore, biomonitoring can be used as a community health tool to track changes in the composition of WW by treating it with a population of microorganisms that is well-established (Novoveská et al., 2023).

The cultivation of microalgae with NPs holds great potential for maximizing biomass yield and producing high-value products. To determine the most effective way to use NPs for microalgae cultivation—single addition, composites, and PBRs—more research is required. Studying how NPs affect microalgae development and harvesting will drive these processes toward greater energy efficiency and thus greater environmental, economic, and sustainable friendliness (Vargas-Estrada et al., 2020).

12.13 CONCLUSION

Algal production combined with WW treatment is a potential approach for long-term resource recovery and environmentally acceptable wastewater bioremediation. WW reuse is a cost-effective and long-term solution to water scarcity and quality degradation. WW treatment, in conjunction with nutrient removal and the manufacture of biofuel and other economically essential high-value products from algae, is an appealing and expanding sector aimed at WW valorization and environmental and economic sustainability. When compared to alternative generating feedstocks, algae as a bioenergy feedstock offers incredible benefits. WW may be used as a culture medium, which has the potential to help fight environmental pollution, climate change, and a lack of energy. More studies in this area would present excellent prospects to expand algae as a platform for integrated water treatment, biofuel production, and the manufacture of energy storage technologies.

REFERENCES

Abdelfattah, A., Ali, S.S., Ramadan, H., El-Aswar, E.I., Eltawab, R., Ho, S.H., Elsamahy, T., Li, S., El-Sheekh, M.M., Schagerl, M., Kornaros, M., Sun, J. Microalgae-based wastewater treatment: Mechanisms, challenges, recent advances, and future prospects. *Environ. Sci. Ecotechnol.* 2023, 13. https://doi.org/10.1016/j.ese.2022.100205

Abusweireh, R.S., Rajamohan, N, Sonne, C., Vasseghian, Y. Algae biogas production focusing on operating conditions and conversion mechanisms: A review. *Heliyon. Elsevier Ltd* 2023;9(7):e17757. https://doi.org/10.1016/j.heliyon.2023.e17757

Alazaiza, M.Y.D., Albahnasawi, A., Ahmad, Z., Bashir, M.J.K., Al-Wahaibi, T., Abujazar, M.S.S., Abu Amr, S.S., Nassani, D.E. Potential use of algae for the bioremediation of different types of wastewater and contaminants: Production of bioproducts and biofuel for green circular economy. *J. Environ. Manage.* 2022;324:116415. https://doi.org/10.1016/j.jenvman.2022.116415

Alfonsín, V., Maceiras, R., Gutiérrez, C. Bioethanol production from industrial algae waste. *Waste Manag.* 2019;87:791–797. https://doi.org/https://doi.org/10.1016/j.wasman.2019.03.019

Ali, S.S., El-Sheekh, M., Manni, A., Ruiz, H.A., Elsamahy, T., Sun, J., Schagerl, M. Microalgae-mediated wastewater treatment for biofuels production: A comprehensive review. *Microbiol. Res.* 2022;265:127187.

Amaro, H.M., Salgado, E.M., Nunes, O.C., Pires, J.C.M., Esteves, A.F. Microalgae systems: Environmental agents for wastewater treatment and further potential biomass valorisation. *J. Environ. Manage.* 2023;337. https://doi.org/10.1016/j.jenvman.2023.117678

Ananthi, V., Raja, R., Carvalho, I.S., Brindhadevi, K., Pugazhendhi, A., Arun, A. A realistic scenario on microalgae based biodiesel production: Third generation biofuel. *Fuel* 2021;284:118965.

Arora, K., Kumar, P., Bose, D., Li, X., Kulshrestha, S. Potential applications of algae in biochemical and bioenergy sector. *3 Biotech.* 2021;11:296. https://doi.org/10.1007/s13205-021-02825-5

Baldev, E., Mubarak Ali, D., Pugazhendhi, A., Thajuddin, N. Wastewater as an economical and ecofriendly green medium for microalgal biofuel production. *Fuel* 2021;294:120484. https://doi.org/10.1016/j.fuel.2021.120484

Bhatt, P., Bhandari, G., Turco, R.F., Aminikhoei, Z., Bhatt, K., Simsek, H. Algae in wastewater treatment, mechanism, and application of biomass for production of value-added product. *Environ. Pollut.* 2022;309:119688. https://doi.org/10.1016/j.envpol.2022.119688

Biris-Dorhoi, E.S., Michiu, D., Pop, C.R., Rotar, A.M., Tofana, M., Pop, O.L., Socaci, S.A., Farcas, A.C. Macroalgae: A sustainable source of chemical compounds with biological activities. *Nutrients* 2020;12:1–23. https://doi.org/10.3390/nu12103085

Brar, A., Kumar, M., Pareek, N. Comparative appraisal of biomass production, remediation, and bioenergy generation potential of microalgae in dairy wastewater. *Front. Microbiol.* 2019;10;1–12. https://doi.org/10.3389/fmicb.2019.00678

Chaieb, K., Kouidhi, B., Ayed, L., Bakr Hosawi, S., Abdulbaqi Abdulhakim, J., Hajri, A., Altayb, H.N. Enhanced textile dye removal from wastewater using natural biosorbent and Shewanella algae B29: Application of Box Behnken design and genomic approach. *Bioresour. Technol.* 2023;374:128755. https://doi.org/10.1016/j.biortech.2023.128755

Chavarría-Hernández, J.C., Falcón-Valdés, J.A., Pedrero-Alvarez, A.E., Toledano-Thompson, T., Barahona-Pérez, L.F., Alzate-Gaviria, L., Tapia-Tusell, R., Polanco-Lugo, E., Valdez-Ojeda, R. Obtaining liquid biofuels through hydrodeoxygenation and transesterification of lipid extracts from Scenedesmus sp. Cultivated in synthetic domestic wastewater. *Fuel* 2022;330. https://doi.org/10.1016/j.fuel.2022.125639

Chinnasamy, S., Bhatnagar, A., Hunt, R.W., Das, K.C. Microalgae cultivation in a wastewater dominated by carpet mill effluents for biofuel applications. *Bioresour. Technol.* 2010;101:3097–3105. https://doi.org/10.1016/j.biortech.2009.12.026

Choi, H.-J. Parametric study of brewery wastewater effluent treatment using Chlorella vulgaris microalgae. *Environ. Eng. Res.* 2016;21:401–408. https://doi.org/10.4491/eer.2016.024

Chojnacka, K., Chojnacki, A., Górecka, H. Trace element removal by Spirulina sp. from copper smelter and refinery effluents. *Hydrometallurgy* 2004;73:147–153. https://doi.org/10.1016/j.hydromet.2003.10.003

Crini, G., Lichtfouse, E. Advantages and disadvantages of techniques used for wastewater treatment. *Environ. Chem. Lett.* 2019;17:145–155. https://doi.org/10.1007/s10311-018-0785-9

Darko, C.N.S., Agyei-Tuffour, B., Faloye, D.F., Goosen, N.J., Nyankson, E., Dodoo-Arhin, D. Biomethane production from residual algae biomass (*Ecklonia maxima*): Effects of inoculum acclimatization on yield. *Waste Biomass Valorization* 2022;13:497–509. https://doi.org/10.1007/s12649-021-01497-9

de Mendonça, H.V., Otenio, M.H., Marchão, L., Lomeu, A., de Souza, D.S., Reis, A. Biofuel recovery from microalgae biomass grown in dairy wastewater treated with activated sludge: The next step in sustainable production. *Sci. Total Environ.* 2022;824:153838. https://doi.org/10.1016/j.scitotenv.2022.153838

Díaz, V., Leyva-Díaz, J.C., Almécija, M.C., Poyatos, J.M., del Mar Muñío, M., Martín-Pascual, J. Microalgae bioreactor for nutrient removal and resource recovery from wastewater in the paradigm of circular economy. *Bioresour. Technol.* 2022;363. https://doi.org/10.1016/j.biortech.2022.127968

Dinka, M.O. Safe drinking water: Concepts, benefits, principles and standards. In: Glavan, M. (Ed.), *Water Challenges of an Urbanizing World.* IntechOpen, Rijeka, 2018. https://doi.org/10.5772/intechopen.71352

Do, J.-M., Jo, S.-W., Kim, I.-S., Na, H., Lee, J.H., Kim, H.S., Yoon, H.-S. A feasibility study of wastewater treatment using domestic microalgae and analysis of biomass for potential applications. *Water* 2019;11. https://doi.org/10.3390/w11112294

El-Sheekh, M.M., El-Chaghaby, G.A., Rashad, S. Bioenergy production from algae: Biomass sources and applications. In: Shah, M.P. (Ed.), *Green Approach to Alternative Fuel for a Sustainable Future*. Elsevier, 2023 a, pp. 59–69. https://doi.org/10.1016/B978-0-12-824318-3.00032-1

El-Sheekh, M.M., Galal, H.R., Mousa, A.S.H., Farghl, A.A.M. Coupling wastewater treatment, biomass, lipids, and biodiesel production of some green microalgae. *Environ. Sci. Pollut. Res.* 2023 b;30:35492–35504. https://doi.org/10.1007/s11356-023-25628-y

Ge, S., Champagne, P. Cultivation of the marine macroalgae Chaetomorpha linum in municipal wastewater for nutrient recovery and biomass production. *Environ. Sci. Technol.* 2017;51:3558–3566. https://doi.org/10.1021/acs.est.6b06039

Ghaffar, I., Deepanraj, B., Sundar, L.S., Vo, D.V.N., Saikumar, A., Hussain, A. A review on the sustainable procurement of microalgal biomass from wastewaters for the production of biofuels. *Chemosphere* 2023;311:137094. https://doi.org/10.1016/j.chemosphere.2022.137094

Gondi, R., Kavitha, S., Yukesh Kannah, R., Kumar, G., Rajesh Banu, J. Wastewater based microalgae valorization for biofuel and value-added products recovery. *Sustain. Energy Technol. Assessments* 2022;53:102443. https://doi.org/10.1016/j.seta.2022.102443

Gouveia, L., Graça, S., Sousa, C., Ambrosano, L., Ribeiro, B., Botrel, E.P., Neto, P.C., Ferreira, A.F., Silva, C.M. Microalgae biomass production using wastewater: Treatment and costs. Scale-up considerations. *Algal Res.* 2016;16:167–176. https://doi.org/10.1016/j.algal.2016.03.010

Hamidian, N., Zamani, H. Potential of Chlorella sorokiniana cultivated in dairy wastewater for bioenergy and biodiesel production. *BioEnergy Res.* 2021;15:1–12. https://doi.org/10.1007/s12155-021-10338-5

Hamouda, R.A., Alhumairi, A.M., Saddiq, A.A. Simultaneous bioremediation of petroleum hydrocarbons and production of biofuels by the micro-green alga, cyanobacteria, and its consortium. *Heliyon* 2023;9:e16656. https://doi.org/10.1016/j.heliyon.2023.e16656

Han, M., Zhang, C., Ho, S.H. Immobilized microalgal system: An achievable idea for upgrading current microalgal wastewater treatment. *Environ. Sci. Ecotechnol.* 2023;14:100227. https://doi.org/10.1016/j.ese.2022.100227

Harini, A.B., Sarangi, N.V., Nisha, N., Rajkumar, R. Cultivation of marine diatom, Amphora sp. in municipal wastewater for enhancing lipids toward sustainable biofuel production. *South African J. Bot.* 2023;155:288–297. https://doi.org/10.1016/j.sajb.2023.02.007

Houria, E.B., Miloudia, S., Jamal, M., Aamri, F.E., Meryem, E.Y. Improvement of fatty acid composition by a strain of Chaetoceros sp. isolated from Moroccan seawater of Sidi Moussa for biodiesel production. *Euro-Mediterranean J. Environ. Integr.* 2023;8:507–515. https://doi.org/10.1007/s41207-023-00392-0

Ijaz, A., Anwar, Z., Zafar, M. Screening of wastewater Oedogonium oblangum algae for hyper-oil transformation into biodiesel by response surface methodology. *Kuwait J. Sci.* 2023;50:627–638. https://doi.org/10.1016/j.kjs.2023.03.005

Jabłońska-Trypuć, A., Wołejko, E., Ernazarovna, M.D., Głowacka, A., Sokołowska, G., Wydro, U. Using algae for biofuel production: A review. *Energies* 2023;16. https://doi.org/10.3390/en16041758

Kaloudas, D., Pavlova, N., Penchovsky, R. Phycoremediation of wastewater by microalgae: A review. *Environ. Chem. Lett.* 2021;19:2905–2920. https://doi.org/10.1007/s10311-021-01203-0

Kandasamy, S., Narayanan, M., Raja, R., Devarayan, K., Kavitha, R. The current state of algae in wastewater treatment and energy conversion: A critical review. *Curr. Opin. Environ. Sci. Heal.* 2023;33:100469. https://doi.org/10.1016/j.coesh.2023.100469

Khan, S., Ullah, A., Ayaz, T., Aziz, A., Aman, K., Habib, M., Yilmaz, S., Farid, A., Yasmin, H., Ali, Q. Phycoremediation of industrial wastewater using Vaucheria debaryana and Cladophora glomerata. *Environ. Monit. Assess.* 2023;195:825. https://doi.org/10.1007/s10661-023-11357-9

Kothari, R., Pathak, V.V., Kumar, V., Singh, D.P. Experimental study for growth potential of unicellular alga Chlorella pyrenoidosa on dairy waste water: An integrated approach for treatment and biofuel production. *Bioresour. Technol.* 2012;116:466–470. https://doi.org/10.1016/j.biortech.2012.03.121

Koul, B., Sharma, K., Shah, M.P. Phycoremediation: A sustainable alternative in wastewater treatment (WWT) regime. *Environ. Technol. Innov.* 2022;25:102040. https://doi.org/10.1016/j.eti.2021.102040

Kumar, C.S.P., Sylas, V.P., Mechery, J., Ambily, V., Kabeer, R., Sunila, C.T. Phycoremediation of cashew nut processing wastewater and production of biodiesel using Planktochlorella nurekis and Chlamydomonas reinhardtii. *Algal Res.* 2023;69:102924. https://doi.org/10.1016/j.algal.2022.102924

Lakatos, G. Biological wastewater treatment. In: Holm, M.D., Lund, M.C. (Eds.), *Wastewater and Water Contamination: Sources, Assessment and Remediation.* Nova Science Publishers, 2018, pp. 105–128. https://doi.org/10.1201/b18368-4

Li, P., Zou, T., Sun, X., Liu, X., Wan, H., Zhang, Y. Bioremediation of domestic wastewaters integrated with enhanced biodiesel production using Desmodesmus sp. mutated by atmospheric and room temperature plasma. *J. Environ. Chem. Eng.* 2023;11:110957. https://doi.org/10.1016/j.jece.2023.110957

Liu, J., Pemberton, B., Scales, P.J., Martin, G.J.O. Ammonia tolerance of filamentous algae Oedogonium, Spirogyra, Tribonema and Cladophora, and its implications on wastewater treatment processes. *Algal Res.* 2023;72:103126. https://doi.org/10.1016/j.algal.2023.103126

Mahapatra, S., Kumar, D., Singh, B., Sachan, P.K. Biofuels and their sources of production: A review on cleaner sustainable alternative against conventional fuel, in the framework of the food and energy nexus. *Energy Nexus* 2021;4:100036. https://doi.org/10.1016/j.nexus.2021.100036

Maity, J.P., Hou, C.-P., Majumder, D., Bundschuh, J., Kulp, T.R., Chen, C.-Y., Chuang, L.-T., Nathan Chen, C.-N., Jean, J.-S., Yang, T.-C., Chen, C.-C. The production of biofuel and bioelectricity associated with wastewater treatment by green algae. *Energy* 2014;78:94–103. https://doi.org/10.1016/j.energy.2014.06.023

Mohsenpour, S.F., Hennige, S., Willoughby, N., Adeloye, A., Gutierrez, T. Integrating micro-algae into wastewater treatment: A review. *Sci. Total Environ.* 2021;752:142168. https://doi.org/10.1016/j.scitotenv.2020.142168

Muradov, N., Taha, M., Miranda, A.F., Wrede, D., Kadali, K., Gujar, A., Stevenson, T., Ball, A.S., Mouradov, A. Fungal-assisted algal flocculation: Application in wastewater treatment and biofuel production. *Biotechnol. Biofuels* 2015;8. https://doi.org/10.1186/s13068-015-0210-6

Muthuraman, V.S., Kasianantham, N. Valorization opportunities and adaptability assessment of algae based biofuels for futuristic sustainability: A review. *Process Safety Environ. Protect.* 2023;174:694–721. https://doi.org/10.1016/j.psep.2023.04.043

Nirmalakhandan, N., Selvaratnam, T., Henkanatte-Gedera, S.M., Tchinda, D., Abeysiriwardana-Arachchige, I.S.A., Delanka-Pedige, H.M.K., Munasinghe-Arachchige, S.P., Zhang, Y., Holguin, F.O., Lammers, P.J. Algal wastewater treatment: Photoautotrophic vs. mixotrophic processes. *Algal Res.* 2019;41:101569. https://doi.org/10.1016/J.ALGAL.2019.101569

Novoveská, L., Nielsen, S.L., Eroldoğan, O.T., Haznedaroglu, B.Z., Rinkevich, B., Fazi, S., Robbens, J., Vasquez, M., Einarsson, H. Overview and challenges of large-scale cultivation of photosynthetic microalgae and cyanobacteria. *Mar. Drugs.* 2023;21(8). https://www.mdpi.com/1660-3397/21/8/445

Papadopoulos, K.P., Economou, C.N., Stefanidou, N., Moustaka-Gouni, M., Genitsaris, S., Aggelis, G., Tekerlekopoulou, A.G., Vayenas, D.V. A semi-continuous algal-bacterial wastewater treatment process coupled with bioethanol production. *J. Environ. Manage.* 2023;326:116717. https://doi.org/10.1016/j.jenvman.2022.116717

Phang, S. Mustafa, E. M., Ambati, R. R. Meriam, n., Sulaiman, N., Lim, P., Majid, N.A., Dommange, X. Checklist of microalgae collected from different habitats in Peninsular Malaysiam for selection of algal biofuel feed-stocks. *Malaysian Journal of Science.* 2015. 34(2):141–167.

Pittman, J.K., Dean, A.P., Osundeko, O. The potential of sustainable algal biofuel production using wastewater resources. *Bioresour. Technol.* 2011;102:17–25. https://doi.org/10.1016/j.biortech.2010.06.035

Pouresmaieli, M., Ataei, M., Nouri Qarahasanlou, A., Barabadi, A. Integration of renewable energy and sustainable development with strategic planning in the mining industry. *Results Eng.* 2023;20:101412. Available from: https://www.sciencedirect.com/science/article/pii/S259012302300539X

Prabakaran, P., Karthikeyan, S. Algae biofuel: A futuristic, sustainable, renewable and green fuel for I.C. engines. *Mater. Today Proc.* 2023. https://doi.org/10.1016/j.matpr.2023.03.579

Prandini, J.M., da Silva, M.L.B., Mezzari, M.P., Pirolli, M., Michelon, W., Soares, H.M. Enhancement of nutrient removal from swine wastewater digestate coupled to biogas purification by microalgae Scenedesmus spp. *Bioresour. Technol.* 2016;202:67–75. https://doi.org/10.1016/j.biortech.2015.11.082

Rafa, N., Ahmed, S.F., Badruddin, I.A., Mofijur, M., Kamangar, S. Strategies to produce cost-effective third-generation biofuel from microalgae. *Front. Energy Res.* 2021;9. https://www.frontiersin.org/articles/10.3389/fenrg.2021.749968

Rahman, A., Agrawal, S., Nawaz, T., Pan, S., Selvaratnam, T. A review of algae-based produced water treatment for biomass and biofuel production. *Water* 2020;12:2351. https://doi.org/10.3390/w12092351

Ramachandra, T.V., Durga Madhab, M., Shilpi, S., Joshi, N.V. Algal biofuel from urban wastewater in India: Scope and challenges. *Renew. Sustain. Energy Rev.* 2013;21:767–777. https://doi.org/10.1016/j.rser.2012.12.029

Ramesh, B., Saravanan, A., Senthil Kumar, P., Yaashikaa, P.R., Thamarai, P., Shaji, A., Rangasamy, G. A review on algae biosorption for the removal of hazardous pollutants from wastewater: Limiting factors, prospects and recommendations. *Environ. Pollut.* 2023;327:121572. https://doi.org/10.1016/j.envpol.2023.121572

Rashad, S., El-Chaghaby, G.A. Marine algae in Egypt: Distribution, phytochemical composition and biological uses as bioactive resources (a review). *Egypt. J. Aquat. Biol. Fisheries* 2020;24(5):147–160.

Rashid, R., Shafiq, I., Akhter, P., Iqbal, M.J., Hussain, M. A state-of-the-art review on wastewater treatment techniques: The effectiveness of adsorption method. *Environ. Sci. Pollut. Res.* 2021;28:9050–9066. https://doi.org/10.1007/s11356-021-12395-x

Rather, S., Davoodbasha, M., Bamufleh, H.S., Alhumade, H., Saeed, U., Taimoor, A.A., Sulaimon, A.A., Al-Alaya, W., Shariff, A.M. Utilization of wastewater as a nutritional source for the production of algal biomass. *Int. J. Energy Res.* 2023:1–8. https://doi.org/10.1155/2023/9939252

Reyimu, Z., Özçimen, D. Batch cultivation of marine microalgae Nannochloropsis oculata and Tetraselmis suecica in treated municipal wastewater toward bioethanol production. *J. Clean. Prod.* 2017;150:40–46. https://doi.org/10.1016/j.jclepro.2017.02.189

Sakarika, M., Kornaros, M. *Chlorella vulgaris* as a green biofuel factory: Comparison between biodiesel, biogas and combustible biomass production. *Bioresour. Technol.* 2019;273:237–243. https://doi.org/10.1016/j.biortech.2018.11.017

Salman, J.M., Majrashi, N., Hassan, F.M., Al-Sabri, A., Abdul-Adel Jabar, E., Ameen, F. Cultivation of blue green algae (Arthrospira platensis Gomont, 1892) in wastewater for biodiesel production. *Chemosphere* 2023;335:139107. https://doi.org/10.1016/j.chemosphere.2023.139107

Satheesh, S., Pugazhendi, A., Al-Mur, B.A., Balasubramani, R. Biohydrogen production coupled with wastewater treatment using selected microalgae. *Chemosphere* 2023;334:138932. https://doi.org/10.1016/j.chemosphere.2023.138932

Saxena, R., Rodríguez-Jasso, R.M., Chávez-González, M.L., Aguilar, C.N., González-Fernández, C., Quijano, G., Ruiz, H.A. A plausible scenario for the third generation of biofuels from microalgae. In: Shah, M.P. (Ed.), *Green Approach to Alternative Fuel for a Sustainable Future.* [Internet]. Elsevier, 2023, pp. 107–21. Available from: https://www.sciencedirect.com/science/article/pii/B9780128243183000138

Senousy, H.H., El-Sheekh, M.M., Khairy, H.M., El-Sayed, H.S., Mahmoud, G.A.-E., Hamed, A.A. Biodiesel production from the marine alga nannochloropsis oceanica grown on yeast wastewater and the effect on its biochemical composition and gene expression. *Plants* 2023;12. https://doi.org/10.3390/plants12162898

Sharma, J., Kumar, V., Kumar, S.S., Malyan, S.K., Mathimani, T., Bishnoi, N.R., Pugazhendhi, A. Microalgal consortia for municipal wastewater treatment – Lipid augmentation and fatty acid profiling for biodiesel production. *J. Photochem. Photobiol. B Biol.* 2020;202:111638. https://doi.org/10.1016/j.jphotobiol.2019.111638

Siddiqua, A., Hahladakis, J.N., Al-Attiya, W.A.K.A. An overview of the environmental pollution and health effects associated with waste landfilling and open dumping. *Environ. Sci. Pollut. Res.* 2022;58514–58536. https://doi.org/10.1007/s11356-022-21578-z

Soffian, M.S., Abdul Halim, F.Z., Aziz, F., Rahman, M.A., Mohamed Amin, M.A., Awang Chee, D.N. Carbon-based material derived from biomass waste for wastewater treatment. *Environ. Adv.* 2022;9:100259. https://doi.org/10.1016/j.envadv.2022.100259

Sydney, E.B., da Silva, T.E., Tokarski, A., Novak, A.C., de Carvalho, J.C., Woiciecohwski, A.L., Larroche, C., Soccol, C.R. Screening of microalgae with potential for biodiesel production and nutrient removal from treated domestic sewage. *Appl. Energy* 2011;88:3291–3294. https://doi.org/10.1016/j.apenergy.2010.11.024

Taher, Z.A., Mahfouz, A.Y., Elhenawy, S.B., Rabea, H.A., Sidkey, N.M. Optimization of cultural and nutritional conditions for selection of the most promising microalgae intended for wastewater treatment and biodiesel production by direct transesterification method. *Egypt. J. Chem.* 2022;65:895–907. https://doi.org/10.21608/EJCHEM.2022.132259.5818

Tarlan, E., Dilek, F.B., Yetis, U. Effectiveness of algae in the treatment of a wood-based pulp and paper industry wastewater. *Bioresour. Technol.* 2002;84:1–5. https://doi.org/10.1016/S0960-8524(02)00029-9

Udayan, A., Pandey, A.K., Sirohi, R., Sreekumar, N., Sang, B.-I., Sim, S.J., Pandey, A. Production of microalgae with high lipid content and their potential as sources of nutraceuticals. *Phytochem. Rev.* 2023;22(4):833–860. https://doi.org/10.1007/s11101-021-09784-y

V, G.S., M, D.K., Pugazhendi, A., Bajhaiya, A.K., Gugulothu, P., J, R.B.

Godvin Sharmila V., Dinesh Kumar M., Arulazhagan Pugazhendi, Amit Kumar Bajhaiya, Poornachander Gugulothu, Rajesh Banu J. Biofuel production from macroalgae: Present scenario and future scope. *Bioengineered* 2021;12:9216–9238. https://doi.org/10.1080/21655979.2021.1996019

Vargas-Estrada, L., Torres-Arellano, S., Longoria, A., Arias, D.M., Okoye, P.U., Sebastian, P.J. Role of nanoparticles on microalgal cultivation: A review. *Fuel* 2020;280:118598.

Wang, J., Yin, Y. Fermentative hydrogen production using pretreated microalgal biomass as feedstock. *Microb. Cell Fact.* 2018;17:22. https://doi.org/10.1186/s12934-018-0871-5

Wang, K., Khoo, K.S., Chew, K.W., Selvarajoo, A., Chen, W.-H., Chang, J.-S., Show, P.L. Microalgae: The future supply house of biohydrogen and biogas. *Front. Energy Res.* 2021;9. https://doi.org/10.3389/fenrg.2021.660399

Wollmann, F., Dietze, S., Ackermann, J.U., Bley, T., Walther, T., Steingroewer, J., Krujatz, F. Microalgae wastewater treatment: Biological and technological approaches. *Eng. Life Sci.* 2019;19:860–871. https://doi.org/10.1002/elsc.201900071

Wu, Y.-H., Yang, J., Hu, H.-Y., Yu, Y. Lipid-rich microalgal biomass production and nutrient removal by Haematococcus pluvialis in domestic secondary effluent. *Ecol. Eng.* 2013;60:155–159. https://doi.org/10.1016/j.ecoleng.2013.07.066

Yaashikaa, P.R., Keerthana Devi, M., Senthil Kumar, P., Pandian, E. A review on biodiesel production by algal biomass: Outlook on lifecycle assessment and techno-economic analysis. *Fuel* 2022;324:124774. https://doi.org/10.1016/j.fuel.2022.124774

13 Biomethane Production from the Anaerobic Digestion of Wastewater
An Introduction

Srinivas Kasulla and S. J. Malik

13.1 INTRODUCTION

The metabolites of organic waste, which would be converted through anaerobic digestion, enable the production of biomethane. It is considered to be a relatively effective renewable resource. This could be attributed to the application of this very process. Biomethane production is the process through which this goal is achieved. It is characterized by several defining features and includes notable methane concentration. Due to its inherent characteristics, biomethane is considered a substantial alternative to fossil fuels. This is because it is a source of energy, being renewable by nature. Anaerobic digestion is a good method for the treatment of wastewater, giving the added advantage of energy recovery as biomethane (Cheng et al., 2021). This is a major advantage. This provides a significant advantage against other water treatment techniques. Biomethane is a rich grade of biogas that contains methane content of over 70%. Biomethane is a higher grade of biogas (Piechota et al., 2023). Besides their dual roles in waste management and energy production, these processes are highly related to the goals of several global initiatives aimed at curbing carbon emissions, enhancing better waste management practices, and shifting toward a more circular economy. Both pathways show concurrence at a very large degree. These two pathways show high concordance concerning their alignment.

Traditionally, the primary goal of wastewater treatment was the removal of contaminants, namely, organic compounds, nutrients, and pathogenic contaminants. Such pollutants are dangerous in themselves because their mere presence can be highly threatening to human health. This chapter describes a variety of themes, which include the technologies, processes, and microbial systems applied to the production of biomethane from wastewater sources. Modern WWTPs are increasingly embracing technologies focused on achieving either net energy neutrality or net energy surplus. AD is one of the technologies contributing toward this objective. Besides, the paper discusses several barriers and potential benefits likely to be realized when this technology is rolled out to a wider population. The identified

DOI: 10.1201/9781003459286-13

challenges and benefits are probably going to be faced. These challenges, therefore, with all their inherent advantages, are most likely to be faced in the future (Uddin and Wright, 2023; Nleya et al., 2023).

The objectives of this chapter are to describe the production of biomethane by the anaerobic digestion process, outline the most important microbial steps involved in the process, present various technologies for biogas upgrading, and summarize environmental and economic impacts and advantages. Besides, it tries to demarcate the problems in scaling up biomethane production and propose future potential prospects and innovations that can overcome existing barriers.

13.2 ANAEROBIC DIGESTION: THEORETICAL BACKGROUND

13.2.1 Definition and Overview

Anaerobic digestion (Obileke et al., 2021) is a biological process by which organic matter, in the absence of oxygen, is degraded into a gas mixture, both in terms of volume and moles, dominated by methane (CH_4) and carbon dioxide (CO_2). Anaerobic digestion of wastewater is indeed a very efficient method for the treatment of waste streams with high organic content and for the simultaneous production of biogas, which, after upgrading, goes under the name biomethane. This biomethane can be utilized for the generation of electricity and heat or as fuel for vehicles, catering to energy sustainability goals (Ometto et al., 2019).

The factors affecting the biodegradable organic matter (Meegoda et al., 2018) removal efficiency is:

1. The nature and composition of organic matter to be removed.
2. Appropriate environmental factors.
3. Sludge retention time in the reactor.
4. Intensity of mixing, hence contact between bacterial biomass and organic matter.
5. Organic matter-specific loading concerning the bacterial sludge mass and retention time.

Essentially, factors (1) and (2) are contingent upon the properties of the wastewater, while factors (3) to (5) are associated with the specific type and design of the treatment system. Metabolic conversion of intricate macromolecules of organic material found in wastewater into biogas necessitates the involvement of many microbial groups. Anaerobic digestion of proteins, carbohydrates, and lipids requires distinct sequencing of steps.

13.2.2 The Biochemical Pathway of Anaerobic Digestion

The overall conversion process of organic matter to biogas (Wellinger, Murphy, and Baxter, 2013; Piechota et al., 2023) can be divided into four distinct phases, as the anaerobic digestion process passes through four crucial stages:

1. **Hydrolysis**: Large organic molecules, which include carbohydrates, proteins, and lipids, are split into simpler molecules, such as sugars, amino acids, and fatty acids, by hydrolytic enzymes (Li et al., 2014; Karthikeyan, Selvam, and Wong, 2016).
2. **Acidogenesis**: Acidogenic bacteria ferment the products of hydrolysis into volatile fatty acids (VFAs), alcohols, hydrogen (H_2), and CO_2 (Carneiro et al., 2020; Duan et al., 2018).
3. **Acetogenesis**: Acetogenic bacteria cause the VFAs to be converted into acetic acid, H_2, and CO_2, which provide the methanogens with substrates in the final step (Fu, Conrad, and Blaser, 2018).
4. **Methanogenesis**: With methanogenic archaea, acetic acid and H_2 are converted into CH_4 and water, the latter resulting in the production of biogas (Zeikus, 1977; Lyu et al., 2018).

Various environmental conditions, such as temperature, pH, organic load, and nutrient balancing in the wastewater, will vary at every stage. Thus, these need to be optimized to maximize the yield of biogas production (Jiang et al., 2020).

13.2.3 Wastewater in Anaerobic Digestion

Wastewater, municipal, industrial, or agricultural, is proven to contain organic materials that can be a very suitable substrate for anaerobic digestion. Of the three, wastewater has the highest BOD and COD values, therefore proving to have great potential for the generation of biomethane. Compared to other organic waste streams, wastewater offers several advantages to the process of anaerobic digestion, the foremost being continuity in feedstock supply, amenability to integration with existing treatment infrastructure, and environmental advantages related to water pollution reduction (Jenicek et al., 2010).

13.3 OVERVIEW OF THE ANAEROBIC DEGRADATION PROCESS

The anaerobic biological transformation of organic waste into CH_4 is a complicated process that involves different microbial communities with substrate and product affinities peculiar to the respective groups (Ometto et al., 2019). The complete conversion mechanism can be characterized as having both direct and indirect symbiotic interactions between diverse bacterial groups. The product formed by one type of bacterium serves as the substrate for another; therefore, it has to balance bacterial populations with levels of substrate. Biochemical transformations of organic matter occur tri-phasically (Obileke et al., 2021). The first stage of this process is hydrolysis (Menzel, Neubauer, and Junne, 2020) of high molecular weight materials to conversion to materials that can be used as a source of energy and carbon for cellular activities. The following stage, known as acidogenesis (Carneiro et al., 2020), involves the conversion of these products of the first stage by bacteria into recognizable lower-molecular-weight intermediate products. Short-chain VFAs, produced during acidogenesis, are further degraded by a particular group of bacteria called acetogens, which produce acetate. The final stage, methanogenesis (Lyu et al., 2018), consists of the bacterial breakdown of these intermediate products into even more elementary end products such as CH_4 and CO_2. Several nomenclatures have been proposed for these three steps. Speece and McCarty, in 1962, called the first and the second steps the constant BOD phase and the third step, the reducing BOD phase, since only the CH_4 formation in the third step brings about the reduction of BOD or COD throughout the process.

13.4 TYPES OF WASTEWATERS SUITABLE FOR ANAEROBIC DIGESTION

13.4.1 Municipal Wastewater

Anaerobic digestion feedstock is also available in municipal wastewater (Azevedo et al., 2023). This source typically comprises a mix of domestic sewage and minor industrial effluent, holding a high amount of organic material that can be easily converted into biogas. Anaerobic digestion of sewage sludge (Calbry-Muzyka et al., 2022) is already widely applied in many WWTPs. Biomethane generation in such systems can balance the energy requirements of the plant or even lead to excess energy injected into the grid.

13.4.2 Industrial Wastewater

Industrial wastewater is one of the significant and heterogeneous sources of organic matter amenable to anaerobic digestion (Kong et al., 2019). Many industries, like food and beverage, dairy, pulp and paper, and petrochemical sectors, produce wastewater flows that have high organic strength. Such examples:

- **Food and Beverage Wastewater**: Due to high concentrations of sugars, starches, and oils, waste effluents from food and beverage processing are highly suitable for anaerobic digestion (Ankathi et al., 2024).
- **Dairy Industry**: Dairy processing wastewater contains proteins, fats, and carbohydrates, which provide high CH_4 yields during digestion (Nleya et al., 2023).
- **Pulp and Paper Wastewater**: Though it is challenging because of the presence of lignocellulosic materials, the effluent from the pulp and paper industries is very rich in biodegradable organic matter. This could be a prime source for biogas generation (Bakraoui et al., 2019).

13.4.3 Agricultural Wastewater

Agricultural activities generate a significant amount of wastewater, primarily in the forms of slurry, manure, and runoff from agricultural fields. The major effluent streams are nutrient-rich in both nitrogen (N_2) and phosphorus as well as organic substances. Anaerobic digestion of these wastes can produce considerable amounts of biomethane and, at the same time, minimize odors, pathogens, and nutrient losses to become a source of water pollution (Hernandez and Jimenez, 2018).

13.4.4 Domestic Wastewater

Graywater and blackwater, unlike industrial or agricultural wastewaters, are normally less concentrated. However, they still contain organic matter, fats, and oils that can easily be converted into biogas. Decentralized anaerobic systems have, nowadays, become an increasing pursuit for the treatment of domestic wastewater with energy recovery in rural and remote locations (Porto et al., 2023).

13.5 BIOGAS COMPOSITION AND QUALITY

13.5.1 Components of Biogas

Typically, biogas will contain 50%–75% CH_4, 25%–50% CO_2, and trace amounts of such gases as hydrogen sulfide (H_2S), N_2, ammonia (NH_3), and trace elements (Kasulla and Malik, 2020). CH_4, extracted from CO_2, has gained the name biomethane because of its high energy content.

13.5.2 Quality of Biogas from Wastewater

The composition of the biogas produced from wastewater varies with the feedstock nature and prevailing environmental conditions within the digester. For example, industrial wastewater, whose organic content is higher compared to municipal wastewater, may have higher percentages of CH_4 (Nganyira et al., 2023). Temperature, pH, as well as organic loading rate (OLR), influence the gas composition as well as the cumulative yield obtained (Roubík et al., 2018; Mel et al., 2015).

13.6 TECHNOLOGICAL ASPECTS OF ANAEROBIC DIGESTION

13.6.1 Reactor Technologies

Proper choice of reactor technology (Obileke et al., 2021) is fundamental to the success of anaerobic digestion in treatment processes. Among the most popular reactor configurations used in wastewater treatment are:

Upflow Anaerobic Sludge Blanket (UASB) Reactor: The UASB reactor has been widely applied to the anaerobic treatment of both municipal and industrial wastewater. This reactor's working principle involves the upward flow of wastewater through a sludge blanket in which natural anaerobic microorganisms break down the organic material and produce biogas.

Continuous Stirred-Tank Reactor (CSTR): In a CSTR, the wastewater and biomass are in continuous contact to allow adequately mixed environments with equal distribution of microorganisms. This technology is often used in high-strength wastewater streams (Finalis et al., 2022).

Expanded Granular Sludge Bed (EGSB) Reactor: The EGSB reactor is an advanced type of UASB reactor that is used to enhance the operation of wastewater treatment plants at higher hydraulic loading rates (Cruz-Salomón et al., 2019).

This chapter focuses on critical operating conditions and process control variables pertinent to the efficiency and effectiveness of anaerobic digestion in wastewater treatment. This section addresses those elements deemed critical for the optimization of biomass production, including OLR, hydraulic retention time (HRT), and nutrient balances. Finally, discussion on different reactor configurations is made complete with which a needed understanding of the dynamic associated with wastewater AD is ensured.

13.7 OPERATING CONDITIONS AND PROCESS CONTROL

Its effectiveness depends highly on careful regulation of a variety of operational parameters influencing microbial performance, biogas generation, and the quality of the effluent. The primary variables (Sarker et al., 2019) involved are OLR, HRT, temperature, pH, and finally, the balance of nutrients in the system.

Organic Loading Rate (OLR): OLR refers to the amount of organic matter (expressed as COD or BOD) introduced into the reactor per unit volume and time. The OLR must be balanced to avoid overloading the microbial community, which can lead to system failure, process inhibition, or low CH_4 yields. Higher OLRs often enhance biogas production, but only up to the point where microbial communities can handle the load. The optimal OLR is generally specific to the type of wastewater and the reactor configuration, but typical values range from 1.5 to 6 kg COD/m³/day in conventional systems (Mel et al., 2015; Parajuli et al., 2022).

Hydraulic Retention Time (HRT): HRT refers to the amount of time that the wastewater remains in the digester. It is a key factor in ensuring that sufficient time is available for the microbial community to degrade the organic matter fully. In conventional systems, HRT typically ranges from 10 to 30 days, depending on the reactor type and wastewater composition. Shorter HRT may lead to incomplete digestion, while longer HRT allows for more complete organic matter breakdown and higher CH_4 production (Parajuli et al., 2022; Haryanto, Triyono, and Wicaksono, 2018).

Nutrient Balance: Maintaining an optimal balance of nutrients such as N_2 (Ahmad et al., 2021), phosphorus, and trace metals is crucial for supporting microbial growth and activity. A balanced C: N ratio (typically 100:10:1) is often cited as ideal for microbial health and CH_4 production. Wastewater from certain industries may require nutrient supplementation to achieve this balance. Anaerobic wastewater treatment processes are often used for industrial waste with only a minor amount of nutrients present. This might result in nutrient deficiency unless additional nutrients are supplemented. Often the COD/N ratio and COD/N/P ratio are used to describe the nutrient requirements (Kasulla and Malik, 2020). The optimum N/P ratio can be considered to be 7. The theoretical minimum COD/N ratio is considered to be 350/7. A value around 400/7 is considered reasonable for high-rate anaerobic processes (0.8–1.2 kg COD/kg VSS per day). Other than N_2 and phosphorus, trace metals (Kasulla and Malik, 2020) are also essential for anaerobic processes. The presence of trace metals such as molybdenum, selenium, tungsten, and nickel is probably necessary for the activity of several enzyme systems. When these trace elements are not present in the wastewater, the addition of nickel, cobalt, and molybdenum can increase CH_4 production and allow greater volumes of wastewater to be effectively treated by decreasing the reactor residence time (Xu et al., 2018; Sommer et al., 2015; Werkneh and Gebru, 2023).

Temperature: Another important parameter within the process of anaerobic digestion is temperature. Most of the AD processes operate within the mesophilic (35°C–40°C) or thermophilic (50°C–60°C) temperature ranges. Thermophilic systems usually have high biogas production rates and short retention times; however, they are more susceptible to process disturbances. Being a biological process, much like any other, anaerobic digestion is temperature-dependent. As the temperature increases, the microbial activities increase up to the maximum temperature. Further elevation of temperature beyond its optimal value causes a steeper fall in activity. Anaerobic processes can take place in an extremely wide range of temperatures (4°C–60°C). Once an effective temperature is achieved, small fluctuations may lead to a process upset. Although most of the sludge digesters are operated in the mesophilic range (30°C–40°C), methanogenesis can take place at temperatures as low as 12°C–15°C. Increasing temperature has a very strong influence on biochemical reaction rate in the range of 4°C–25°C.

The optimal temperature for the growth of anaerobic microorganisms is 35°C or more. Although anaerobic digesters have been reported to operate at substantially lower temperatures, like 20°C, anaerobic growth under these temperature conditions is slow, requiring a long start-up time and difficulties in operation. If the reactor's operating temperature is low, then initiation of start-up at about 35°C will be beneficial. At temperatures less than 25°C, the rate of digestion drops steeply, and conventional anaerobic reactors working at ambient temperatures in colder climates may need detention times as high as 12 weeks for sewage sludge treatment. Most industrial digester systems function in the mesophilic range of 30°C–40°C. Probably, an increase in rates of microbial reaction at the high temperatures used in thermophilic processes (50°C–60°C), and hence a reduction in SRT, may be advantageous under some circumstances. However, instability in thermophilic municipal waste treatment can occur. Thermophilic digestion becomes most viable when the wastewater stream to be treated is discharged at an extremely high temperature and the digester is sited in the same location. Anaerobic digestion requires uniformity of temperature throughout all of the contents of the vessel when it operates in the psychrophilic, mesophilic, or thermophilic ranges. A change in temperature of a few degrees can cause a marked upset in microbial metabolism, and rapid changes in reactions in the reactor may be necessary with at least several days for recovery. Adequate mixing of the reactor by paddle agitation, gas sparging, or fluid flow over the heat exchangers should ensure an appropriate temperature distribution in the entire system (Nie et al., 2021; Ibeaja, Nwaogazie, and Udeh, 2023; Jiang et al., 2020; Nekhubvi and Tinarwo, 2022; Wang et al., 2019).

pH Control: Usually, methanogens have optimal pH ranges of 6.6–7.6. Extreme fluctuation in pH will repress CH_4 production or make the system fail. Always, stable pH levels have to be maintained through the addition of either buffering or chemical dosing during wastewater digestion when the latter has fluctuating acidity or alkalinity. The minute changes in pH values are susceptible to methanogenic microorganisms. The optimal pH value for methanogenic bacteria that are sensitive to changes in their surroundings is between 6.6 and 7.6. The smaller genus group of non-methanogenic bacteria is not very sensitive to the environment and can therefore work at any pH value between 5 and 8.5. Such pH values within the reactor are a result of both the CO_2-bicarbonate buffering system and volatile acids-NH_3 produced through the process. Suppression of an accumulation of acids to such a concentration that may inhibit the activity of methanogenic bacteria is required. Regarding this, there should be enough buffering capacity in the reactor so that souring is avoided. Although sodium calcium carbonates and bicarbonates are introduced into digesters so that they can buffer, lime, and more specifically calcium hydroxide, are mainly used for buffering purposes. Toxicity happens only if the concentration of unionized volatile acids falls into the range of 30–60 mg/L (Hussain, Filiatrault, and Guiot, 2017).

Inhibitory Substances: The anaerobic digestion process can experience inhibition, which may be influenced to differing extents by toxic substances located within the system. These compounds can originate from either the influent wastewater or as byproducts generated through metabolic processes in the digester. Among the inhibitory toxic agents are sulfides, which play a significant role in the treatment of waste derived from sources such as molasses fermentation, petroleum refining, and the tanning sector. The accumulation of volatile acids and various microbial metabolites may impair the buffering capacity of reactors. Inhibitions may also arise from the elevated concentrations of NH_3, alkali, alkaline earth metals, and heavy metals present in the system (Li et al., 2013; Ilangovan and Kanchinadham, 2023).

Volatile Acids Inhibition: The instability of anaerobic reactors is generally marked by a sharp and marked increase in the concentrations of VFAs; this is generally an indicator of the collapse of the methanogenic community, which may result from a myriad of environmental shocks, such as shock loadings, nutrient deficiencies, or inhibitory compounds. Among these volatile acids, acetate is considered the least harmful, while propionate is often linked with considerable adverse effects on digester performance (Shyan et al., 2023).

The volatile acids are supposed to inhibit acidic pH values, which can be attributed to the presence of unionized VFAs in large quantities within the system.

The undissociated nature of these acids allows them to penetrate the bacterial cell membrane more effectively than their ionized counterparts, and once assimilated, they induce an intracellular decrease in pH and hence a decrease in microbial metabolic rate. The VFA concentration at any point in time in the reactor should be maintained below 500 mg/L at any point in time and preferably below 200 mg/L for optimum performance (Hussain, Filiatrault, and Guiot, 2017).

Ammonia – Nitrogen Inhibition: While NH_3 serves a critical function as a buffer in anaerobic processes, elevated concentrations of NH_3 can significantly contribute to operational failures. Shock loadings of high NH_3 concentration usually cause immediate production of VFAs such that the buffering capacity of the reactor system may not be able to compensate for the decrease in pH. Further depression of alkalinity and reduction of pH could lead to failure of the reactor system. Inhibition is characterized by a reduction in gas production accompanied by an escalation in the formation of volatile acids (Moestedt et al., 2016; Yellezuome et al., 2022).

Sulfide Inhibition: Under the conditions prevailing in anaerobic digesters, sulfate and other oxidized forms of sulfur are readily reduced to form sulfide. Sulfur-containing amino acids incorporated into proteins may also be metabolized to produce sulfide. These are particularly relevant for consideration of anaerobic treatment options in industrial processes that produce significant volumes of sulfides in their waste effluent. These sulfides formed by the activity of reactor microorganisms may be soluble or insoluble, depending upon their associated cations. When the salts formed are insoluble, they have negligible effects on digestion. Iron addition, for example, can suppress sulfide inhibition by removing S_2- ions from the solution by precipitation. Desulfovibrio, among other sulfate-reducing genera, transforms sulfates into sulfides, and some of the fermentative microorganisms utilize sulfur-containing amino acids to produce sulfides (Rinzema and Lettinga, 1988).

The concentration of sulfide over 200 mg/L in a digester at 35°C, with continuous feeding and mixing, produced severe inhibitory effects, including the complete cessation of gas production. All the heavy metals, except for chromium, form insoluble sulfide salts and thus can be removed from the solution by sulfide present in the system by precipitation. Free sulfide can also be removed as H_2S by intense gas formation (Abdel-Shafy and Mansour, 2014).

Heavy Metals Inhibition: Heavy metals are the most significant prevailing factors that cause inhibition and dysfunction in sewage sludge digesters. Generally, soluble heavy metals pose higher reactor toxicity compared to their insoluble counterparts. Moreover, the anaerobic digestion process has the potential to reduce some of the valence states of the heavy metals. For example, both copper and iron can be reduced from a trivalent state to a divalent state. This reduces the amount of the precipitating agent, such as sulfide, required to remove the metal ion from the solution. The heavy metals can be removed from anaerobic systems by adsorption. Those digesters, such as the CSTR configuration, which are tolerant to the wastes containing high levels of suspended solids, are effective in metal removal, provided sufficient adsorption sites are present. Examples of metals that may cause toxicity within the reactor include copper, chromium, nickel, and lead, and their tolerable concentration levels vary in the wastewater samples to be treated (Kadam et al., 2022; Abdel-Shafy and Mansour, 2014; Zhang et al., 2023).

13.8 BIOMETHANE UPGRADING TECHNOLOGIES

Although raw biogas is an energy-dense mixture, its CH_4 content typically falls in the 50%–75% range, with the rest thereof being CO_2 and trace contaminants. Upgrading biomethane is conducted to clean biogas to biomethane standards by removing CO_2, H_2S, water vapor, and other impurities. The value for upgraded biomethane is realized when its CH_4 fraction goes above 90%; it can then be blended into natural gas distribution networks or used as a vehicle fuel (Malik et al., 2018).

13.8.1 Biogas to Biomethane: An Overview

Technologies applied to convert biogas to biomethane are several, and each bears its distinct advantage relating to the application. Subsequently, four main technologies (Malik et al., 2018) are often used for biogas upgrading: pressure swing adsorption (PSA), membrane separation, water scrubbing, and chemical absorption.

Pressure Swing Adsorption (PSA): PSA is a technology in which biogas is compressed, which allows CO_2 and other impurities to be attracted onto a solid substrate—a fixed bed consisting of activated carbon, zeolites, or else. Under these conditions, the CH_4 component, being not reactive with the adsorbent, passes through the system, where it is recovered afterward as biomethane. Although PSA is one of the most effective technologies, it should be acknowledged that the process involves high energy consumption (Canevesi et al., 2018; Augelletti, Conti, and Annesini, 2017).

Membrane Separation: The primary principles behind the membrane-separated technologies are the size and solubility of the gas molecules that will be used to partition CO_2 from the target gas, CH_4. The compact membrane method, being highly efficient, allows operation at a lower pressure compared to PSA systems (Chen et al., 2015; Peppers et al., 2019).

Water Scrubbing: In this process, biogas is passed through water; in that water, CO_2 and H_2S dissolve more readily than CH_4, leaving CH_4 to be separated. Water scrubbing is relatively simple and inexpensive, but it requires large quantities of water and needs to be regenerated often (Paolini et al., 2021; Werkneh and Gebru, 2023).

Chemical Absorption: Some solvents, including amines, can separate CO_2 from biogas. Although this system represents good efficiency, it may come with a higher cost because of the requirements for chemical regeneration and solvent handling (Abdeen et al., 2016).

13.9 MICROBIAL COMMUNITIES IN ANAEROBIC DIGESTION

Anaerobic digestion is a complex microbiological process involving a diverse population of microorganisms and successive biochemical reactions. Understanding their roles, especially methanogens, will determine the scope to which biomethane yields can be improved. AD is a biochemical process for the decomposition of organic matter without oxygen, producing biogas consisting mainly of CH_4 and CO_2. This entire process is based on an intricate and intensely linked ecosystem of microorganisms. The microbiological system sequentially progresses through specific phases to break down organic substances and produce biomethane. Gaining insight into the functions performed by these microorganisms, namely, methanogens, is essential for optimizing the efficiency of biomethane generation from wastewater. The effectiveness of this anaerobic digestion process is determined by the intricate equilibrium and viability of this microbial community (Nguyen, Nguyen, and Nghiem, 2019; Lim, Ge, and Tong, 2018).

13.9.1 Roles of Different Microorganisms

The process of anaerobic digestion involves several stages performed by distinct microbial groups, each accountable for particular biochemical changes. The primary groups implicated are:

Hydrolytic Microorganisms: Hydrolytic microorganisms are microorganisms that decompose complex organic substances (carbohydrates, fats, and proteins) into simpler compounds including sugars, fatty acids, and amino acids.

Function: Anaerobic digestion begins with hydrolysis, a process in which intricate organic polymers like carbohydrates, fats, and proteins are enzymatically degraded into more easily soluble molecules, including sugars, fatty acids, and amino acids. Hydrolytic microorganisms, particularly bacteria that release extracellular enzymes, catalyze this process (Menzel, Neubauer, and Junne, 2020).

Enzymes Involved: Enzymes, including proteases, lipases, cellulases, and amylases, catalyze the hydrolysis of complex molecules into smaller, more easily metabolizable substrates, thus facilitating subsequent microbial activity.

Importance: Hydrolysis is the step that limits the rate of anaerobic digestion, especially for substrates that have a high concentration of lignocellulosic material, such as agricultural residues. The efficiency of subsequent stages of digestion would be compromised in the absence of effective hydrolysis (Blank and Hoffmann, 2011).

Acidogenic Bacteria: Acidogenic bacteria transform hydrolysis products into VFAs, alcohols, H_2, and CO_2 (Carneiro et al., 2020).

Function: After hydrolysis generates basic organic compounds, acidogenic bacteria assume control by transforming these products into VFAs, alcohols, H_2, and CO_2. This stage, referred to as acidogenesis, is the initial substantial decomposition of organic matter into intermediate compounds (Duan et al., 2018).

Metabolic Byproducts: The major byproducts of this stage consist of short-chain fatty acids, including acetic acid, propionic acid, butyric acid, and ethanol. Furthermore, H_2 and CO_2 are liberated, acting as substrates for the following stages in anaerobic digestion (Karthikeyan, Selvam, and Wong, 2016).

Significance: Acidogenic bacteria play a vital role in supplying the necessary intermediate substrates (VFAs) for the optimal functioning of subsequent microbial groups. Nevertheless, an over-buildup of VFAs can result in the acidification of the digester, disrupting the functioning of methanogens.

Acetogenic Bacteria: The acetogenic bacteria metabolize VFAs to generate acetic acid, H_2, and CO_2, which act as the substrates for methanogens.

Function: Acetogenic bacteria are responsible for the conversion of VFAs and other intermediates from the acidogenic phase into acetic acid, H_2, and CO_3. Acetogenesis is a crucial process in producing the necessary substrates for methanogenesis, the definitive phase of anaerobic digestion (Fu, Conrad, and Blaser, 2018).

Hydrogen Regulation: An essential role is played by acetogens in preserving the balance of H_2 concentration in the digester. An overabundance of H_2 impedes the synthesis of acetic acid and can disrupt the entire digestion process. The preservation of this balance is crucial as methanogens derive energy from H_2, so controlling H_2 levels and promoting effective acetogenesis.

Methanogens: Methanogens are a distinct category of archaea that carry out the last stage of anaerobic digestion by transforming acetic acid, H_2, and CO_2 into CH_4 and water. The methanogens can be classified into two primary categories: acetoclastic methanogens, which degrade acetate, and hydrogenotrophic methanogens, which utilize H_2 and CO_2 to generate CH_4 (Lyu et al., 2018).

Function: Methanogens are specialized unicellular organisms that participate in the last and most crucial phase of anaerobic digestion, known as methanogenesis. This phase involves the direct conversion of acetic acid, H_2, and CO_2 into CH_4 and water by methanogens. CH_4 is the main constituent of biogas and the main byproduct generated during anaerobic digestion (Amin et al., 2021).

Types of Methanogens: There are two major types of methanogens:

Acetoclastic Methanogens: Acetoclastic methanogens are bacterial species by which acetic acid is decomposed into CH_4 and carbon monoxide (CO). They function as the primary methanogens in numerous digesters since around 70% of CH_4 generation is derived from the decomposition of acetate (Hao et al., 2012).

Hydrogenotrophic Methanogens: The hydrogenotrophic methanogen is a specific form of methanogen that generates CH_4 through the utilization of H_2 and carbon monoxide (CO_2). Their major role is to regulate the chemical equilibrium in the digester by preventing the buildup of H_2, which might otherwise hinder the growth of acetogenic bacteria (Hao et al., 2012).

Importance of Methanogens: Methanogens are very important in stabilizing the process of anaerobic digestion. Methanogenesis is the step where CH_4 is produced, which happens to be the primary component of biogas. Generally, the other populations of microbes grow much faster and are much less sensitive to disruptions in process conditions such as a sharp shift in temperature, pH, or the composition of the substrate. CH_4 production is very closely dependent on the health and stability of methanogenic communities.

13.10 INFLUENCE OF WASTEWATER COMPOSITION ON MICROBIAL ACTIVITY

The microbiological populations in an anaerobic digester are often affected by the properties of the wastewater (Jiang et al., 2020). More specifically, industrial effluent with a high concentration of fatty acids or certain toxins may hinder the functioning of microorganisms. Therefore, by ensuring a balanced integration of substrate, nutritional level, and environmental factors, one can ensure a strong population and achieve optimal production of biomethane (Zhang et al., 2019).

Methanogens are the basis of the microbes that perform the ultimate methanogenesis steps. Without methanogenesis, the anaerobic digestion process cannot be considered to be fully operational. The success of methanogenesis impacts not only the volume but also the quality of the biogas produced. Unlike other microbial communities contributing to anaerobic digestion, the growth rate of methanogens is slower. They are particularly vulnerable to changes in environmental parameters and process disturbances (Šmigáň et al., 1990).

Sensitivity to Environmental Conditions: Methanogens prefer operating within a limited range of environmental parameters, such as pH and temperature, and substrate composition. Generally, they favor operating over a neutral pH from 6.5 to 7.5 and are thus rather sensitive to changes in acidity. Sudden changes in temperature or substrate would hinder the growth and may even cause system failure (Kiener and Leisinger, 1983).

Slow Growth Rate: Methanogens grow relatively slowly compared to other bacteria within the AD process; hence, their risk for washout within the systems with high hydraulic loading rates is probable. The slow growth further implies that the recovery of methanogenic populations will take some time after a disturbance; therefore, biogas production will also decline with disturbances (Pasalari et al., 2021; Nguyen, Nguyen and Nghiem, 2019).

Role in Stabilizing the AD Process: Methanogens stabilize the whole process of anaerobic digestion. This is by converting intermediate products like acetic acid, H_2, and CO_2 into CH_4, hence preventing their inhibition of other microbial groups. Through this, they ensure proper progression in the digestion process and high-quality biogas production (Obileke et al., 2021).

Because of this, methanogens are the main obstacle in anaerobic digestion, and any factor that inhibits their activity directly impacts the synthesis of biomethane. It is important to know the biology and ecology of methanogens to improve the effectiveness of anaerobic digesters and optimize CH_4 production.

13.11 ENVIRONMENTAL AND ECONOMIC BENEFITS

Wastewater anaerobic digestion for the production of biomethane has several environmental and economic benefits (Lombardi and Francini, 2020), matching and going beyond the goals of true sustainability and the transition to a circular economy.

13.11.1 GHG Emission Reduction

Producing biomethane from wastewater reduces greenhouse gas (GHG) emissions in two main ways:

Direct Emission Reduction: The anaerobic digestion process efficiently sequesters CH_4 that would otherwise be released into the atmosphere due to untreated waste or poorly controlled landfill environments.

Offsetting Fossil Fuels: Using biomethane as a renewable energy source displaces fossil fuels, further reducing carbon emissions (Roubík et al., 2020).

13.11.2 Circular Economy and Resource Recovery

The anaerobic digestion of wastewater fits well with the basic principles of a circular economy, as it converts waste materials into useful resources. In addition to the production of renewable energy (Bose et al., 2022), digestion systems also produce nutrient-dense digestate, which may be used as a soil enhancer, thus recovering critical nutrients such as N_2 and phosphorus.

13.11.3 Economic Benefits: Energy Savings and Reduced Sludge Handling Costs

Anaerobic digesters help WWTPs have large energy savings because of biogas generation for onsite heat and power (Lombardi and Francini, 2020). Sometimes, WWTPs can be energy-positive—that is, they produce more energy than they consume for anaerobic digestion. In addition, anaerobic digestion reduces the volume of wastewater sludge resulting from the treatment process, which decreases the cost of sludge handling and disposal.

13.12 CHALLENGES IN BIOMETHANE PRODUCTION FROM WASTEWATERS

Although the benefits are evident, technical, economic, and regulatory barriers limit large-scale production of biomethane from wastewater.

13.12.1 Technical Challenges: Process Inhibition and Variability in Wastewater Composition

Process inhibition is one of the major problems in the anaerobic digestion process when wastewater contains toxic substances (Cheng et al., 2021), such as heavy metals, or has high NH_3 concentrations. Variation in wastewater composition also causes uncertain biogas yields and technical problems (Uddin and Wright, 2023).

13.12.2 Economic Viability: Initial Investment and Maintenance Costs

While AD offers long-term savings, the high upfront investment in infrastructure (Lombardi and Francini, 2020), current upgrading technologies (Malik et al., 2018), and the costs of maintenance (Malik et al., 2019) might be extremely cost-prohibitive for smaller facilities.

13.12.3 Regulatory Barriers and Policy Gaps

The main challenges against the general use of biomethane would include the absence in most countries of clear regulatory frameworks regarding the production and injection of biomethane into the grid. Policies that promote renewable energy production and offer financial support for initial capital investments will be a step in establishing biomethane infrastructure (Capodaglio, Callegari, and Lopez, 2016).

13.12.4 Future Prospects and Innovations

Prospects for biomethane production from wastewater appear encouraging (Budzianowski, 2016), as various new technologies and policy advancements are being developed to tackle existing obstacles and enhance efficiency.

13.12.5 Advanced Technologies: Electrochemical Enhancement and Co-Digestion

The electrochemical enhancement of the processes involved in anaerobic digestion and co-digestion of different sources of organic wastes may hold promise for using these techniques in biomethane production, thereby potentially improving system efficiency.

13.12.6 Policy Support and Incentives for Biomethane

Policies that bring financial support, tax benefits, and assured feed-in tariffs for biomethane should be put in place for governmental bodies around the world to realize the promising role that biomethane can play under the banner of renewable energy sources (Lowes, Woodman, and Fitch-Roy, 2019).

13.12.7 Role of Industry and Academia in Driving Research

Effective collaboration between industrial enterprises and academic institutions is crucial for advancing research in anaerobic digestion, biomethane enhancement, and system optimization. Through collaborative efforts, innovation can be fostered, and technical barriers that limit the full use of biomethane reduced.

13.12.8 Long-Term Sustainability: Energy Transition and Climate Goals

Biogenic CH_4 plays a crucial role in the worldwide transition to sustainable energy (Ankathi et al., 2024) sources and the reduction of carbon emissions. The value of biomethane is anticipated to increase as nations strive to fulfill their climate commitments by substituting fossil fuels with sustainable energy sources that promote environmental and economic objectives.

13.13 CONCLUSION

The production of biomethane through the anaerobic digestion of wastewater is a supremely sustainable and efficient way of managing waste and ensuring renewable energy. In the course of this chapter, we have critically reviewed the issues involved in anaerobic digestion: biochemical processes, microbial communities, technologies, and challenges influencing biomethane production.

The diverse community of microorganisms that significantly degrades organic substances into CH_4 and CO_2 is part of the process of anaerobic digestion. On the hydrolysis initiated by hydrolytic microorganisms, acidogenic and acetogenic bacteria will continue to metabolize these products into VFAs, H_2, and CO_2, which methanogens will, in turn, utilize. Methanogens, a unique class of archaea, catalyze the final step of the process by converting the intermediate products into CH_4, the main component of biogas. The system's balance and health are determining factors for the biomethane production stability and efficiency.

We investigated operating conditions and process control parameters, including rates of organic loading, durations of hydraulic retention, and nutrient balances. The production of maximum yields of CH_4 and the avoidance of process inhibition are dependent on the optimal conditions for microbial activity. Any fluctuation in the substrate composition or environmental factors can greatly influence the performance of the microbes involved, especially the sensitive methanogens.

For the upgrading of biomethane toward its use as a renewable energy source, numerous technologies are in use, which feature PSA, membrane separation, water scrubbing, and chemical absorption. These provide the purification of raw biogas by its impurities- CO_2 and H_2S. As a result of this purification process, the resulting biomethane meets all the quality requirements for injection into natural gas networks or use as a fuel for transport.

If considered from the environmental perspective, this resource of biomethane generated from wastewater delivers outstanding depositions of GHG and, through its utilization in various sectors, aids significantly toward attaining a circular economy. This not only presents renewable energy production but also recovers nutrient elements from wastewater, which then become very useful in fertilizers. Applying biomethane would help bring strong economic efficiency benefits for its users, particularly regarding energy efficiency and reduced sludge management expenditures at a wastewater treatment facility.

Empirical studies of highly successful biomethane ventures around the world have demonstrated the realizability and scalability of this technology. Yet, there remain many issues presently facing biomethane ventures, are all technical concerns related to varying composition of wastewater, economic barriers linked to high start-up costs, and regulatory shortcomings preventing full use of all biomethane capabilities. Novel techniques such as electrochemical enhancement and co-digestion hold great potential for increasing biomethane production in the next years. Furthermore, the provision of policy support and incentives for biomethane production will be crucial in overcoming existing obstacles. Further research and collaboration between industry and academia will contribute to the long-term viability of those endeavors. The prospective long-term survival of biomethane in the light of global energy transition seems promising, as it aligns with international climate goals and serves as a pathway toward a more sustainable and secure energy infrastructure.

Biomethane production from wastewater anaerobic digestion will thus meet the fast-growing need for renewable energy production while simultaneously being an environmentally and financially sound method of waste management. Conclusion: With advancements in technology as well as supporting policies, biomethane will take on an increasingly pivotal role in global change toward cleaner energy as well as circular resource use.

REFERENCES

Abdeen, F.R.H. et al. (2016) A review of chemical absorption of carbon dioxide for biogas upgrading. *Chinese Journal of Chemical Engineering*, 24(6). https://doi.org/10.1016/j.cjche.2016.05.006

Abdel-Shafy, H.I. and Mansour, M.S.M. (2014) Biogas production as affected by heavy metals in the anaerobic digestion of sludge. *Egyptian Journal of Petroleum*, 23(4). https://doi.org/10.1016/j.ejpe.2014.09.009

Ahmad, F. et al. (2021) Assessing integrated nitrogen and planting density on growth, yield component and financial analysis of maize crops (Zea mays). *Journal of Hygienic Engineering and Design*, 37:125.

Amin, F.R. et al. (2021) Functions of bacteria and archaea participating in the bioconversion of organic waste for methane production. *Science of the Total Environment*, 763. https://doi.org/10.1016/j.scitotenv.2020.143007

Ankathi, S.K. et al. (2024) Sustainability of biogas production from anaerobic digestion of food waste and animal manure. *Applied Microbiology*. https://doi.org/10.3390/applmicrobiol4010029

Augelletti, R., Conti, M., and Annesini, M.C. (2017) Pressure swing adsorption for biogas upgrading. A new process configuration for the separation of biomethane and carbon dioxide. *Journal of Cleaner Production*, 140. https://doi.org/10.1016/j.jclepro.2016.10.013

Azevedo, A. et al. (2023) Opportunities and challenges in the anaerobic co-digestion of municipal sewage sludge and fruit and vegetable wastes: A review. *Energy Nexus*. https://doi.org/10.1016/j.nexus.2023.100202

Bakraoui, M. et al. (2019) Experimental biogas production from recycled pulp and paper wastewater by biofilm technology. *Biotechnology Letters*, 41(11). https://doi.org/10.1007/s10529-019-02735-w

Blank, A. and Hoffmann, E. (2011) Upgrading of a co-digestion plant by implementation of a hydrolysis stage. *Waste Management and Research*, 29(11). https://doi.org/10.1177/0734242X11423954

Bose, A. et al. (2022) Evaluation of a biomethane, food and biofertiliser polygeneration system in a circular economy system. *Renewable and Sustainable Energy Reviews*, 170. https://doi.org/10.1016/j.rser.2022.112960

Budzianowski, W.M. (2016) A review of potential innovations for production, conditioning and utilization of biogas with multiple-criteria assessment. *Renewable and Sustainable Energy Reviews*. https://doi.org/10.1016/j.rser.2015.10.054

Calbry-Muzyka, A. et al. (2022) Biogas composition from agricultural sources and organic fraction of municipal solid waste. *Renewable Energy*, 181. https://doi.org/10.1016/j.renene.2021.09.100

Canevesi, R.L.S. et al. (2018) Pressure swing adsorption for biogas upgrading with carbon molecular sieve. *Industrial and Engineering Chemistry Research*, 57(23). https://doi.org/10.1021/acs.iecr.8b00996

Capodaglio, A.G., Callegari, A., and Lopez, M.V. (2016) European framework for the diffusion of biogas uses: Emerging technologies, acceptance, incentive strategies, and institutional-regulatory support. *Sustainability (Switzerland)*. https://doi.org/10.3390/su8040298

Carneiro, R.B. et al. (2020) Acidogenesis is a key step in the anaerobic biotransformation of organic micropollutants. *Journal of Hazardous Materials*, 389. https://doi.org/10.1016/j.jhazmat.2019.121888

Chen, X.Y. et al. (2015) Membrane gas separation technologies for biogas upgrading. *RSC Advances*. https://doi.org/10.1039/c5ra00666j

Cheng, Q. et al. (2021) Challenges of anaerobic digestion in China. *International Journal of Environmental Science and Technology*. https://doi.org/10.1007/s13762-020-03087-z

Cruz-Salomón, A. et al. (2019) Expanded granular sludge bed bioreactor in wastewater treatment. *Global Journal of Environmental Science and Management*. https://doi.org/10.22034/gjesm.2019.01.10

Duan, X. et al. (2018) Acidogenic bacteria assisted biodegradation of nonylphenol in waste activated sludge during anaerobic fermentation for short-chain fatty acids production. *Bioresource Technology*, 268. https://doi.org/10.1016/j.biortech.2018.08.053

Finalis, E.R. et al. (2022) Development of bio-CSTR design for bio-H_2 from POME as renewable fuel. *Evergreen*, 9(2). https://doi.org/10.5109/4794177

Fu, B., Conrad, R., and Blaser, M. (2018) Potential contribution of acetogenesis to anaerobic degradation in methanogenic rice field soils. *Soil Biology and Biochemistry*, 119. https://doi.org/10.1016/j.soilbio.2017.10.034

Hao, X. et al. (2012) Distinguishing activity decay and cell death from bacterial decay for two types of methanogens. *Water Research*, 46(4). https://doi.org/10.1016/j.watres.2011.12.029

Haryanto, A., Triyono, S., and Wicaksono, N.H. (2018) Effect of hydraulic retention time on biogas production from cow dung in a semi continuous anaerobic digester. *International Journal of Renewable Energy Development*, 7(2). https://doi.org/10.14710/ijred.7.2.93-100

Hernandez, S.C. and Jimenez, L.D. (2018) The potential for biogas production from agriculture wastes in Mexico. In: K. Biernat (ed.), *Biofuels: State of Development*. IntechOpen. https://doi.org/10.5772/intechopen.75457

Hussain, A., Filiatrault, M., and Guiot, S.R. (2017) Acidogenic digestion of food waste in a thermophilic leach bed reactor: Effect of pH and leachate recirculation rate on hydrolysis and volatile fatty acid production. *Bioresource Technology*, 245. https://doi.org/10.1016/j.biortech.2017.08.130

Ibeaja, A.C.E., Nwaogazie, I.L., and Udeh, N.U. (2023) Temperature effect on biogas production from co-digestion of food waste, potash and cow dung. *Journal of Engineering Research and Reports*, 24(6). https://doi.org/10.9734/jerr/2023/v24i6821

Ilangovan, A. and Kanchinadham, S.B.K. (2023) The formation, influence and recovery of inhibitory compounds in anaerobic digestion. In: R.B. Jeyakumar, K. Sankarapandian, and Y.K. Ravi (eds.), *A Closer Look at Anaerobic Digestion*. Nova Science Publishers. https://doi.org/10.52305/SKCZ3668

Jenicek, P. et al. (2010) Advantages of anaerobic digestion of sludge in microaerobic conditions. *Water Science and Technology*, 62(2). https://doi.org/10.2166/wst.2010.305

Jiang, J. et al. (2020) Comparison of microbial communities during anaerobic digestion of kitchen waste: Effect of substrate sources and temperatures. *Bioresource Technology*, 317. https://doi.org/10.1016/j.biortech.2020.124016

Kadam, R. et al. (2022) Occurrence, fate, and implications of heavy metals during anaerobic digestion: A review. *Energies*. https://doi.org/10.3390/en15228618

Karthikeyan, O.P., Selvam, A., and Wong, J.W.C. (2016) Hydrolysis-acidogenesis of food waste in solid-liquid-separating continuous stirred tank reactor (SLS-CSTR) for volatile organic acid production. *Bioresource Technology*, 200. https://doi.org/10.1016/j.biortech.2015.10.017

Kasulla, S. and Malik, S.J. (2020) Enhancement in biogas production from press mud. *International Journal of Engineering Applied Sciences and Technology*, 5(5). https://doi.org/10.33564/ijeast.2020.v05i05.036

Kiener, A. and Leisinger, T. (1983) Oxygen sensitivity of methanogenic bacteria. *Systematic and Applied Microbiology*, 4(3). https://doi.org/10.1016/S0723-2020(83)80017-4

Kong, Z. et al. (2019) Challenges and prospects for the anaerobic treatment of chemical-industrial organic wastewater: A review. *Journal of Cleaner Production*. https://doi.org/10.1016/j.jclepro.2019.05.233

Li, G. et al. (2014) Sewage sludge thermal hydrolysis process. *Advanced Materials Research*, 827: 368–373. Available at: https://doi.org/10.4028/www.scientific.net/AMR.827.368

Li, W. et al. (2013) Inhibitory mechanism of different treated manure biogas slurry on some fungi. In *ICMREE 2013 - Proceedings: 2013 International Conference on Materials for Renewable Energy and Environment*, Chengdu, China. https://doi.org/10.1109/ICMREE.2013.6893831

Lim, J.W., Ge, T., and Tong, Y.W. (2018) Monitoring of microbial communities in anaerobic digestion sludge for biogas optimisation. *Waste Management*, 71. https://doi.org/10.1016/j.wasman.2017.10.007

Lombardi, L. and Francini, G. (2020) Techno-economic and environmental assessment of the main biogas upgrading technologies. *Renewable Energy*, 156. https://doi.org/10.1016/j.renene.2020.04.083

Lowes, R., Woodman, B., and Fitch-Roy, O. (2019) Policy change, power and the development of Great Britain's renewable heat incentive. *Energy Policy*, 131. https://doi.org/10.1016/j.enpol.2019.04.041

Lyu, Z. et al. (2018) Methanogenesis. *Current Biology*. https://doi.org/10.1016/j.cub.2018.05.021

Malik, S.J. et al. (2019) Techno commercial aspects of bio-CNG from 100 TPD press mud plant. In *DSTA 65th Annual Convention*, Pune, Maharashtra.

Malik, S.J., Gunjal, B.B., and Kasulla, S. (2018) Study of upgradation technologies to treat biogas from spentwash to BioCNG. In *DSTA 64th Annual Convention*, Gandhinagar, Gujarat [Preprint].

Meegoda, J.N. et al. (2018) A review of the processes, parameters, and optimization of anaerobic digestion. *International Journal of Environmental Research and Public Health*. https://doi.org/10.3390/ijerph15102224

Mel, M. et al. (2015) Effect of organic loading rate (OLR) of slurry on biogas production quality. *Advanced Materials Research*, 1115. https://doi.org/10.4028/www.scientific.net/amr.1115.325

Menzel, T., Neubauer, P., and Junne, S. (2020) Role of microbial hydrolysis in anaerobic digestion. *Energies*. https://doi.org/10.3390/en13215555

Moestedt, J. et al. (2016) Ammonia threshold for inhibition of anaerobic digestion of thin stillage and the importance of organic loading rate. *Microbial Biotechnology*, 9(2). https://doi.org/10.1111/1751-7915.12330

Nekhubvi, V. and Tinarwo, D. (2022) Prediction of slurry operating temperature and biogas production rate using ambient temperature forecast as input parameter for underground brick-built biogas digesters. *Cogent Engineering*, 9(1). https://doi.org/10.1080/23311916.2022.2034375

Nganyira, P.D. et al. (2023) Quality of biogas generated through co-digestion of Brewer's spent grain and cattle dung. *Energy Reports*, 10. https://doi.org/10.1016/j.egyr.2023.09.012

Nguyen, L.N., Nguyen, A.Q., and Nghiem, L.D. (2019) Microbial community in anaerobic digestion system: Progression in microbial ecology. In: A.K. Agarwal (ed.), *Energy, Environment, and Sustainability*. Springer Nature. https://doi.org/10.1007/978-981-13-3259-3_15

Nie, E. et al. (2021) How does temperature regulate anaerobic digestion? *Renewable and Sustainable Energy Reviews*. https://doi.org/10.1016/j.rser.2021.111453

Nleya, Y. et al. (2023) Opportunities and challenges for anaerobic digestion of farm dairy effluent. *ChemBioEng Reviews*. https://doi.org/10.1002/cben.202300021

Obileke, K.C. et al. (2021) Anaerobic digestion: Technology for biogas production as a source of renewable energy: A review. *Energy and Environment*. https://doi.org/10.1177/0958305X20923117

Ometto, F. et al. (2019) Anaerobic digestion: An engineered biological process. In: M. Materazzi and P.U. Foscolo (eds.), *Substitute Natural Gas from Waste: Technical Assessment and Industrial Applications of Biochemical and Thermochemical Processes*. Academic Press. https://doi.org/10.1016/B978-0-12-815554-7.00005-2

Paolini, V. et al. (2021) Water scrubbing for biogas upgrading: Developments and innovations. In: N. Aryal (ed.), *Emerging Technologies and Biological Systems for Biogas Upgrading*. Academic Press. https://doi.org/10.1016/B978-0-12-822808-1.00001-5

Parajuli, A. et al. (2022) Effect of hydraulic retention time and organic-loading rate on two-staged, semi-continuous mesophilic anaerobic digestion of food waste during start-up. *Fermentation*, 8(11). https://doi.org/10.3390/fermentation8110620

Pasalari, H. et al. (2021) Perspectives on microbial community in anaerobic digestion with emphasis on environmental parameters: A systematic review. *Chemosphere*. https://doi.org/10.1016/j.chemosphere.2020.128618

Peppers, J. et al. (2019) Performance analysis of membrane separation for upgrading biogas to biomethane at small scale production sites. *Biomass and Bioenergy*, 128. https://doi.org/10.1016/j.biombioe.2019.105314

Piechota, G. et al. (2023) An introduction to biogas and biomethane. In: A. Yousuf and L. Melville (eds.), *Biogas to Biomethane: Engineering, Production, Sustainability*. Woodhead Publishing. https://doi.org/10.1016/B978-0-443-18479-6.00008-9

Porto, P.M.B.V. et al. (2023) Energy recovery from biogas in domestic waste water treatment plant in the last 5 decades: A bibliometric analysis. *Journal of Scientometric Research*, 12(2). https://doi.org/10.5530/jscires.12.2.031

Rinzema, A. and Lettinga, G. (1988) The effect of sulphide on the anaerobic degradation of propionate. *Environmental Technology Letters*, 9(2). https://doi.org/10.1080/09593338809384544

Roubík, H. et al. (2018) Biogas quality across small-scale biogas plants: A case of central Vietnam. *Energies*, 11(7). https://doi.org/10.3390/en11071794

Roubík, H. et al. (2020) Emission reduction potential of household biogas plants in developing countries: The case of central Vietnam. *Journal of Cleaner Production*, 270. https://doi.org/10.1016/j.jclepro.2020.122257

Sarker, S. et al. (2019) A review of the role of critical parameters in the design and operation of biogas production plants. *Applied Sciences (Switzerland)*. https://doi.org/10.3390/app9091915

Shyan, L.L. et al. (2023) Effort to mitigate volatile fatty acid inhibition by using mixed inoculum and compost for the degradation of food waste and the production of biogas. *Sustainability (Switzerland)*, 15(2). https://doi.org/10.3390/su15021185

Šmigáň, P. et al. (1990) Methanogenic bacteria as a key factor involved in changes of town gas stored in an underground reservoir. *FEMS Microbiology Letters*, 73(3). https://doi.org/10.1016/0378-1097(90)90733-7

Sommer, S.G. et al. (2015) Pig slurry characteristics, nutrient balance and biogas production as affected by separation and acidification. *Journal of Agricultural Science*, 153(1). https://doi.org/10.1017/S0021859614000367

Uddin, M.M. and Wright, M.M. (2023) Anaerobic digestion fundamentals, challenges, and technological advances. *Physical Sciences Reviews*, 8(9). https://doi.org/10.1515/psr-2021-0068

Wang, S. et al. (2019) Influence of temperature on biogas production efficiency and microbial community in a two-phase anaerobic digestion system. *Water (Switzerland)*, 11(1). https://doi.org/10.3390/w11010133

Wellinger, A., Murphy, J., and Baxter, D. (2013) *The Biogas Handbook: Science, Production and Applications*. Woodhead Publishing Limited. https://doi.org/10.1533/9780857097415

Werkneh, A.A. and Gebru, S.B. (2023) Development of ecological sanitation approaches for integrated recovery of biogas, nutrients and clean water from domestic wastewater. *Resources, Environment and Sustainability*. https://doi.org/10.1016/j.resenv.2022.100095

Xu, R. et al. (2018) A critical review on the interaction of substrate nutrient balance and microbial community structure and function in anaerobic co-digestion. *Bioresource Technology*. https://doi.org/10.1016/j.biortech.2017.09.095

Yellezuome, D. et al. (2022) Mitigation of ammonia inhibition in anaerobic digestion of nitrogen-rich substrates for biogas production by ammonia stripping: A review. *Renewable and Sustainable Energy Reviews*. https://doi.org/10.1016/j.rser.2021.112043

Zeikus, J.G. (1977) The biology of methanogenic bacteria. *Bacteriological Reviews*. https://doi.org/10.1128/mmbr.41.2.514-541.1977

Zhang, Q. et al. (2019) High variations of methanogenic microorganisms drive full-scale anaerobic digestion process. *Environment International*, 126. https://doi.org/10.1016/j.envint.2019.03.005

Zhang, X. et al. (2023) Tolerance of anaerobic digestion sludge to heavy metals: COD removal, biogas production, and microbial variation. *Journal of Water Process Engineering*, 55. https://doi.org/10.1016/j.jwpe.2023.104157

14 Biomethane Production from the Anaerobic Digestion of Wastewater

Process, Mechanism, and Cost Analysis

Sameh S. Ali, Rania Al-Tohamy, Majid Al-Zahrani, Tamer Elsamahy, and Yehia A.-G. Mahmoud

14.1 INTRODUCTION

Anaerobic digestion is a natural biological process that involves the decomposition of organic matter in the absence of oxygen (Ali et al., 2022, 2019). It is an ancient process that has been harnessed by humans for thousands of years, primarily for waste treatment and energy recovery. During anaerobic digestion, complex organic compounds are broken down by a consortium of microorganisms into simpler compounds, such as biogas, which primarily consists of methane (CH_4) and carbon dioxide (CO_2) (Yadav et al., 2022). The process of anaerobic digestion occurs in different stages, each facilitated by specific groups of microorganisms. These microorganisms convert complex organic matter, such as wastewater, agricultural residues, and organic waste, into biogas through a series of biochemical reactions (Yadav et al., 2022; Zamri et al., 2021). Anaerobic digestion is a versatile and sustainable technology that not only treats organic waste but also provides an opportunity to recover valuable energy in the form of biomethane (Iglesias et al., 2021).

Biomethane, also known as renewable natural gas, is a methane-rich gas produced through the anaerobic digestion process. It has gained significant attention as a renewable energy source due to its multiple benefits (Pavičić et al., 2022). Biomethane has similar properties to natural gas and can be used directly as a fuel for heating, electricity generation, and transportation. It is a versatile energy carrier that can be easily integrated into existing natural gas infrastructure, making it an attractive option for decarbonizing various sectors (Kabeyi and Olanrewaju, 2022). The production of biomethane offers several advantages over other renewable energy sources. Firstly, it provides a renewable and sustainable alternative to fossil fuels, helping to reduce greenhouse gas emissions and combat climate change (Kabeyi and Olanrewaju, 2022). Secondly, biomethane production can contribute to waste management by treating organic waste and converting it into a valuable

DOI: 10.1201/9781003459286-14

resource. Additionally, biomethane production can provide energy security by diversifying the energy mix and reducing dependence on imported fossil fuels (Kabeyi and Olanrewaju, 2022; Richards and Al Zaili, 2020).

Wastewater, generated from various sources such as domestic, industrial, and agricultural activities, represents a significant and abundant feedstock for anaerobic digestion. Wastewater contains high concentrations of organic matter, including carbohydrates, proteins, and lipids, which can be efficiently converted into biomethane (Bhatia et al., 2020). Utilizing wastewater as a feedstock for anaerobic digestion not only addresses the environmental concerns associated with wastewater treatment but also presents an opportunity to recover energy and resources.

The anaerobic digestion of wastewater offers several benefits. Firstly, it provides a sustainable and cost-effective solution for wastewater treatment, minimizing the release of pollutants into the environment. Secondly, the process produces biogas, which can be upgraded to biomethane and used as a renewable energy source (Huang et al., 2019). This dual benefit of wastewater treatment and biomethane production makes anaerobic digestion an attractive option for sustainable wastewater management.

Hence, this chapter aims to provide a comprehensive exploration of biomethane production from the anaerobic digestion of wastewater. By delving into the fundamental principles of anaerobic digestion, the significance of biomethane production, and the potential of wastewater as a valuable feedstock, this chapter aims to shed light on the multifaceted benefits and applications of this sustainable technology. Furthermore, the chapter aims to discuss the key aspects of biomethane production from wastewater, including process optimization strategies, biogas upgrading techniques, and techno-economic and environmental assessments. Through this comprehensive analysis, the chapter aims to enhance the knowledge and insights necessary to understand the potential of biomethane production from wastewater as a means of achieving sustainable wastewater management and contributing to the global transition toward renewable energy sources.

14.2 ANAEROBIC DIGESTION PROCESS

14.2.1 Basic Principles of Anaerobic Digestion

As shown in Figure 14.1, anaerobic digestion is a complex biological process that occurs in the absence of oxygen and involves a series of interconnected reactions. The process can be divided into four main stages: hydrolysis, acidogenesis, acetogenesis, and methanogenesis (Ali et al., 2019). In the first stage, hydrolysis, complex organic compounds present in the wastewater are broken down into simpler compounds by hydrolytic bacteria (Wainaina et al., 2019). These compounds include carbohydrates, proteins, and lipids, which are converted into smaller molecules such as sugars, amino acids, and fatty acids. Ali et al. (2019) stated that hydrolysis is a crucial step as it makes the organic matter more accessible for microbial degradation. The second stage, acidogenesis, involves the conversion of the hydrolyzed products into volatile fatty acids (VFAs) and alcohols. Acidogens, which are acid-producing bacteria, play a vital role in this step. They ferment the hydrolyzed products, producing

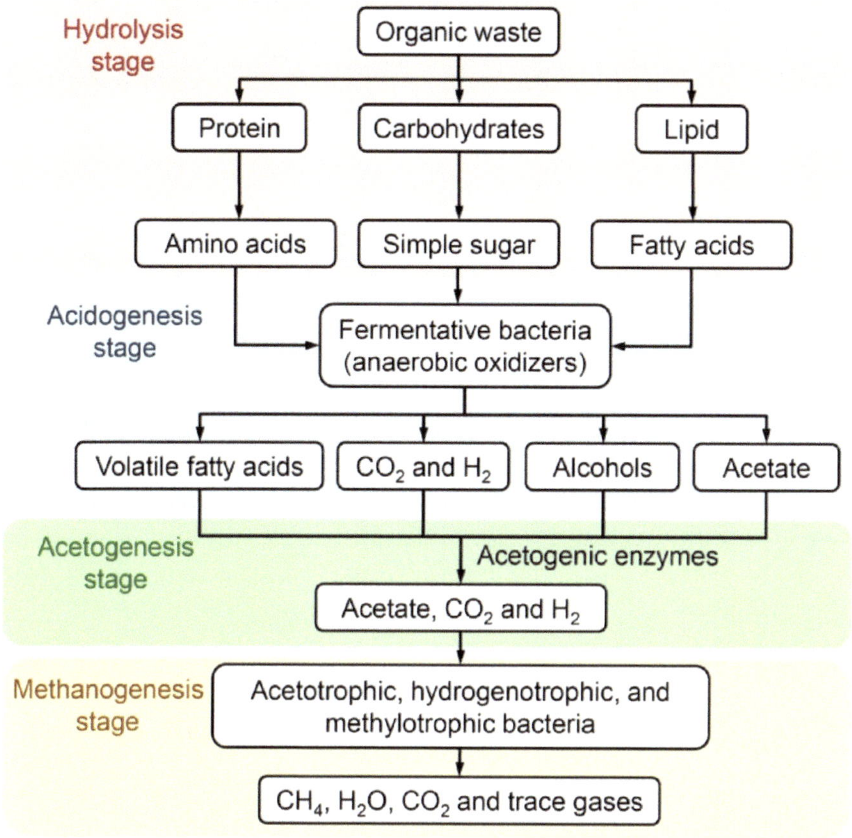

FIGURE 14.1 Biogas and methane production stages during the anaerobic digestion process.

VFAs such as acetic acid, propionic acid, and butyric acid, as well as alcohols like ethanol (Wainaina et al., 2019). The VFAs and alcohols serve as key substrates for subsequent stages of the process.

In the third stage, acetogenesis, the VFAs produced during acidogenesis are further metabolized by acetogens. These acetogenic bacteria convert the VFAs into acetic acid, H_2, and CO_2 through a process known as acetogenesis (Liu et al., 2021). The production of acetic acid is particularly important as it serves as a precursor for the final stage of anaerobic digestion (Figure 14.1). The fourth and final stage, methanogenesis, involves the conversion of acetic acid, hydrogen, and carbon dioxide into CH_4 and CO_2 by methanogenic archaea (Berghuis et al., 2019; Wainaina et al., 2019). Methanogens are specialized microorganisms that thrive in the absence of oxygen and utilize the carbon sources produced in the previous stages to produce methane. Methanogenesis can occur through three pathways (Ahuja et al., 2024). As shown in Figure 14.2, in acetoclastic methanogenesis, methanogens directly convert acetic acid into methane and CO_2, and in hydrogenotrophic methanogenesis, methanogens use

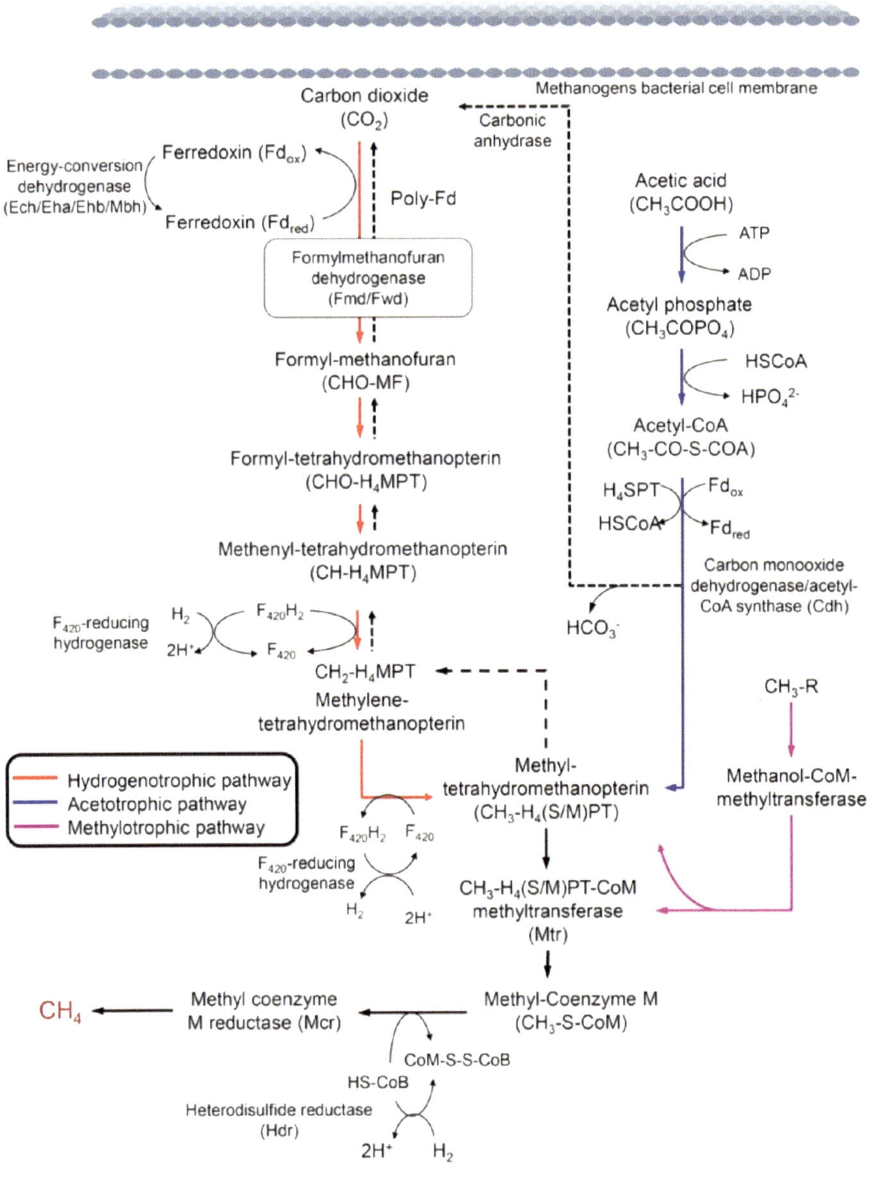

FIGURE 14.2 Different biomethane production pathways in methanogens and the involved enzymes.

hydrogen and CO_2 to produce methane (Ni et al., 2022). Acetoclastic methanogenesis is the most well-known pathway of methanogenesis and involves the conversion of acetate into CH_4 and CO_2. This pathway is primarily carried out by a group of microorganisms called acetoclastic methanogens (Xiao et al., 2020b). These methanogens belong to the archaeal domain and are typically found in anaerobic environments such as wetlands, sediments, and the digestive tracts of animals (Söllinger and Urich, 2019). The process begins with acetate (CH_3COO^-) uptake by acetoclastic methanogens. Inside these microorganisms, acetate is enzymatically converted to CH_4 and carbon dioxide CO_2 (Sahni et al., 2023). The reaction is as follows:

$$CH_3COO^- \rightarrow CH_4 + CO_2$$

Hydrogenotrophic methanogenesis is another significant pathway of methanogenesis. In this pathway, methanogens convert H_2 and CO_2 into CH_4. These methanogens are called hydrogenotrophic methanogens and are also members of the archaeal domain. The process occurs in two steps. First, hydrogenotrophic methanogens use H_2 as an electron donor and reduce CO_2 to form formate ($HCOO^-$) (Berghuis et al., 2019). The enzyme hydrogenase catalyzes this reaction:

$$CO_2 + 4H_2 \rightarrow HCOO^- + 2H_2O$$

In the second step, formate is further reduced to methane (CH_4) by the enzyme formate dehydrogenase:

$$HCOO^- \rightarrow CH_4 + H_2O$$

Methylotrophic methanogenesis is a less common pathway compared to the previous two. It involves the conversion of methyl compounds, such as methanol (CH_3OH) and methylamines, into CH_4 (Conrad, 2020). Methylotrophic methanogens, also belonging to the archaeal domain, carry out this pathway. The process begins with the uptake of methyl compounds by methylotrophic methanogens. Inside these microorganisms, methyl compounds are sequentially transformed into methane CH_4 through several enzymatic reactions involving intermediate compounds (Ali et al., 2023a; Conrad, 2020). The exact steps of this pathway can vary depending on the specific methyl compound being utilized. Overall, methanogenesis is a vital process in nature, contributing to methane production in various environments. The three pathways—acetoclastic methanogenesis, hydrogenotrophic methanogenesis, and methylotrophic methanogenesis—demonstrate the diversity of microorganisms and metabolic strategies involved in the production of methane gas.

The anaerobic digestion process is driven by a complex microbial consortium consisting of various bacterial and archaeal species. These microorganisms work in synergy, each performing specific metabolic functions to ensure the complete degradation of organic matter and the production of methane. The microbial communities involved in anaerobic digestion are highly diverse and can vary depending on the characteristics of the wastewater feedstock, operating conditions, and reactor configuration (Ali et al., 2019; Pasalari et al., 2021). The microbial consortia involved in anaerobic digestion are composed of both bacteria and archaea. Bacteria play a crucial role in the initial stages of organic matter degradation, including hydrolysis and acidogenesis. Different groups of bacteria are responsible for breaking down specific organic compounds, such as carbohydrates, proteins, and lipids, into simpler molecules.

On the other hand, methanogenic archaea are responsible for the final stage of anaerobic digestion and methanogenesis. They convert the intermediate products, such as acetic acid and hydrogen, into CH_4 and CO_2 (Li et al., 2019). Methanogens are classified into two main groups: acetoclastic methanogens and hydrogenotrophic methanogens. Acetoclastic methanogens utilize acetic acid as a substrate and directly convert it into CH_4 and CO_2 (Xiao et al., 2020b). Hydrogenotrophic methanogens utilize hydrogen and CO_2 as substrates to produce CH_4 (Berghuis et al., 2019; Ni et al., 2022). The microbial consortia involved in anaerobic digestion are highly diverse and dynamic, with different species adapting and thriving under varying environmental conditions. The composition and abundance of microbial populations can be influenced by factors such as temperature, pH, organic loading rate, and the type of organic matter being digested. Microbial diversity and stability are key factors in ensuring the efficiency and stability of the anaerobic digestion process (Zamri et al., 2021).

Understanding the microbial consortia involved in anaerobic digestion is crucial for optimizing the process and enhancing methane production. By characterizing and manipulating the microbial communities, it becomes possible to improve process stability, increase methane yield, and enhance the degradation of specific organic compounds (Ali et al., 2021). Additionally, Che and Men (2019) reported that advances in molecular biology techniques, such as DNA sequencing and metagenomics, have provided valuable insights into the diversity and functional potential of the microbial consortia, enabling researchers to better understand and harness their capabilities.

In summary, the anaerobic digestion process relies on a complex microbial consortium consisting of bacteria and archaea. These microorganisms work synergistically, each performing specific metabolic functions, to degrade organic matter and produce methane. Understanding the composition and dynamics of these microbial consortia is essential for optimizing anaerobic digestion systems and maximizing biomethane production.

14.2.2 Reactor Configurations for Biomethane Production

Anaerobic digestion can be carried out in various reactor configurations, each offering unique advantages and considerations for biomethane production (Banerjee et al., 2021). The choice of reactor configuration depends on factors such as the characteristics of the wastewater feedstock, desired process efficiency, and operational requirements. Some commonly used reactor configurations for biomethane production include batch reactors, continuous stirred-tank reactors (CSTRs), up-flow anaerobic sludge blanket (UASB) reactors, and anaerobic sequencing batch reactors (ASBRs) (Banerjee et al., 2021). Figure 14.3 shows different bioreactor types used for biomethane production from different wastewaters.

Batch reactors are simple and cost-effective systems that operate in a discontinuous mode. Wastewater is placed in a closed vessel, where anaerobic digestion occurs over a specific period (Zhao et al., 2019). Once the digestion process is complete, the reactor is emptied, and a new batch is introduced. While batch reactors are relatively easy to operate, they have limitations in terms of scalability and continuous operation (Ruiz et al., 2020). However, CSTRs are widely used for anaerobic digestion due to their ability to maintain a constant volume and operate in a steady state (Wang et al., 2020). In CSTRs, wastewater is continuously fed into the reactor, and the digested material, or effluent, is continuously withdrawn. The reactor contents

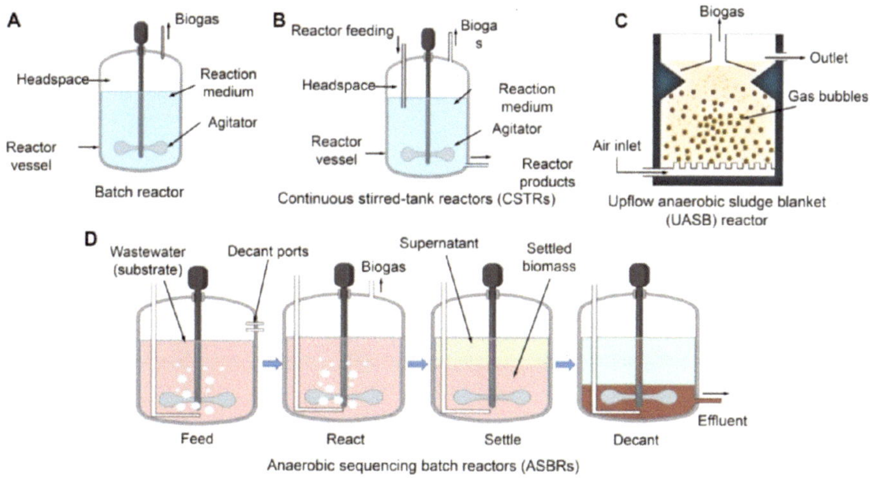

FIGURE 14.3 Different bioreactor types used for biomethane production from different wastewater.

are wellmixed, ensuring uniform conditions for microbial activity. CSTRs offer better process stability and can handle a wide range of feedstock, making them suitable for large-scale biomethane production (Mohiddin et al., 2021).

Moreover, UASB reactors are another commonly employed configuration for anaerobic digestion. In UASB reactors, wastewater is introduced from the bottom and flows upwards through a bed of granular sludge (Mainardis et al., 2020). The sludge bed acts as both a support medium for microbial growth and a means for solids retention. UASB reactors are known for their high organic loading capacity and efficient biogas production. They are particularly suitable for wastewater with high organic content and suspended solids (Cecconet et al., 2022).

Also, ASBRs operate in batch mode but with intermittent mixing and settling phases. The wastewater is sequentially introduced into the reactor and undergoes a series of stages, including filling, reaction, settling, and decanting (Lauzurique et al., 2023; Zhan et al., 2023). ASBRs offer flexibility in terms of process control and can handle variable organic loading rates. They are suitable for small to medium-scale applications and can be used for the digestion of complex wastewater compositions (Pereira et al., 2022).

Recent research has focused on advancements in reactor configurations due to the various disadvantages of different applied bioreactors as shown in Table 14.1, such as hybrid systems combining different reactor types, novel materials for biofilm development, and advanced process control strategies (Gunes, 2021). These developments aim to improve reactor performance, increase biogas yields, and enhance the overall efficiency of biomethane production. Thus, selecting an appropriate reactor configuration is crucial for successful biomethane production. Each configuration has its advantages and limitations, which must be carefully considered based on the specific requirements of the application. Further research and development efforts should focus on addressing the challenges associated with reactor configurations and exploring innovative approaches to enhance biomethane-production efficiency.

TABLE 14.1
Different Reactor Types and Their Advantages and Disadvantages

Reactor Type	Operate Mode	Advantages	Disadvantages
Batch reactor	Operates in batches	• Simple operation and control • Flexibility for different feedstocks • Easy to start up	• Lower biogas production compared to continuous reactors • Longer treatment time • Manual operation and labor-intensive
Continuous stirred-tank reactor (CSTR)	Operates continuously	• Steady-state operation • High-biogas production • Shorter retention time	• Higher complexity in design and operation • Sensitivity to hydraulic and organic loading variations
Up-flow anaerobic sludge blanket (UASB)	Uses a blanket of sludge as the biomass	• High-loading rates • Compact design	• Prone to sludge washout • Sensitivity to fluctuations in influent characteristics
Anaerobic sequencing batch reactor (ASBR)	Operates in batch mode with sequential steps	• High removal efficiency • Versatile for different substrates • Reduced sludge production	• Complex operation and control • Longer cycle times • Higher energy consumption

14.2.3 Factors Affecting Anaerobic Digestion Performance

Several factors influence the performance of anaerobic digestion and the production of biomethane. These factors include substrate characteristics, environmental conditions, reactor design, and operational parameters. Substrate characteristics play a critical role in anaerobic digestion performance (Rocha-Meneses et al., 2022). Nwokolo et al. (2020) stated that the composition and concentration of organic matter, nutrient availability, and the presence of inhibitory substances can significantly impact the process. Easily degradable substrates, such as sugars and starches, are readily converted into biogas, while complex compounds may require longer digestion times. High concentrations of toxic compounds, such as heavy metals or certain chemicals, can inhibit microbial activity and reduce methane production (Ali et al., 2021).

Environmental conditions, such as temperature and pH, are crucial for the activity and stability of the microbial consortia involved in anaerobic digestion. Methanogenic bacteria are sensitive to temperature, and different temperature ranges favor specific microbial groups. Mesophilic anaerobic digestion typically carried out at temperatures around 35°C–40°C, is the most common temperature range for biomethane production. Thermophilic anaerobic digestion, conducted at temperatures of 50°C–60°C, offers faster degradation rates and higher pathogen reduction but requires additional energy for heating (Akbar et al., 2020; Zamri et al., 2021). Moreover, the design and operation of the anaerobic digestion reactor also influence performance. Factors such as reactor configuration, mixing intensity, hydraulic retention time (HRT), and organic loading rate (OLR) affect the microbial activity and the efficiency of the process (Banerjee et al., 2021). Proper mixing ensures the distribution of substrate and microorganisms, while the HRT determines the residence time of the wastewater in the reactor (Zhan et al., 2023). The OLR indicates the amount of organic matter fed into the system per unit volume and time. Balancing these parameters is crucial to maintaining stability and maximizing the methane production of the anaerobic digestion process (Nkuna et al., 2021).

Pasalari et al. (2021) reported that other operational parameters, such as alkalinity, trace element supplementation, and inoculum selection, can also impact anaerobic digestion performance. Adequate alkalinity helps to buffer pH fluctuations and maintain favorable conditions for microbial activity. Trace elements, such as iron, cobalt, and nickel, are essential for the growth and activity of certain microorganisms involved in anaerobic digestion (Ali et al., 2023b). Inoculum selection, including the use of specific microbial consortia or pre-acclimatized biomass, can enhance process performance and stability.

Understanding and optimizing these factors are essential for efficient biomethane production from anaerobic digestion. By carefully controlling substrate characteristics, environmental conditions, reactor design, and operational parameters, it is possible to achieve higher methane yields, improve process stability, and enhance the overall performance of anaerobic digestion systems.

14.3 WASTEWATER CHARACTERISTICS

Wastewater is a valuable resource that can be effectively treated and utilized for biomethane production through anaerobic digestion. The characteristics of wastewater play a crucial role in determining its suitability for the process and influencing the overall efficiency of biomethane production (Kaur et al., 2019). Various types of wastewater are suitable for anaerobic digestion, including municipal wastewater, industrial wastewater, agricultural wastewater, and agro-industrial residues. Municipal wastewater, generated from households, commercial establishments, and urban areas, contains a diverse range of organic matter and nutrients, making it an ideal feedstock for anaerobic digestion (Zamri et al., 2021). Industrial wastewater, generated from industries such as food processing, breweries, and pharmaceuticals, can also be effectively treated through anaerobic digestion, provided that they are compatible with the process and does not contain inhibitory substances (Serrano-Meza et al., 2020). Agricultural wastewater, such as animal manure and slurries, is rich in organic matter and nutrients, making it highly suitable for anaerobic digestion. The high concentration of organic material in agricultural wastewater can contribute to the production of significant amounts of biogas (Giwa et al., 2023). Agro-industrial residues, such as crop residues, food processing residues, and distillery waste, can also be utilized as feedstock for anaerobic digestion, contributing to waste management and renewable energy production (Valenti et al., 2020).

14.3.1 Composition and Characteristics of Wastewaters

The composition and characteristics of wastewater can vary significantly depending on the source and the specific activities generating the wastewater, as shown in Table 14.2. Wastewater typically contains organic matter, nutrients, suspended solids, and other compounds that can influence the anaerobic digestion process (Gutu et al., 2021). Organic matter in wastewater is primarily composed of carbohydrates, proteins, lipids, and other complex organic compounds. The concentration and composition of organic matter determine the methane potential and the degradation kinetics of the wastewater (Xiao et al., 2020a). High concentrations of easily degradable organic matter contribute to higher methane production rates, while complex compounds may require longer digestion times and additional pretreatment (Li et al., 2019).

Nutrients (e.g., organic matter, nitrogen, and phosphorus) are essential for microbial growth and metabolism. The availability and balance of nutrients in wastewater

TABLE 14.2
Composition and Characteristics of Wastewater Suitable for Biogas Production

Wastewater Type	Major Constituents	COD (mg/L)	Suspended Solids (mg/L)	Dissolved Solids (mg/L)	Nitrogen (N)	Phosphorus (P)	pH	C/N Ratio
Municipal wastewater	Water (99.9%) suspended and dissolved organic and inorganic solids	200–400	200–400	800–1,200	20–40	5–15	6.5–8.5	5–15
Industrial wastewater	Varies depending on the industry	200–800	100–500	500–1,500	10–50	1–10	6.0–9.0	10–30
Agricultural wastewater	Water (99.9%) suspended and dissolved organic and inorganic solids	300–600	100–300	400–800	30–60	5–20	6.0–8.0	10–20
Agro-industrial residues	Varies depending on the industry	400–1,000	200–600	600–2,000	20–80	5–30	6.0–8.0	10–40

can affect the microbial activity and the overall process efficiency (Wang et al., 2022a). Imbalances in nutrient concentrations can lead to process instabilities and the accumulation of intermediate metabolites. Suspended solids in wastewater can include organic particles, inorganic solids, and microbial biomass (Wang et al., 2022b). The presence of suspended solids affects the settling characteristics of the wastewater and can influence the design and operation of anaerobic digestion systems (Jover-Smet et al., 2017). High concentrations of suspended solids may require additional pretreatment steps, such as solids separation, to ensure optimal process performance.

Salama et al. (2019) reported that other compounds present in wastewater, such as fats, oils, and greases (FOGs), heavy metals, and toxic chemicals, can impact anaerobic digestion performance. FOGs can form scum layers and reduce the availability of organic matter for microbial degradation. Also, Guo et al. (2019) stated that heavy metals and toxic chemicals, if present in high concentrations, can inhibit microbial activity and hinder the anaerobic digestion process. The presence of these compounds necessitates proper characterization and pretreatment to ensure process stability and prevent operational issues.

14.3.2 Pretreatment of Wastewater to Enhance Anaerobic Digestion

Pretreatment plays a crucial role in enhancing the efficiency and effectiveness of anaerobic digestion, a widely utilized process for converting organic waste into methane gas (Ali et al., 2023a, 2023b). By subjecting wastewater to pretreatment, complex organic compounds can be broken down into simpler forms, facilitating the subsequent anaerobic digestion process. Pretreatment of wastewater offers several key benefits in enhancing anaerobic digestion (Aili Hamzah et al., 2023; Lomwongsopon and Aramrueang, 2022). Pretreatment methods increase the availability of easily digestible organic matter, leading to higher biogas yields. By breaking down complex compounds, such as lignocellulosic materials, fats, and proteins, pretreatment enhances substrate accessibility and promotes microbial activity (Poddar et al., 2021). Certain pretreatment techniques, such as chemical or thermal pretreatment, can accelerate the degradation of organic matter, reducing the required HRT in anaerobic digesters. This enables higher treatment capacities and improved process efficiency (Zamri et al., 2021). However, these techniques have their advantages and disadvantages, as shown in Table 14.3.

Pretreatment can facilitate the recovery of valuable nutrients from wastewater. This not only reduces the environmental impact of nutrient discharge but also provides opportunities for resource recycling and utilization (Perera et al., 2019). Several pretreatment methods are employed to optimize anaerobic digestion performance. Physical methods involve thermal or mechanical processes such as screening, grinding, and solid-liquid separation (Atelge et al., 2020). These techniques remove large solids, debris, and non-biodegradable materials from the wastewater, preventing equipment clogging and damage. Physical pretreatment also increases the surface area for microbial digestion, enhancing biogas production. However, it may have limited effectiveness in breaking down complex organic compounds (Abraham et al., 2020). Thermal pretreatment has also been studied for its effect on the anaerobic

TABLE 14.3
Advantages and Disadvantages of Different Pretreatment Technologies for Biomethane Production

Pretreatment Category	Methods	Advantages	Disadvantages
Physical pretreatment	Screening, grinding, solid-liquid separation	• Removes large solids and debris, preventing clogging and damage to equipment • Increases the surface area for microbial digestion, enhancing biogas production • Improves the overall efficiency of the anaerobic digestion process • Can be easily implemented and requires less maintenance compared to other methods	• Limited effectiveness in breaking down complex organic compounds • Does not remove dissolved or fine particulate matter effectively
Chemical pretreatment	Coagulation, oxidation	• Efficiently breaks down complex organic compounds, enhancing biogas production • Removes heavy metals, pathogens, and certain toxic substances from the wastewater • Can improve the biodegradability and stability of the wastewater • May require less hydraulic retention time in anaerobic digestion, leading to higher treatment capacity	• Requires careful dosage and control of chemicals to avoid adverse impacts on the environment and microbial activity • Generation of chemical sludge, which requires proper handling and disposal
Biological pretreatment	Microaeration, anaerobic, or aerobic treatment	• Breaks down complex organic compounds into simpler forms, enhancing biogas production • Utilizes naturally occurring microorganisms to degrade organic matter, reducing the need for chemicals • Can achieve high removal efficiencies for organic pollutants and nutrients • Supports a sustainable and environmentally friendly approach to wastewater treatment	• Longer treatment duration compared to physical or chemical methods • Requires careful control of environmental conditions (e.g., temperature, pH, nutrient levels) to maintain microbial activity • Potential for process instability if not properly managed

(*Continued*)

TABLE 14.3 (*Continued*)
Advantages and Disadvantages of Different Pretreatment Technologies for Biomethane Production

Pretreatment Category	Methods	Advantages	Disadvantages
Integrated pretreatment systems	Combination of physical, chemical, and biological methods	• Offers the advantages of multiple pretreatment methods, optimizing the overall efficiency of the process • Provides flexibility to tailor the treatment approach based on the specific characteristics of the wastewater • Can achieve higher removal efficiencies for a wide range of pollutants and organic compounds	• Requires careful design and operational control to ensure proper integration and synergy among different pretreatment stages • Increased complexity and capital investment compared to individual pretreatment methods
Advanced pretreatment technologies	Membrane filtration, advanced oxidation processes	• Provides superior removal efficiency for dissolved solids, fine particles, and recalcitrant pollutants • Enhances biogas production by improving the digestibility of complex organic compounds • Enables the recovery of valuable resources from the wastewater	• Requires higher capital and operational costs compared to conventional pretreatment methods • Requires skilled operation and maintenance due to the complexity of the technology

digestion of dewatered sludge, showing increased biomethane yield and promoting the growth of hydrolytic bacteria (Kim et al., 2022). Another pretreatment method that has gained prominence is hydrothermal subcritical water pretreatment. This environmentally friendly method uses water as the main medium and has shown the potential to enhance biogas production through anaerobic digestion (Aili Hamzah et al., 2023). Several examples are listed in Table 14.4 showing the efficiency of different pretreatment methods to enhance biogas and methane production from different wastewaters.

Chemical methods involve pH adjustment, coagulation, oxidation, or the addition of chemicals such as acids or bases. Chemical pretreatment efficiently breaks down complex organic compounds, improving biogas production (Mitraka et al., 2022). It also removes heavy metals, pathogens, and certain toxic substances from the wastewater (Elalami et al., 2019). In this regard, Lomwongsopon and Aramrueang (2022) stated that 2.0% NaOH pretreatment solubilized up to 54% of the chemical oxygen demand and achieved the maximum methane yield of 324 ± 5 mL/gVS. However, a proper concentration and control of chemicals are required to avoid adverse

TABLE 14.4
Biogas and Methane Production Potential from Different Wastewater after Treatment Process

Type of Wastewater	Pretreatment Process	Biogas Production	Remarks	References
Waste-activated sludge	Digestion liquid-based alkaline pretreatment by Reflux proper of digestion liquid into waste-activated sludge (pH 9.5 for 24 hours)	• Refluxing digestion liquid into waste-activated sludge increases methane production • Ammonium in digestion liquid promotes sludge solubilization and methane production	• Methane production from anaerobic digestion (AD) of waste-activated sludge increased significantly • Refluxing digestion liquid into the pretreatment unit improved methane yield • Increased methane production through reutilization of biodegradable organics	He et al. (2021)
Wastewater sludge	• Mechanical cutting was used as the pretreatment process • A sludge disintegration degree of 6.7% was achieved after 8 minutes of mechanical cutting	• Mechanical cutting pretreatment increased biogas production • Biogas production was 3.06 times higher in the pretreatment group	• Mechanical pretreatment increased sludge fermentation and AD • Soluble COD production and biogas yield were significantly higher in the pretreatment group	Wang et al. (2021)
Vinegar residue wastewater	• Alkali pretreatment and microwave pretreatment were used to enhance the degradation • Alkaline hydrogen peroxide and microwave co-pretreatment showed the best results	• Alkali-microwave co-pretreatment increased methane yield by 34.6% • Solid-state AD showed potential for efficient methane production	• Alkali-microwave co-pretreatment increased cumulative methane yield by 34.6% • The lignocellulosic structure of vinegar residue was transformed after co-pretreatment	Jin et al. (2022)
Municipal and industrial wastewaters–Dewatered sludge from small wastewater treatment facilities	• Thermal pretreatment was conducted at varying temperatures (40°C–220°C) and durations (1–3 hours) • The most economical pretreatment condition is 60°C for 3 hours	• Thermal pretreatment increased biomethane yield by 81% compared to control • The most economical pretreatment condition for biogas production is 60°C for 3 hours	• The highest biomethane yield was achieved at 140°C for 3 hours • The most economical pretreatment condition was 60°C for 3 hours	Kim et al. (2022)

(Continued)

TABLE 14.4 (*Continued*)
Biogas and Methane Production Potential from Different Wastewater after Treatment Process

Type of Wastewater	Pretreatment Process	Biogas Production	Remarks	References
Various organic wastes/ wastewater, including sewage sludge	• Bioelectrochemical systems (BESs) can be used as a pretreatment process in AD • BESs help in treating inhibitory compounds, including ammonia, in AD processes	• Bioelectrochemical systems (BESs) can improve biogas production in AD processes • BES operations aid in the upgradation of biogas to valuable chemicals and fuels	• The paper provides a comprehensive summary of bioelectrochemical systems (BES) technologies and operations • It suggests upscaling and future opportunities for BESAD systems	Nagendranatha Reddy et al. (2022)
Food waste and sewage sludge mixture (FW-SS mixture)	• Hydrothermal pretreatment (HTP) is used for co-digestion of food waste and sewage sludge mixture • HTP process modifies the physical properties of the mixture to enhance the solubilization of organic solids	HTP enhances the biogas production rate and the maximum increase in biogas production of 50% is achieved with a HTP temperature of 140°C	HTP enhances the solubilization of organic solids increasing biogas production	Park et al. (2020)
ND	• Biological treatment using microaeration pretreatment was used in the study • Microaeration led to the release of dissolved organic matter and the generation of volatile fatty acids	Microaeration increased biogas production by 16.4% and enhanced methane content in biogas	Microaeration increased biogas production and improved sludge dewaterability which promoted enrichment of hydrolytic and fermentative bacteria in AD reactor	Ruan et al. (2019)

(*Continued*)

TABLE 14.4 (*Continued*)
Biogas and Methane Production Potential from Different Wastewater after Treatment Process

Type of Wastewater	Pretreatment Process	Biogas Production	Remarks	References
Sewage sludge	• The thermal hydrolysis process (THP) is used as the pretreatment for sludge AD • Post-THP treatment (AD-THP) is an improved AD configuration for sludge treatment	• Post-THP treatment (AD-THP) resulted in further improvement in methane production • The maximum specific methane production of the AD-THP system was 344.25 mL CH_4/g VSadd	• Post-THP treatment improved methane production and volatile solid reduction • The dewaterability of THP-treated sludge was significantly better	Yang et al. (2019)
• The wastewater used in the study is a mixture of food waste and sewage sludge • The study does not specify the type of wastewater used	• Thermal-alkali pretreatment with specific conditions (temperature, NaOH concentration, reaction time) • Ammonia stripping performed to remove NH_4^+ after pretreatment	• Optimum pretreatment conditions increased methane yield by 84% • Ammonia stripping increased methane yield by 7%	Ammonia stripping pretreatment conditions improved solubilization efficiency and methane yield	Lee et al. (2019)
Thickened waste-activated sludge (TWAS)	• Free nitrous acid (FNA) was used as a pretreatment chemical • FNA dose ranged from 0.4 to 2.8 mg N/L	• FNA pretreatment increased methane production by 20% • Soluble chemical oxygen demand (SCOD) increased by 140%	• Soluble chemical oxygen demand (SCOD) increased by 140% • Methane production increased by 20% with an FNA dose of 1.4 mg N/L	Elbeshbishy et al. (2017)

(*Continued*)

TABLE 14.4 (*Continued*)
Biogas and Methane Production Potential from Different Wastewater after Treatment Process

Type of Wastewater	Pretreatment Process	Biogas Production	Remarks	References
Wastewater-activated sludge	• Physicochemical pretreatments used: thermal, thermo-alkaline, and ozonation • Best performance obtained with thermo-alkaline pretreatment	• Thermo-alkaline pretreatment resulted in biogas yields analogous to mesophilic conditions • Biogas yields were 0.30–0.36 m^3/kg VS in standard conditions	• Thermo-alkaline pretreatment showed the best performance in improving anaerobic digestion • Thermo-alkaline pretreatment made AD at 20°C economically profitable for a 20,000 PE WWTP	Chiappero et al. (2019)
Supernatant rich in carbon and nitrogen from the organic fraction of municipal solidwaste treatment	• Combination of the low-temperature thermal process (75°C) and ammonia stripping (1.3 L air/L supernatant/min) • Pretreatment contact time of 4 or 8 hours	• Anaerobic postdigestion of supernatant improved biogas production by 18% • Ammonia stripping maintained ammonia concentration for high-methanogenic activity	• 8-hour pretreatment contact time improved methanation by 18% • Ammonia stripping maintained ammonia concentration in the digester below inhibitory values	Pedizzi et al. (2018)
ND	Five typical chemicals (solubilizer, alkaline, peroxide, biological, and chemical surfactants) were used for pretreating waste-activated sludge	• Biogas production was mainly associated with NaOH and SDS treatments • The microbial community in NaOH and SDS treatments was related to biogas production	• CD treatment achieved the highest efficiency in WAS digestion • Microbial community structure shifted in all chemical-pretreated waste-activated sludge	Zhou et al. (2017)

(Continued)

TABLE 14.4 (*Continued*)
Biogas and Methane Production Potential from Different Wastewater after Treatment Process

Type of Wastewater	Pretreatment Process	Biogas Production	Remarks	References
• The wastewater mentioned in the text is from a wastewater treatment plant • The specific type of wastewater is not mentioned in the text	• Alkaline pretreatment: adjusted pH to 10.0 for 2 minutes, solubilization at ambient temperature • Thermal pretreatment: sludge pretreated at 120°C for 0.5 hours	• Anaerobic digestion is chosen as an emerging approach for biogas production • More methane is produced from pretreated sludge with a higher volatile to total solids (VS/TS) ratio	• Methane production from sludge increased with a volatile to total solids (VS/TS) ratio • Average methane yield improved from 2.8 L/day to 4.7 L/day with VS/TS ratio increase	Wang et al. (2016)
Oxidation ditches wastewater from a health-promoting hospital in Thailand	• The alkaline pretreatment process was used to increase the biodegradability of waste-activated sludge • The pH value of 8 was found to be the optimal condition for alkaline pretreatment	• PTAN removed 50.64% of COD and provided higher efficiency in gas production	• Alkaline pretreatment increased the biodegradability of waste-activated sludge • Alkaline pretreatment in thermophilic anaerobic condition (PTAN) showed higher removal efficiency and gas production	Wonglertarak and Wichitsathian (2014)

(*Continued*)

TABLE 14.4 (*Continued*)
Biogas and Methane Production Potential from Different Wastewater after Treatment Process

Type of Wastewater	Pretreatment Process	Biogas Production	Remarks	References
N/A	• Low-temperature thermal pretreatment of sludge was used in the study • The optimal pretreatment condition was a temperature of 70°C for 30 minutes	• Biogas production increased by 7.3%, 15.6%, and 24.4% with thermal pretreatment at 60°C, 70°C, and 80°C, respectively • In the semi-continuous pilot scale HSAD experiment, biogas yield increased by 11% compared to no pretreatment	• Thermal pretreatment increased biogas production and digestion rate • Optimal pretreatment condition: 70°C for 30 minutes	Liao et al. (2016)
Dairy waste-activated sludge (WAS) wastewater	• Two-step sono-alkalization pretreatment using different alkaline agents, pH, and sonic reaction times • Combination of alkaline and sonication pretreatments to enhance sludge solubilization	• Biogas production increased with lower SRTs and pretreated sludge • Methane content in biogas varied from 65 to 70	• Two-step sono-alkalization pretreatment increased COD solubilization, SS reduction, and biogas production • Combining pretreatment with anaerobic digestion led to significant SS and VS reduction and improved biogas production	Rani et al. (2014)

impacts on the environment and microbial activity (Kim, 2018). Additionally, alkali and microwave pretreatments have been used to enhance the degradation of vinegar residue in solid-state anaerobic digestion (Jin et al., 2022). The generation of chemical sludge is another consideration in chemical pretreatment.

Biological methods utilize naturally occurring microorganisms to degrade organic matter (Ali et al., 2023b). This approach can achieve high removal efficiencies for organic pollutants and nutrients. Biological pretreatment supports a sustainable and environmentally friendly approach to wastewater treatment. However, it requires careful control of environmental conditions, such as temperature, pH, and nutrient levels, to maintain microbial activity (Babu Ponnusami et al., 2023). Longer treatment durations and potential process instability are other considerations. Microaeration is one of the biological pretreatment methods that has been investigated. It has been found to accelerate hydrolysis, improve microbial diversity, and enhance enzymatic production (Huiliñir et al., 2023).

Song et al. (2021) stated that integrated systems combine multiple pretreatment methods to optimize the overall efficiency of the process. By integrating physical, chemical, and biological pretreatment stages, these systems can achieve higher removal efficiencies for a wide range of pollutants and organic compounds. However, they require careful design and operational control to ensure proper integration and synergy among different pretreatment stages. Integrated systems are more complex and involve higher capital investment compared to individual pretreatment methods (Khan et al., 2022).

Hence, the characteristics of wastewaters significantly influence their suitability for anaerobic digestion and the efficiency of biomethane production. Understanding the composition, nutrient content, suspended solids, and potential inhibitory substances in the wastewater is crucial for process optimization. Pretreatment of wastewater plays a critical role in enhancing anaerobic digestion performance. It improves biogas production, reduces HRT, and enables nutrient recovery. Physical, chemical, and biological pretreatment methods offer various advantages and disadvantages, and the selection of a specific method depends on the characteristics of the wastewater and treatment goals. Integrated pretreatment systems and advanced technologies provide additional options for optimizing anaerobic digestion. By carefully assessing the wastewater characteristics and implementing appropriate pretreatment strategies, the anaerobic digestion process can be optimized for maximum methane yield and stable operation.

14.4 BIOMETHANE POTENTIAL ASSESSMENT

14.4.1 Methods for Determining Biomethane Potential

The determination of biomethane potential is crucial for assessing the gas production potential of organic materials and optimizing anaerobic digestion processes. Several methods are available for determining biomethane potential, including batch tests, continuous digestion tests, and biochemical methane potential (BMP) assays (Koch et al., 2020; Sillero et al., 2022). Batch tests involve placing a known quantity of substrate in a sealed container, such as a glass bottle or a laboratory-scale reactor, and monitoring the biogas production over a specific period (Nahar et al., 2022). The accumulated gas volume and composition are measured to determine the biomethane potential. On the other hand, batch tests are relatively simple and can provide valuable information about the gas production potential of different substrates. However, they may not accurately represent the conditions of continuous digestion or long-term operation (Manyi-Loh and Lues, 2023).

Continuous digestion tests involve the operation of a CSTR or a similar system under controlled conditions. The substrate is continuously fed into the reactor, and the produced biogas is continuously collected and measured (Szilágyi et al., 2021). Continuous digestion tests provide a more realistic representation of the long-term gas production potential and can help assess the stability and performance of anaerobic digestion systems (Zamri et al., 2021). However, they require longer experimental durations and more complex setups.

BMP assays are commonly used to determine the maximum methane yield of a substrate (Koch et al., 2020). BMP tests involve incubating the substrate in controlled laboratory conditions, typically using inoculum from an anaerobic digester, and measuring the methane production over an extended period (Koch et al., 2020). BMP assays provide a comprehensive assessment of the substrate's biogas production potential and can be used to compare different feedstocks or evaluate the effects of pretreatment. However, BMP tests require specialized equipment, longer experimental durations, and careful monitoring of various parameters (Filer et al., 2019).

14.4.2 Experimental Techniques for Biomethane Potential Assessment

Experimental techniques are commonly used to determine the biomethane potential from wastewater. These methods involve conducting laboratory-scale anaerobic digestion tests to measure the methane production potential of the substrate (Nikpour et al., 2021). Batch tests involve sealing a sample of wastewater in a closed vessel, such as a batch reactor or a bottle, and monitoring the gas production over a specific period. The accumulated methane is measured, and the methane yield is calculated based on the volume of gas produced per unit volume or mass of the substrate (Bekele et al., 2022). CSTR experiments simulate the continuous flow of wastewater in a controlled environment. The reactor is operated under stable conditions, and the biogas production is measured over an extended period (Banerjee et al., 2021). The methane yield is calculated based on the biogas composition and production rate.

Additionally, ASBR experiments involve cyclic operation of the reactor with different stages, including feeding, reaction, settling, and decanting (Lauzurique et al., 2023). The methane production is measured during each stage, and the overall methane yield is determined. Also, inhibition studies are conducted to assess the potential inhibitory effects of certain compounds or substances present in the wastewater on the biomethane production process (Zhan et al., 2023). By varying the concentration of these inhibitory substances, the threshold concentration for inhibition can be determined, enabling the identification of potential process limitations.

14.4.3 Mathematical Modeling for Predicting Biomethane Yield

Mathematical modeling is an alternative approach to assessing the biomethane potential from wastewater. These models use mathematical equations to predict the methane yield based on the composition and characteristics of the wastewater (Morais et al., 2021). Anaerobic digestion models, such as the Anaerobic Digestion Model No. 1 (ADM1), consider various biochemical and physicochemical processes occurring during anaerobic digestion (Frunzo et al., 2019). These models describe the transformations of different organic compounds and predict methane production based on substrate characteristics, temperature, pH, and other factors (Li et al., 2020). Statistical models use regression analysis and other statistical techniques to establish relationships between the wastewater characteristics and the methane yield. These models are developed based on experimental data and can provide valuable insights into the impact of specific parameters on biomethane production (Li et al., 2020).

ANN models are data-driven models that simulate the human neural network to predict complex relationships between input and output variables (Tufaner and Demirci, 2020). ANN models have been applied to predict biomethane yield by training the network using large datasets of wastewater characteristics and corresponding methane production data (Dominguillo-Ramírez et al., 2023). Computational fluid dynamics CFD models simulate fluid flow, heat transfer, and mass transfer within anaerobic digestion reactors. These models can predict the distribution of substrate, gas, and microorganisms within the reactor, providing insights into process efficiency, mixing, and substrate utilization (Caillet et al., 2023).

Moreover, the Monod kinetic model is a widely used empirical model for predicting biomethane yield. It is based on the Monod equation, which describes the growth rate of microorganisms as a function of substrate concentration (Abilmazhinov et al., 2023). In the context of biomethane production, the Monod model relates the specific methane production rate to the concentration of volatile solids (VS) or chemical oxygen demand (COD) in the feedstock. The Monod model equation can be represented as:

$$dM/dt = \frac{\mu^* X^* S}{K_s + S}$$

Where dM/dt is the specific methane production rate, μ is the maximum specific growth rate of methanogens (1/day), X is the concentration of methanogenic biomass (g VS or g COD/L), S is the concentration of substrate (g VS or g COD/L), and K_s is the saturation constant (g VS or g COD/L).

The Modified Gompertz model is a curve-fitting model commonly used to predict cumulative biomethane production over time (Awadalla et al., 2023). It is particularly useful for analyzing batch or semi-continuous anaerobic digestion experiments. The model assumes that biomethane production follows a sigmoidal growth pattern and incorporates three parameters: lag phase duration, maximum methane production rate, and time to reach the maximum methane production rate (Zhao et al., 2019). The Modified Gompertz model equation can be represented as:

$$V = A * \exp\left(-\exp\left(\left(\mu_{\max} * \exp(1)/A\right) * (\lambda - t) + 1\right)\right)$$

Where V is the cumulative volume of methane production at time t, A is the methane volume production potential, μ_{max} is the maximum volume of methane production rate, and λ is the lag phase duration by days.

These mathematical models can assist in process optimization, design, and scale-up of anaerobic digestion systems. They offer a cost-effective and time-efficient approach to assessing the biomethane potential from wastewaters, enabling operators to make informed decisions and maximize biogas production. Therefore, a combination of experimental techniques and mathematical modeling is employed to assess the biomethane potential from wastewater. Experimental methods provide direct measurements of methane production, while mathematical models offer predictive capabilities based on wastewater characteristics. These approaches contribute to the development of efficient and sustainable anaerobic digestion processes for wastewater treatment and resource recovery.

14.5 PROCESS OPTIMIZATION

Biomethane production from wastewater can be optimized through various strategies and process enhancements. By carefully controlling operating parameters, implementing effective process control mechanisms, and exploring co-digestion options, the methane yield can be maximized, leading to improved energy recovery and resource management.

14.5.1 Operating Parameters and Process Control

Optimizing operating parameters and implementing robust process control strategies are crucial for maximizing biomethane production from wastewater (Wu et al., 2021). Anaerobic digestion is highly temperature-dependent, and optimizing the temperature can significantly impact methane production. Mesophilic digestion, typically conducted at temperatures around 35°C–40°C, is common for wastewater treatment. However, thermophilic digestion operated at temperatures around 50°C–60°C, can achieve higher methane yields due to faster microbial activity (Akbar et al., 2020). The selection of the appropriate temperature regime depends on factors such as wastewater characteristics, system design, and energy requirements.

Maintaining the optimal pH range is essential for maintaining microbial activity and ensuring efficient biomethane production. The pH range for anaerobic digestion is typically between 6.5 and 8.5 (Lackner et al., 2020). pH control can be achieved through the addition of alkaline substances, such as lime or sodium hydroxide, or by implementing pH feedback control systems (Lackner et al., 2020). HRT refers to the average duration that wastewater remains in the anaerobic digester. Optimizing the HRT based on the specific characteristics of the wastewater can enhance methane production. A balance must be struck between minimizing HRT for higher turnover rates and ensuring sufficient time for the complete degradation of organic matter (Zhan et al., 2023). Yu et al. (2020) stated that OLR represents the amount of organic matter fed into the digester per unit volume or weight per day. Properly managing the OLR is crucial for maintaining stable microbial activity and avoiding process inhibition. Gradually increasing the OLR while monitoring digester performance allows the microbial community to adapt to higher organic loads (Yu et al., 2020).

Moreover, ensuring an adequate supply of essential nutrients, such as nitrogen and organic carbon content, is important for maintaining microbial activity and optimizing anaerobic digestion (Zamri et al., 2021). Nutrient supplementation, either through wastewater composition or external addition, should be considered to prevent nutrient limitations and optimize methane production.

14.5.2 Maximizing Methane Yield from Wastewaters through Co-Digestion

Co-digestion is a process that involves the simultaneous anaerobic digestion of multiple organic feedstocks, such as wastewater sludge, agricultural waste, food waste, and other organic residues (Karki et al., 2021). Maximizing methane yield from wastewater through co-digestion is an effective strategy to enhance biogas production and energy recovery. The co-digestion process offers several advantages when compared to the digestion of wastewater sludge alone. By combining different organic feedstocks, co-digestion can optimize the nutrient balance, improve process stability, and enhance the biodegradability of the overall substrate (Elalami et al., 2019; Lee et al., 2019). This results in a higher biogas yield and increased methane content in the produced biogas.

The success of co-digestion relies on selecting compatible and complementary feedstocks that can enhance the overall digestion process (Kaur et al., 2019; Park et al., 2020; Zhan et al., 2023). Wastewater sludge can be combined with various organic residues, such as animal manure, crop residues, or food waste, to create a diverse and balanced substrate. Different feedstocks have varying nutrient compositions (Pereira et al., 2022). Co-digesting wastewaters with other organic materials helps achieve a better balance of nutrients, such as carbon, nitrogen, and phosphorus, which can promote optimal microbial activity and biogas production. Adjusting the feedstock mixture can ensure that essential nutrients are available in the appropriate proportions for microbial growth and methane production (Ghaleb et al., 2021).

Co-digestion can improve process stability by reducing the risk of process inhibition or failure. By blending different feedstocks, the co-digestion system becomes more robust and resilient to fluctuations in feedstock characteristics, such as pH, OLR, and nutrient content (Ali et al., 2022; Elalami et al., 2019; Lee et al., 2019). Hence, co-digestion can prevent process upsets ensure consistent and efficient biogas production, and increase the methane content of the produced biogas (Ali et al., 2019; Salama et al., 2019). Certain organic residues, such as agricultural waste or food waste, tend to have higher carbon-to-nitrogen (C/N) ratios than wastewater sludge. Combining these feedstocks with wastewater sludge can help balance the C/N ratio, favoring the growth of methanogenic bacteria and increasing methane production (Ghaleb et al., 2021). Additionally, Kunatsa and Xia (2022) stated that to maximize methane yield, it is crucial to continuously optimize and monitor the co-digestion process. This includes adjusting the feedstock mixture, maintaining optimal process conditions (such as temperature, pH, and HRT), and monitoring key process parameters (such as biogas production, methane content, VFAs, and alkalinity) to ensure efficient and stable operation.

Therefore, optimizing biomethane production from wastewater involves a combination of controlling operating parameters, implementing effective process control mechanisms, and exploring co-digestion options. By fine-tuning the anaerobic digestion process and maximizing the utilization of organic substrates, the methane yield can be enhanced, leading to improved energy recovery and sustainable wastewater treatment. These strategies contribute to the development of efficient and economically viable biogas production systems, promoting environmental sustainability and resource management. Also, maximizing methane yield from wastewater through co-digestion has significant environmental and economic benefits. It enables the efficient utilization of organic waste streams, reduces greenhouse gas emissions, and generates renewable energy in the form of biogas. Additionally, co-digestion can contribute to the circular economy by transforming waste into valuable resources and promoting sustainable waste management practices.

14.6 TECHNO-ECONOMIC AND ENVIRONMENTAL ASSESSMENT

Techno-economic and environmental assessments play a crucial role in evaluating the feasibility, economic viability, and sustainability of biomethane production from wastewater (Mahmod et al., 2021). These assessments consider the cost implications, life cycle impacts, and environmental benefits associated with the production and utilization of biomethane.

14.6.1 Cost Analysis of Biomethane Production from Wastewaters

Cost analysis of biomethane production from wastewater involves assessing the expenses associated with the entire process, from wastewater treatment to the production of biomethane (Silva et al., 2024). Biomethane, also known as renewable natural gas (RNG), is a valuable energy source derived from organic waste materials, including wastewater (Skorek-Osikowska et al., 2020). Analyzing the costs involved in its production helps to evaluate the economic feasibility and competitiveness of biomethane as a renewable energy option (Silva et al., 2024).

The first step in biomethane production is wastewater treatment. The costs associated with wastewater treatment include infrastructure investment, operation and maintenance, chemicals, energy consumption, and labor (Lyu et al., 2020). Advanced treatment methods might involve higher capital and operational costs but can yield higher-quality biogas suitable for biomethane production (Bhojwani et al., 2019). Anaerobic digestion is a crucial process in biomethane production. It involves breaking down organic matter in wastewater to produce biogas. The costs include the capital investment for anaerobic digesters, process monitoring and control systems, mixing equipment, and maintenance. The size and efficiency of the digesters are significant cost factors (Đurđević and Hulenić, 2020).

Biogas produced from anaerobic digestion contains impurities, such as CO_2, H_2S, and moisture. Upgrading the biogas to biomethane requires removing these impurities. Common biogas upgrading methods include pressure swing adsorption (PSA), water scrubbing, and membrane separation (Ahmed et al., 2021). The costs include equipment, chemicals, energy consumption, and operational and maintenance expenses.

After upgrading, biomethane can be injected into the natural gas grid or used as transportation fuel. Injecting biomethane into the grid requires additional infrastructure, including compressors, gas pipelines, and injection stations. The costs depend on the distance to the existing gas grid and the required capacity. If biomethane is used as transportation fuel, additional costs for refueling stations and vehicle modifications need to be considered (Keogh et al., 2022).

In analyzing the cost of biomethane production, it is essential to consider revenue streams and incentives (D'Adamo et al., 2023). The sale of biomethane can generate revenue, either through gas grid injection or direct sales to end-users. In some regions, governments provide incentives, such as feed-in tariffs, tax credits, or renewable energy certificates, which can offset production costs. Regular operation and maintenance costs are significant factors in the overall cost analysis. These include labor, monitoring, repairs, and replacements of equipment (Ren et al., 2021). Effective maintenance practices can improve the efficiency and longevity of the biomethane production facility, reducing long-term costs. The scale of the biomethane production facility affects the overall cost. Larger facilities typically benefit from economies of scale, as they can spread out fixed costs over a larger production volume. However, smaller decentralized systems can be advantageous in terms of reducing transportation costs and utilizing localized waste streams.

While not directly related to financial costs, it is important to consider the environmental costs and benefits of biomethane production (D'Adamo et al., 2023). Biomethane production from wastewater can provide significant environmental benefits by reducing greenhouse gas emissions, mitigating odors from wastewater treatment plants, and reducing dependence on fossil fuels (Marconi and Rosa, 2023).

Therefore, the cost analysis of biomethane production from wastewater involves evaluating expenses related to wastewater treatment, anaerobic digestion, biogas upgrading, biomethane injection, operation and maintenance, revenue streams, and incentives. Assessing these costs is crucial for determining the economic viability of biomethane production and its competitiveness as a renewable energy option. A comprehensive cost analysis provides insights into the economic feasibility of biomethane production from wastewaters, helping stakeholders make informed investment decisions and identify potential areas for cost optimization.

14.6.2 Life Cycle Assessment of Biomethane Production from Wastewaters

Life Cycle Assessment (LCA) is a comprehensive methodology used to evaluate the environmental impacts associated with the entire life cycle of a product or process, including raw material extraction, production, use, and end-of-life disposal. When applied to biomethane production from wastewaters, LCA provides valuable insights into the environmental performance of this renewable energy option (Amato et al., 2023).

The LCA of biomethane production from wastewater, as shown in Figure 14.4 is initiated by the acquisition of the wastewater feedstock. This stage considers the environmental impacts associated with the collection and transportation of wastewater to the treatment facility. It also evaluates the potential environmental benefits of diverting waste materials from landfills or other disposal methods. The next stage

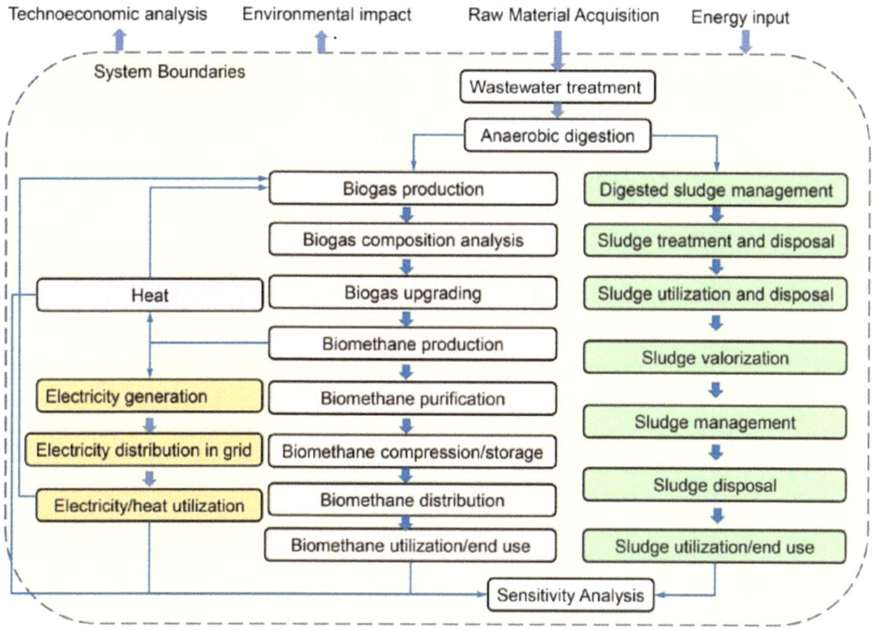

FIGURE 14.4 Life cycle assessment of biomethane production from wastewaters.

involves the treatment of wastewater to remove impurities and facilitate the production of biomethane. Li et al. (2021) reported that LCA assesses the environmental impacts of various treatment processes, such as energy consumption, chemical usage, and emissions to air, water, and soil. Advanced treatment methods, such as membrane bioreactors or anaerobic digestion, can result in different environmental profiles (Robles et al., 2021).

Anaerobic digestion is a key process in biomethane production, where organic matter is converted into biogas. LCA evaluates the environmental impacts of anaerobic digestion, including greenhouse gas emissions (e.g., methane and carbon dioxide), energy consumption, and potential air and water pollution (Ingrao et al., 2019; Valenti et al., 2020). The efficiency and management of the anaerobic digestion process play a significant role in determining environmental performance (Duan et al., 2020). LCA analyzes the environmental impacts associated with biogas upgrading processes, which remove impurities from the biogas to produce biomethane (Ingrao et al., 2019; Omrany et al., 2021; Zhang et al., 2021). This stage considers energy consumption, emissions from the upgrading process, and the environmental benefits of utilizing renewable biomethane as a replacement for fossil fuels (Valli et al., 2017). The environmental impacts of utilizing biomethane as a renewable energy source are evaluated in this stage of the LCA. It includes the transportation and distribution of biomethane, as well as its use as a fuel in different applications such as heat generation, electricity production, or transportation (D'Adamo et al., 2019). Comparisons with conventional fossil fuel alternatives can provide insights into the environmental benefits of biomethane utilization (Richards and Al Zaili, 2020; Valli et al., 2017).

LCA considers the environmental impacts associated with the end-of-life stage of the biomethane production system. This includes the disposal or reuse of waste by-products generated during the wastewater treatment and biogas upgrading processes (Ingrao et al., 2019). Zhang et al. (2021) stated that proper management of waste materials can reduce potential environmental impacts. Additionally, defining the system boundaries is essential in LCA to ensure a comprehensive assessment (Omrany et al., 2021). It involves determining the functional unit (e.g., energy produced or emissions avoided) and establishing the scope of the study, including the inclusion or exclusion of upstream and downstream processes, such as raw material production or waste treatment after use. LCA often includes a sensitivity analysis to assess the influence of different parameters and assumptions on the results (Bamber et al., 2019). This analysis helps identify the key factors that significantly affect the environmental performance of biomethane production and highlights areas for improvement or optimization.

By applying LCA to biomethane production from wastewaters, decision-makers can gain a holistic understanding of the environmental impacts associated with the entire life cycle of this renewable energy option. This information can guide the development and implementation of more sustainable and environmentally friendly biomethane production systems.

14.6.3 Environmental Benefits and Sustainability Considerations

Biomethane production from wastewater offers several environmental benefits and contributes to sustainable resource management:

- Biomethane production reduces greenhouse gas emissions by capturing and utilizing methane, a potent greenhouse gas, which would otherwise be released into the atmosphere. The utilization of biomethane as a renewable energy source displaces the use of fossil fuels, further reducing carbon dioxide emissions (Valli et al., 2017).
- Biomethane production provides a renewable energy source that can be used for electricity generation, heating, or transportation. By replacing fossil fuels, it contributes to the transition to a low-carbon energy system and reduces reliance on finite resources (Amato et al., 2023; D'Adamo et al., 2019).
- Anaerobic digestion of wastewater reduces the organic load and pollutants in the wastewater, improving water quality. The process also helps in the management of organic waste, diverting it from landfills and reducing methane emissions from decomposing organic matter (Logan and Visvanathan, 2019).
- Biomethane production from wastewaters aligns with the principles of the circular economy by utilizing organic wastes as feedstock, closing the waste loop, and promoting resource recovery. The digestate produced as a by-product can be used as a nutrient-rich fertilizer, completing the cycle of nutrient recycling (Ellacuriaga et al., 2021).
- Biomethane can generate economic benefits at the local level. It creates job opportunities, supports local industries, and reduces dependence on imported energy sources (D'Adamo et al., 2019). The revenue generated

from biomethane production can contribute to the local economy and foster regional development.

- Biomethane production enhances energy security by diversifying the energy mix and reducing reliance on fossil fuel imports. It provides a decentralized energy source that can be produced and utilized locally, reducing vulnerability to geopolitical risks and price fluctuations (Richards and Al Zaili, 2020).

Zhu et al. (2019) stated that sustainability considerations in biomethane production include ensuring the sustainable sourcing of feedstock, minimizing environmental impacts throughout the life cycle, promoting social acceptance, and complying with regulatory frameworks and standards.

By assessing the techno-economic aspects and evaluating the environmental benefits and sustainability considerations, stakeholders can make informed decisions about biomethane production from wastewater. This holistic approach ensures that biomethane projects are economically viable and environmentally sustainable, and contribute to a more sustainable and resilient energy and waste management system.

14.7 FUTURE PERSPECTIVES AND EMERGING TRENDS

Biomethane production from wastewater has gained significant attention as a sustainable solution for renewable energy generation and waste management. As the technology continues to evolve, several future perspectives and emerging trends are shaping the field of biomethane production from wastewaters:

1. Ongoing research and development efforts are focused on improving anaerobic digestion processes, such as reactor design, microbial communities, and process control. These advancements aim to enhance process efficiency, increase biomethane yields, and improve the stability and reliability of the system.
2. Pretreatment technologies are being explored to enhance the biodegradability of wastewater and improve the overall performance of anaerobic digestion. Techniques such as thermal hydrolysis, ultrasound, and chemical pretreatment can increase the solubilization of organic matter, resulting in higher biogas production.
3. Co-digestion involves the simultaneous digestion of multiple feedstocks, such as agricultural residues, food waste, and energy crops, along with wastewater. This approach diversifies the feedstock portfolio, improves process stability, and optimizes biogas production by utilizing a wider range of organic materials.
4. Gas upgrading technologies, such as PSA and membrane separation, are being integrated into biomethane production systems to purify the biogas and upgrade it to biomethane quality. These technologies enable the injection of biomethane into the natural gas grid or its use as a transportation fuel.
5. There is a growing trend toward decentralized and on-site biomethane production. Small-scale biomethane plants can be located near the source of wastewater generation, reducing transportation costs and enabling the direct utilization of the produced biomethane for local energy needs.

6. Power-to-gas technology integrates renewable electricity sources, such as solar or wind, with biomethane production. Excess renewable electricity is used for electrolysis to produce hydrogen, which is then combined with CO_2 from biogas to produce biomethane via a methanation process. This approach provides a means for energy storage and grid balancing while producing renewable biomethane.
7. Biomethane production can be integrated with other circular economy concepts, such as nutrient recovery from digestate and the utilization of digestate as a soil amendment. By closing the loop on nutrient recycling and waste management, biomethane production becomes an integral part of a circular and sustainable system.
8. Governments and regulatory bodies across the globe are recognizing the potential of biomethane production from wastewater and implementing supportive policies and incentives. These policies encourage the development of biomethane projects, facilitate grid injection, and promote the use of biomethane as a renewable energy source.
9. Rapid advancements in technology, such as artificial intelligence, automation, and remote monitoring, are being integrated into biomethane production systems. These innovations improve process control, optimize performance, and enable real-time monitoring and data analysis, leading to more efficient and reliable operations.
10. Collaboration among researchers, industry players, and policymakers is crucial for accelerating the development and adoption of biomethane production from wastewater. International knowledge sharing, research networks, and best practice sharing platforms facilitate the exchange of ideas and experiences, driving innovation and standardization.

These future perspectives and emerging trends pave the way for a more sustainable and efficient biomethane production sector. As technologies mature and awareness of the environmental and economic benefits of biomethane increases, the role of biomethane in the global energy transition is expected to grow, contributing to a more sustainable and low-carbon future.

14.8 CONCLUSION

Biomethane production from wastewater is a promising solution for renewable energy generation, waste management, and environmental sustainability. The chapter's techno-economic and environmental assessments provide valuable insights into the feasibility, economic viability, and sustainability considerations of this technology. Cost analysis helps stakeholders understand the financial implications and optimize costs, while revenue streams and local economic benefits enhance project viability. LCA identifies environmental impacts, such as greenhouse gas emissions, and supports sustainable decision-making. Biomethane production aligns with circular economy principles by utilizing organic waste as feedstock, closing the waste loop, and promoting resource recovery. It also contributes to energy security by diversifying the energy mix and reducing reliance on fossil fuels. Future trends include advancements in anaerobic digestion technology, pretreatment integration, and co-digestion with multiple feedstocks, leading to process efficiency and optimization. Gas upgrading technologies, decentralized production, and power-to-gas systems further enhance sustainability and efficiency. Policy support, technological innovations, and international collaboration are critical for the development and adoption of biomethane production. These factors create an enabling environment for the sector, facilitating a more sustainable and low-carbon future. Biomethane production offers significant potential for renewable energy generation, waste management, and environmental protection. Stakeholders can use the assessments in this chapter to make informed decisions and contribute to a sustainable energy and waste management system. Continued advancements and supportive policies will further establish biomethane production from wastewaters as a vital player in the global energy transition. This technology holds the key to a greener and more sustainable future by harnessing renewable energy, managing waste effectively, and reducing environmental impact.

REFERENCES

Abilmazhinov, Y., Shakerkhan, K., Meshechkin, V., Shayakhmetov, Y., Nurgaliyev, N., Suychinov, A., 2023. Mathematical modeling for evaluating the sustainability of biogas generation through anaerobic digestion of livestock waste. *Sustainability* 15, 5707. https://doi.org/10.3390/su15075707

Abraham, A., Mathew, A.K., Park, H., Choi, O., Sindhu, R., Parameswaran, B., Pandey, A., Park, J.H., Sang, B.-I., 2020. Pretreatment strategies for enhanced biogas production from lignocellulosic biomass. *Bioresour. Technol.* 301, 122725. https://doi.org/10.1016/j.biortech.2019.122725

Ahmed, S.F., Mofijur, M., Tarannum, K., Chowdhury, A.T., Rafa, N., Nuzhat, S., Kumar, P.S., Vo, D.-V.N., Lichtfouse, E., Mahlia, T.M.I., 2021. Biogas upgrading, economy and utilization: A review. *Environ. Chem. Lett.* 19, 4137–4164. https://doi.org/10.1007/s10311-021-01292-x

Ahuja, V., Sharma, C., Paul, D., Dasgupta, D., Saratale, G.D., Banu, J.R., Yang, Y., Bhatia, S.K., 2024. Unlocking the power of synergy: Cosubstrate and coculture fermentation for enhanced biomethane production. *Biomass Bioenergy* 180, 106996. https://doi.org/10.1016/j.biombioe.2023.106996

Aili Hamzah, A.F., Hamzah, M.H., Che Man, H., Jamali, N.S., Siajam, S.I., Show, P.L., 2023. Subcritical water pretreatment for anaerobic digestion enhancement: A review. *Pertanika J. Sci. Technol.* 31, 1011–1034. https://doi.org/10.47836/pjst.31.2.19

Akbar, M., Khan, M.F.S., Qian, L., Wang, H., 2020. Degradation of polyacrylamide (PAM) and methane production by mesophilic and thermophilic anaerobic digestion: Effect of temperature and concentration. *Front. Environ. Sci. Eng.* 14. https://doi.org/10.1007/s11783-020-1277-2

Ali, S.S., Al-Tohamy, R., Manni, A., Luz, F.C., Elsamahy, T., Sun, J., 2019. Enhanced digestion of bio-pretreated sawdust using a novel bacterial consortium: Microbial community structure and methane-producing pathways. *Fuel* 254. https://doi.org/10.1016/j.fuel.2019.06.012

Ali, S.S., Elsamahy, T., Abdelfattah, A., Mustafa, A.M., Khalil, M.A., Giannis Mastropetros, S., Kornaros, M., Sun, J., Azab, M., 2022. Exploring the potential of anaerobic co-digestion of water hyacinth and cattle dung for enhanced biomethanation and techno-economic feasibility. *Fuel* 329, 125397. https://doi.org/10.1016/j.fuel.2022.125397

Ali, S.S., Jiao, H., El-Sapagh, S., Sun, J., 2023a. Biodegradation of willow sawdust by novel cellulase-producing bacterial consortium from wood-feeding termites for enhancing methane production. *Bioresour. Technol.* 383, 129232. https://doi.org/10.1016/j.biortech.2023.129232

Ali, S.S., Mustafa, A.M., Kornaros, M., Sun, J., Khalil, M., El-Shetehy, M., 2021. Biodegradation of creosote-treated wood by two novel constructed microbial consortia for the enhancement of methane production. *Bioresour. Technol.* 323, 124544. https://doi.org/10.1016/j.biortech.2020.124544

Ali, S.S., Zagklis, D., Kornaros, M., Sun, J., 2023b. Cobalt oxide nanoparticles as a new strategy for enhancing methane production from anaerobic digestion of noxious aquatic weeds. *Bioresour. Technol.* 368, 128308. https://doi.org/10.1016/j.biortech.2022.128308

Amato, A., Tsigkou, K., Becci, A., Beolchini, F., Ippolito, N.M., Ferella, F., 2023. Life cycle assessment of biomethane vs. fossil methane production and supply. *Energies* 16, 4555. https://doi.org/10.3390/en16124555

Atelge, M.R., Atabani, A.E., Banu, J.R., Krisa, D., Kaya, M., Eskicioglu, C., Kumar, G., Lee, C., Yildiz, Y.Ş., Unalan, S., Mohanasundaram, R., Duman, F., 2020. A critical review of pretreatment technologies to enhance anaerobic digestion and energy recovery. *Fuel* 270, 117494. https://doi.org/10.1016/j.fuel.2020.117494

Awadalla, O.A., Atawy, W.A., Bedaiwy, M.Y., Ali, S.S., Mahmoud, Y.A.-G., 2023. Anaerobic digestion of lignocellulosic waste for enhanced methane production and biogas-digestate utilization. *Ind. Crops Prod.* 195, 116420. https://doi.org/10.1016/j.indcrop.2023.116420

Babu Ponnusami, A., Sinha, S., Ashokan, H., Paul, M.V., Hariharan, S.P., Arun, J., Gopinath, K.P., Hoang Le, Q., Pugazhendhi, A., 2023. Advanced oxidation process (AOP) combined biological process for wastewater treatment: A review on advancements, feasibility and practicability of combined techniques. *Environ. Res.* 237, 116944. https://doi.org/10.1016/j.envres.2023.116944

Bamber, N., Turner, I., Arulnathan, V., Li, Y., Zargar Ershadi, S., Smart, A., Pelletier, N., 2019. Comparing sources and analysis of uncertainty in consequential and attributional life cycle assessment: Review of current practice and recommendations. *Int. J. Life Cycle Assess.* 25, 168–180. https://doi.org/10.1007/s11367-019-01663-1

Banerjee, S., Prasad, N., Selvaraju, S., 2021. Reactor design for biogas production-A short review. *J. Energy Power Technol.* 4, 1. https://doi.org/10.21926/jept.2201004

Bekele, W., Guinguina, A., Zegeye, A., Simachew, A., Ramin, M., 2022. Contemporary methods of measuring and estimating methane emission from ruminants. *Methane* 1, 82–95. https://doi.org/10.3390/methane1020008

Berghuis, B.A., Yu, F.B., Schulz, F., Blainey, P.C., Woyke, T., Quake, S.R., 2019. Hydrogenotrophic methanogenesis in archaeal phylum Verstraetearchaeota reveals the shared ancestry of all methanogens. *Proc. Natl. Acad. Sci. U. S. A.* 116, 5037–5044. https://doi.org/10.1073/pnas.1815631116

Bhatia, R.K., Sakhuja, D., Mundhe, S., Walia, A., 2020. Renewable energy products through bioremediation of wastewater. *Sustainability* 12, 7501. https://doi.org/10.3390/su12187501

Bhojwani, S., Topolski, K., Mukherjee, R., Sengupta, D., El-Halwagi, M.M., 2019. Technology review and data analysis for cost assessment of water treatment systems. *Sci. Total Environ.* 651, 2749–2761. https://doi.org/10.1016/j.scitotenv.2018.09.363

Caillet, H., Bastide, A., Adelard, L., 2023. Advances in computational fluid dynamics modeling of anaerobic digestion process for renewable energy production: A review. *Clean. Waste Syst.* 6, 100124. https://doi.org/10.1016/j.clwas.2023.100124

Cecconet, D., Callegari, A., Capodaglio, A.G., 2022. UASB performance and perspectives in urban wastewater treatment at sub-mesophilic operating temperature. *Water* 14, 115. https://doi.org/10.3390/w14010115

Che, S., Men, Y., 2019. Synthetic microbial consortia for biosynthesis and biodegradation: Promises and challenges. *J. Ind. Microbiol. Biotechnol.* 46, 1343–1358. https://doi.org/10.1007/s10295-019-02211-4

Chiappero, M., Demichelis, F., Lin, X., Liu, C., Frigon, D., Fiore, S., 2019. Investigation of pre-treatments improving low-temperature anaerobic digestion of waste activated sludge. *Process Saf. Environ. Prot.* 131, 28–37. https://doi.org/10.1016/j.psep.2019.08.034

Conrad, R., 2020. Importance of hydrogenotrophic, aceticlastic and methylotrophic methanogenesis for methane production in terrestrial, aquatic and other anoxic environments: A mini review. *Pedosphere* 30, 25–39. https://doi.org/10.1016/s1002-0160(18)60052-9

D'Adamo, I., Falcone, P.M., Ferella, F., 2019. A socio-economic analysis of biomethane in the transport sector: The case of Italy. *Waste Manag.* 95, 102–115. https://doi.org/10.1016/j.wasman.2019.06.005

D'Adamo, I., Ribichini, M., Tsagarakis, K.P., 2023. Biomethane as an energy resource for achieving sustainable production: Economic assessments and policy implications. *Sustain. Prod. Consum.* 35, 13–27. https://doi.org/10.1016/j.spc.2022.10.014

Dominguillo-Ramírez, D., Aburto, J., Hugo Leon-Santiesteban, H., Martinez-Hernandez, E., 2023. Neural network model for predicting the biomethane yield in an anaerobic digester using biomass composition profiles. *Fuel* 344, 128053. https://doi.org/10.1016/j.fuel.2023.128053

Duan, N., Zhang, D., Khoshnevisan, B., Kougias, P.G., Treu, L., Liu, Z., Lin, C., Liu, H., Zhang, Y., Angelidaki, I., 2020. Human waste anaerobic digestion as a promising low-carbon strategy: Operating performance, microbial dynamics and environmental footprint. *J. Clean. Prod.* 256, 120414. https://doi.org/10.1016/j.jclepro.2020.120414

Đurđević, D., Hulenić, I., 2020. Anaerobic digestate treatment selection model for biogas plant costs and emissions reduction. *Processes* 8, 142. https://doi.org/10.3390/pr8020142

Elalami, D., Carrere, H., Monlau, F., Abdelouahdi, K., Oukarroum, A., Barakat, A., 2019. Pretreatment and co-digestion of wastewater sludge for biogas production: Recent research advances and trends. *Renew. Sustain. Energy Rev.* 114, 109287. https://doi.org/10.1016/j.rser.2019.109287

Elbeshbishy, E., Okoye, F., Ho, D., Sarathy, S., Parker, W., 2017. Free nitrous acid pretreatment of thickened waste activated sludge improves anaerobic degradability. *Proc. Water Environ. Fed.* 2017, 75–84. https://doi.org/10.2175/193864717821495591

Ellacuriaga, M., García-Cascallana, J., Gómez, X., 2021. Biogas production from organic wastes: Integrating concepts of circular economy. *Fuels* 2, 144–167. https://doi.org/10.3390/fuels2020009

Filer, J., Ding, H.H., Chang, S., 2019. Biochemical methane potential (BMP) assay method for anaerobic digestion research. *Water* 11, 921. https://doi.org/10.3390/w11050921

Frunzo, L., Fermoso, F.G., Luongo, V., Mattei, M.R., Esposito, G., 2019. ADM1-based mechanistic model for the role of trace elements in anaerobic digestion processes. *J. Environ. Manage.* 241, 587–602. https://doi.org/10.1016/j.jenvman.2018.11.058

Ghaleb, A.A.S., Kutty, S.R.M., Salih, G.H.A., Jagaba, A.H., Noor, A., Kumar, V., Almahbashi, N.M.Y., Saeed, A.A.H., Saleh Al-dhawi, B.N., 2021. Sugarcane bagasse as a co-substrate with oil-refinery biological sludge for biogas production using batch mesophilic anaerobic co-digestion technology: Effect of carbon/nitrogen ratio. *Water* 13, 590. https://doi.org/10.3390/w13050590

Giwa, A.S., Maurice, N.J., Luoyan, A., Liu, X., Yunlong, Y., Hong, Z., 2023. Advances in sewage sludge application and treatment: Process integration of plasma pyrolysis and anaerobic digestion with the resource recovery. *Heliyon* 9, e19765–e19765. https://doi.org/10.1016/j.heliyon.2023.e19765

Gunes, B., 2021. A critical review on biofilm-based reactor systems for enhanced syngas fermentation processes. *Renew. Sustain. Energy Rev.* 143, 110950. https://doi.org/10.1016/j.rser.2021.110950

Guo, Q., Majeed, S., Xu, R., Zhang, K., Kakade, A., Khan, A., Hafeez, F.Y., Mao, C., Liu, P., Li, X., 2019. Heavy metals interact with the microbial community and affect biogas production in anaerobic digestion: A review. *J. Environ. Manage.* 240, 266–272. https://doi.org/10.1016/j.jenvman.2019.03.104

Gutu, L., Basitere, M., Harding, T., Ikumi, D., Njoya, M., Gaszynski, C., 2021. Multi-integrated systems for treatment of abattoir wastewater: A review. *Water* 13, 2462. https://doi.org/10.3390/w13182462

He, D., Xiao, J., Wang, D., Liu, X., Fu, Q., Li, Y., Du, M., Yang, Q., Liu, Y., Wang, Q., Ni, B.-J., Song, K., Cai, Z., Ye, J., Yu, H., 2021. Digestion liquid based alkaline pretreatment of waste activated sludge promotes methane production from anaerobic digestion. *Water Res.* 199, 117198. https://doi.org/10.1016/j.watres.2021.117198

Huang, B.-C., Li, W.-W., Wang, X., Lu, Y., Yu, H.-Q., 2019. Customizing anaerobic digestion-coupled processes for energy-positive and sustainable treatment of municipal wastewater. *Renew. Sustain. Energy Rev.* 110, 132–142. https://doi.org/10.1016/j.rser.2019.04.064

Huiliñir, C., Pagés-Díaz, J., Vargas, G., Vega, S., Lauzurique, Y., Palominos, N., 2023. Microaerobic condition as pretreatment for improving anaerobic digestion: A review. *Bioresour. Technol.* 384, 129249. https://doi.org/10.1016/j.biortech.2023.129249

Iglesias, R., Muñoz, R., Polanco, M., Díaz, I., Susmozas, A., Moreno, A.D., Guirado, M., Carreras, N., Ballesteros, M., 2021. Biogas from anaerobic digestion as an energy vector: Current upgrading development. *Energies* 14, 2742. https://doi.org/10.3390/en14102742

Ingrao, C., Bacenetti, J., Adamczyk, J., Ferrante, V., Messineo, A., Huisingh, D., 2019. Investigating energy and environmental issues of agro-biogas derived energy systems: A comprehensive review of life cycle assessments. *Renew. Energy* 136, 296–307. https://doi.org/10.1016/j.renene.2019.01.023

Jin, Y., Feng, J.-Y., Song, C., Wang, L.-G., Chen, C., Liu, G.-Q., 2022. Alkali-microwave co-pretreatment to enhance the solid-state anaerobic digestion performance of vinegar residue. *J. Biobased Mater. Bioenergy* 16, 481–487. https://doi.org/10.1166/jbmb.2022.2186

Jover-Smet, M., Martín-Pascual, J., Trapote, A., 2017. Model of suspended solids removal in the primary sedimentation tanks for the treatment of urban wastewater. *Water* 9, 448. https://doi.org/10.3390/w9060448

Kabeyi, M.J.B., Olanrewaju, O.A., 2022. Biogas production and applications in the sustainable energy transition. *J. Energy* 2022, 1–43. https://doi.org/10.1155/2022/8750221

Karki, R., Chuenchart, W., Surendra, K.C., Shrestha, S., Raskin, L., Sung, S., Hashimoto, A., Kumar Khanal, S., 2021. Anaerobic co-digestion: Current status and perspectives. *Bioresour. Technol.* 330, 125001. https://doi.org/10.1016/j.biortech.2021.125001

Kaur, G., Luo, L., Chen, G., Wong, J.W.C., 2019. Integrated food waste and sewage treatment: A better approach than conventional food waste-sludge co-digestion for higher energy recovery via anaerobic digestion. *Bioresour. Technol.* 289, 121698. https://doi.org/10.1016/j.biortech.2019.121698

Keogh, N., Corr, D., O'Shea, R., Monaghan, R.F.D., 2022. The gas grid as a vector for regional decarbonisation: A techno economic case study for biomethane injection and natural gas heavy goods vehicles. *Appl. Energy* 323, 119590. https://doi.org/10.1016/j.apenergy.2022.119590

Khan, M.U., Usman, M., Ashraf, M.A., Dutta, N., Luo, G., Zhang, S., 2022. A review of recent advancements in pretreatment techniques of lignocellulosic materials for biogas production: Opportunities and limitations. *Chem. Eng. J. Adv.* 10, 100263. https://doi.org/10.1016/j.ceja.2022.100263

Kim, D., 2018. Physico-chemical conversion of lignocellulose: Inhibitor effects and detoxification strategies: A mini review. *Molecules* 23, 309. https://doi.org/10.3390/molecules23020309

Kim, G.-B., Cayetano, R.D.A., Park, J., Jo, Y., Jeong, S.Y., Lee, M.Y., Pandey, A., Kim, S.-H., 2022. Impact of thermal pretreatment on anaerobic digestion of dewatered sludge from municipal and industrial wastewaters and its economic feasibility. *Energy* 254, 124345. https://doi.org/10.1016/j.energy.2022.124345

Koch, K., Hafner, S.D., Weinrich, S., Astals, S., Holliger, C., 2020. Power and limitations of biochemical methane potential (BMP) tests. *Front. Energy Res.* 8. https://doi.org/10.3389/fenrg.2020.00063

Kunatsa, T., Xia, X., 2022. A review on anaerobic digestion with focus on the role of biomass co-digestion, modelling and optimisation on biogas production and enhancement. *Bioresour. Technol.* 344, 126311. https://doi.org/10.1016/j.biortech.2021.126311

Lackner, N., Wagner, A.O., Markt, R., Illmer, P., 2020. pH and phosphate induced shifts in carbon flow and microbial community during thermophilic anaerobic digestion. *Microorganisms* 8, 286. https://doi.org/10.3390/microorganisms8020286

Lauzurique, Y., Meza, A., Huiliñir, C., Salazar-González, R., Vargas-Morales, G., 2023. Evaluation of the anaerobic digestion of winery wastewater: Effect of fly ash in anaerobic sequencing batch reactor. *J. Environ. Chem. Eng.* 11, 110997. https://doi.org/10.1016/j.jece.2023.110997

Lee, W., Park, S., Cui, F., Kim, M., 2019. Optimizing pre-treatment conditions for anaerobic co-digestion of food waste and sewage sludge. *J. Environ. Manage.* 249, 109397. https://doi.org/10.1016/j.jenvman.2019.109397

Li, H., Chen, Z., Fu, D., Wang, Y., Zheng, Y., Li, Q., 2020. Improved ADM1 for modelling C, N, P fates in anaerobic digestion process of pig manure and optimization approaches to biogas production. *Renew. Energy* 146, 2330–2336. https://doi.org/10.1016/j.renene.2019.08.086

Li, Y., Chen, Y., Wu, J., 2019. Enhancement of methane production in anaerobic digestion process: A review. *Appl. Energy* 240, 120–137. https://doi.org/10.1016/j.apenergy.2019.01.243

Li, Y., Xu, Y., Fu, Z., Li, W., Zheng, L., Li, M., 2021. Assessment of energy use and environmental impacts of wastewater treatment plants in the entire life cycle: A system meta-analysis. *Environ. Res.* 198, 110458. https://doi.org/10.1016/j.envres.2020.110458

Liu, C., Ren, L., Yan, B., Luo, L., Zhang, J., Awasthi, M.K., 2021. Electron transfer and mechanism of energy production among syntrophic bacteria during acidogenic fermentation: A review. *Bioresour. Technol.* 323, 124637. https://doi.org/10.1016/j.biortech.2020.124637

Logan, M., Visvanathan, C., 2019. Management strategies for anaerobic digestate of organic fraction of municipal solid waste: Current status and future prospects. *Waste Manag. Res. J. Sustain. Circ. Econ.* 37, 27–39. https://doi.org/10.1177/0734242x18816793

Lomwongsopon, P., Aramrueang, N., 2022. Mild chemical pretreatment of cassava pulp for enhancing high-load anaerobic digestion. *Bioresour. Technol. Reports* 17, 100896. https://doi.org/10.1016/j.biteb.2021.100896

Lyu, Y., Ye, H., Zhao, Z., Tian, J., Chen, L., 2020. Exploring the cost of wastewater treatment in a chemical industrial park: Model development and application. *Resour. Conserv. Recycl.* 155, 104663. https://doi.org/10.1016/j.resconrec.2019.104663

Mahmod, S.S., Jahim, J.M., Abdul, P.M., Luthfi, A.A.I., Takriff, M.S., 2021. Techno-economic analysis of two-stage anaerobic system for biohydrogen and biomethane production from palm oil mill effluent. *J. Environ. Chem. Eng.* 9, 105679. https://doi.org/10.1016/j.jece.2021.105679

Mainardis, M., Buttazzoni, M., Goi, D., 2020. Up-flow anaerobic sludge blanket (UASB) technology for energy recovery: A review on state-of-the-art and recent technological advances. *Bioeng. (*Basel, Switzerland*)* 7, 43. https://doi.org/10.3390/bioengineering7020043

Manyi-Loh, C.E., Lues, R., 2023. Anaerobic digestion of lignocellulosic biomass: Substrate characteristics (challenge) and innovation. *Fermentation* 9, 755. https://doi.org/10.3390/fermentation9080755

Marconi, P., Rosa, L., 2023. Role of biomethane to offset natural gas. *Renew. Sustain. Energy Rev.* 187, 113697. https://doi.org/10.1016/j.rser.2023.113697

Mitraka, G.-C., Kontogiannopoulos, K.N., Batsioula, M., Banias, G.F., Zouboulis, A.I., Kougias, P.G., 2022. A comprehensive review on pretreatment methods for enhanced biogas production from sewage sludge. *Energies* 15, 6536. https://doi.org/10.3390/en15186536

Mohiddin, M.N.B., Tan, Y.H., Seow, Y.X., Kansedo, J., Mubarak, N.M., Abdullah, M.O., Chan, Y.S., Khalid, M., 2021. Evaluation on feedstock, technologies, catalyst and reactor for sustainable biodiesel production: A review. *J. Ind. Eng. Chem.* 98, 60–81. https://doi.org/10.1016/j.jiec.2021.03.036

Morais, N.W.S., Coelho, M.M.H., de Sousa e Silva, A., Silva, F.S.S., Ferreira, T.J.T., Pereira, E.L., dos Santos, A.B., 2021. Biochemical potential evaluation and kinetic modeling of methane production from six agro-industrial wastewaters in mixed culture. *Environ. Pollut.* 280, 116876. https://doi.org/10.1016/j.envpol.2021.116876

Nagendranatha Reddy, C., Kondaveeti, S., Mohanakrishna, G., Min, B., 2022. Application of bioelectrochemical systems to regulate and accelerate the anaerobic digestion processes. *Chemosphere* 287, 132299. https://doi.org/10.1016/j.chemosphere.2021.132299

Nahar, G., Rajput, S., Grasham, O., Dalvi, V.H., Dupont, V., Ross, A.B., Pandit, A.B., 2022. Technoeconomic analysis of biogas production using simple and effective mechanistic model calibrated with biomethanation potential experiments of water lettuce (Pistia stratiotes) inoculated by buffalo dung. *Energy* 244, 122911. https://doi.org/10.1016/j.energy.2021.122911

Ni, R., Xu, C., Shi, X., Yang, S., Li, L., Peng, X., Song, L., 2022. Acetoclastic methanogenesis pathway stability despite the high microbial taxonomic variability in the transition from acidogenesis to methanogenesis during food waste anaerobic digestion. *J. Clean. Prod.* 372, 133758. https://doi.org/10.1016/j.jclepro.2022.133758

Nikpour, B., Jalilzadeh Yengejeh, R., Takdastan, A., Hassani, A.H., Zazouli, M.A., 2021. Studying the effluent quality of enhanced modified ludzack ettinger-oxic settling anaerobic process (E-MLE-OSA) for treating real municipal wastewater. *J. Adv. Environ. Heal. Res.* 9, 345–360. https://doi.org/10.32598/jaehr.9.4.1243

Nkuna, R., Roopnarain, A., Rashama, C., Adeleke, R., 2021. Insights into organic loading rates of anaerobic digestion for biogas production: A review. *Crit. Rev. Biotechnol.* 42, 1–21. https://doi.org/10.1080/07388551.2021.1942778

Nwokolo, N., Mukumba, P., Obileke, K., Enebe, M., 2020. Waste to energy: A focus on the impact of substrate type in biogas production. *Processes* 8, 1224. https://doi.org/10.3390/pr8101224

Omrany, H., Soebarto, V., Zuo, J., Chang, R., 2021. A comprehensive framework for standardising system boundary definition in life cycle energy assessments. *Buildings* 11, 230. https://doi.org/10.3390/buildings11060230

Park, S., Han, S.K., Song, E., Kim, H., Kim, M., Lee, W., 2020. Effect of hydrothermal pre-treatment on physical properties and co-digestion from food waste and sewage sludge mixture. *Waste Manag. Res.* 38, 546–553. https://doi.org/10.1177/0734242x19897123

Pasalari, H., Gholami, M., Rezaee, A., Esrafili, A., Farzadkia, M., 2021. Perspectives on microbial community in anaerobic digestion with emphasis on environmental parameters: A systematic review. *Chemosphere* 270, 128618. https://doi.org/10.1016/j.chemosphere.2020.128618

Pavičić, J., Novak Mavar, K., Brkić, V., Simon, K., 2022. Biogas and biomethane production and usage: Technology development, advantages and challenges in Europe. *Energies* 15, 2940. https://doi.org/10.3390/en15082940

Pedizzi, C., Lema, J.M., Carballa, M., 2018. A combination of ammonia stripping and low temperature thermal pre-treatment improves anaerobic post-digestion of the supernatant from organic fraction of municipal solid waste treatment. *Waste Manag.* 78, 271–278. https://doi.org/10.1016/j.wasman.2018.05.051

Pereira, E., Borges, A., da Silva, G., 2022. Effect of the progressive increase of organic loading rate in an anaerobic sequencing batch reactor for biodiesel wastewater treatment. *Water* 14, 223. https://doi.org/10.3390/w14020223

Perera, M.K., Englehardt, J.D., Dvorak, A.C., 2019. Technologies for recovering nutrients from wastewater: A critical review. *Environ. Eng. Sci.* 36, 511–529. https://doi.org/10.1089/ees.2018.0436

Poddar, B.J., Nakhate, S.P., Gupta, R.K., Chavan, A.R., Singh, A.K., Khardenavis, A.A., Purohit, H.J., 2021. A comprehensive review on the pretreatment of lignocellulosic wastes for improved biogas production by anaerobic digestion. *Int. J. Environ. Sci. Technol.* 19, 3429–3456. https://doi.org/10.1007/s13762-021-03248-8

Ren, Z., Verma, A.S., Li, Y., Teuwen, J.J.E., Jiang, Z., 2021. Offshore wind turbine operations and maintenance: A state-of-the-art review. *Renew. Sustain. Energy Rev.* 144, 110886. https://doi.org/10.1016/j.rser.2021.110886

Richards, S.J., Al Zaili, J., 2020. Contribution of encouraging the future use of biomethane to resolving sustainability and energy security challenges: The case of the UK. *Energy Sustain. Dev.* 55, 48–55. https://doi.org/10.1016/j.esd.2019.12.003

Robles, Á., Serralta, J., Martí, N., Ferrer, J., Seco, A., 2021. Anaerobic membrane bioreactors for resource recovery from municipal wastewater: A comprehensive review of recent advances. *Environ. Sci. Water Res. Technol.* 7, 1944–1965. https://doi.org/10.1039/d1ew00217a

Rocha-Meneses, L., Zannerni, R., Inayat, A., Abdallah, M., Shanableh, A., Ghenai, C., Kamil, M., Kikas, T., 2022. Current progress in anaerobic digestion reactors and parameters optimization. *Biomass Convers. Biorefinery.* https://doi.org/10.1007/s13399-021-02224-z

Ruan, D., Zhou, Z., Pang, H., Yao, J., Chen, G., Qiu, Z., 2019. Enhancing methane production of anaerobic sludge digestion by microaeration: Enzyme activity stimulation, semi-continuous reactor validation and microbial community analysis. *Bioresour. Technol.* 289, 121643. https://doi.org/10.1016/j.biortech.2019.121643

Ruiz, H.A., Conrad, M., Sun, S.-N., Sanchez, A., Rocha, G.J.M., Romaní, A., Castro, E., Torres, A., Rodríguez-Jasso, R.M., Andrade, L.P., Smirnova, I., Sun, R.-C., Meyer, A.S., 2020. Engineering aspects of hydrothermal pretreatment: From batch to continuous operation, scale-up and pilot reactor under biorefinery concept. *Bioresour. Technol.* 299, 122685. https://doi.org/10.1016/j.biortech.2019.122685

Sahni, N., Hans, M., Kumar, S., 2023. Fermentative approaches in wastewater treatment for harnessing renewable energy. In P. Dheeran, S. Kumar (eds.), *Extremophiles*. CRC Press. https://doi.org/10.1201/9781003335221-4

Salama, E.-S., Saha, S., Kurade, M.B., Dev, S., Chang, S.W., Jeon, B.-H., 2019. Recent trends in anaerobic co-digestion: Fat, oil, and grease (FOG) for enhanced biomethanation. *Prog. Energy Combust. Sci.* 70, 22–42. https://doi.org/10.1016/j.pecs.2018.08.002

Serrano-Meza, A., Garzón-Zuñiga, M.A., Barragán-Huerta, B.E., Estrada-Arriaga, E.B., Almaraz-Abarca, N., García-Olivares, J.G., 2020. Anaerobic digestion inhibition indicators and control strategies in processes treating industrial wastewater and wastes. *Rev. Mex. Ing. Química* 19, 29–44. https://doi.org/10.24275/rmiq/ia1221

Sillero, L., Solera, R., Perez, M., 2022. Biochemical assays of potential methane to test biogas production from dark fermentation of sewage sludge and agricultural residues. *Int. J. Hydrogen Energy* 47, 13289–13299. https://doi.org/10.1016/j.ijhydene.2022.02.080

Silva, J., Gonçalves, J.C., Rocha, C., Vilaça, J., Madeira, L.M., 2024. Biomethane production from biogas obtained in wastewater treatment plants: Process optimization and economic analysis. *Renew. Energy* 220, 119469. https://doi.org/10.1016/j.renene.2023.119469

Skorek-Osikowska, A., Martín-Gamboa, M., Dufour, J., 2020. Thermodynamic, economic and environmental assessment of renewable natural gas production systems. *Energy Convers. Manag.* X 7, 100046. https://doi.org/10.1016/j.ecmx.2020.100046

Söllinger, A., Urich, T., 2019. Methylotrophic methanogens everywhere — Physiology and ecology of novel players in global methane cycling. *Biochem. Soc. Trans.* 47, 1895–1907. https://doi.org/10.1042/bst20180565

Song, B., Lin, R., Lam, C.H., Wu, H., Tsui, T.-H., Yu, Y., 2021. Recent advances and challenges of inter-disciplinary biomass valorization by integrating hydrothermal and biological techniques. *Renew. Sustain. Energy Rev.* 135, 110370. https://doi.org/10.1016/j.rser.2020.110370

Szilágyi, Á., Bodor, A., Tolvai, N., Kovács, K.L., Bodai, L., Wirth, R., Bagi, Z., Szepesi, Á., Markó, V., Kakuk, B., Bounedjoum, N., Rákhely, G., 2021. A comparative analysis of biogas production from tomato bio-waste in mesophilic batch and continuous anaerobic digestion systems. *PLoS One* 16, e0248654–e0248654. https://doi.org/10.1371/journal.pone.0248654

Tufaner, F., Demirci, Y., 2020. Prediction of biogas production rate from anaerobic hybrid reactor by artificial neural network and nonlinear regressions models. *Clean Technol. Environ. Policy* 22, 713–724. https://doi.org/10.1007/s10098-020-01816-z

Valenti, F., Liao, W., Porto, S.M.C., 2020. Life cycle assessment of agro-industrial by-product reuse: A comparison between anaerobic digestion and conventional disposal treatments. *Green Chem.* 22, 7119–7139. https://doi.org/10.1039/d0gc01918f

Valli, L., Rossi, L., Fabbri, C., Sibilla, F., Gattoni, P., Dale, B.E., Kim, S., Ong, R.G., Bozzetto, S., 2017. Greenhouse gas emissions of electricity and biomethane produced using the Biogasdoneright™ system: Four case studies from Italy. *Biofuels Bioprod. Biorefining* 11, 847–860. https://doi.org/10.1002/bbb.1789

Wainaina, S., Lukitawesa, Kumar Awasthi, M., Taherzadeh, M.J., 2019. Bioengineering of anaerobic digestion for volatile fatty acids, hydrogen or methane production: A critical review. *Bioengineered* 10, 437–458. https://doi.org/10.1080/21655979.2019.1673937

Wang, N., Zhao, K., Li, F., Peng, H., Lu, Y., Zhang, L., Pan, J., Jiang, S., Chen, A., Yan, B., Luo, L., Huang, H., Li, H., Wu, G., Zhang, J., 2022a. Characteristics of carbon, nitrogen, phosphorus and sulfur cycling genes, microbial community metabolism and key influencing factors during composting process supplemented with biochar and biogas residue. *Bioresour. Technol.* 366, 128224. https://doi.org/10.1016/j.biortech.2022.128224

Wang, X., Duan, X., Chen, J., Fang, K., Feng, L., Yan, Y., Zhou, Q., 2016. Enhancing anaerobic digestion of waste activated sludge by pretreatment: Effect of volatile to total solids. *Environ. Technol.* 37, 1520–1529. https://doi.org/10.1080/09593330.2015.1120783

Wang, X., Gao, C., Qi, X., Zhang, Y., Chen, T., Xie, Y., Zhang, A., Gao, J., 2021. Enhancing sludge fermentation and anaerobic digestion by mechanical cutting pretreatment. *J. Water Process Eng.* 40, 101812. https://doi.org/10.1016/j.jwpe.2020.101812

Wang, Y., Zhang, S., Song, J., Sheng, C., Shang, Z., Wang, R., Wang, X., Yang, G., Feng, Y., Ren, G., 2022b. Strategies to improve production of biomethane from organic wastes with anaerobic co-digestion: A systematic review. *Biofuels Bioprod. Biorefining* 16, 1388–1411. https://doi.org/10.1002/bbb.2367

Wang, Z., Ma, T., Xing, L., 2020. Process performance and microbial interaction in two-stage continuously stirred tank reactors for sludge anaerobic digestion operated at different temperatures. *Biochem. Eng. J.* 161, 107682. https://doi.org/10.1016/j.bej.2020.107682

Wonglertarak, W., Wichitsathian, B., 2014. Alkaline pretreatment of waste activated sludge in anaerobic digestion. *J. Clean Energy Technol.* 2, 118–121. https://doi.org/10.7763/jocet.2014.v2.104

Wu, D., Peng, X., Li, L., Yang, P., Peng, Y., Liu, H., Wang, X., 2021. Commercial biogas plants: Review on operational parameters and guide for performance optimization. *Fuel* 303, 121282. https://doi.org/10.1016/j.fuel.2021.121282

Xiao, K., Abbt-Braun, G., Horn, H., 2020a. Changes in the characteristics of dissolved organic matter during sludge treatment: A critical review. *Water Res.* 187, 116441. https://doi.org/10.1016/j.watres.2020.116441

Xiao, L., Liu, F., Lichtfouse, E., Zhang, P., Feng, D., Li, F., 2020b. Methane production by acetate dismutation stimulated by Shewanella oneidensis and carbon materials: An alternative to classical CO_2 reduction. *Chem. Eng. J.* 389, 124469. https://doi.org/10.1016/j.cej.2020.124469

Yadav, M., Joshi, C., Paritosh, K., Thakur, J., Pareek, N., Masakapalli, S.K., Vivekanand, V., 2022. Organic waste conversion through anaerobic digestion: A critical insight into the metabolic pathways and microbial interactions. *Metab. Eng.* 69, 323–337. https://doi.org/10.1016/j.ymben.2021.11.014

Yang, D., Hu, C., Dai, L., Liu, Z., Dong, B., Dai, X., 2019. Post-thermal hydrolysis and centrate recirculation for enhancing anaerobic digestion of sewage sludge. *Waste Manag.* 92, 39–48. https://doi.org/10.1016/j.wasman.2019.04.044

Yu, Q., Feng, L., Zhen, X., 2020. Effects of organic loading rate and temperature fluctuation on the microbial community and performance of anaerobic digestion of food waste. *Environ. Sci. Pollut. Res.* 28, 13176–13187. https://doi.org/10.1007/s11356-020-11548-8

Zamri, M.F.M.A., Hasmady, S., Akhiar, A., Ideris, F., Shamsuddin, A.H., Mofijur, M., Fattah, I.M.R., Mahlia, T.M.I., 2021. A comprehensive review on anaerobic digestion of organic fraction of municipal solid waste. *Renew. Sustain. Energy Rev.* 137, 110637. https://doi.org/10.1016/j.rser.2020.110637

Zhan, Y., Zhu, J., Xiao, Y., Wu, S., Ndeddy Aka, R.J., 2023. Efficient methane production from anaerobic co-digestion of poultry litter with wheat straw in sequencing batch reactor: Effects of carbon-to-nitrogen ratio, total solids, and hydraulic retention time. *Bioresour. Technol.* 381, 129127. https://doi.org/10.1016/j.biortech.2023.129127

Zhang, J., Qin, Q., Li, G., Tseng, C.-H., 2021. Sustainable municipal waste management strategies through life cycle assessment method: A review. *J. Environ. Manage.* 287, 112238. https://doi.org/10.1016/j.jenvman.2021.112238

Zhao, X., Li, L., Wu, D., Xiao, T., Ma, Y., Peng, X., 2019. Modified anaerobic digestion model no. 1 for modeling methane production from food waste in batch and semi-continuous anaerobic digestions. *Bioresour. Technol.* 271, 109–117. https://doi.org/10.1016/j.biortech.2018.09.091

Zhou, A., Zhang, J., Varrone, C., Wen, K., Wang, G., Liu, W., Wang, A., Yue, X., 2017. Process assessment associated to microbial community response provides insight on possible mechanism of waste activated sludge digestion under typical chemical pretreatments. *Energy* 137, 457–467. https://doi.org/10.1016/j.energy.2017.02.166

Zhu, T., Curtis, J., Clancy, M., 2019. Promoting agricultural biogas and biomethane production: Lessons from cross-country studies. *Renew. Sustain. Energy Rev.* 114, 109332. https://doi.org/10.1016/j.rser.2019.109332

15 Biohydrogen
A Green Fuel for Sustainable Energy

Chetna Kumari, Sudesh Kumar, Nirmala Kumari Jangid, and Navjeet Kaur

15.1 INTRODUCTION

15.1.1 Economy of Renewable Energy

Governments around the world are increasingly recognizing the importance of sustainable development, which encompasses not only industrial progress but also economic, social, and climatic considerations [1]. According to Davidson, Sustainable energy is described as energy that provides services that are reasonable, easily available, and trustworthy, while meeting the social, economic, and climatic needs of society. This definition also emphasizes the importance of equitable distribution in fulfilling those needs. The foundational elements of sustainable development — energy, economy, and environment — are interconnected, forming a triangle. While renewable energy is crucial for sustainable energy, its significance to the economy and environment must also be recognized [2]. Figure 15.1 shows the important key requirements for sustainable energy [2]. In 2013, the second generation of biofuel feedstock accounted for approximately 3% of the world energy requirement [2]. These biofuels are considered favorable as they do not compete with land and food resources. However, their adoption is hindered by technical challenges, primarily low-energy conversion efficiencies. The third generation of biofuels utilizes microalgae as feedstock [3,4].

15.1.2 Global Warming Effect and Current Global Energy Situation

Energy is the backbone of physical, economic, and social activities. However, meeting the increasing global demand poses a significant challenge. According to the International Energy Outlook, it is projected that the world's energy needs will rise by 50% from 2018 to 2050 [5]. Historically, this demand has been met mainly through the use of fossil fuels. Fossil fuels, which are carbon-rich energy sources, are formed through photosynthesis and the breakdown of organic matter over the years. Their efficient energy transferability has played a crucial role in driving a big change in the industrial sector [6]. On the flip side, the combustion of fossil fuels leads to global warming and climate change. This pressing issue has far-reaching consequences, such as rising temperatures, which have a severe impact on agriculture and food security [7,8]. Additionally, the repercussions might result in warfare,

 DOI: 10.1201/9781003459286-15

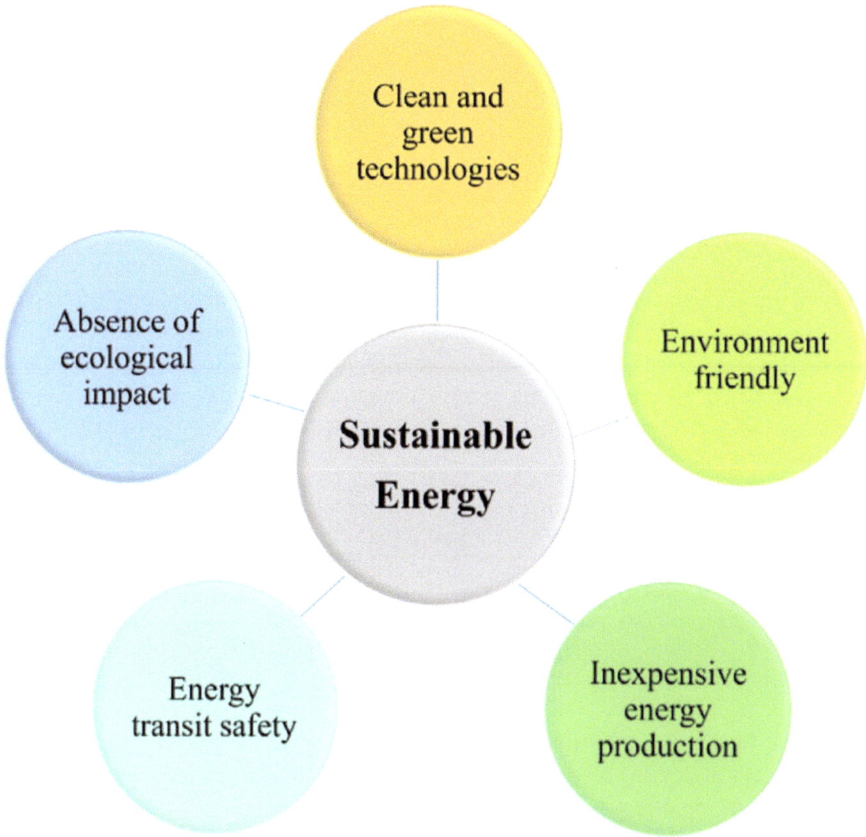

FIGURE 15.1 Important requirements for sustainable energy. (Adapted from [2].)

political and social instability, and even large-scale immigration. Acidified oceans and the rise in sea levels are also significant problems that arise from this situation [9,10]. Furthermore, the impact of global warming is emphasizing the need for green and eco-friendly energy solutions [11–13].

15.1.3 Economy of Hydrogen

Biohydrogen has become a potential biofuel in today's world due to its clean and efficient nature as an energy carrier. Hydrogen (H_2) offers numerous advantages, including its high-energy content per unit mass of 142 kJ/g [14]. As a result, H_2 is utilized in fuel cells and the transport sector [15]. H_2 is also used as a starting material in different processes and applications [16]. However, the economic sustainability of using H_2 as a fuel source faces challenges such as high-production costs, storage issues, transportation limitations, and underdeveloped hydrogen infrastructure [16–20]. Biohydrogen can be produced through different methods like biological and thermochemical [21–27]. The ecological benefits and lower energy intensity of biological processes make them preferable, especially in reducing global carbon dioxide emissions [17,19,28]. Additionally, using raw biomass as feedstock, such as organic matter and municipal, industrial, and agricultural wastewater, hydrogen production can be economical [17,19]. Biohydrogen production methods include photofermentation, dark fermentation, microbial electrolysis cells (MECs), and water biophotolysis. Dark fermentation stands out due to its advantages, as it uses less energy and can be done at atmospheric conditions of temperature and pressure [29].

15.2 BIOHYDROGEN PRODUCTION TECHNOLOGY

Biohydrogen production involves both biological and chemical methods, relying on biomass generated through photosynthesis. This approach offers significant benefits, including abundant raw material storage, energy efficiency, and strong environmental performance. As a result, it has gained considerable attention as a research focus in the area of hydrogen generation. Figure 15.2 depicts the different hydrogen generation methods.

15.2.1 Biological Methods

Biological hydrogen production uses microbes for hydrogen generation. Photofermentation, Dark-photo co-fermentation, Dark fermentation, and water photolysis are the four main methods that come under biological hydrogen production (Figure 15.2).

15.2.1.1 Photofermentation

Photofermentation is a process that generates hydrogen in anaerobic and light environments. Photosynthetic bacteria catalyze an anaerobic fermentation reaction using small organic matter and light energy, resulting in hydrogen production [30]. Currently, common substrates for photosynthetic bacteria to produce hydrogen include starch wastewater, dairy wastewater, and soybean wastewater [31]. The most extensively studied photosynthetic bacteria for hydrogen production include *Rhodospifillum rubrum*, *Rhodopseudomonas*, *Rhodopseudomonas capsulata*, and

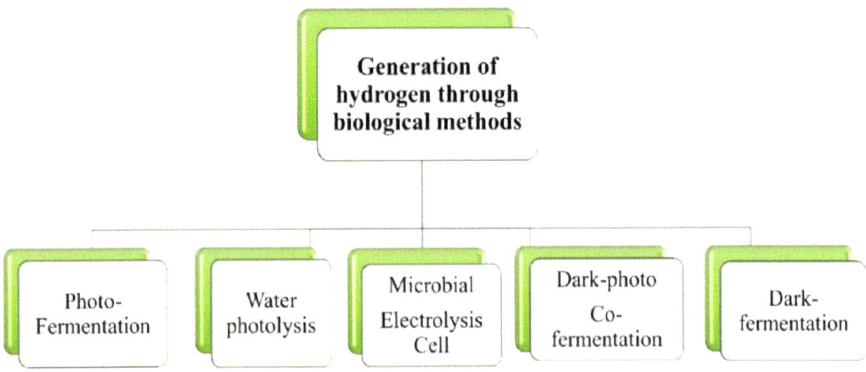

FIGURE 15.2 Different hydrogen generation methods.

Thiocapsa roseopersicina [32]. Numerous studies have demonstrated that the nitrogenase dehydrogenation mechanism is used for H_2 generation for photosynthetic bacteria [33]. Photosynthetic bacteria only possess PS I, and the electron donor is typically organic. As a result, H_2 production takes place in the absence of O_2 [34]. In comparison to the H_2 generation by water photolysis, the absence of oxygen release eliminates the need to separate hydrogen from oxygen, simplifying the production process significantly [35]. The biochemical pathways involved in hydrogen production through photofermentation can be summarized as follows:

$$(CH_2O)x \rightarrow \text{Ferredoxin} \rightarrow \text{Dinitrogenase} \rightarrow H_2$$

In the absence of O_2 and N_2, photosynthetic bacteria typically utilize butyric acid, acetic acid, malic acid, and lactic acid as substrates for H_2 generation [36]. The reaction for producing hydrogen from lactic acid is shown below:

$$C_6H_6O_3 + 3H_2O \rightarrow 6H_2 + 3CO_2 \tag{15.1}$$

Photosynthetic bacteria can utilize not only organic acids but also polysaccharides and monosaccharides for the production of H_2. The chemical reaction for this process is depicted below:

$$C_6H_6O_6 + 12H_2O \rightarrow 12H_2 + 6CO_2 \tag{15.2}$$

The production rate of hydrogen in photofermentation is influenced by factors such as the concentration of the substrate, pH value of the solution, intensity of light, and type of microorganism [37]. Organic acids, produced by anaerobic microorganisms like Clostridium during metabolism, can enhance hydrogen production efficiency. To improve efficiency, researchers have cultivated composite formulas combining Clostridium and photosynthetic bacteria [38]. For example, *Rhodobacter sphaeroides* KUPB, a salt-tolerant photosynthetic bacterium, was produced by Kawagoshi using mixed acid as a substrate and continuous light [39]. Wu et al. cultured Anabaena sp. CH_3 and *Rhodopseudomonas palustris* WP3-5 use fructose and acetic acid as sources of carbon, resulting in nearly twice the cumulative hydrogen production compared to individual cultures [40]. Sun et al. found that the rate of H_2 generation was enhanced when *Clostridium acidisoli* and *Rhodobacter sphaeroides* were cultured together under specific conditions of initial pH, the concentration of the medium, and the inoculum amount [41]. Similarly, mixing *Lactobacillus delbrueckiiNBRC13953* with photosynthetic bacteria *Rhodobacter sphaeroides* RV increased the maximum hydrogen production rate [42]. Table 15.1 presents the H_2 generation results through hybrid culture studies for fermented bacteria [43].

In conclusion, enhancing hydrogen production efficiency through gene technology or combining photosynthetic bacteria with other microorganisms holds great potential in photofermentation hydrogen production technology. This method can utilize diverse organic products like organic acids and organic wastewater [33]. The availability of these sources is extensive, and the synthesis method does not require oxygen [35]. However, the reliance on light for photofermentation poses challenges in conducting amplification tests [44].

TABLE 15.1
Hydrogen Generation Results through Hybrid Culture Studies for Fermented Bacteria

Carbon Source	Microorganism 1	Microorganism 11	Yield of H_2 (mol H_2/mol hexose)
Glucose	*Clostridium butyricum*	*Rhodobacter sp.M-19*	6.6
Glucose	*Escherichia coli HD701*	*Rhodobacter sphaeroides O.U.001*	2.4
Sucrose	*Microflora*	*Rhodobacter sphaeroides SH2C*	3.32
Glucose	*Clostridium butyricum*	*Rhodobacter sphaeroides*	5.6
Glucose	*Rhodopseudomonas palustris P4*	*Rhodopseudomonas palustris P4*	4.8–5.6

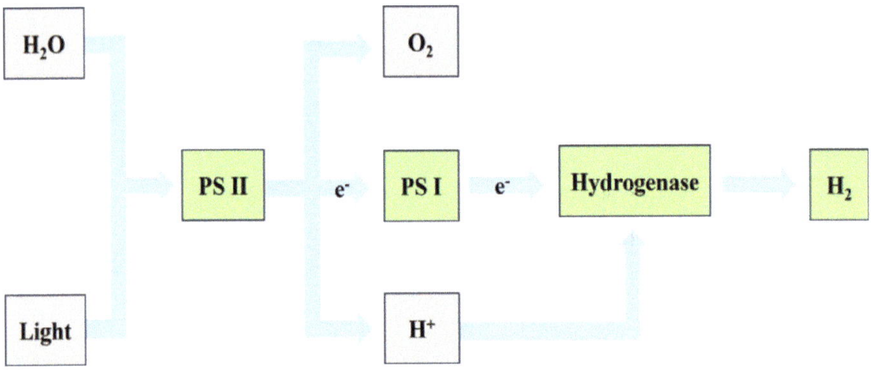

FIGURE 15.3 Mechanism of the water photolysis method. (Adapted from [47].)

15.2.1.2 Water Photolysis

Water photolysis uses microorganisms such as blue and green algae to decompose the water [45,46]. These microorganisms can decompose water through photosynthesis, using light, air, water, and simple inorganic salts. Production of H_2 can be done by: photosynthetic system I and photosynthetic system II. Initially, photosynthetic system II decomposes water in light and generates electrons (e^-), hydrogen ions (H^+), and O_2. On the other hand, photosystem I produce a reducing agent to capture carbon dioxide. Electrons (e^-) generated by photosystem II are transferred to hydrogenase through ferredoxin from both photosystems. This transfer catalyzes the production of H_2. The method is depicted in Figure 15.3. Water photolysis is a potential green and biological method for hydrogen production [47].

In 1939, Gafforn found that *Scenedesmus obliquus* can produce hydrogen through its metabolism [48]. Numerous subsequent studies have confirmed the ability of various green algae, like *Chlorellafiscal* [49], and *Chlamydomonas reinhardtii* [50,51], to produce hydrogen. Cyanobacteria, a prokaryote, is capable of producing oxygen through photosynthesis and exists in different forms. In the *Nostoc* and *Anabaena* genera, cyanobacteria's ability to produce hydrogen has been well explored [52].

Currently, the development of techniques to improve hydrogenase enzymes' response to oxygen is the subject of study, thereby prolonging hydrogen production and increasing yields [53]. The benefit of the water photolysis method for the production of H_2 is that it can occur in an aqueous environment at standard temperature and pressure, utilizing water as the source [37]. Several variables, including the catalyst, raw material, light intensity, and temperature affect the efficiency of H_2 production through the water photolysis method [37]. Yan [54] discovered that using $Ni_3P/Ni/G\text{-}C_3N_4$ composite as the raw material resulted in a hydrogen generation rate of up to 203.3 μmol/(g·h). By using NixPy/CdS composite as the basic material and adding other additives, Yan [54] achieved an improved rate of approximately 601.5 μmol/(g·h). Li et al. [55] achieved a maximum rate of H_2 production of 1,061.87 mmol/(g·h) by using foamy carbon as the raw material and employing different catalysts. Effective utilization of catalysts can significantly enhance the H_2 yield by water photolysis [56,57]. However, several technical challenges still need to be addressed for industrial applications, such as low efficiency, complex synthesis methods, and high costs [58].

15.2.1.3 Dark-Photo Co-fermentation

Dark-photo co-fermentation method combines the strengths of dark-fermentation and photofermentation bacteria to increase the efficiency of H_2 generation and substrate conversion [59,60]. The process involves pre-treating biomass raw materials, dark fermenting them to generate H_2 and releasing tail liquid, and then using small organic acids in the tail liquid as electron donors for photofermentation to generate H_2 [47, 61,62] (Figure 15.4). This technique significantly increases the total amount of hydrogen, making it efficient and cost-effective [63,64]. Researchers have explored the use of different bacteria and substrates in dark-photo co-fermentation, such as

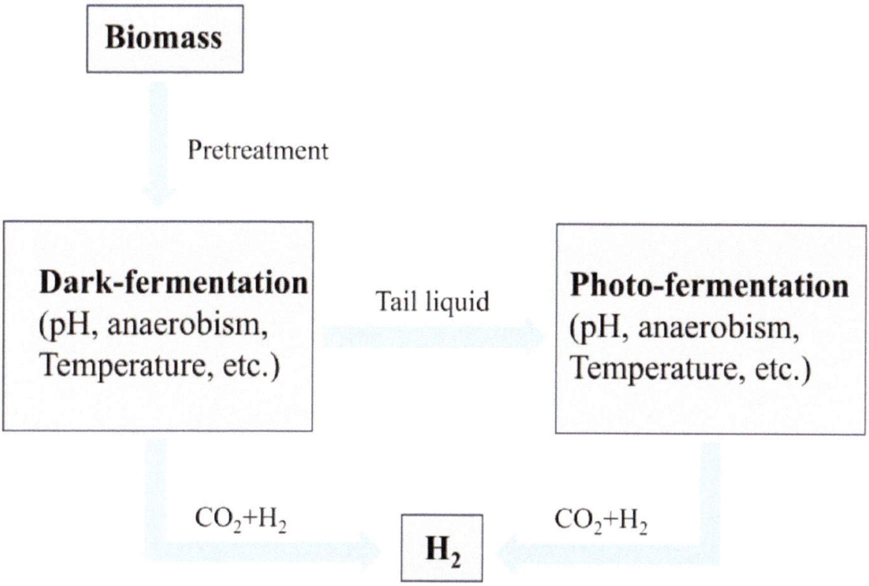

FIGURE 15.4 Production of H_2 by dark-photo co-fermentation method. (Adapted from [47].)

Rhodobacter sphaeroides O.U.001 with *Enterobacter cloacae DM11* metabolites and sucrose as a substrate, resulting in higher H_2 content and production [65,66]. However, there may be limitations or challenges associated with this method [67,68]. The earlier tables show that dark-photo co-fermentation is more effective in producing hydrogen. In a study by Zhang et al. [59], they examined the conversion rate of energy and rate of H_2 generation of dark-photo co-fermentation, dark fermentation, and photofermentation with corn straw. The dark-photo co-fermentation showed a higher rate of H_2 generation and a good conversion rate of energy.

15.2.1.4 Dark Fermentation

Dark fermentation or anaerobic fermentation is a biological method of H_2 production, that produces organic acids from organic waste using heterotrophic anaerobic bacteria in the absence of O_2 via the fermentation method [69,70]. Various species, such as glucose, carbohydrates, starch, galactose, cellulose, etc. can be used and decomposed by fermentative bacteria for the production of H_2 [71,72]. From organic waste, anaerobic bacteria can produce hydrogen and can be classified as obligate and facultative bacteria [73]. Obligate anaerobic bacteria include *Clostridium*, *Desulfovibrio*, and others, while *Escherichia*, *Klebsiella*, and *Enterobacter* are the most common facultative anaerobic bacteria [74]. *Enterobacter*, *Enterobacter escherichia*, *Clostridium*, and *Bacillus* have been extensively researched and have shown good H_2 production rates [59,74]. Fermented bacteria can convert various substrates like cellulose, formic acid, etc. into H_2 using hydrogenase enzyme or nitrogenase in the dark fermentation process.

The reaction for producing hydrogen from glucose is shown below:

$$C_6H_{12}O_6 + 2H_2O \rightarrow 4H_2 + 2CO_2 + 2CH_3COOH \tag{15.3}$$

Numerous studies have been conducted on H_2 generation by the dark fermentation method. Xu et al. studied that the phosphate amount influenced the rate of H_2 generation and growth of *E. harbinenseB49*, with the optimal phosphate concentration being 50 mmol/L [75]. Chen et al. achieved a yield of H_2 of 4.52 mol/mol using a CSTR reactor with maltose as the substrate [76]. Fang et al. discovered that rates of H_2 generation and glucose usage were maximum at a pH of 5.5 [77]. Delavar and Wang observed that the rate of H_2 production was increased with an injection of acid by 40% [74]. Table 15.2 provides the production of H_2 through dark fermentation using microorganisms [78,79].

TABLE 15.2

Production of Hydrogen through Dark Fermentation Using Microorganisms

Carbon Source	Microorganism	Yield of H_2 (mol H_2/molhexose)
Glucose	*Citrobacter sp. Y19*	2.49
Glucose	*Enterobacter cloacae IIT-BT08*	6.0
Glucose	*Enterobacter cloacae IIT-BT08*	2.28

Dark fermentation offers benefits such as stability, versatility in substrates, eco-friendliness, continuous production of H_2 without a light source, and non-hazardous reaction conditions. However, it has drawbacks like low-raw material utilization, product inhibition, and environmental pollution from the tail fluid. The H_2 production through dark fermentation still requires in-depth cost analysis [80–83].

15.2.1.5 Microbial Electrolysis Cell

When a small amount of soluble organic matter present in wastewater is subjected to an electric current, microbially catalyzed electrolysis cells create hydrogen during the fermentation process [84,85]. Acetate is the most straightforward carbon source for MECs to produce H_2. Nonetheless, it is thought to be advantageous for MECs to produce hydrogen when volatile fatty acids are present in the biohydrogen production effluent through the dark process [75,86]. Figure 15.5 shows the four essential components of the MEC: the anode and cathode, the external electrical connection,

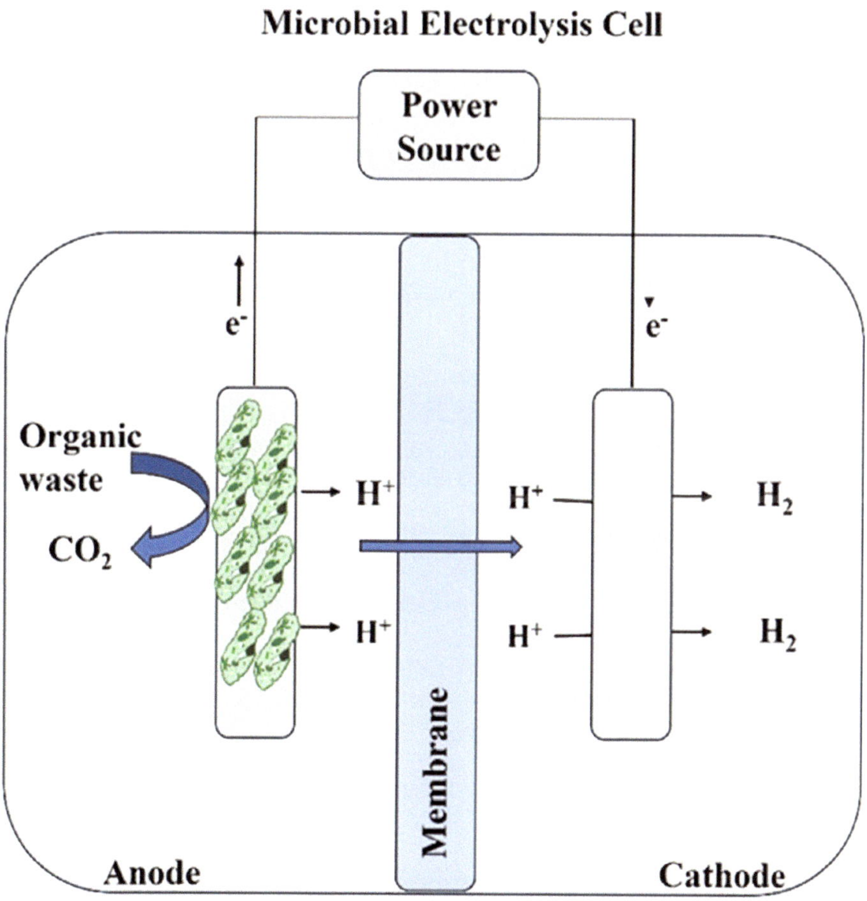

FIGURE 15.5 Diagrammatic representation of a microbial electrolysis cell used to produce hydrogen. (Adapted from [2].)

and the cationic exchange membrane that separates the electrodes. Through a series of redox reactions, bacteria living on the anode surface of the cell oxidize the carbon sources of waste to produce protons, electrons, and carbon dioxide.

The bacteria transport the electrons they have created to the anode's surface. In addition, the proton contributes to the reduction reactions taking place on the cathode electrode surface as it readily diffuses into the solution and migrates via the cation exchange membrane to the cathode compartment. Concurrently, an external power source transports the electrons to the cathode electrode, where they couple with the free proton in the electrolyte solution to generate H_2 gas [87]. To power the MEC, an AC/DC electrical supply is needed. MEC uses only 0.11 V of energy, which is less than water electrolysis's 1.8–2.0 V [88]. With limited restrictions, including decreased mass transfer and energy waste, this technology is therefore thought to have great potential to develop. MECs encounter five primary obstacles: metabolic diversity, the loss of methanogenic electrons, electrode resistance (anode), power supply, intricate wiring in cell design, membrane issues, and cathode-related problems (high-catalyst cost, long-term stability, side reaction, and electrode resistance). Progress in addressing the above-mentioned problems will inform future research endeavors in this domain [89]. When paired with dark fermentation methods, MECs function effectively. Compared to the single-stage method, the MEC–DF combined process yields around 98% more H_2 [90]. Table 15.3 provides a comparison of biohydrogen production such as process reaction, advantages, and disadvantages [91–94].

15.3 HYDROGEN PRODUCTION FROM ALGAE: A SUSTAINABLE METHOD

Small photosynthetic organisms known as microalgae are present in freshwater surface waters and play a crucial role in water turbidity in lagoons and aquariums. They are considered abundant renewable resources and are even classified as aquatic weeds. While their photosynthetic mechanism is similar to terrestrial plants, microalgae have the advantage of growing in an aquatic environment with ample water supply, resulting in a higher rate of conversion of solar energy into biomass compared to other plants. Microalgae-derived biofuels are less harmful to the environment as compared to biofuels made from other sources [95–97]. Microalgae are a plentiful renewable resource. They contribute to water turbidity in lagoons and aquariums and are considered aquatic weeds. The fat in their cellular structure can be used to produce biofuel [98]. Microalgae can absorb CO_2, grow rapidly, and can be cultivated year-round. They can even be grown in sewage without harming water resources [99,100].

15.3.1 Cultivation of Microalgae

To generate hydrogen from algae, a significant amount of algae biomass is needed. Algae require light, carbon dioxide, water, and mineral salts for growth. Some algae can grow heterotrophically, using organic carbon sources and not relying on light. The ideal range of temperature range for their development and growth is 15°C–30°C.

TABLE 15.3
Comparison of Three Different Biohydrogen Production Systems

Process	Process Reaction	Advantages	Disadvantages
Biophotolysis	$2H_2O + light \rightarrow 2H_2 + O_2$	Direct production of H_2 in the absence of sunlight and water Water in abundance for the substrate Substrate is not required as a nutrition	Low-H yields due to low-light conversion efficiency Energy required to provide light is restricted by day-and-night cycles when using solar energy, and extra expenses when using other light sources
Dark fermentation	$C_6H_{12}O_6 + 2H_2O \rightarrow 2CH_3COOH + 2CO_2 + 4H_2$	Generates the largest H_2 yields when compared to alternative methods of producing biohydrogen. No requirement for light, which H_2 can generate continuously Various carbon sources, including garbage, can be utilized as substrates	Process expenses are increased since a separation unit is needed to separate the CO_2 and H_2 product mixture
Photofermentation	$CH_3COOH + 2H_2O + light \rightarrow 4H_2 + 2CO_2$	The diverse light energy spectrum is useful for bacteria In this process, wastewater can produce H_2 H_2 and CO_2 are entirely transformed from organic molecules	Low-H_2 yields are caused by low-light conversion efficiency Energy required to provide light is restricted by day-and-night cycles when using solar energy, and extra expenses when using other light sources

Algae growth requires essential nutrients such as phosphorus, nitrogen, silicon, and iron. To achieve large-scale production, algae cells need to be continuously mixed and the biomass prevented from settling. Different cultivation methods include photobioreactors, open raceway ponds, and heterotrophic systems [101,102].

15.3.1.1 Open Raceway Ponds

The open raceway pond method considered the easiest and most ancient for algae cultivation, involves tanks with a depth of 30 cm where algae are grown under equal conditions [103]. A flow design and gear ensure the mixing and circulation of algae cells with nutrients, preventing biomass settling. This method has lower costs compared to photobioreactors, but it relies on evaporation for cooling, which can limit carbon dioxide utilization by algae due to water scarcity. Additionally, in this process, microbial contamination is unavoidable, limiting biomass production [104].

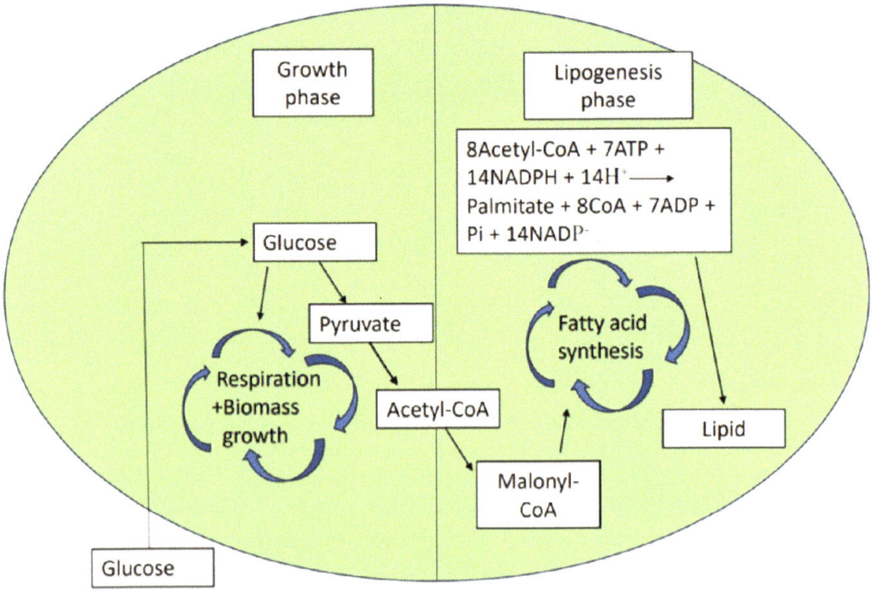

FIGURE 15.6 Heterotropic cultivation of microalgae. (Adapted from [106].)

15.3.1.2 Heterotrophic

Certain algae can thrive in heterotrophic conditions, where they don't require a light source and obtain carbon from organic sources in their environment [105]. These algae tend to generate less protein and more fat compared to photosynthetic algae as shown in Figure 15.6 [106]. Heterotrophic cultures need to be fully sterile and rich in nutrients. An environment rich in organic carbon sources, such as carbohydrates and organic acids, is ideal for the growth of algae. However, only some can thrive in heterotrophic environments, and fat production in this system depends on factors such as the culture medium's vitality, nutrient availability, and environmental factors like temperature, pH, and salinity [105,106].

15.3.2 Mechanism of H_2 Production from Microalgae

Microalgal H_2 production can be achieved through three mechanisms: mixing, heterotrophic, and photoautotrophic. The photoautotrophic approach consists of both closed and open-pond photobioreactor methods [107]. In the open-pond method, artificial ponds or natural water sources can be used. Artificial systems often consist of circular closed loops, around 0.2–0.5 m deep, with equipment like agitators to avoid sedimentation and give algae the necessary CO_2. This method is cost-effective, requires less energy, and is easier to clean compared to the closed method [108].

15.3.2.1 Mixed Production System

This approach combines elements of both the open and closed methods. The initial step takes place in photobioreactors to maintain controlled reaction conditions.

TABLE 15.4
Value-Added Products Produced by Distinct Algae Strains throughout the Biohydrogen Conversion Process

Algae	Final Products	Application
Porphyridium cruentum	Polysaccharide	Medical application
Arthrospira, Spirulina sp.	Biomass, phycocianin, protein, vitamin B12	Anti-oxidants, cosmetics
Chlorella sp.	Carbohydrates	Substitutes for food
Phaedactylum tricomutum	Lipid, fatty acid	Biofuels
Haematococcus pluvialis	Astaxanthin, carotenoids	Medical application

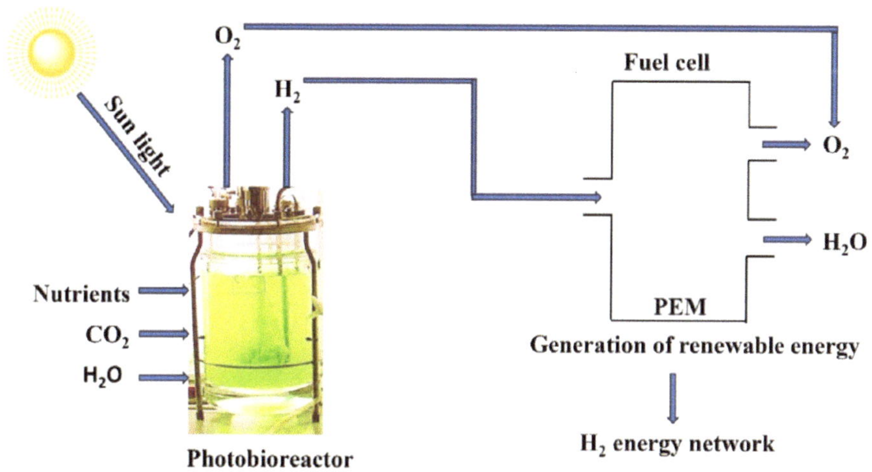

FIGURE 15.7 The production of biohydrogen from microalgae in photobioreactors. (Adapted from [15].)

The microalgae are moved to open ponds in the second stage, subjecting the cells to nutritional stress and boosting lipid synthesis. Table 15.4 provides an overview of the value-added products produced by distinct algae strains throughout the biohydrogen conversion process [109–111].

15.3.2.2 Photobioreactors

Photobioreactors are more effective and address difficulties with evaporation and pollution. The attainment of high-hydrogen generation efficiencies at relatively cheap prices is a prerequisite for the commercial production of hydrogen. Photobioreactors designed specifically for the growing of microalgae are required for the industrial production of biohydrogen using microalgae [16,112]. A photobioreactor used to produce biohydrogen from microalgae is depicted in Figure 15.7 [15]. Agitation and tank depth are the two main parameters influencing PBR design [113]. The majority of the time, open ponds are used for commercial algae farming. Open ponds are ineffective for producing hydrogen through microalgal fermentation since hydrogen is a gas. The production of biohydrogen can be better controlled by the use of closed PBRs, which come in a variety of designs for large-scale microalgae growing [112,114]. Light penetration, which is typically provided by sunlight, is necessary for all photobioreactors. Indoor farming does, however, also make use of artificially produced regulated light. In general, the strain's properties are taken into account while designing microalga PBRs. The primary variables that impact the efficiency of photobioreactors are temperature, gas exchange, agitation, volume ratio, and deep light penetration. Hydrogen is produced using many types of PBRs. These photobioreactors are categorized as flat panel, vertical column, and tubular types. A specific control system governs the temperature, pH, optical density, and hydrogen generation in this photo-bioreactor [112]. Furthermore, these photobioreactors have the capacity to achieve significant photosynthetic activity. Nevertheless, flat panel PBRs can only be used in labs and little enterprises [115]. Comparing flat panel photobioreactors to other types, they are thought to be more cost-effective. Vertical photobioreactors are made up of a clear glass column surrounded by sheets of water. By moving water and allowing extensive light penetration into the suspension, it enables temperature control. Long transparent tubes make up tubular-type photobioreactors. These tubes are used by airlift pumps to transport photosynthetic microalga cells [116].

15.4 FACTORS AFFECTING THE H_2 GENERATION

Several factors influence the rate of hydrogen production and its yield in both dark and photofermentation processes. These factors include:

15.4.1 Effect of Initial pH

In the dark fermentation method, the initial pH is an important factor, as each microbe has different optimal pH conditions. For efficient hydrogen production, the effective pH range is 5–8, with production halving when the pH drops below 5 [117]. In batch fermentation, a pH range of 5–9 has been used, but an initial pH range of 5–6 is reported as optimal [118]. Volatile fatty acids produced during fermentation lower the pH, so an initial pH of 6–7 is set to compensate [119]. The pH also affects hydrogen production by green algae, with an improved yield of 2.4% observed at a pH of 6.5 [120]. Purple non-sulfur bacteria (PNSB) also require a suitable pH, with studies suggesting a pH of 7 for efficient electron transport to nitrogenase enzymes during photofermentation for hydrogen production [121].

15.4.2 Effect of Nutrients

Macronutrients like sulfur, phosphorus, and nitrogen are crucial for bacteria growth and H_2 production [122]. Sulfate, commonly present in organic matter, is converted into sulfide by sulfate-reducing bacteria during fermentation. However, high-sulfide levels can be toxic and inhibit H_2 production, while also reducing the availability of trace elements [123]. Nitrogen, particularly in the form of ammonia, can also decrease bacterial activity and H_2 production. Protein degradation in the fermentation medium can lead to high-ammonia levels [124,125]. Phosphorus is another important nutrient for enhanced H_2 production, with a rate increase obtained at a concentration of 600 mg/L K_2HPO_4 [126,127]. A 40% rise in H_2 generation was seen.

15.4.3 Pretreatment Methods

Using various food wastes as raw materials offers useful nutrients and organic compounds that enhance the rate of H_2 production. However, the presence of inhibitory compounds can negatively impact production and yield [128]. To improve the utilization of raw materials for the production of hydrogen, different pretreatment methods have been explored. Preheating and hydrolysis methods are the most commonly used among all. Alkaline hydrolysis for 6 hours at pH 12 can increase H_2 generation by 206 times, while acid hydrolysis for 12 hours at pH 2 can triple H_2 production. However, these methods require large quantities of chemicals and extra steps to balance the pH [119,129]. Ultrasonication of food waste, combined with hydrolysis, can increase H_2 yield by 75%–88% [130]. Nevertheless, the commercialization of this method faces challenges due to equipment investment and energy costs. Preheating food waste is another effective pretreatment method [131]. The findings showed that preheating the waste at 90°C for a minimum of 20 minutes before fermentation can enhance the yield of H_2.

15.4.4 Effect of Metal Ions

Moderate amounts of metal ions are used to promote microbial growth and optimize enzyme activity in dark- and photofermentation. However, high concentrations of these metal ions can inhibit fermentation by damaging the cell membrane, disrupting ion and nutrient transmission, and causing intracellular metal accumulation [132,133]. Iron (Fe) is important for metabolic changes and hydrogenase enzyme expression, but excessive Fe concentration can lead to cell clumping and reduced mass transfer activity [118]. Magnesium, calcium, and sodium ions are useful in the growth of bacteria, but their large amounts can slow down growth and become toxic. Researchers found that hydrogen production was high when sodium was not present [134]. Sodium, in particular, can raise osmotic pressure and negatively affect bacterial activity. It's recommended to keep sodium concentration below 20 g/L for optimal hydrogen production [135]. Calcium (Ca^{2+}) and magnesium (Mg^{2+}) are also essential trace elements for bacterial growth and function [136]. Nickel (Ni^{2+}) at a level of 0.1 mg/L does not inhibit H_2 yield. Currently, there are no significant measures to regulate the inhibition of metal during the fermentation process, but some pretreatments may be explored [137].

15.4.5 Effect of Concentration of Substrate

Optimal substrate concentration is crucial for efficient H_2 production in dark fermentation. Higher substrate concentrations can create unfavorable conditions, leading to changes in pH, H_2 partial pressure, and volatile fatty acid concentrations. To minimize substrate inhibition, it is important to arrange the initial substrate concentration appropriately [138]. While inhibition by substrate concentration has been extensively studied for carbohydrate sources, there is limited literature on using wastewater and organic waste as substrates [139]. Fed-batch reactors can help mitigate substrate inhibition, and certain bacteria like *Enterobacter aerogens* can stimulate microbial activity for H_2 production, reducing inhibition [138]. The influence of substrate concentration on the production of hydrogen was also assessed for *Clostridium butyricum* and *Lactobacillus casei*, with glucose and galactose. Under low-glucose concentrations, *L. casei* outcompeted *C. butyricum*, resulting in lower hydrogen production [140].

15.4.6 Effect of Temperature

The temperature has an important impact on the production of hydrogen and enzyme activity [141,142]. In dark fermentation, thermophilic bacteria show higher hydrogen synthesis at high-temperatures in comparison to mesophilic bacteria. The optimal temperature range for enhanced biohydrogen production is reported to be 30°C–55°C, with inhibition of H_2 producers at temperatures above 60°C [142,143]. Only hyperthermophilic bacteria like *Thermotoga maritime* and *Pyrococcus furiosus* can generate H_2 at temperatures above 80°C [144]. PNSBs are also temperature-sensitive. A study on photofermentation showed that the rate of production of H_2 was increased to 30°C, but hydrogen production declined above 30°C due to enzyme inhibition by higher temperatures [121].

15.4.7 Effect of Light Intensity

Light intensity plays a crucial role in H_2 production by PNSB. Increasing the light intensity can enhance PNSB performance, but excessive increase can reduce growth and H_2 production. Higher light intensity demands more ATP and reduces power for the photosynthetic system and nitrogenase enzyme, limiting H_2 production [145]. Photoinhibition occurs when light intensity exceeds certain levels, with studies showing decreased H_2 production above 200 W/m^2 or at a light intensity of 6,000 lx. LED light with a wavelength range of 770–920 nm is favorable for PNSB activity due to its cost-effectiveness and energy efficiency [146,147].

15.5 DIFFICULTIES WITH H_2 GENERATION

The earlier section covered different techniques/methods to produce hydrogen. Although biohydrogen is a cleaner alternative to fossil fuels, there are still challenges to overcome in scaling up production and improving yield for the complete replacement of fossil-based fuels. This section will address these challenges and the need for commercial feasibility and sustainability in biohydrogen production.

15.5.1 Microbial Electrolysis Cell

The performance of MECs depends on factors like MEC architecture, electrode materials, type of inocula, substrate nature, applied potential, CH_4 production, and cost considerations. However, there are challenges to overcome before scaling up and commercializing MEC technology [148]. Optimizing EAB as biocatalysts and understanding electron transfer mechanisms are important. The methanogenic activity contributes to low-H_2 yield, so strategies to inhibit methanogens need further investigation [149]. Exploring low-cost, highly conductive electrode materials is essential. MEC technology faces the challenge of high-capital costs for cathode materials and catalysts, which make up a significant portion (47%–85%) of the total expenses [150]. The high-capital cost of cathode materials and catalysts, especially platinum, is a challenge. Cheaper and eco-friendly alternatives are being sought [151]. Microbial cathodes made from carbon are a potential option [152]. For scaling up, an MEC design that combines high-H_2 yields with low costs is highly desirable. This technique merges the metabolic processes of microorganisms with the electrochemical cell. A MEC consists of two terminal electrodes —positive and negative. Within the MEC, specific bacteria known as exoelectrogenic bacteria (also referred to as electro-active bacteria) transform organic substances into protons, CO_2, and electrons. These electrons are transferred to the positive terminal of MEC, while the protons go to the electrolytic solution. Subsequently, the electrons move from the anodic electrode to the cathodic electrode through an outside electrical pathway, where they unite with available protons to generate hydrogen within the cathode, which operates in an oxygen-free environment. To facilitate this electron flow, an external electrical potential is necessary, imposed on the MEC in the presence of biological assistance. Nevertheless, the electric potential required for MECs is notably lesser compared to that needed for electrolysis. The MECs can utilize either a pure strain or a mixed consortium of microorganisms as an inoculum. The preference for mixed cultures often aligns with these advantages. These microbial communities can be cultured from sources like cultured microbial sediment, effluent from wastewater, residues from sewage treatment, or by-products of fermentation. The substrates employed in this process encompass a range, starting from sugars like glucose, extending to compounds like acetate and fermentation by-products, and even including effluent from wastewater. This indicates that MECs possess comparable potential for both remediation and the generation of green energy, akin to the biohydrogen production processes through fermentation. Numerous studies have employed acetate, a side product of the fermentation process, as a starting material for MECs. Initially, the design of MECs consisted of a dual-electrode configuration isolated by an ion exchange partition within a binary-compartment arrangement. The frequently employed membrane was the Proton Exchange Membrane (PEM). This setup facilitated the distinct trapping of carbon dioxide at the anodic terminal while capturing Hydrogen at the cathode. It also prevented the utilization of hydrogen by the bacteria at the anode and enabled the gathering of purer hydrogen, along with preventing cathode fouling due to bacteria. Nonetheless, this design led to the notable loss of energy and pH alterations that prohibited the production of hydrogen, subsequently resulting in a shift toward MECs with a single chamber. Consequently, chambers without membranes exhibit lower investment costs compared to their double-chambered MECs. Furthermore, the absence of a membrane lowers the resistance, thereby enhancing the efficiency of hydrogen generation in chambers without membranes. Nevertheless, a key drawback of single-chamber MECs is the reduced hydrogen production, primarily attributed to the presence of methanogenic activity [153].

15.5.2 Fermentative Biohydrogen Production

Switching to agricultural, food, and livestock wastes as substrates reduces operational costs in fermentative biohydrogen production. Using sewage sludge and wastewater as substrates addresses bioremediation and energy production. Pretreatment of complex substrates and inocula improves hydrogen yields. Selection of the appropriate microorganism is crucial for optimal yield. Purification steps are necessary to separate hydrogen from the biogas mixture and capture CO_2 for carbon neutrality. The DF process is faster but has a low-hydrogen yield due to the formation of by-products such as butyric acid, lactic acid, acetic acid, methanol, butanol, acetone, methane, and hydrogen sulfide [154]. In addition to hydrogen (H_2) and carbon dioxide (CO_2), the fermentative process also produces by-products like butyric acid, lactic acid, acetic acid, methanol, butanol, acetone, methane, and hydrogen sulfide. The composition of biogas depends on the microorganism and substrate used [155]. While these by-products can be used in other hydrogen production methods, their formation reduces the H_2 yield by 1–2.5 mol of H_2 per mole of glucose [156]. Selecting the right microorganisms, and pretreatment technologies, and optimizing process parameters can enhance H_2 yield and production rate.

Organic waste materials consist of intricate, lengthy chains of polymer. Consequently, the fermentation of organic substances typically requires an extended period and frequently yields a limited amount of hydrogen. To accelerate the reaction rate and enhance the quantity of produced hydrogen, it becomes imperative to subject the organic wastes to pretreatment methods. The fundamental aim of pretreatment techniques is to disintegrate the intricate structure of organic wastes, rendering them more amenable to fermentation. These pretreatment methods can be broadly categorized into physical approaches (such as milling, microwave, grinding, thermal treatments, and sonication), chemical methods (including acidic solutions, alkaline solutions, and ozonation reactions), physicochemical procedures (like ammonia fiber expansion and steam explosion), as well as biological interventions (employing bacteria and fungi) [157].

Microorganisms utilized in fermentation can be cultivated as either a monoculture, a cooperative culture, or a heterogeneous culture. In co-cultures, the selection of microorganisms is meticulously carried out under the synergy of their functions. For example, combining cellulolytic and hydrogen-generating bacteria in a co-culture accelerates the transformation of cellulose into simpler sugars by the microbes. Subsequently, these simpler sugars are utilized by the anaerobic microorganisms to generate hydrogen [158]. Mixed cultures typically comprise three or more distinct varieties of microorganisms. These cultures can be sourced from materials such as effluent from wastewater, residues from sewage treatment, or by-products of fermentation. Mixed cultures offer multiple advantages compared to pure strains, including variability of microorganisms, adaptable substrate utilization, and the absence of sterilization prerequisites.

15.5.3 Biophotolysis

Biophotolysis for hydrogen production has a low yield due to sensitivity to oxygen and low-light conversion efficiency. To control oxygen, O_2-binding proteins can be used or O_2 production can be inhibited. Sparging the reactor with an inert gas also reduces O_2 levels but increases costs. Nutrient deprivation can improve yield by reducing O_2 evolution. Metabolic engineering can prevent H_2 consumption by hydrogenases [159]. Genetically modified algae and cyanobacteria have shown increased light conversion efficiency [160,161]. However, biophotolysis raises safety concerns and requires artificial illumination in closed systems, increasing costs [162]. Other factors affecting yield include light-capturing and CO_2 fixation efficiency [163]. Approaches to improve biophotolysis include enhancing light-utilization efficiency [164], metabolic and genetic engineering [165], bioreactor design, bioprospecting [166], and optimizing culture and process parameters [167].

15.6 NANOTECHNOLOGY AND BIOHYDROGEN

Nanotechnology, focusing on nm-sized particles, has various applications in biosensors, medicine, immobilization, and biofuel production [168]. Nanoparticles (NPs) have been found to influence microbial activities and enhance biohydrogen production under aerobic conditions [169]. NPs made from activated carbon, gold, nickel, copper, iron, palladium, silver, titanium, and iron, etc. have been used to increase H_2 production by providing a larger surface area for electron adsorption and stimulating hydrogen-producing enzymes [168,170–173].

15.6.1 Use of Metallic Nanoparticles

Zhang's group studied the enhancement of H_2 biosynthesis utilizing gold nanoparticles (NPs) [173]. They utilized artificial wastewater and observed that the NPs increased microbial metabolic activity, resulting in higher H_2 production rates in comparison to the control. Gold nanoparticles of 5 nm showed the highest yield [173]. Another research used mixed culture and silver nanoparticles to produce hydrogen from glucose [174]. The H_2 production was increased with the increasing Ag-NPs concentration up to 20 nM. However, higher concentrations did not further increase the hydrogen production rate. The obtained yield of H_2 was 67.6% at 20 nM Ag-NPs. Additionally, the Ag-NPs boosted cell biomass and shortened the lag phase in H_2 production.

Han et al. used an anaerobic fed-batch procedure to investigate the effects of hematite nanoparticles (NPs) and the starting pH on the generation of hydrogen in mixed bacteria. The highest H_2 yield was obtained at 3.21 mol H_2/mol sucrose. Transmission electron microscopy confirmed the slow release of hematite NPs and their impact on bacteria shape. Another study explored the use of biogenic palladium NPs and ions [170]. Palladium NPs synthesized from *Cortandrum sattvum* leaf extract achieved the highest yield of hydrogen of 1.48 mol H_2/mol glucose at a concentration of 5 mg/L, due to increased hydrogenase enzyme activity. However, palladium ions did not show a favorable impact on H_2yield and lag phase. Iron NPs, specifically iron ions, and Fe-NPs, were investigated for H_2 production [175]. The favorable impact on hydrogen yield was shown by both in comparison to the control, but Fe-NPs led to a 75% reduction in propionate production compared to a 35% reduction with Fe^{2+} ions. Phytogenic iron NPs and ions were also studied, resulting in a 100% increase in hydrogen production (1.9 mol H_2/mol hexose) with 100 mg/L Fe-NPs compared to the control (0.95 mol H_2/mol glucose) [176]. Other studies have also documented the use of various methods to prepare Fe, Fe_2O_3, and Fe_3O_4 nanoparticles to improve hydrogen biosynthesis [177].

Researchers have also studied the rate of hydrogen production using ZnO nanoparticles [178]. These nanoparticles, synthesized through precipitation, in comparison to the control were found to decrease the yield of hydrogen. However, metallic nanoparticles such as titanium dioxide, copper, and nickel have been shown to increase the rate of hydrogen production [179–182]. While there are numerous studies on increasing the rate of H_2 generation in dark fermentation processes using metallic nanoparticles, there are fewer studies on their impact on photo-fermentative hydrogen production. For example, the activity of photosynthetic bacteria was enhanced with TiO_2 nanoparticles, which resulted in a high yield of H_2 while it reduced the activity of hydrogenase enzyme uptake [181]. Pandey et al. conducted a study that showed a similar enhancement effect of TiO_2 nanoparticles on photo-fermentative hydrogen production [182]. Additionally, Kanwal et al. utilized the photo-fermentative method for H_2 generation by the use of a Phyto-fabricated nanoscale iron complex [183]. Carbon nanotubes have also been explored for enhancing H_2 production using bacteria [184].

15.7 FUTURE OUTLOOK

The depletion of fossil fuels and climate change are driving nations to explore renewable and eco-friendly energy sources. Hydrogen is becoming a potential source of energy, with fuel cells gaining attention. Japan has been at the forefront of promoting a hydrogen society, showcasing initiatives like the Harumi Flag and hydrogen-powered vehicles. California, the first US state to launch fuel cell electric vehicles (FCEVs), offers incentives to attract buyers. South Korea has a Hydrogen Economy Roadmap and is establishing demonstration cities powered by hydrogen. Notably, Hyundai's Nexo, Toyota's Mirai, and Honda's Clarity Fuel Cell Electric Vehicle are active in the market. China has also recognized the role of FCEVs in decarbonizing its transport sector. The document also advocates for increased research and development in fuel cell technology. We can expect a growing number of FCEVs and refueling stations in the coming years. China has formed a partnership with Norway to promote alkaline electrolyzers. In Europe, there is a strong focus on research and development to enhance fuel cell and hydrogen technologies. European countries like Norway, Austria, the Netherlands, and Germany have made significant investments in large-scale hydrogen production through electrolysis. Currently, electrolysis is the most popular method for producing clean hydrogen, but it requires a substantial amount of energy. However, the cost of hydrogen production may improve with the use of renewable energy sources like wind and solar power. As technology advances, the costs of electrolyzers and renewable electricity are expected to decrease, making hydrogen from renewables the most affordable source of clean hydrogen in the future, according to IRENA. In their study, Hossienzadeh et al. evaluated various H_2-producing techniques based on techno-economic and life cycle assessments.

When combined with photofermentation and MECs, the efficacy of the dark fermentation process increased. This combined approach yielded approximately 1 L of H_2 per gram of organic waste. In terms of cost-effectiveness, the dark-fermentative process was the most economical at 2.3 US $/g, followed by MECs at 2.8 US $/g, and photofermentation at 3.5 US $/g. Furthermore, both fermentation processes demonstrated lower greenhouse gas emissions compared to MECs. According to their assessment, MEC technology and hybrid dark fermentation have utilized 1.3 billion metric tons of food waste to produce 105 million tons of H_2 gas, surpassing the actual demand in 2020 by 120%.

Biohydrogen production is the most promising method and deserves our attention. Despite being less polluting, challenges remain regarding gas production rate and cost. Here are some main recommendations for H_2 production:

1. Hydrogen benefits the environment but its production requires improved control of reaction conditions. Exploring advanced hydrogen production methods and optimizing the reaction conditions can enhance the efficiency of H_2 production.
2. The rate of hydrogen production can be increased by enhancing the activity of bacteria and by improving the efficiency of the conversion of biomass.
3. Incorporating biomass technology, like biohydrogen carriers, can effectively reduce costs and improve the efficiency of hydrogen production.

15.8 CONCLUSIONS

Hydrogen energy is a key focus in modern energy research as a potential replacement for fossil fuels. Biomass green hydrogen production is gaining attention because traditional methods are expensive and non-eco-friendly. Biohydrogen production utilizes a wide range of raw materials, resulting in less pollution. Chemical methods yield more hydrogen but require higher temperatures and incur certain costs. Biological methods offer stable production rates and have the potential for sustainable green energy. However, as it can lower greenhouse gas emissions while meeting future energy demands, producing hydrogen in this way may be very advantageous. The various categories of solid and liquid waste feedstock, microorganisms, and enzymes involved in biohydrogen production were discussed. Examples of these routes of biological hydrogen production are photolysis, photofermentation, dark fermentation, and MECs. Furthermore, a number of variables impact the rate of production, such as pH, temperature, and substrate concentration.

CONFLICT OF INTEREST

No conflict of interest.

ACKNOWLEDGMENT

The authors are grateful to Banasthali Vidyapith for providing all the necessities to complete this work.

REFERENCES

1. Vidadili, Nurtaj, Elchin Suleymanov, Cihan Bulut, and Ceyhun Mahmudlu. "Transition to renewable energy and sustainable energy development in Azerbaijan." *Renewable and Sustainable Energy Reviews* 80 (2017): 1153–1161. https://doi.org/10.1016/j.rser.2017.05.168
2. Kanwal, Fariha, and Angel A. J. Torriero. "Biohydrogen—A green fuel for sustainable energy solutions." *Energies* 15, no. 20 (2022): 7783. https://doi.org/10.3390/en15207783.
3. Rodionova, Margarita V., Roshan Sharma Poudyal, Indira Tiwari, Roman A. Voloshin, Sergei K. Zharmukhamedov, Hong Gil Nam, Bolatkhan K. Zayadan, Barry D. Bruce, Harvey J. M. Hou, and Suleyman I. Allakhverdiev. "Biofuel production: challenges and opportunities." *International Journal of Hydrogen Energy* 42, no. 12 (2017): 8450–8461. https://doi.org/10.1016/j.ijhydene.2016.11.125
4. Voloshin, Roman A., Margarita V. Rodionova, Sergey K. Zharmukhamedov, T. Nejat Veziroglu, and Suleyman I. Allakhverdiev. "Biofuel production from plant and algal biomass." *International Journal of Hydrogen Energy* 41, no. 39 (2016): 17257–17273. https://doi.org/10.1016/j.ijhydene.2016.07.084
5. Chaves, Luiz Inácio, Carlos Alberto Favarin Murari, Carlos Alberto de Castro Júnior, Marcelo José da Silva, Marcelo Nanni, Pedro de Assis Sobreira Júnior, Samuel Nelson Melegari de Souza, and Carlos Eduardo Camargo Nogueira. "Current status and perspectives of the electric sector in Brazil." *IEEE Latin America Transactions* 19, no. 4 (2021): 660–668.
6. Bayro-Kaiser, Vinzenz, and Nathan Nelson. "Feasibility of sustainable photosynthetic hydrogen production." In Jian-Ren Shen, Kimiyuki Satoh, and Suleyman I. Allakhverdiev (eds.), *Photosynthesis: Molecular Approaches to Solar Energy Conversion*, pp. 567–587. Cham: Springer International Publishing (2021). https://doi.org/10.1007/978-3-030-67407-6_21.
7. von Braun, Joachim. "Climate change risks for agriculture, health, and nutrition." In Wael Al-Delaimy, Veerabhadran Ramanathan, and Marcelo Sánchez Sorondo (eds.), *Health of People, Health of Planet and Our Responsibility: Climate Change, Air Pollution and Health*, pp. 135–148. Cham: Springer Open (2020).
8. Tai, Amos P. K., Maria Val Martin, and Colette L. Heald. "Threat to future global food security from climate change and ozone air pollution." *Nature Climate Change* 4, no. 9 (2014): 817–821. https://doi.org/10.1038/nclimate2317
9. Schaeffer, Michiel, William Hare, Stefan Rahmstorf, and Martin Vermeer. "Long-term sea-level rise implied by 1.5C and 2C warming levels." *Nature Climate Change* 2, no. 12 (2012): 867–870. https://doi.org/10.1038/nclimate1584
10. Kroeker, Kristy J., Rebecca L. Kordas, Ryan Crim, Iris E. Hendriks, Laura Ramajo, Gerald S. Singh, Carlos M. Duarte, and Jean-Pierre Gattuso. "Impacts of ocean acidification on marine organisms: quantifying sensitivities and interaction with warming." *Global Change Biology* 19, no. 6 (2013): 1884–1896. https://doi.org/10.1111/gcb.12179
11. Corlett, Richard T., and David A. Westcott. "Will plant movements keep up with climate change?" *Trends in Ecology & Evolution* 28, no. 8 (2013): 482–488.
12. Franks, Steven J., Jennifer J. Weber, and Sally N. Aitken. "Evolutionary and plastic responses to climate change in terrestrial plant populations." *Evolutionary Applications* 7, no. 1 (2014): 123–139. https://doi.org/10.1111/eva.12112
13. Moritz, Craig, and Rosa Agudo. "The future of species under climate change: resilience or decline?" *Science* 341, no. 6145 (2013): 504–508. https://doi.org/10.1126/science.1237190
14. Mitra, Romit, G. Balachandar, Vaishali Singh, Pallavi Sinha, and Debabrata Das. "Improvement in energy recovery by dark fermentative biohydrogen followed by biobutanol production process using obligate anaerobes." *International Journal of Hydrogen Energy* 42, no. 8 (2017): 4880–4892. https://doi.org/10.1016/j.ijhydene.2017.01.183

15. Khetkorn, Wanthanee, Rajesh P. Rastogi, Aran Incharoensakdi, Peter Lindblad, Datta Madamwar, Ashok Pandey, and Christian Larroche. "Microalgal hydrogen production–a review." *Bioresource Technology* 243 (2017): 1194–1206. https://doi.org/10.1016/j.biortech.2017.07.085
16. Oey, Melanie, Anne Linda Sawyer, Ian Lawrence Ross, and Ben Hankamer. "Challenges and opportunities for hydrogen production from microalgae." *Plant Biotechnology Journal* 14, no. 7 (2016): 1487–1499. https://doi.org/10.1111/pbi.12516
17. Bundhoo, M. A. Zumar, and Romeela Mohee. "Inhibition of dark fermentative bio-hydrogen production: a review." *International Journal of Hydrogen Energy* 41, no. 16 (2016): 6713–6733. https://doi.org/10.1016/j.ijhydene.2016.03.057
18. Chu, Steven, and Arun Majumdar. "Opportunities and challenges for a sustainable energy future." *Nature* 488, no. 7411 (2012): 294–303. https://doi.org/10.1038/nature11475
19. Ghimire, Anish, Luigi Frunzo, Francesco Pirozzi, Eric Trably, Renaud Escudie, Piet N. L. Lens, and Giovanni Esposito. "A review on dark fermentative biohydrogen production from organic biomass: process parameters and use of by-products." *Applied Energy* 144 (2015): 73–95. https://doi.org/10.1016/j.apenergy.2015.01.045
20. Rashid, Naim, Muhammad Saif Ur Rehman, Sheeraz Memon, Zia Ur Rahman, Kisay Lee, and Jong-In Han. "Current status, barriers and developments in biohydrogen production by microalgae." *Renewable and Sustainable Energy Reviews* 22 (2013): 571–579. https://doi.org/10.1016/j.rser.2015.01.051
21. Rahman, Siti Najibah Abd, Mohd Shahbudin. Masdar, Masli Irwan. Rosli, Edy Herianto Majlan, Teuku Husaini, Siti Kartom Kamarudin, and Wan Ramli Wan Daud. "Overview biohydrogen technologies and application in fuel cell technology." *Renewable and Sustainable Energy Reviews* 66 (2016): 137–162. https://doi.org/10.1016/j.rser.2016.07.047
22. Torzillo, Giuseppe, Alberto Scoma, Cecilia Faraloni, and Luca Giannelli. "Advances in the biotechnology of hydrogen production with the microalga Chlamydomonas reinhardtii." *Critical Reviews in Biotechnology* 35, no. 4 (2015): 485–496. https://doi.org/10.3109/07388551.2014.900734
23. Rumpel, Sigrun, Judith F. Siebel, Christophe Farès, Jifu Duan, Edward Reijerse, Thomas Happe, Wolfgang Lubitz, and Martin Winkler. "Enhancing hydrogen production of microalgae by redirecting electrons from photosystem I to hydrogenase." *Energy & Environmental Science* 7, no. 10 (2014): 3296–3301. https://doi.org/10.1039/C4EE01444H
24. Demirbas, Ayhan. "Hydrogen-rich gas from fruit shells via supercritical water extraction." *International Journal of Hydrogen Energy* 29, no. 12 (2004): 1237–1243. https://doi.org/10.1016/j.ijhydene.2003.11.012
25. Pipitone, Giuseppe, Giulia Zoppi, Raffaele Pirone, and Samir Bensaid. "A critical review on catalyst design for aqueous phase reforming." *International Journal of Hydrogen Energy* 47, no. 1 (2022): 151–180. https://doi.org/10.1016/j.ijhydene.2021.09.206
26. Tak, Sakshi S., Omkar Shetye, Omkar Muley, Harsh Jaiswal, and Sameena N. Malik. "Emerging technologies for hydrogen production from wastewater." *International Journal of Hydrogen Energy* 47, no. 88 (2022): 37282–37301. https://doi.org/10.1016/j.ijhydene.2022.06.225
27. Zoppi, Giulia, Giuseppe Pipitone, Camilla Galletti, Andrea Maria Rizzo, David Chiaramonti, Raffaele Pirone, and Samir Bensaid. "Aqueous phase reforming of lignin-rich hydrothermal liquefaction by-products: a study on catalyst deactivation." *Catalysis Today* 365 (2021): 206–215. https://doi.org/10.1016/j.cattod.2020.08.013
28. Liu, Hong, Hongqiang Hu, Jeremy Chignell, and Yanzhen Fan. "Microbial electrolysis: novel technology for hydrogen production from biomass." *Biofuels* 1, no. 1 (2010): 129–142. https://doi.org/10.4155/bfs.09.9

29. Lakshmidevi, Rajendran, and Karuppan Muthukumar. "Enzymatic saccharification and fermentation of paper and pulp industry effluent for biohydrogen production." *International Journal of Hydrogen Energy* 35, no. 8 (2010): 3389–3400. https://doi.org/10.1016/j.ijhydene.2009.12.165
30. Su, Huibo, Jun Cheng, Junhu Zhou, Wenlu Song, and Kefa Cen. "Hydrogen production from water hyacinth through dark-and photo-fermentation." *International Journal of Hydrogen Energy* 35, no. 17 (2010): 8929–8937. https://doi.org/10.1016/j.ijhydene.2010.06.035
31. Wang, Hanxi, Jianling Xu, Lianxi Sheng, Xuejun Liu, Yue Lu, and Wei Li. "A review on bio-hydrogen production technology." *International Journal of Energy Research* 42, no. 11 (2018): 3442–3453. https://doi.org/10.1002/er.4044
32. Ren, Nanqi, Wanqian Guo, Bingfeng Liu, Guangli Cao, and Jie Ding. "Biological hydrogen production by dark fermentation: challenges and prospects towards scaled-up production." *Current Opinion in Biotechnology* 22, no. 3 (2011): 365–370. https://doi.org/10.1016/j.copbio.2011.04.022
33. Argun, Hidayet, and Fikret Kargi. "Bio-hydrogen production by different operational modes of dark and photo-fermentation: an overview." *International Journal of Hydrogen Energy* 36, no. 13 (2011): 7443–7459. https://doi.org/10.1016/j.ijhydene.2011.03.116
34. Hallenbeck, Patrick C., and Yuan Liu. "Recent advances in hydrogen production by photosynthetic bacteria." *International Journal of Hydrogen Energy* 41, no. 7 (2016): 4446–4454. https://doi.org/10.1016/j.ijhydene.2015.11.090
35. Li, Yong Feng, Nan Qi Ren, Ying Chen, and Guo Xiang Zheng. "Ecological mechanism of fermentative hydrogen production by bacteria." *International Journal of Hydrogen Energy* 32, no. 6 (2007): 755–760. https://doi.org/10.1016/j.ijhydene.2006.08.004
36. Kapdan, Ilgi K., Fikret Kargi, Rukiye Oztekin, and Hidayet Argun. "Bio-hydrogen production from acid hydrolyzed wheat starch by photo-fermentation using different Rhodobacter sp." *International Journal of Hydrogen Energy* 34, no. 5 (2009): 2201–2207. https://doi.org/10.1016/j.ijhydene.2009.01.017
37. Akhlaghi, Neda, and Ghasem Najafpour-Darzi. "A comprehensive review on biological hydrogen production." *International Journal of Hydrogen Energy* 45, no. 43 (2020): 22492–22512. https://doi.org/10.1016/j.ijhydene.2020.06.182
38. Lu, Lu, Defeng Xing, Bingfeng Liu, and Nanqi Ren. "Enhanced hydrogen production from waste activated sludge by cascade utilization of organic matter in microbial electrolysis cells." *Water Research* 46, no. 4 (2012): 1015–1026. https://doi.org/10.1016/j.watres.2011.11.073
39. Kawagoshi, Yasunori, Yukinori Oki, Issei Nakano, Aya Fujimoto, and Hirokazu Takahashi. "Biohydrogen production by isolated halotolerant photosynthetic bacteria using long-wavelength light-emitting diode (LW-LED)." *International Journal of Hydrogen Energy* 35, no. 24 (2010): 13365–13369. https://doi.org/10.1016/j.ijhydene.2009.11.121
40. Wu, Siang Chen, Pi Fen Lu, Yu Chien Lin, Pei Chung Chen, and Chi Mei Lee. "Bio-hydrogen production enhancement by co-cultivating Rhodopseudomonas palustris WP3-5 and Anabaena sp. CH3." *International Journal of Hydrogen Energy* 37, no. 3 (2012): 2231–2238. https://doi.org/10.1016/j.ijhydene.2011.10.066
41. Sun, Qin, Wei Xiao, Dan Xi, Jiping Shi, Xing Yan, and Zhihua Zhou. "Statistical optimization of biohydrogen production from sucrose by a co-culture of Clostridium acidisoli and Rhodobacter sphaeroides." *International Journal of Hydrogen Energy* 35, no. 9 (2010): 4076–4084. https://doi.org/10.1016/j.ijhydene.2010.01.145
42. Asada, Toichiro, Pu Chen, Carl Chiarella, and Peter Flaschel. "Keynesian dynamics and the wage–price spiral: a baseline disequilibrium model." *Journal of Macroeconomics* 28, no. 1 (2006): 90–130. https://doi.org/10.1016/j.jmacro.2005.10.007

43. Hitam, Che Nor Che, and Aishah Abdul Jalil. "A review on biohydrogen production through photo-fermentation of lignocellulosic biomass." *Biomass Conversion and Biorefinery* 13 (2020): 1–19. https://doi.org/10.1007/s13399-020-01140-y
44. Rai, Pankaj K., and S. P. Singh. "Integrated dark-and photo-fermentation: recent advances and provisions for improvement." *International Journal of Hydrogen Energy* 41, no. 44 (2016): 19957–19971. https://doi.org/10.1016/j.ijhydene.2016.08.084
45. Shi, Ying, An-Fei Yang, Chun-Shuai Cao, and Bin Zhao. "Applications of MOFs: recent advances in photocatalytic hydrogen production from water." *Coordination Chemistry Reviews* 390 (2019): 50–75. https://doi.org/10.1016/j.ccr.2019.03.012
46. Ghirardi, Maria L., Paul W. King, David W. Mulder, Carrie Eckert, Alexandra Dubini, Pin-Ching Maness, and Jianping Yu. "Hydrogen production by water biophotolysis." In Davide Zannoni, and Roberto De Philippis (eds.), *Microbial BioEnergy: Hydrogen Production*, pp. 101–135. Dordrecht: Springer (2014). https://doi.org/10.1007/978-94-017-8554-9_5
47. Xu, Xianxian, Quan Zhou, and Dehai Yu. "The future of hydrogen energy: Bio-hydrogen production technology." *International Journal of Hydrogen Energy* 47, no. 79 (2022): 33677–33698. https://doi.org/10.1016/j.ijhydene.2022.07.261
48. Gaffron, Hans, and Jack Rubin. "Fermentative and photochemical production of hydrogen in algae." *The Journal of General Physiology* 26, no. 2 (1942): 219–240. https://doi.org/10.1085/jgp.26.2.219
49. Kessler, Erich. "Effect of anaerobiosis on photosynthetic reactions and nitrogen metabolism of algae with and without hydrogenase." *Archiv für Mikrobiologie* 93 (1973): 91–100. https://doi.org/10.1007/BF00424940
50. Llama, María J., Juan L. Serra, K. Krishna Rao, and David O. Hall. "Isolation and characterization of the hydrogenase activity from the non-heterocystous cyanobacterium Spirulina maxima." *FEBS Letters* 98, no. 2 (1979): 342–346. https://doi.org/10.1016/0014-5793(79)80213-6
51. Maione, Theodore E., and Martin Gibbs. "Hydrogenase-mediated activities in isolated chloroplasts of Chlamydomonas reinhardii." *Plant Physiology* 80, no. 2 (1986): 360–363. https://doi.org/10.1104/pp.80.2.360
52. Belkin, Shimshon, and Etana Padan. "Hydrogen metabolism in the facultative anoxygenic cyanobacteria (blue-green algae) Oscillatoria limnetica and Aphanothece halophytica." *Archives of Microbiology* 116 (1978): 109–111. https://doi.org/10.1007/BF00408741
53. Melo, Marcos de Oliveira, and Luciana Almeida Silva. "Photocatalytic production of hydrogen: an innovative use for biomass derivatives." *Journal of the Brazilian Chemical Society* 22 (2011): 1399–1406. https://doi.org/10.1590/S0103-50532011000800002
54. Yan, R. *Nickel Phosphide-Based Composites for Photocatalytic Hydrogen Production D]*. Liaoning: Dalian Univ of Technol; 2021.
55. Li, Guosheng, Zhipeng Xie, Shuming Chai, Xiong Chen, and Xinchen Wang. "A facile one-step fabrication of holey carbon nitride nanosheets for visible-light-driven hydrogen evolution." *Applied Catalysis B: Environmental* 283 (2021): 119637. https://doi.org/10.1016/j.apcatb.2020.119637
56. Girginer, Burcu, Giancarlo Galli, Emo Chiellini, and Niyazi Bicak. "Preparation of stable CdS nanoparticles in aqueous medium and their hydrogen generation efficiencies in photolysis of water." *International Journal of Hydrogen Energy* 34, no. 3 (2009): 1176–1184. https://doi.org/10.1016/j.ijhydene.2008.10.086
57. Feng, Xiang, Yuhai Zhao, Dakuo Liu, Yasi Mo, Yibin Liu, Xiaobo Chen, Wenjuan Yan et al. "Towards high activity of hydrogen production from ammonia borane over efficient non-noble Ni5P4 catalyst." *International Journal of Hydrogen Energy* 43, no. 36 (2018): 17112–17120. https://doi.org/10.1016/j.ijhydene.2018.07.055

58. Çelik, Dicle, and Meltem Yıldız. "Investigation of hydrogen production methods in accordance with green chemistry principles." *International Journal of Hydrogen Energy* 42, no. 36 (2017): 23395–23401. https://doi.org/10.1016/j.ijhydene.2017.03.104
59. Zhang, Tian, Danping Jiang, Huan Zhang, Yanyan Jing, Nadeem Tahir, Yang Zhang, and Quanguo Zhang. "Comparative study on bio-hydrogen production from corn stover: photo-fermentation, dark-fermentation and dark-photo co-fermentation." *International Journal of Hydrogen Energy* 45, no. 6 (2020): 3807–3814. https://doi.org/10.1016/j.ijhydene.2019.04.170
60. Ren, Nan-Qi, Wan-Qian Guo, and Bing-Feng Liu. "Development and application prospect of bio-hydrogen production technology." *Harbin GongyeDaxueXuebao (Journal of Harbin Institute of Technology)* 42, no. 6 (2010): 855–863.
61. Dawood, Furat, Martin Anda, and G. M. Shafiullah. "Hydrogen production for energy: an overview." *International Journal of Hydrogen Energy* 45, no. 7 (2020): 3847–3869. https://doi.org/10.1016/j.ijhydene.2019.12.059
62. Momirlan, Magdalena, and Veziroğlu T. Nejat. "Recent directions of world hydrogen production." *Renewable and Sustainable Energy Reviews* 3, no. 2–3 (1999): 219–231. https://doi.org/10.1016/S1364-0321(98)00017-3
63. Nath, Kaushik, and Debabrata Das. "Improvement of fermentative hydrogen production: various approaches." *Applied Microbiology and Biotechnology* 65 (2004): 520–529. https://doi.org/10.1007/s00253-004-1644-0
64. Hawkes, Freda R., Richard Dinsdale, Dennis L. Hawkes, and Ines Hussy. "Sustainable fermentative hydrogen production: challenges for process optimisation." *International Journal of Hydrogen Energy* 27, no. 11–12 (2002): 1339–1347. https://doi.org/10.1016/S0360-3199(02)00090-3
65. Nath, Kaushik, Anish Kumar, and Debabrata Das. "Hydrogen production by Rhodobactersphaeroides strain OU 001 using spent media of Enterobacter cloacae strain DM11." *Applied Microbiology and Biotechnology* 68 (2005): 533–541. https://doi.org/10.1007/s00253-005-1887-4
66. Tao, Yongzhen, Yang Chen, Yongqiang Wu, Yanling He, and Zhihua Zhou. "High hydrogen yield from a two-step process of dark-and photo-fermentation of sucrose." *International Journal of Hydrogen Energy* 32, no. 2 (2007): 200–206. https://doi.org/10.1016/j.ijhydene.2006.06.034
67. Asada, Yasuo, Masaru Tokumoto, Yasuyuki Aihara, Masayo Oku, Katsuhiro Ishimi, Tatsuki Wakayama, Jun Miyake, Masamitsu Tomiyama, and Hideki Kohno. "Hydrogen production by co-cultures of Lactobacillus and a photosynthetic bacterium, Rhodobactersphaeroides RV." *International Journal of Hydrogen Energy* 31, no. 11 (2006): 1509–1515. https://doi.org/10.1016/j.ijhydene.2006.06.017
68. Chen, Chun-Yen, Mu-Hoe Yang, Kuei-Ling Yeh, Chien-Hung Liu, and Jo-Shu Chang. "Biohydrogen production using sequential two-stage dark and photo fermentation processes." *International Journal of Hydrogen Energy* 33, no. 18 (2008): 4755–4762. https://doi.org/10.1016/j.ijhydene.2008.06.055
69. Lin, Chiu-Yue, Chyi-How Lay, Biswarup Sen, Chen-Yeon Chu, Gopalakrishnan Kumar, Chin-Chao Chen, and Jo-Shu Chang. "Fermentative hydrogen production from wastewaters: a review and prognosis." *International Journal of Hydrogen Energy* 37, no. 20 (2012): 15632–15642. https://doi.org/10.1016/j.ijhydene.2012.02.072
70. Brennan, Liam, and Philip Owende. "Biofuels from microalgae: a review of technologies for production, processing, and extractions of biofuels and co-products." *Renewable and Sustainable Energy Reviews* 14, no. 2 (2010): 557–577. https://doi.org/10.1016/j.rser.2009.10.009

71. Kumar, Gopalakrishnan, Guangyin Zhen, Takuro Kobayashi, Periyasamy Sivagurunathan, Sang Hyoun Kim, and Kai Qin Xu. "Impact of pH control and heat pre-treatment of seed inoculum in dark H_2 fermentation: a feasibility report using mixed microalgae biomass as feedstock." *International Journal of Hydrogen Energy* 41, no. 7 (2016): 4382–4392. https://doi.org/10.1016/j.ijhydene.2015.08.069
72. Kumar, Gopalakrishnan, Periyasamy Sivagurunathan, Arivalagan Pugazhendhi, Ngoc Bao Dung Thi, Guangyin Zhen, Kuppam Chandrasekhar, and Abudukeremu Kadier. "A comprehensive overview on light independent fermentative hydrogen production from wastewater feedstock and possible integrative options." *Energy Conversion and Management* 141 (2017): 390–402. https://doi.org/10.1016/j.enconman.2016.09.087
73. Nandi, Rakhi, and S. Sengupta. "Microbial production of hydrogen: an overview." *Critical Reviews in Microbiology* 24, no. 1 (1998): 61–84. https://doi.org/10.1080/10408419891294181
74. Delavar, Mojtaba Aghajani, and Junye Wang. "Numerical investigation of pH control on dark fermentation and hydrogen production in a microbioreactor." *Fuel* 292 (2021): 120355. https://doi.org/10.1016/j.fuel.2021.120355
75. Xu, Ji-Fei, Yuan-Ting Mi, and Nan-Qi Ren. "Buffering action of acetate on hydrogen production by Ethanoligenens harbinense B49." *Electronic Journal of Biotechnology* 23 (2016): 7–11. https://doi.org/10.1016/j.ejbt.2016.07.002
76. Chen, Chinchao, and Chiuyue Lin. "Using sucrose as a substrate in an anaerobic hydrogen-producing reactor." *Advances in Environmental Research* 7, no. 3 (2003): 695–699. https://doi.org/10.1016/S1093-0191(02)00035-7
77. Fang, Herbert H. P., and Hong Liu. "Effect of pH on hydrogen production from glucose by a mixed culture." *Bioresource Technology* 82, no. 1 (2002): 87–93. https://doi.org/10.1016/S0960-8524(01)00110-9
78. Kumar, Narendra, and Debabrata Das. "Enhancement of hydrogen production by Enterobacter cloacae IIT-BT 08." *Process Biochemistry* 35, no. 6 (2000): 589–593. https://doi.org/10.1016/S0032-9592(99)00109-0
79. Oh, You-Kwan, Eun-Hee Seol, Jung Rae Kim, and Sunghoon Park. "Fermentative biohydrogen production by a new chemoheterotrophic bacterium Citrobacter sp. Y19." *International Journal of Hydrogen Energy* 28, no. 12 (2003): 1353–1359. https://doi.org/10.1016/S0360-3199(03)00024-7
80. Sivagurunathan, Periyasamy, Gopalakrishnan Kumar, Jeong-Hoon Park, Jong-Hun Park, Hee-Deung Park, Jeong-Jun Yoon, and Sang-Hyoun Kim. "Feasibility of enriched mixed cultures obtained by repeated batch transfer in continuous hydrogen fermentation." *International Journal of Hydrogen Energy* 41, no. 7 (2016): 4393–4403. https://doi.org/10.1016/j.ijhydene.2015.06.133
81. Moon, Chungman, Sujin Jang, Yeo-Myeong Yun, Mo-Kwon Lee, Dong-Hoon Kim, Won-Seok Kang, Seung-Shin Kwak, and Mi-Sun Kim. "Effect of the accuracy of pH control on hydrogen fermentation." *Bioresource Technology* 179 (2015): 595–601. https://doi.org/10.1016/j.biortech.2014.10.128
82. Yue, Xiao Fang, Hong Yuan Sun, Xu Xin Zhao, and Li Qing Zhao. "Research progress of food waste fermentation for bio-hydrogen production." *Advanced Materials Research* 550 (2012): 569–573. https://doi.org/10.4028/www.scientific.net/AMR.550-553.569
83. Pareek, Alka, Rekha Dom, Jyoti Gupta, Jyothi Chandran, Vivek Adepu, and Pramod H. Borse. "Insights into renewable hydrogen energy: recent advances and prospects." *Materials Science for Energy Technologies* 3 (2020): 319–327. https://doi.org/10.1016/j.mset.2019.12.002

84. Modestra, J. Annie, M. Lenin Babu, and S. Venkata Mohan. "Electro-fermentation of real-field acidogenic spent wash effluents for additional biohydrogen production with simultaneous treatment in a microbial electrolysis cell." *Separation and Purification Technology* 150 (2015): 308–315.
85. Zhang, Yifeng, and Irini Angelidaki. "Microbial electrochemical systems and technologies: it is time to report the capital costs." *Environmental Science & Technology* 50 (2016): 5432–5433.
86. Rivera, Isaac, Germán Buitrón, Péter Bakonyi, Nándor Nemestóthy, and Katalin Bélafi-Bakó. "Hydrogen production in a microbial electrolysis cell fed with a dark fermentation effluent." *Journal of Applied Electrochemistry* 45 (2015): 1223–1229.
87. Kadier, Abudukeremu, Pratiksha Jain, Bin Lai, Mohd Sahaid Kalil, Sanath Kondaveeti, Khulood Fahad Saud Alabbosh, Ibrahim M. Abu-Reesh, and Gunda Mohanakrishna. "Biorefinery perspectives of microbial electrolysis cells (MECs) for hydrogen and valuable chemicals production through wastewater treatment." *Biofuel Research Journal* 7 (2020): 1128.
88. Kadier, Abudukeremu, Raghuveer Singh, Dongsheng Song, Farshid Ghanbari, Nur Syamimi Zaidi, Putu Teta Prihartini Aryanti, Dipak A. Jadhav et al. "A novel pico-hydro power (PHP)-Microbial electrolysis cell (MEC) coupled system for sustainable hydrogen production during palm oil mill effluent (POME) wastewater treatment." *International Journal of Hydrogen Energy* 48, no. 55 (2023): 21066–21087.
89. Hosseinzadeh, Ahmad, John L. Zhou, Xiaowei Li, Morteza Afsari, and Ali Altaee. "Techno-economic and environmental impact assessment of hydrogen production processes using bio-waste as renewable energy resource." *Renewable and Sustainable Energy Reviews* 156 (2022): 111991.
90. Lee, Hyung-Sool, Wang Xin, Ranaprathap Katakojwala, S. Venkata Mohan, and Noori M. D. Tabish. "Microbial electrolysis cells for the production of biohydrogen in dark fermentation-a review." *Bioresource Technology* 363 (2022): 127934.
91. Kosourov, Sergey, Anatoly Tsygankov, Michael Seibert, and Maria L. Ghirardi. "Sustained hydrogen photoproduction by Chlamydomonas reinhardtii: effects of culture parameters." *Biotechnology and Bioengineering* 78, no. 7 (2002): 731–740.
92. Melis, Anastasios, Liping Zhang, Marc Forestier, Maria L. Ghirardi, and Michael Seibert. "Sustained photobiological hydrogen gas production upon reversible inactivation of oxygen evolution in the green alga Chlamydomonas reinhardtii." *Plant Physiology* 122, no. 1 (2000): 127–136.
93. Mizuno, Osamu, Richard Dinsdale, Freda R. Hawkes, Dennis L. Hawkes, and Tatsuya Noike. "Enhancement of hydrogen production from glucose by nitrogen gas sparging." *Bioresource Technology* 73, no. 1 (2000): 59–65.
94. Lay, Jiunn-Jyi. "Modeling and optimization of anaerobic digested sludge converting starch to hydrogen." *Biotechnology and Bioengineering* 68, no. 3 (2000): 269–278.
95. Show, Kuan-Yeow, Yuegen Yan, Ming Ling, Guoxiang Ye, Ting Li, and Duu-Jong Lee. "Hydrogen production from algal biomass–advances, challenges and prospects." *Bioresource Technology* 257 (2018): 290–300. https://doi.org/10.1016/j.biortech.2018.02.105
96. Michael, Maria, Arun Meyyazhagan, Krishnakumar Velayudhannair, Manikantan Pappuswamy, Anu Maria, Vincent Xavier, Balamuralikrishnan Balasubramanian et al. "The content of heavy metals in cigarettes and the impact of their leachates on the aquatic ecosystem." *Sustainability* 14, no. 8 (2022): 4752. https://doi.org/10.3390/su14084752
97. Vasseghian, Yasser, Priya Arunkumar, Sang-Woo Joo, Lalitha Gnanasekaran, Hesam Kamyab, Saravanan Rajendran, Deepanraj Balakrishnan, Shreeshivadasan Chelliapan, and Jiří Jaromír Klemeš. "Metal-organic framework-enabled pesticides are an emerging tool for sustainable cleaner production and environmental hazard reduction." *Journal of Cleaner Production* 373 (2022): 133966. https://doi.org/10.1016/j.jclepro.2022.133966

98. Cheng, Jun, Liangchen Yue, Lingkan Ding, Yu-You Li, Qing Ye, Junhu Zhou, Kefa Cen, and Richen Lin. "Improving fermentative hydrogen and methane production from an algal bloom through hydrothermal/steam acid pretreatment." *International Journal of Hydrogen Energy* 44, no. 12 (2019): 5812–5820. https://doi.org/10.1016/j.ijhydene.2019.01.046
99. Ban, Shidong, Weitie Lin, Fangyan Wu, and Jianfei Luo. "Algal-bacterial cooperation improves algal photolysis-mediated hydrogen production." *Bioresource Technology* 251 (2018): 350–357. https://doi.org/10.1016/j.biortech.2017.12.072
100. Abd Majid, Muhd Zaimi, Yao Bigah, Ali Keyvanfar, Arezou Shafaghat, Jahangir Mirza, and Hesam Kamyab. "Green highway development features to control stormwater runoff pollution." *Journal of Environmental Treatment Techniques* 4, no. 4 (2015): 173–175.
101. Ibrahim, Amr F. M., Kodanda Phani Raj Dandamudi, Shuguang Deng, and Jerry Y. S. Lin. "Pyrolysis of hydrothermal liquefaction algal biochar for hydrogen production in a membrane reactor." *Fuel* 265 (2020): 116935. https://doi.org/10.1016/j.fuel.2019.116935
102. Hosur Shivaramaiah, Naveen Kumar, Subhaschandra Kattimani, Mahdi Shariati, and Trung Nguyen-Thoi. "Geometrically nonlinear behavior of two-directional functionally graded porous plates with four different materials." *Proceedings of the Institution of Mechanical Engineers, Part C: Journal of Mechanical Engineering Science* 236, no. 22 (2022): 11008–11023. https://doi.org/10.1177/09544062221111038
103. Maroušek, Josef, Otakar Strunecký, Vojtěch Bartoš, and Marek Vochozka. "Revisiting competitiveness of hydrogen and algae biodiesel." *Fuel* 328 (2022): 125317. https://doi.org/10.1016/j.fuel.2022.125317
104. Pankratz, Stan, Mayank Kumar, AdetoyeseOlajireOyedun, Eskinder Gemechu, and Amit Kumar. "Environmental performances of diluents and hydrogen production pathways from microalgae in cold climates: open raceway ponds and photobioreactors coupled with thermochemical conversion." *Algal Research* 47 (2020): 101815. https://doi.org/10.1016/j.algal.2020.101815
105. Show, Kuan-Yeow, Yue-Gen Yan, and Duu-Jong Lee. "Biohydrogen production from algae: perspectives, challenges, and prospects." In Ashok Pandey, Jo-Shu Chang, Carlos Ricardo Soccol, Duu-Jong Lee, and Yusuf Chisti (eds.), *Biofuels from Algae*, pp. 325–343. Elsevier (2019). https://doi.org/10.1016/B978-0-444-64192-2.00013-5
106. Mohan, S. Venkata, M. V. Rohit, P. Chiranjeevi, Rashmi Chandra, and B. Navaneeth. "Heterotrophic microalgae cultivation to synergize biodiesel production with waste remediation: progress and perspectives." *Bioresource Technology* 184 (2015): 169–178. https://doi.org/10.1016/j.biortech.2014.10.056
107. Mona, Sharma, and Anubha Kaushik. "Screening metal-dye-tolerant photoautotrophic microbes from textile wastewaters for biohydrogen production." *Journal of Applied Phycology* 27 (2015): 1185–1194. https://doi.org/10.1007/s10811-014-0396-1
108. Sharma, Amit Kumar, Praveen Kumar Ghodke, Suvendu Manna, and Wei-Hsin Chen. "Emerging technologies for sustainable production of biohydrogen production from microalgae: a state-of-the-art review of upstream and downstream processes." *Bioresource Technology* 342 (2021): 126057. https://doi.org/10.1016/j.biortech.2021.126057
109. Mathimani, Thangavel, and Arivalagan Pugazhendhi. "Utilization of algae for biofuel, bio-products and bio-remediation." *Biocatalysis and Agricultural Biotechnology* 17 (2019): 326–330. https://doi.org/10.1016/j.bcab.2018.12.007
110. Pulz, Otto, and Wolfgang Gross. "Valuable products from biotechnology of microalgae." *Applied Microbiology and Biotechnology* 65 (2004): 635–648. https://doi.org/10.1007/s00253-004-1647-x

111. Abu-Ghosh, Said, Zvy Dubinsky, Vitor Verdelho, and David Iluz. "Unconventional high-value products from microalgae: a review." *Bioresource Technology* 329 (2021): 124895. https://doi.org/10.1016/j.biortech.2021.124895
112. Skjånes, Kari, Uno Andersen, Thorsten Heidorn, and Stig A. Borgvang. "Design and construction of a photobioreactor for hydrogen production, including status in the field." *Journal of Applied Phycology* 28 (2016): 2205–2223.
113. de Jesus, Sérgio S., and Rubens Maciel Filho. "Potential of algal biofuel production in a hybrid photobioreactor." *Chemical Engineering Science* 171 (2017): 282–292.
114. Geada, Pedro, Vítor Vasconcelos, A. Vicente, and B. Fernandes. "Microalgal biomass cultivation." In Rajesh Prasad Rastogi, Datta Madamwar, and Ashok Pandey (eds.), *Algal Green Chemistry: Recent Progress in Biotechnology*, pp. 257–284. Elsevier (2017).
115. Bolatkhan, Kenzhegul, Bekzhan D. Kossalbayev, Bolatkhan K. Zayadan, Tatsuya Tomo, T. Nejat Veziroglu, and Suleyman I. Allakhverdiev. "Hydrogen production from phototrophic microorganisms: reality and perspectives." *International Journal of Hydrogen Energy* 44, no. 12 (2019): 5799–5811.
116. Oncel, Suphi, and Ayşe Kose. "Comparison of tubular and panel type photobioreactors for biohydrogen production utilizing Chlamydomonas reinhardtii considering mixing time and light intensity." *Bioresource Technology* 151 (2014): 265–270.
117. Dong-Jie, Niu, Wang Jing-Yuan, Wang Bao-Ying, and Zhao You-Cai. "Effect of Mo-containing additives on biohydrogen fermentation from cassava's stillage." *International Journal of Hydrogen Energy* 36, no. 9 (2011): 5289–5295. https://doi.org/10.1016/j.ijhydene.2011.01.139
118. Elbeshbishy, Elsayed, Bipro Ranjan Dhar, George Nakhla, and Hyung-Sool Lee. "A critical review on inhibition of dark biohydrogen fermentation." *Renewable and Sustainable Energy Reviews* 79 (2017): 656–668. https://doi.org/10.1016/j.rser.2017.05.075
119. Jarunglumlert, Teeraya, Chattip Prommuak, Namtip Putmai, and Prasert Pavasant. "Scaling-up bio-hydrogen production from food waste: feasibilities and challenges." *International Journal of Hydrogen Energy* 43, no. 2 (2018): 634–648. https://doi.org/10.1016/j.ijhydene.2017.10.013
120. Maneeruttanarungroj, Cherdsak, and Saranya Phunpruch. "Effect of pH on biohydrogen production in green alga Tetraspora sp. CU2551." *Energy Procedia* 138 (2017): 1085–1092. https://doi.org/10.1016/j.egypro.2017.10.122
121. Wang, Bing-Nan, Chu-Fang Yang, and Chi-Mei Lee. "The factors influencing direct photohydrogen production and anaerobic fermentation hydrogen production combination bioreactors." *International Journal of Hydrogen Energy* 36, no. 21 (2011): 14069–14077. https://doi.org/10.1016/j.ijhydene.2011.04.134
122. Hwang, Jae-Hoon, Jeong-A. Choi, You-Kwan Oh, Reda A. I. Abou-Shanab, Hocheol Song, Booki Min, Yunchul Cho, Jeong-Geol Na, Jakon Koo, and Byong-Hun Jeon. "Hydrogen production from sulfate-and ferrous-enriched wastewater." *International Journal of Hydrogen Energy* 36, no. 21 (2011): 13984–13990. https://doi.org/10.1016/j.ijhydene.2011.03.156
123. Dhar, Bipro Ranjan, Elsayed Elbeshbishy, and George Nakhla. "Influence of iron on sulfide inhibition in dark biohydrogen fermentation." *Bioresource Technology* 126 (2012): 123–130. https://doi.org/10.1016/j.biortech.2012.09.043
124. Ahmadi-Pirlou, Mansour, Mohammadali Ebrahimi-Nik, Mehdi Khojastehpour, and Seyed Hadi Ebrahimi. "Mesophilic co-digestion of municipal solid waste and sewage sludge: effect of mixing ratio, total solids, and alkaline pretreatment." *International Biodeterioration & Biodegradation* 125 (2017): 97–104. https://doi.org/10.1016/j.ibiod.2017.09.004

125. del Pilar Anzola-Rojas, Mélida, Samantha Gonçalves da Fonseca, Cynthia Canedo da Silva, Valeria Maia de Oliveira, and Marcelo Zaiat. "The use of the carbon/nitrogen ratio and specific organic loading rate as tools for improving biohydrogen production in fixed-bed reactors." *Biotechnology Reports* 5 (2015): 46–54. https://doi.org/10.1016/j.btre.2014.10.010
126. Zhou, Peiqing, Elsayed Elbeshbishy, and George Nakhla. "Optimization of biological hydrogen production for anaerobic co-digestion of food waste and wastewater biosolids." *Bioresource Technology* 130 (2013): 710–718. https://doi.org/10.1016/j.biortech.2012.12.069
127. Lin, Chiuyue, and Chyi How Lay. "Carbon/nitrogen-ratio effect on fermentative hydrogen production by mixed microflora." *International Journal of Hydrogen Energy* 29, no. 1 (2004): 41–45. https://doi.org/10.1016/S0360-3199(03)00083-1
128. Ghosh, Shiladitya, Umme Kulsoom Dairkee, Ranjana Chowdhury, and Pinaki Bhattacharya. "Hydrogen from food processing wastes via photofermentation using purple non-sulfur bacteria (PNSB): a review." *Energy Conversion and Management* 141 (2017): 299–314. https://doi.org/10.1016/j.enconman.2016.09.001
129. Jang, Sujin, Dong-Hoon Kim, Yeo-Myeong Yun, Mo-Kwon Lee, Chungman Moon, Won-Seok Kang, Seung-Shin Kwak, and Mi-Sun Kim. "Hydrogen fermentation of food waste by alkali-shock pretreatment: microbial community analysis and limitation of continuous operation." *Bioresource Technology* 186 (2015): 215–222. https://doi.org/10.1016/j.biortech.2015.03.031
130. Elbeshbishy, Elsayed, Hisham Hafez, and George Nakhla. "Viability of ultrasonication of food waste for hydrogen production." *International Journal of Hydrogen Energy* 37, no. 3 (2012): 2960–2964. https://doi.org/10.1016/j.ijhydene.2011.01.008
131. Kim, Dong-Hoon, Sang-Hyoun Kim, and Hang-Sik Shin. "Hydrogen fermentation of food waste without inoculum addition." *Enzyme and Microbial Technology* 45, no. 3 (2009): 181–187. https://doi.org/10.1016/j.enzmictec.2009.06.013
132. Lemire, Joseph A., Joe J. Harrison, and Raymond J. Turner. "Antimicrobial activity of metals: mechanisms, molecular targets and applications." *Nature Reviews Microbiology* 11, no. 6 (2013): 371–384. https://doi.org/10.1038/nrmicro3028
133. Le, Diep Thi Hong, and RachnarinNitisoravut. "Modified hydrotalcites for enhancement of biohydrogen production." *International Journal of Hydrogen Energy* 40, no. 36 (2015): 12169–12176. https://doi.org/10.1016/j.ijhydene.2015.06.142
134. Junghare, Madan, Sanjukta Subudhi, and Banwari Lal. "Improvement of hydrogen production under decreased partial pressure by newly isolated alkaline tolerant anaerobe, Clostridium butyricum TM-9A: optimization of process parameters." *International Journal of Hydrogen Energy* 37, no. 4 (2012): 3160–3168. https://doi.org/10.1016/j.ijhydene.2011.11.043
135. Cao, Xianyan, and Youcai Zhao. "The influence of sodium on biohydrogen production from food waste by anaerobic fermentation." *Journal of Material Cycles and Waste Management* 11 (2009): 244–250. https://doi.org/10.1007/s10163-009-0237-5
136. Yu, Hanqing, Joohwa Tay, and Herbert H. P. Fang. "The roles of calcium in sludge granulation during UASB reactor start-up." *Water Research* 35, no. 4 (2001): 1052–1060. https://doi.org/10.1016/S0043-1354(00)00345-6
137. Singhania, Reeta Rani, Anil Kumar Patel, Gwendoline Christophe, Pierre Fontanille, and Christian Larroche. "Biological upgrading of volatile fatty acids, key intermediates for the valorization of biowaste through dark anaerobic fermentation." *Bioresource Technology* 145 (2013): 166–174. https://doi.org/10.1016/j.biortech.2012.12.137
138. Kothari, Richa, Virendra Kumar, Vinayak V. Pathak, Shamshad Ahmad, Ochieng Aoyi, and V. V. Tyagi. "A critical review on factors influencing fermentative hydrogen production." *Frontiers in Bioscience* 22, no. 8 (2017): 1195–1220.

139. Kim, Eui-Jin, Mi-Sun Kim, and Jeong K. Lee. "Hydrogen evolution under photoheterotrophic and dark fermentative conditions by recombinant Rhodobacter sphaeroides containing the genes for fermentative pyruvate metabolism of Rhodospirillum rubrum." *International Journal of Hydrogen Energy* 33, no. 19 (2008): 5131–5136.
140. Park, Jeong-Hoon, Do-Hyung Kim, Sang-Hyoun Kim, Jeong-Jun Yoon, and Hee-Deung Park. "Effect of substrate concentration on the competition between Clostridium and Lactobacillus during biohydrogen production." *International Journal of Hydrogen Energy* 43, no. 25 (2018): 11460–11469. https://doi.org/10.1016/j.ijhydene.2017.08.150
141. Guo, Xin Mei, Eric Trably, Eric Latrille, Hélène Carrère, and Jean-Philippe Steyer. "Hydrogen production from agricultural waste by dark fermentation: a review." *International Journal of Hydrogen Energy* 35, no. 19 (2010): 10660–10673. https://doi.org/10.1016/j.ijhydene.2010.03.008
142. Sinha, Pallavi, and Anjana Pandey. "An evaluative report and challenges for fermentative biohydrogen production." *International Journal of Hydrogen Energy* 36, no. 13 (2011): 7460–7478. https://doi.org/10.1016/j.ijhydene.2011.03.077
143. Saady, Noori M. Cata. "Homoacetogenesis during hydrogen production by mixed cultures dark fermentation: unresolved challenge." *International Journal of Hydrogen Energy* 38, no. 30 (2013): 13172–13191. https://doi.org/10.1016/j.ijhydene.2015.07.122
144. Pawar, Sudhanshu S., and Ed W. J. van Niel. "Thermophilic biohydrogen production: how far are we?" *Applied Microbiology and Biotechnology* 97 (2013): 7999–8009.
145. Assawamongkholsiri, Thitirut, and Alissara Reungsang. "Photo-fermentational hydrogen production of Rhodobacter sp. KKU-PS1 isolated from an UASB reactor." *Electronic Journal of Biotechnology* 18, no. 3 (2015): 221–230.
146. Kim, Mi-Sun, Jin-Sook Baek, and Jeong K. Lee. "Comparison of H_2 accumulation by Rhodobacter sphaeroides KD131 and its uptake hydrogenase and PHB synthase deficient mutant." *International Journal of Hydrogen Energy* 31, no. 1 (2006): 121–127. https://doi.org/10.1016/j.ijhydene.2004.10.023
147. Akroum-Amrouche, Dahbia, Nadia Abdi, Hakim Lounici, and Nabil Mameri. "Effect of physico-chemical parameters on biohydrogen production and growth characteristics by batch culture of Rhodobacter sphaeroides CIP 60.6." *Applied Energy* 88, no. 6 (2011): 2130–2135. https://doi.org/10.1016/j.apenergy.2010.12.044
148. Kadier, Abudukeremu, Yong Jiang, Bin Lai, Pankaj Kumar Rai, Kuppam Chandrasekhar, Azah Mohamed, and Mohd Sahaid Kalil. "Biohydrogen production in microbial electrolysis cells from renewable resources." In Ozcan Konur (ed.), *Bioenergy and Biofuels*, pp. 331–356. Boca Raton, FL: CRC Press (2018).
149. Chae, Kyu-Jung, Mi-Jin Choi, Kyoung-Yeol Kim, F. F. Ajayi, In-Seop Chang, and In S. Kim. "Selective inhibition of methanogens for the improvement of biohydrogen production in microbial electrolysis cells." *International Journal of Hydrogen Energy* 35, no. 24 (2010): 13379–13386. https://doi.org/10.1016/j.ijhydene.2009.11.114
150. Rozendal, René A., Hubertus V. M. Hamelers, Korneel Rabaey, Jurg Keller, and Cees J. N. Buisman. "Towards practical implementation of bioelectrochemical wastewater treatment." *Trends in Biotechnology* 26, no. 8 (2008): 450–459. https://doi.org/10.1016/j.tibtech.2008.04.008
151. Jeremiasse, Adriaan W., Hubertus V. M. Hamelers, and Cees J. N. Buisman. "Microbial electrolysis cell with a microbial biocathode." *Bioelectrochemistry* 78, no. 1 (2010): 39–43. https://doi.org/10.1016/j.bioelechem.2009.05.005
152. Croese, Elsemiek, Maria Alcina Pereira, Gert-Jan W. Euverink, Alfons J. M. Stams, and Jeanine S. Geelhoed. "Analysis of the microbial community of the biocathode of a hydrogen-producing microbial electrolysis cell." *Applied Microbiology and Biotechnology* 92 (2011): 1083–1093. https://doi.org/10.1007/s00253-011-3583-x

153. Ferraren-De Cagalitan, Decien Dee., and Michael Lochinvar Sim Abundo. "A review of biohydrogen production technology for application towards hydrogen fuel cells." *Renewable and Sustainable Energy Reviews* 151 (2021): 111415. https://doi.org/10.1016/j.rser.2021.111413
154. Chandrasekhar, Kuppam, Yong-Jik Lee, and Dong-Woo Lee. "Biohydrogen production: strategies to improve process efficiency through microbial routes." *International Journal of Molecular Sciences* 16, no. 4 (2015): 8266–8293. https://doi.org/10.3390/ijms16048266
155. Sarangi, Prakash K., and Sonil Nanda. "Biohydrogen production through dark fermentation." *Chemical Engineering & Technology* 43, no. 4 (2020): 601–612. https://doi.org/10.1002/ceat.201900452
156. Łukajtis, Rafał, Iwona Hołowacz, Karolina Kucharska, Marta Glinka, Piotr Rybarczyk, Andrzej Przyjazny, and Marian Kamiński. "Hydrogen production from biomass using dark fermentation." *Renewable and Sustainable Energy Reviews* 91 (2018): 665–694. https://doi.org/10.1016/j.rser.2018.04.043
157. Bundhoo, M. A. Zumar, Romeela Mohee, and M. Ali Hassan. "Effects of pretreatment technologies on dark fermentative biohydrogen production: a review." *Journal of Environmental Management* 157 (2015): 20–48. https://doi.org/10.1016/j.jenvman.2015.04.006
158. Hitit, Zeynep Yilmazer, Carolina Zampol Lazaro, and Patrick C. Hallenbeck. "Single stage hydrogen production from cellulose through photo-fermentation by a co-culture of Cellulomonas fimi and Rhodopseudomonas palustris." *International Journal of Hydrogen Energy* 42, no. 10 (2017): 6556–6566. https://doi.org/10.1016/j.ijhydene.2016.12.035
159. Ding, Chang, Kunlin Yang, and Jianzhong He. "Biological and fermentative production of hydrogen." In Rafael Luque, Carol Sze Ki Lin, Karen Wilson, and Chenyu Du (eds.), *Handbook of Biofuels Production*, pp. 303–333. Woodhead Publishing (2016). https://doi.org/10.1016/B978-0-08-100455-5.00011-4
160. Volgusheva, Alena, Stenbjörn Styring, and Fikret Mamedov. "Increased photosystem II stability promotes H_2 production in sulfur-deprived Chlamydomonas reinhardtii." *Proceedings of the National Academy of Sciences* 110, no. 18 (2013): 7223–7228. https://doi.org/10.1073/pnas.1220645110
161. Ghirardi, Maria L. "Implementation of photobiological H_2 production: the O_2 sensitivity of hydrogenases." *Photosynthesis Research* 125, no. 3 (2015): 383–393. https://doi.org/10.1007/s11120-015-0158-1
162. Das, Debabrata, Namita Khanna, and Chitralekha Nag Dasgupta. *Biohydrogen Production: Fundamentals and Technology Advances.* Boca Raton, FL: CRC Press (2014).
163. Nagarajan, Dillirani, Duu-Jong Lee, Akihiko Kondo, and Jo-Shu Chang. "Recent insights into biohydrogen production by microalgae–from biophotolysis to dark fermentation." *Bioresource Technology* 227 (2017): 373–387. https://doi.org/10.1016/j.biortech.2016.12.104
164. Kosourov, Sergey, Gayathri Murukesan, Michael Seibert, and Yagut Allahverdiyeva. "Evaluation of light energy to H_2 energy conversion efficiency in thin films of cyanobacteria and green alga under photoautotrophic conditions." *Algal Research* 28 (2017): 253–263. https://doi.org/10.1016/j.algal.2017.09.027
165. Khetkorn, Wanthanee, Namita Khanna, Aran Incharoensakdi, and Peter Lindblad. "Metabolic and genetic engineering of cyanobacteria for enhanced hydrogen production." *Biofuels* 4, no. 5 (2013): 535–561. https://doi.org/10.4155/bfs.13.41

166. Kumar, Sunil, Sushma Sharma, Sapna Thakur, Tanuja Mishra, Puneet Negi, Shashank Mishra, Abd El-Latif Hesham, Ali A. Rastegari, Neelam Yadav, and Ajar Nath Yadav. "Bioprospecting of microbes for biohydrogen production: current status and future challenges." In Gustavo Molina, Vijai Kumar Gupta, Brahma N. Singh, and Nick Gathergood (eds.), *Bioprocessing for Biomolecules Production*, pp. 443–471. Hoboken, NJ: Wiley (2019). https://doi.org/10.1002/9781119434436.ch22
167. Tamburic, Bojan, PongsathornDechatiwongse, Fessehaye W. Zemichael, Geoffrey C. Maitland, and Klaus Hellgardt. "Process and reactor design for biophotolytic hydrogen production." *Physical Chemistry Chemical Physics* 15, no. 26 (2013): 10783–10794. https://doi.org/10.1039/C3CP51866C
168. Elreedy, Ahmed, Eman Ibrahim, Nazly Hassan, Ali El-Dissouky, Manabu Fujii, Chihiro Yoshimura, and Ahmed Tawfik. "Nickel-graphene nanocomposite as a novel supplement for enhancement of biohydrogen production from industrial wastewater containing mono-ethylene glycol." *Energy Conversion and Management* 140 (2017): 133–144. https://doi.org/10.1016/j.enconman.2017.02.080
169. Beckers, Laurent, Serge Hiligsmann, Stéphanie D. Lambert, Benoît Heinrichs, and Philippe Thonart. "Improving effect of metal and oxide nanoparticles encapsulated in porous silica on fermentative biohydrogen production by Clostridium butyricum." *Bioresource Technology* 133 (2013): 109–117. https://doi.org/10.1016/j.biortech.2012.12.168
170. Han, Hongliang, Maojin Cui, Liling Wei, Haijun Yang, and Jianquan Shen. "Enhancement effect of hematite nanoparticles on fermentative hydrogen production." *Bioresource Technology* 102, no. 17 (2011): 7903–7909. https://doi.org/10.1016/j.biortech.2011.05.089.
171. Mohanraj, Sundaresan, Krishnasamy Anbalagan, Palanisamy Rajaguru, and Velan Pugalenthi. "Effects of phytogenic copper nanoparticles on fermentative hydrogen production by Enterobacter cloacae and Clostridium acetobutylicum." *International Journal of Hydrogen Energy* 41, no. 25 (2016): 10639–10645. https://doi.org/10.1016/j.ijhydene.2016.04.197
172. Gadhe, Abhijit, Shriram S. Sonawane, and Mahesh N. Varma. "Influence of nickel and hematite nanoparticle powder on the production of biohydrogen from complex distillery wastewater in batch fermentation." *International Journal of Hydrogen Energy* 40, no. 34 (2015): 10734–10743. https://doi.org/10.1016/j.ijhydene.2015.05.198
173. Zhang, Yongfang, and Jianquan Shen. "Enhancement effect of gold nanoparticles on biohydrogen production from artificial wastewater." *International Journal of Hydrogen Energy* 32, no. 1 (2007): 17–23. https://doi.org/10.1016/j.ijhydene.2006.06.004
174. Zhao, Wei, Yongfang Zhang, Bin Du, Dong Wei, Qin Wei, and Yanfang Zhao. "Enhancement effect of silver nanoparticles on fermentative biohydrogen production using mixed bacteria." *Bioresource Technology* 142 (2013): 240–245. https://doi.org/10.1016/j.biortech.2015.05.042
175. Taherdanak, Mohsen, Hamid Zilouei, and Keikhosro Karimi. "The effects of Fe^0 and Ni^0 nanoparticles versus Fe^{2+} and Ni^{2+} ions on dark hydrogen fermentation." *International Journal of Hydrogen Energy* 41, no. 1 (2016): 167–173. https://doi.org/10.1016/j.ijhydene.2015.11.110
176. Nath, Dhrubajyoti, Ajay Kumar Manhar, Kuldeep Gupta, Devabrata Saikia, Shymal Kumar Das, and Manabendra Mandal. "Phytosynthesized iron nanoparticles: effects on fermentative hydrogen production by Enterobacter cloacae DH-89." *Bulletin of Materials Science* 38 (2015): 1533–1538. https://doi.org/10.1007/s12034-015-0974-0

177. Lin, Richen, Jun Cheng, Lingkan Ding, Wenlu Song, Min Liu, Junhu Zhou, and Kefa Cen. "Enhanced dark hydrogen fermentation by addition of ferric oxide nanoparticles using Enterobacter aerogenes." *Bioresource Technology* 207 (2016): 213–219. https://doi.org/10.1016/j.biortech.2016.02.009
178. Zada, Bakht, Tariq Mahmood, Shoaib Ahmad Malik, and Ziya Uddin. "Effect of zinc oxide nanoparticles on hyacinth's fermentation." *International Journal of Enhanced Research in Science Technology & Engineering* 3, no. 11 (2014): 78–92.
179. Mullai, Pandian, M. K. Yogeswari, and K. J. B. T. Sridevi. "Optimisation and enhancement of biohydrogen production using nickel nanoparticles–a novel approach." *Bioresource Technology* 141 (2013): 212–219. https://doi.org/10.1016/j.biortech.2015.03.082
180. Patel, Sanjay K. S., Jung-Kul Lee, and Vipin C. Kalia. "Nanoparticles in biological hydrogen production: an overview." *Indian Journal of Microbiology* 58 (2018): 8–18. https://doi.org/10.1007/s12088-017-0678-9
181. Zhao, Yuxiao, and Yinguang Chen. "Nano-TiO_2 enhanced photofermentative hydrogen produced from the dark fermentation liquid of waste activated sludge." *Environmental Science & Technology* 45, no. 19 (2011): 8589–8595. https://doi.org/10.1021/es2016186
182. Pandey, Anjana, Kiran Gupta, and Ashutosh Pandey. "Effect of nanosized TiO_2 on photofermentation by Rhodobacter sphaeroides NMBL-02." *Biomass and Bioenergy* 72 (2015): 273–279. https://doi.org/10.1016/j.biombioe.2014.10.021
183. Kanwal, Fariha, Arifa Tahir, Syed Abdul Qadir Shah, Takuya Tsuzuki, David Nisbet, Junhong Chen, and Yasir Rehman. "Effect of phyto-fabricated nanoscale organic-iron complex on photo-fermentative hydrogen production by Rhodopseudomonas palustris MP2 and Rhodopseudomonas palustris MP4." *Biomass and Bioenergy* 140 (2020): 105667. https://doi.org/10.1016/j.biombioe.2020.105667
184. Liu, Zhidan, Fengxiang Lv, Hao Zheng, Chong Zhang, Fei Wei, and Xin-Hui Xing. "Enhanced hydrogen production in a UASB reactor by retaining microbial consortium onto carbon nanotubes (CNTs)." *International Journal of Hydrogen Energy* 37, no. 14 (2012): 10619–10626. https://doi.org/10.1016/j.ijhydene.2012.04.057

16 Carbon Sequestration and Wastewater Treatment

An Approach to Reduce Carbon Footprint

Ilangovan Pugazhenthi

16.1 INTRODUCTION

Global warming is a concerning phenomenon that leads to an increase in the average surface temperature of the Earth, primarily caused by the emission of greenhouse gases. The Earth's surface temperature rose by approximately 0.6°C–0.9°C from 1906 to 2005 (Den Elzen et al., 2017). Over the last 50 years, the rate of temperature increase has doubled, largely due to globalization and industrial development. Wastewater treatment plant operations (WWTP) are not only energy-intensive but also result in the emission of significant greenhouse gases, including nitrous oxide (N_2O), methane (CH_4), and carbon dioxide (CO_2) (Alex, 2017). Concurrently, water consumption has been on the rise, driven by population growth. Globally, approximately 1,000 km^3 of wastewater is generated, with 300 km^3 constituting municipal wastewater, while the remaining volume consists of industrial wastewater (Rosso and Stenstrom, 2008). To protect the environment, it's essential to treat wastewater and reduce the concentration of harmful elements before releasing it. A WWTP (Wastewater Treatment Plant) relies heavily on electricity and chemicals for its operation. Approximately 30%–40% of the total global electricity consumption is associated with wastewater treatment. The energy usage in wastewater treatment is contingent upon factors such as flow rate, effluent quality, the processes implemented, and the effluent's overall quality. Within WWTPs, specific components such as the aeration system, lifting pumps, and sludge dewatering units contribute significantly to the overall electricity consumption. Figure 16.1 shows the global energy consumption for the treatment of wastewater around the world. India has one of the largest populations globally, and with a growing population, the demand for water and subsequent wastewater generation also increases. Larger populations often lead to more extensive and energy-intensive wastewater treatment processes to meet the sanitation needs of the people. It is crucial to acknowledge the importance of these energy inputs in wastewater treatment processes, as they play a vital role in ensuring the efficiency and effectiveness of the treatment plant. The aeration system, lifting

DOI: 10.1201/9781003459286-16

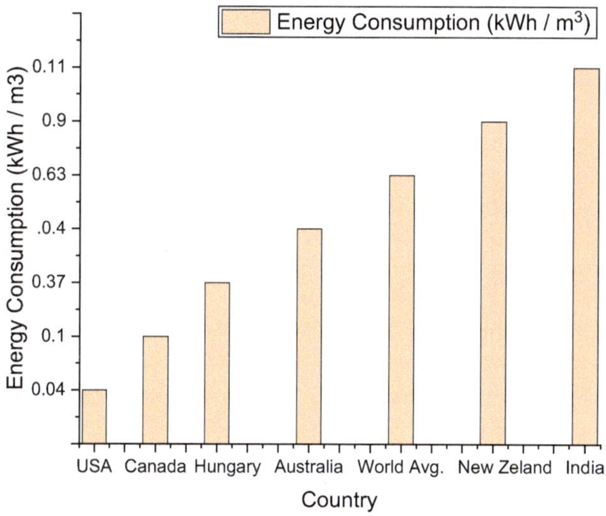

FIGURE 16.1 Energy consumption of the sanitary wastewater system.

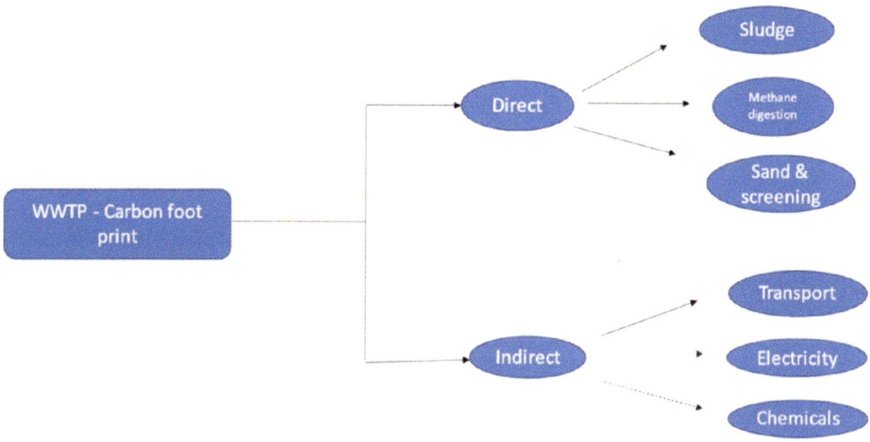

FIGURE 16.2 Carbon footprint of wastewater treatment.

pumps, and sludge dewatering units are essential for various stages of wastewater treatment, such as biological treatment, pumping, and dewatering of sludge.

Efforts to optimize energy usage and explore more energy-efficient technologies in these components can help to reduce the environmental impact of wastewater treatment plants. Additionally, advancements in sustainable practices and the use of alternative energy sources can contribute to mitigating the overall energy consumption associated with wastewater treatment. Assessing the carbon footprint of WWTP involves measuring emissions throughout the collection, treatment, and sludge disposal processes. Figure 16.2 shows carbon footprint in wastewater treatment plants. Sludge, methane removal, and screening are directly related to carbon footprint, whereas transportation of wastewater, electrical power, and chemical treatment are indirectly related. Carbon-footprint assessments serve the dual purpose of reducing environmental impact and enhancing plant efficiency. Emission management can be achieved by implementing various alternative carbon capture techniques. Carbon sequestration can be facilitated through several physical methods, such as the use of membranes, various adsorbents, and cryogenic separation (Allangawi et al., 2023).

This chapter aims to explore the fundamentals, mechanisms, advantages, and disadvantages of various strategies of carbon sequestration techniques. It focuses on non-bio carbon sequestering technologies, namely adsorption, photocatalysts, Electrochemical treatment, Ion exchange, Coagulation/flocculation, Membrane technologies, Hydrodynamic cavitation technology, and Ultrasound technique. In addition, the chapter elaborates on the mechanism of bio-based carbon sequestering techniques such as vermifiltration, Microbial electrolytic carbon capture, Plant microbial fuel cell, Microalgae cultivation, and Biochar production. Furthermore, the future perspectives and challenges of carbon sequestration have been discussed.

16.2 NON-BIOLOGICAL METHODS OF CARBON SEQUESTRATION

16.2.1 Adsorption Process

The adsorption process relies on the interaction of wastewater with a reactive adsorbent that possesses a high-specific surface area is porous, and/or contains specific functional groups. This interaction occurs over a specified equilibrium period. Following water purification, the used adsorbents can be recycled multiple times, with or without regeneration. While the method is straightforward and cost-effective, it may not be suitable for handling large-volumetric flow rates or heavily contaminated wastewater.

16.2.2 Photo-Catalytic Treatment

Semiconductors with wide band gap energy have been utilized for the photo-catalytic reduction of CO_2 in wastewater treatment plants (WWTPs). This process involves the absorption of light and CO_2 on the catalyst's surface, followed by the separation of charges, surface redox reactions, and desorption of the products (Priyadarshini et al., 2022). The absorbed photons on the catalyst's surface generate electron (e^-) and hole (h^+) pairs, which further facilitate the reduction of CO_2 and the oxidation of H_2O. The adsorbed CO_2 is converted into a bi-dentate carbonate intermediate, which is further reduced to end products. Photocatalysts with a high-surface area provide more active sites for the redox reactions. The conduction band and valence band of semiconductors should have a more negative redox potential than CO_2 and a more positive redox potential than H_2O, respectively (Wu et al., 2017).

16.2.3 Metal Oxides

Photocatalysts with suitable band structures, such as oxides of Zr^{4+}, Ta^{5+}, W^{6+}, Ti^{4+}, and Mo^{6+} have been used primarily due to their capability of simultaneously reducing CO_2 and oxidizing H_2O. TiO_2 and ZrO_2 are commonly used photocatalysts for WWTPs due to their low cost, abundance, sustainability, and catalytic activity (Inoue, 2009). Due to the presence of hybridized sp orbitals in their conduction bands, certain other metal cations, including Ga^{3+}, Ge^{4+}, Sn^{4+}, Sb^{5+}, and In^{3+} exhibit catalytic potential in reducing CO_2, facilitating the generation of photo-induced electrons (e^-) with a strong reducing capacity (Feng et al., 2014).

16.2.4 CO_2 Reduction by Carbonaceous Materials

Allotropes of carbon materials such as graphenes, Carbon nanotubes (CNT), graphene sheets, and carbon nanofibers have been utilized for the photo-catalytic reduction of CO_2 exhibited from WWTPs. The biocompatibility, large surface area, and thermal and chemical stability of these carbonaceous materials appeal to great interest in carbon sequestration. The pristine carbonaceous materials have to be doped with heteroatoms to increase their reactivity to the CO_2 or other carbon intermediates. For example, graphitic carbon nitride (g-C_3N_4) possesses significant catalytic ability in the reduction of CO_2.

16.2.5 Electrochemical Treatment

The electrochemical process is one of the robust and simple methods to remove a wide spectrum of organic and inorganic pollutants. The electro-oxidation process is employed in the treatment of industrial wastewater, where it effectively separates and degrades organic pollutants. The emitted CO_2 oxide from WWTPs is reduced at the cathode into formic acid (HCOOH), methane (CH_4), ethylene (C_2H_4), and other hydrocarbons as value-added products. At first, the CO_2 is converted into CO_2• radical which is reduced through the protonation of its oxygen atom to form •COOH or HCOO• (Priyadarshini et al., 2022). The obtained product can be reduced into formate ions (HCOO−). Furthermore, the conversion of carbon dioxide into multi-carbon products is difficult due to poor C^{2+} selectivity at high-CO_2 conversion rates (Xu et al., 2023). Alloy catalysts such as Cu-Ag with bimetal centers have been utilized to enhance the selectivity of CO_2 for C_2H_5OH. The Cu nanoparticles contribute to enhanced binding of the unsaturated intermediates of C_2H_4 whereas Ag nanoparticles decrease the stability of intermediate products. Both Cu and Ag nanoparticles show a synergetic effect on the conversion of CO_2 into ethanol product (Egirani et al., 2020).

16.2.6 Ion Exchange

Ion exchange is a commonly employed water treatment method for the removal of heavy metals, involving the reversible exchange of ions between a solid and a liquid phase. In this process, H^+ ions are released from the acidic groups of the resin, facilitating the displacement of metals by the free functional groups. This technique is known for its speed and cost-effectiveness, particularly when membrane regeneration is straightforward and inexpensive. However, drawbacks include membrane fouling, a potential reduction in overall system efficiency due to the presence of free acid leading to a low affinity for binding, and its unsuitability for treating highly concentrated metal solutions. Another method, precipitation, utilizes chemical agents to induce the formation of insoluble precipitates from pollutants, followed by filtration to eliminate the generated sludge. Widely employed in water and wastewater treatment plants, this technique is lauded for its simplicity and cost-effectiveness. Nonetheless, its efficiency diminishes with high-contaminant loads, and it necessitates the consumption of precipitation agents while generating substantial sludge and exhibiting slow-contact times.

16.2.7 Coagulation/Flocculation

In the coagulation/flocculation process, a coagulant, such as metal salts and polymers, is introduced into contaminated wastewater during a brief period of rapid mixing. This coagulant aids in the aggregation of pollutant particles, causing them to come together and adhere to it. Subsequently, under conditions of low mixing, flocculants, and polymers enable the formation of larger flocs that further aggregate, enhancing the precipitation process. Despite its economic and straightforward nature, this technique faces challenges related to low efficiency, the generation of sludge, and the need for an additional post-treatment stage, namely sedimentation (Moustafa et al., 2021). Several natural materials, including Tamarindus Indica, Moringa oleifera, and the Musa genus, have been proposed as alternative coagulants (Abdullah et al., 2019).

16.2.8 Membrane Technologies

Membrane technology encompasses the use of various membrane processes, including low-pressure-driven membranes such as distillation, microfiltration, and ultrafiltration, high-pressure-driven membranes like reverse osmosis and nanofiltration, and osmotic pressure-driven membrane processes such as electrodialysis and liquid membrane. The efficacy of these processes relies predominantly on membrane characteristics, such as surface charge, pore size, pore distribution, and the degree of hydrophobicity/hydrophilicity, along with the applied pressure (Pan et al., 2019). Membrane structures may comprise polymers, ceramics, carbon, titanium, and titanium oxide-based materials (Dehghani et al., 2019). Classification of membranes is based on their pore size, with microfiltration (0.1–10 μm), ultrafiltration (0.1–0.001 μm), and nanofiltration (0.5–2.0 nm) being common categories.

16.2.9 Ultrasound Technique

Ultrasound (US) is a technique that employs high-oxidant compounds such as $OH^{\bullet}$, $H^{\bullet}$, and H_2O_2, along with hydro-mechanical shear forces, to disintegrate various toxic and organic pollutants into simpler, biodegradable compounds (Suslick et al., 1997). However, this method is associated with high- capital and operating costs, and its standalone application exhibits low efficiency for treating large volumes of wastewater with highly variable compositions. To address these challenges and minimize costs and energy requirements, the US process has been integrated with other techniques, including advanced oxidative processes, nanotechnology, membrane treatment, and adsorption.

16.2.10 Hydrodynamic Cavitation Technology

Hydrodynamic cavitation (HC) technology operates by inducing the formation of microbubbles through a generated pressure drop and an increase in fluid velocity. The collapse of these cavities generates active spots with sufficient energy to dissociate water molecules into potent oxidative radicals, along with high-mechanical shear forces capable of degrading pollutants (Gogate et al., 2020). These combined effects significantly enhance the biodegradability of complex compounds, rendering the treatment more efficient compared to ultrasound (US) techniques (Bis et al., 2015). Notably, employing HC with a venturi tube at 13 bars for 50 minutes resulted in a remarkable 60% increase in the biodegradability index of polluted wastewater. Moreover, a substantial reduction in chemical oxygen demand (COD), total organic carbon (TOC), and color by 32%, 31%, and 48%, respectively, underscored the efficacy of cavitation. The HC process offers advantages such as enhanced biological oxidation (Bis et al., 2015) and a reduction in COD values, attributed to the breakdown of pollutants. Additionally, the HC process facilitates the transformation of complex macro-molecular compounds into micro-molecular ones, leading to an increased solubility of organic matter.

16.3 BIOLOGICAL METHODS

Biological treatment is a widely adopted method in wastewater treatment. This approach encompasses the cultivation, enrichment, and activation of various beneficial microorganisms, including fungi, algae, and bacteria. These microorganisms are either naturally present in the wastewater or introduced from external sources to efficiently degrade a range of biodegradable contaminants in wastewater. Recognized for its environmental friendliness, cost-effectiveness, and energy efficiency, biological treatment can take the form of aerobic processes, yielding carbon dioxide, biomass, and water as primary products, or anaerobic processes, resulting in the production of methane and carbon dioxide. The underlying mechanism involves reactive functional groups, such as carboxyl, phosphate, amino, and hydroxyl, on microorganisms that attract and react with different pollutants, alongside catabolic reactions. However, challenges associated with biological treatment include prolonged treatment times and difficulties in control.

16.3.1 Membrane Bioreactor

A membrane bioreactor (MBR) is a specialized bioreactor that integrates both aeration and membranes in the organic removal process (Ćemanović et al., 2018). This hybrid system is known for its versatility in treating various wastewater types, low-maintenance requirements, small footprint, low-sludge production, and high efficiency (Chang et al., 2002). It offers stability in treatment by retaining biocatalysts and microorganisms. However, a significant drawback is membrane fouling or clogging, necessitating frequent cleaning. The cleaning process not only increases operating costs due to applied pressure but also reduces the lifespan of the membrane (Karabelas et al., 2018).

A photo-catalytic membrane reactor (PMR) is designed to break down complex large organic molecules into smaller ones with the assistance of photocatalysts, resulting primarily in carbon dioxide and water (Moslehyani et al., 2018). This reactor is distinguished by its potential for energy savings, easy separation of photocatalysts from treated water, resistance to fouling, and a reduced carbon footprint. Nevertheless, the reactor still requires post-treatment (coagulation, flocculation, and sedimentation) to separate photocatalysts from the effluent, thereby enhancing water quality.

16.3.2 Vermifiltration

The primary objective of vermifiltration is to deliver high-quality products without compromising environmental sustainability. This involves a biofilter cultivated with earthworms to enhance the decomposition of organics into compost, subsequently used as fertilizer. The process entails the passage of raw wastewater through the activated layer, where organics are transformed into humus-enriched vermicompost. It then passes through filter media with triple actions—clarifying the effluent, supporting microorganisms, and trapping suspended and dissolved solids in the wastewater. The mechanism of vermifiltration is illustrated in Figure 16.3. This biofilter is advantageous as it does not clog, benefiting from earthworms' burrowing action that improves aeration. It can be integrated with other technologies and applied for the treatment of municipal, industrial, and hospital wastewater. From an environmental perspective, vermifiltration is considered an eco-friendly technology (Arora and Saraswat, 2021).

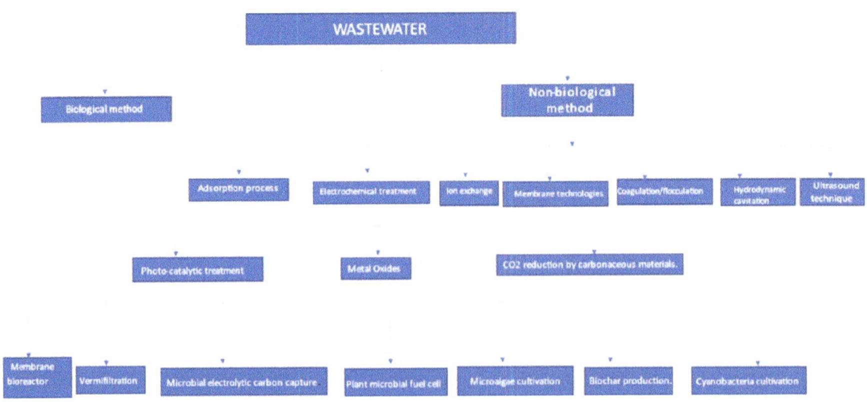

FIGURE 16.3 Diverse strategies for carbon sequestration in wastewater treatment.

Studies on vermifiltration have reported impressive removal rates, including 99.9% for bacteria, 96.9% for Salmonella, and 99.3% for E. coli (Arora and Kazmi, 2015). Additionally, it achieved substantial removal percentages for COD (87.6%), Biological Oxygen Demand (BOD) (91.3%), and total suspended solids (TSS) (98.4%) (Lourenço and Nunes, 2017). Recent advancements involve equipping vermifilters with other processes to enhance overall efficiencies (Rajpal et al., 2014). These bioreactors have been cultivated with plants like Cyprus rotundus (Tomar and Suthar, 2011) Canna indica (Samal et al., 2017), and Carex frankii (Singh et al., 2021) to improve the system's overall efficiency. Ultimately, the treated effluent from vermifiltration can meet the required standards for irrigation and agriculture.

16.3.3 Microbial Electrolytic Carbon Capture

The method involves treating wastewater by utilizing electroactive bacteria in the anode chamber to oxidize microorganisms (Figure 16.4). Specifically, electroactive bacteria (EAB) in the anode chamber oxidize biodegradable substances in wastewater, generating electrons, protons, and CO_2 as by-products. The electrons produced are accepted by the anode and then transferred through an external circuit to the cathode. At the cathode, these electrons reduce water to produce H_2 and OH^- (Guan et al., 2021). The anolyte, enriched with H^+ ions, can release metal ions (such as Ca^{2+}, Mg^{2+}, etc.) from abundant silicate minerals like wollastonite ($CaSiO_3$) or waste materials such as coal fly ash. As the metal ions migrate to the OH^- catholyte, they combine to form metal hydroxide. The subsequent reaction of metal hydroxide with CO_2 results in the spontaneous capture of CO_2, transforming it into stable carbonate or bicarbonate compounds (Maddalwar et al., 2021).

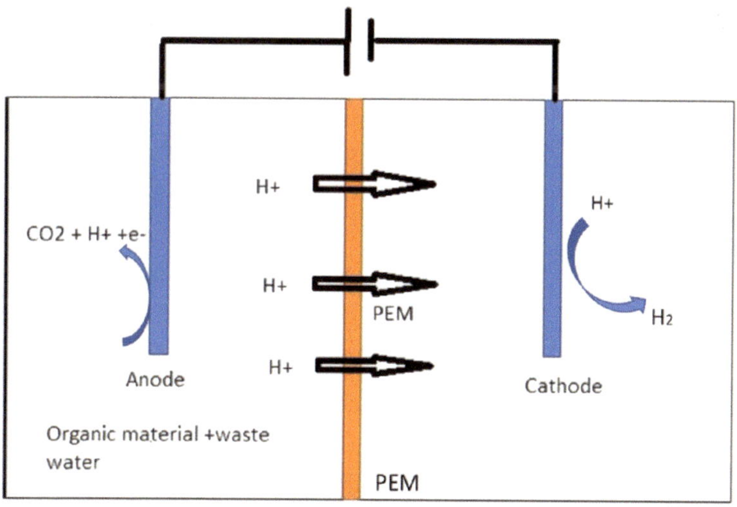

FIGURE 16.4 Mechanism of vermifiltration.

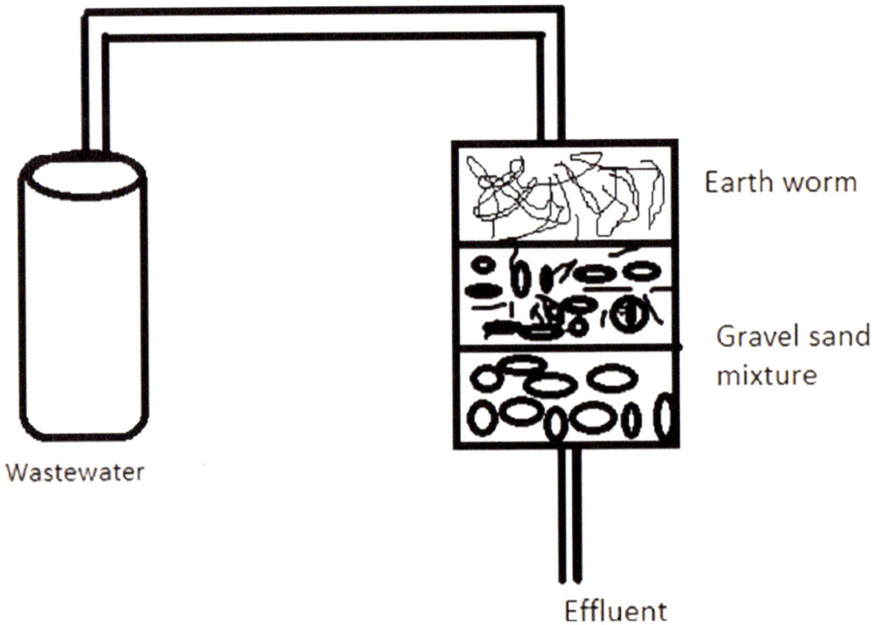

FIGURE 16.5 Mechanism of microbial electrolytic carbon capture.

16.3.4 Plant Microbial Fuel Cell

The plant microbial fuel cell represents an environmentally friendly approach to wastewater treatment, involving the recovery of valuable products and carbon sequestration (Zhu et al., 2022). In this system (Figure 16.5), plants are cultivated in the anodic chamber, providing organic nutrients to rhizobium bacteria in the root nodules. These plants receive additional nutrients from wastewater, which are consumed within the anodic chamber. During photosynthesis, plants release rhizodeposits into the soil through their roots. Electrogenic bacteria in the roots of the anodic chamber oxidize these rhizodeposits and other dissolved organic excretes, producing electrons and protons as by-products. While electrons are reduced near the plant roots, protons migrate from the anode to the cathode through the proton-exchange membrane. The potential difference between the anode and cathode enables electricity generation. The migrated protons and deposited electrons at the cathode react with oxygen to generate water. This entire process showcases a sustainable approach to wastewater treatment, harnessing the power of plants and bacteria for electricity generation while promoting environmental benefits such as carbon sequestration (Figure 16.6).

16.3.5 Microalgae Cultivation

Microalgae cultivation has more attention as it recovers nutrients from wastewater. Many species of microalgae can effectively remove nitrogen, phosphorus, heavy metals, pesticides, organic and inorganic toxins, and pathogens from

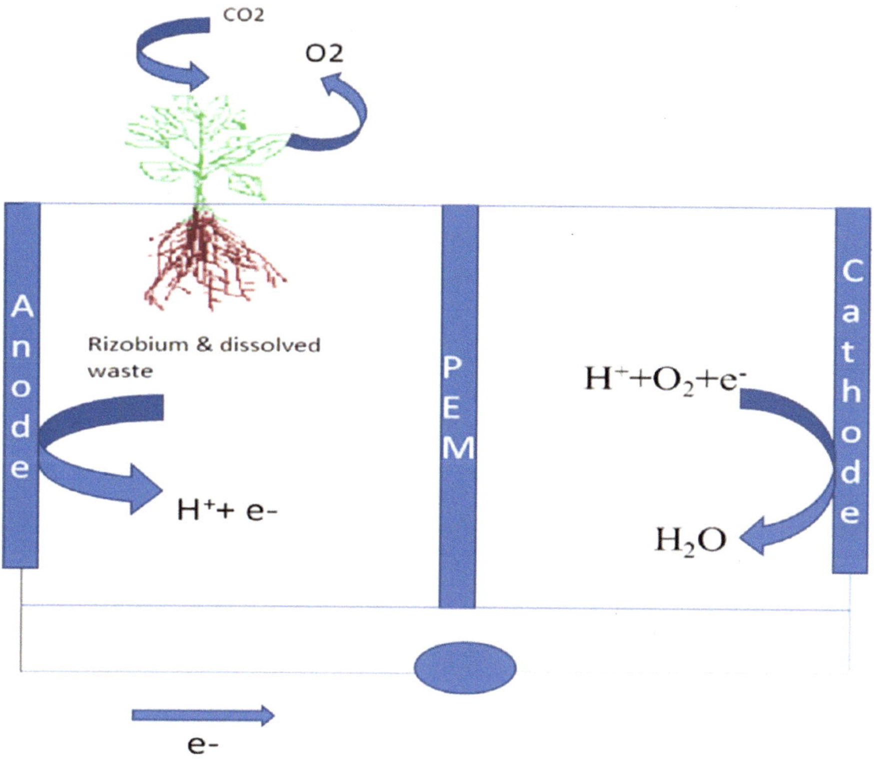

FIGURE 16.6 Mechanism of plant microbial fuel cell.

wastewater (Nitisoravut and Regmi, 2017). The primary mechanism for pollutant removal involves accumulating and utilizing pollutants within the cells. Numerous studies have documented the successful cultivation of various microalgae species, including *Chlorella, Scenedesmus, Phormidium, Botryococcus, Chlamydomonas,* and *Arthrospira*, for wastewater treatment, to demonstrate promising efficacy (Cai et al., 2013). It is essential to highlight that the effectiveness of heavy metal removal is contingent on the algal species. For instance, *Oscillatoria spp.* has demonstrated proficiency in removing chromium, *Chlorella vulgaris* in removing cadmium, copper, and zinc. Moreover, the tolerance to organic pollutants in wastewater varies among species. *Euglena, Oscillatoria, Chlamydomonas, Scenedesmus, Chlorella, Nitzschia, Navicula*, and *Stigeoclonium* have been identified as the most resilient genera to organic pollutants (Cai et al., 2013).

16.3.6 Biochar Production

Biochar, a carbon-rich form of charcoal, is created through pyrolysis in anaerobic conditions, typically being energy self-sufficient during production and yielding surplus energy like heat, bio-oil, or syngas. Although conventional plant residues are commonly used as feedstock, alternative sources such as wastewater sludge, microalgae, and wetland plants have also been employed. When applied as a soil amendment, biochar enhances soil fertility, water retention, and nutrient levels, leading to increased crop yields (Palmer, 1969). Biochar establishes a connection between short-term photosynthesis-mediated carbon cycles and long-term carbon storage in the soil, demonstrating the capacity to suppress soil CH_4 and N_2O emissions, albeit with variability based on biochar and soil properties, crop type, and local climate (Qambrani et al., 2017).

Utilizing wastewater sludge for biochar production offers environmental and economic advantages over current practices like landfilling. Sludge biochar not only eradicates pathogens but also enhances soil structure, adsorbs pesticides, reduces heavy metal leaching, and boosts soil fertility. Studies have shown significant increases in crop production, such as a 64% rise in cherry tomato yield with the application of sludge biochar, attributed to improved nutrient availability and soil conditions (Rogelj et al. 2016). While there is limited literature on biochar production from wastewater microalgae or constructed wetland plants, similar benefits are anticipated. However, due to the high-moisture content in these feedstocks, compared to woody biomass, the conversion process may be more energy-intensive and require additional treatment steps. Therefore, further research is essential in product quality control, life-cycle assessment, and techno-economic analysis (Hossain et al., 2010).

16.3.7 Cyanobacteria Cultivation

The utilization of photosynthetic microorganisms, including eukaryotic microalgae and cyanobacteria, for wastewater treatment was initially suggested by Oswald and Golueke (Lee et al., 2015). The approach involved cultivating cyanobacteria and eukaryotic microalgae in open raceway ponds in California for municipal wastewater treatment, where they coexisted with aerobic heterotrophic bacteria (Borde et al., 2003). In this method, photosynthetic microorganisms play a crucial role by generating molecular oxygen, which serves as an electron acceptor for bacteria engaged in degrading organic matter. Simultaneously, bacteria release CO_2 during the mineralization process, providing a substrate for photosynthesis (Abed et al., 2009). Cyanobacteria, specifically, have diverse applications in wastewater treatment, including nutrient removal, reduction of both COD and BOD, as well as the removal of heavy metals (Honda et al., 2012).

This wastewater treatment approach is considered cost-effective and environmentally friendly, as it minimizes secondary pollution. Provided that the biomass generated can be effectively reused, enabling efficient nutrient recycling, various technologies based on cyanobacteria have demonstrated successful applications in the treatment of municipal, livestock, and industrial wastewaters (Wang et al., 2010). The efficiencies achieved in these applications often surpass those of conventional aerobic-activated sludge processes in terms of nutrient assimilation and pathogen removal, all while requiring much lower energy consumption (de Godos et al., 2009).

16.4 COMPARATIVE STUDIES OF BIOLOGICAL AND NON-BIOLOGICAL METHODS

Table 16.1 shows the comparative analysis of the biological and non-biological treatment of wastewater. Generally, non-biological methods such as Adsorption, Electrochemical treatment, Ion exchange, Coagulation/flocculation, Membrane technologies, HC technology, and Ultrasound techniques are efficient in pollutant removal. However, these methods have challenges including regeneration issues, limitations in handling large volumes of highly polluted water, and potential environmental concerns. These methods are suitable for specific applications but may have drawbacks in terms of sustainability and long-term operation. On the other hand, biological methods such as Membrane bioreactor, Vermifiltration, Microbial electrolytic carbon capture, Plant microbial fuel cell, Microalgae cultivation, and Biochar production are offering a range of advantages including cost-effectiveness, versatility, and ecological friendliness. Common challenges of the biological techniques are membrane fouling, clogging, and limitations in scalability and treatment rates. Bio-based methods demonstrate potential for energy recovery and resource utilization. Choosing between these methods depends on the specific goals, characteristics of the wastewater, and the context of the treatment facility. Integrated approaches combining both biological and non-biological methods may provide enhanced efficiency and sustainability.

16.5 FUTURE PERSPECTIVES

Addressing the urgent global issues of climate change and extensive wastewater generation stemming from human activities is paramount for the well-being of both current and future generations. A pivotal measure in combating climate change involves proactively curbing the release of carbon dioxide (CO_2) and other greenhouse gas emissions into the atmosphere. Simultaneously, a reduction in wastewater production can be achieved by minimizing the usage of harmful chemicals in our daily routines. To enact preventive measures successfully, it is crucial to explore and embrace alternative sustainable practices in our day-to-day lives. This includes advocating for the adoption of environmentally friendly options like compressed natural gas (CNG) or solar-powered vehicles in lieu of petroleum-based ones and choosing natural pesticides over synthetic alternatives.

While recognizing the impracticality of complete replacement overnight, a gradual reduction in the reliance on toxic chemicals and appliances contributing to greenhouse gas emissions can significantly alleviate environmental burdens. Therefore, the most effective strategy for minimizing the impacts of global warming and wastewater generation in the present and future lies in the mindful and responsible utilization of available resources. In conjunction with source monitoring and control, addressing the imperative for CO_2 capture and wastewater treatment is crucial in the contemporary era.

This document extensively explores a variety of both biological and non-biological CO_2 capture technologies. These processes have exhibited promising results in mitigating CO_2 emissions, underscoring the advancements in the field. Despite these

TABLE 16.1
Comparative Studies of Various Carbon Sequestration Method

Method	Process	Advantages	Disadvantages	References
Non-biological treatment	Adsorption	Efficiently removes organic pollutants	Regeneration of the adsorbent Not suitable for large-volumetric flow rates or highly polluted wastewater	Lourenço and Nunes (2017)
	Electrochemical treatment	Increased removal performance remarkably, and decreased operational costs	Formation of toxic-intermediate metabolites, high-energy consumption, and sludge generation	Yakamercan et al. (2023)
	Ion exchange	Low cost, simple equipment, easy operation, and no or little use of organic solvents	It has its drawbacks, such as a long-production cycle, sometimes poor-product quality, and large-pH changes in the production process. On top of this, it is not always possible to find suitable resin	Chen and Lan (2017)
	Coagulation/flocculation	Simple process, good for reclamation or removing pollutants	Produce massive sludge and large particles	Ariffin et al. (2017)
	Membrane technologies	Small-occupation area, high-processing efficiency	Membrane fouling	Zhang et al. (2016)
	HC technology	High-energy efficiency, cost-effective operation, ability to induce chemical reactions, and scale-up possibilities	Cavitation can cause a lot of noise. Cavitation can cause a loss of efficiency in devices like pumps and propellers	Wang et al. (2021)
	Ultrasound technique	Environmentally friendly	Require high-energy	Budiman and Wu (2016)

(Continued)

TABLE 16.1 (*Continued*)
Comparative Studies of Various Carbon Sequestration Method

Method	Process	Advantages	Disadvantages	References
Biological treatment	Membrane bioreactor	Cost-effective, versatile, fast, and ecologically friendly method of removing toxins from the environment	Membrane fouling	Iorhemen et al. (2016)
	Vermifiltration	High-treatment efficiency given the low-space requirement	The major problem of vermifiltration is clogging	Arora et al. (2021)
	Microbial electrolytic carbon capture	The wastewater itself provides electrolytes and is used to dissolve minerals	CO_2 requires a significant amount of electrons for the synthesis of organic compounds due to its thermodynamic stability	Christodoulou et al. (2017)
	Plant microbial fuel cell	More energy-efficient alternative to traditional wastewater treatment plants, as they can generate electricity from the organic matter in wastewater, offsetting the energy needs of the treatment plants	High-power density separator/membrane between the electrodes, which prevents the growth of cell microorganisms on them so as to not degrade their behavior, but this can cause an increase in the internal resistance again	Tsekours et al. (2022)
	Microalgae cultivation	Pollutant removal, support for degrading bacteria, and potential for biodiesel production	Slow-treatment rates, low-toxic tolerance, and difficulties in upscaling bioaugmentation processes	Kumar et al. (2023)
	Biochar production	Less emission of greenhouse gases in the air: Biochar is a stable product that can stay in the soil for many years	Dust is the most serious health hazard associated with biochar	Kumar et al. (2023)

positive developments, several challenges persist that require attention for the successful and practical implementation of these technologies. These obstacles are thoroughly examined below, highlighting the ongoing efforts needed for their seamless integration into mainstream practices. The next generation of technologies will be hugely influenced by the need to reduce carbon emissions combined with rigorous regulations whilepromoting water reuse.

- Integration of anaerobic digestion processes in wastewater treatment facilities can generate bioenergy from organic waste while simultaneously sequestering carbon. This process converts organic matter into biogas and a renewable energy source.
- Implementation of carbon capture technologies within wastewater treatment plants can capture CO_2 emissions before they are released into the atmosphere. This involves the use of absorbents or adsorbents to capture and separate CO_2.
- Advancements in treatment technologies, including membrane-based processes and advanced oxidation methods, can contribute to more efficient carbon removal from wastewater.
- Growing awareness of environmental issues and the implementation of stricter regulations may drive wastewater treatment facilities to adopt more carbon-sequestration-friendly practices.
- Continued collaboration between researchers, industries, and governments can lead to the development and implementation of innovative technologies for carbon sequestration in wastewater treatment.

16.6 CONCLUSION

The exploration of carbon sequestration within the context of wastewater treatment represents a promising approach to mitigate and reduce the carbon footprint associated with industrial processes. This synergy between carbon sequestration and wastewater treatment is crucial in addressing the dual challenges of environmental sustainability and climate change. By strategically integrating carbon sequestration methods into wastewater treatment practices, we not only enhance the efficiency of waste removal but also contribute to the broader goal of carbon-footprint reduction. The adoption of advanced technologies, both biological and non-biological, offers a spectrum of possibilities for capturing and storing carbon dioxide from industrial effluents. Moving forward, continued research, technological innovation, and widespread implementation of these integrated approaches are essential to realizing tangible reductions in the carbon footprint associated with wastewater treatment. By embracing this multifaceted strategy, we pave the way for a evironmentally sustainable solution.

REFERENCES

Abdullah, N., Yusof, N., Lau, W., Jaafar, J., Ismail, A. Recent trends of heavy metal removal from water/wastewater by membrane technologies. *Journal of Industrial and Engineering Chemistry*, 2019, 76, 17–38.

Abed, R. M. M., Dobretsov, S., Sudesh, K. Applications of cyanobacteria in biotechnology. *Journal of Applied Microbiology*, 2009, 106, 1–12.

Alex, S. Tapping sewage as a source of useful materials. *Chemical & Engineering News,* 2017, 95, 30–34.

Allangawi, A., Alzaimoor, E. F. H, Shanaah, H. H., Mohammed, H. A., Saqer, H., El-Fattah, A. A., Kamel, A. H. Carbon capture materials in post-combustion: adsorption and absorption-based processes. Carbon, 2023, 9, 17.

Ariffin, N., Abdullah, M. M. A. B., Mohd Arif, Z., Murshed, M. F., Hariz-Zain, F., Bayuaji, R. Review on adsorption of heavy metal in wastewater by using geopolymer. *MATEC Web of Conferences,* 2017, 97, 01023.

Arora, S., Kazmi, A. The effect of seasonal temperature on pathogen removal efficacy of vermifilter for wastewater treatment. *Water Research,* 2015, 74, 88–99.

Arora, S., Saraswat, S. Vermifiltration as a natural, sustainable and green technology for environmental remediation: a new paradigm for wastewater treatment process. *Current Research in Green and Sustainable Chemistry*, 2021, 4, 100061.

Bis, M., Montusiewicz, A., Ozonek, J., Pasieczna-Patkowska, S. Application of Hydrodynamic cavitation to improve the biodegradability of mature landfill leachate. *Ultrasonics Sonochemistry*, 2015, 26, 378–387.

Borde, X., Guieysse, B., Delgado, O., Muñoz, R., Hatti-Kaul, R., Nugier-Chauvin, C. Synergistic relationships in algal-bacterial microcosms for the treatment of aromatic pollutants. *Bioresource Technology,* 2003, 86(3), 293–300.

Budiman, P. M., Wu, T. Y. Ultrasonication pre-treatment of combined effluents from palm oil, pulp and paper mills for improving photo-fermentative biohydrogen production. *Energy Conversion and Management*, 2016, 119, 142–150.

Cai, T., Park, S. Y., Li, Y. Nutrient recovery from wastewater streams by microalgae: status and prospects. *Renewable and Sustainable Energy Reviews,* 2013, 19, 360–369.

Ćemanović, O. A., Arikan, A., Kizilet, A., Çınar, O. Treatment of industrial wastewaters with anaerobic membrane bioreactors. In *Proceedings of the 4th International Conference on Engineering and Natural Sciences*, Kiev, Ukraine, 2018.

Chang, I. S., Le Clech, P., Jefferson, B., Judd, S. Membrane fouling in membrane bioreactors for wastewater treatment. *Journal of Environmental Engineering*, 2002, 128, 1018–1029.

Chen, H., Lan, W. (eds.). Posttreatment strategies for biomass conversion. In *Technologies for Biochemical Conversion of Biomass*, 2017, pp. 197–217. Academic Press.

Christodoulou, X., Tobechi, O., Simon, P., Sharon, B., Velasquez, O. The use of carbon dioxide in microbial electrosynthesis: advancements, sustainability and economic feasibility. *Journal of* CO_2 *Utilization,* 2017, 18, 390–399.

de Godos, I., Blanco, S., García-Encina, P. A., Becares, E., Muñoz, R. Long-term operation of high-rate algal ponds for the bioremediation of piggery wastewaters at high loading rates. *Bioresource Technology,* 2009, 100(19), 4332–4339.

Dehghani, M. H., Zarei, A., Yousefi, M. Efficiency of ultrasound for degradation of an anionic surfactant from water: surfactant determination using methylene blue active substances method. *MethodsX*, 2019, 6, 805–814.

Den Elzen, M., Höhne, N., Jiang, K.; Cantzler, J.; Drost, P. *The Emissions Gap Report 2017.* Nairobi: United Nations Environment Programme. 2017.

Egirani, D., Poyi, N., Shehata, N. Preparation and characterization of powdered and granular activated carbon from palmae biomass for cadmium removal. *International Journal of Environmental Science and Technology,* 2020, 17, 2443–2454. https://doi.org/10.1007/s13762-020-02652-w

Feng, S., Chen, X., Zhou, Y., Tu, W., Li, P., Zou, Z. $Na_2V_6O_{16}{\cdot}xH_2O$ Nanoribbons: large-scale synthesis and visible-light photocatalytic activity of CO_2 into solar fuels. *Nanoscale* 2014, 6, 1896–1900. https://doi.org/10.1039/c3nr05219b

Gogate, P., Thanekar, P., Oke, A. Strategies to improve biological oxidation of real wastewater using cavitation-based pre-treatment approaches. *Ultrasonics Sonochemistry,* 2020, 64, 105016.

Guan, C.-Y., Yu, C. P. Evaluation of plant microbial fuel cells for urban green roofs in a subtropical *Metropolis. Science of the Total Environment,* 2021, 765, 142786. https://doi.org/10.1016/j.scitotenv.2020.142786

Honda, R., Boonnorat, J., Chiemchaisri, C., Chiemchaisri, W., Yamamoto, K. Carbon dioxide capture and nutrients removal utilizing treated sewage by concentrated microalgae cultivation in a membrane photobioreactor. *Bioresource Technology,* 2012, 125, 59–64.

Hossain, M. K., Strezov, V., Chan, K, Y., Nelson, P. F. Agronomic properties of wastewater sludge biochar and bioavailability of metals in production of cherry tomato (Lycopersicon esculentum). *Chemosphere*, 2010, 78, 1167–1171.

Inoue, Y. Photocatalytic water splitting by RuO_2-loaded metal oxides and nitrides with d^0- and d^{10}-related electronic configurations. *Energy & Environmental Science,* 2009, 2, 364–386.

Iorhemen, O. T., Hamza, R. A., Tay, J. H. Membrane bioreactor (MBR) technology for wastewater treatment and reclamation: membrane fouling. *Membranes,* 2016, 6(2), 33.

Karabelas, A. J., Plakas, K. V., Sarasidis, V. C. How far are we from large-scale PMR applications? In A. Basile, E. Curcio, Inamuddin (eds.), *Current Trends and Future Developments on (Bio-)Membranes*, 2018, pp. 233–295. Elsevier.

Kumar, S, Priyadarshini, M., Azhan, A., Ghangrekar, M. M. Advanced biological and non-biological technologies for carbon sequestration, wastewater treatment, and concurrent valuable recovery: a review. *Journal of CO_2 Utilization*, 2023, 68, 102372.

Lee, C. S., Lee, S. A., Ko, S, R., Oh, H. M., Ahn, C. Y. Effects of photoperiod on nutrient removal, biomass production, and algal-bacterial population dynamics in lab-scale photobioreactors treating municipal wastewater. *Water Research,* 2015, 68, 680–691.

Lourenço, N., Nunes, L. Is filter packing important in a small-scale vermifiltration process of urban wastewater? *International Journal of Environmental Science and Technology,* 2017, 14, 2411–2422.

Maddalwar, S., Kumar Nayak, K., Kumar, M., Singh, L. Plant microbial fuel cell: opportunities, challenges, and prospects. *Bioresource Technology,* 2021, 341, 125772.

Moslehyani, A., Hubadillah, S. K., Othman, M. H. D., Ismail, A. F., Matsuura, T. PMRs in photodegradation of organic contaminants: water and wastewater treatment. In A. Basile, E. Curcio, Inamuddin (eds.), *Current Trends and Future Developments on (Bio-) Membranes,* 2018, pp. 189–208. Elsevier.

Moustafa, D., Mahmoud, R., Abd El-Salam, H. M., Shehata, N. Utilization of residual zinc–iron-layered double hydroxide after methyl orange management as a new sorbent for wastewater treatment. *Applied Nanoscience*, 2021, 11, 709–723.

Nitisoravut, R., Regmi, R. Plant microbial fuel cells: a promising biosystems engineering. *Renewable and Sustainable Energy Reviews*, 2017, 76, 81–89.

Palmer, C. M. A composite rating of algae tolerating organic pollution. *Journal of Phycology*, 1969, 5, 78–82.

Pan, Z., Song, C., Li, L., Wang, H., Pan, Y., Wang, C., Li, J., Wang, T., Feng, X. Membrane Technology coupled with electrochemical advanced oxidation processes for organic wastewater treatment: recent advances and future prospects. *Chemical Engineering Journal,* 2019, 376, 120909.

Priyadarshini, M., Ahmad, A., Das, S., Ghangrekar, M. M. Application of innovative electrochemical and microbial electrochemical technologies for the efficacious removal of emerging contaminants from wastewater: a review. *Journal of Environmental Chemical Engineering,* 2022, 10, 108230.

Qambrani, N. A., Rahman, M. M., Won, S., Shim, S., Ra, C. Biochar properties and eco-friendly applications for climate change mitigation, waste management, and wastewater treatment: a review. *Renewable and Sustainable Energy Reviews,* 2017, 79, 255–273.

Rajpal, A., Arora, S., Bhatia, A., Kumar, T., Bhargava, R., Chopra, A., Kazmi, A. Co-treatment of organic fraction of municipal solid waste (OFMSW) and sewage by vermireactor. *Ecological Engineering*, 2014, 73, 154–161.

Rogelj, J., den Elzen, M., Höhne, N., Fransen, T., Fekete, H., Winkler, H., Schaeffer, R., Sha, F., Riahi, K., Meinshausen, M. Paris agreement climate proposals need a boost to keep warming well below 2°C. *Nature*, 2016, 534, 631–639.

Rosso, D., Stenstrom, M. K. The carbon-sequestration potential of municipal wastewater treatment. *Chemosphere*, 2008, 70, 1468–1475.

Samal, K., Dash, R. R., Bhunia, P. Performance assessment of a Canna indica assisted vermifilter for synthetic dairy wastewater treatment. *Process Safety and Environmental Protection,* 2017, 111, 363–374.

Singh, R., Alessio, M. D., Meneses, Y., Bartelt-Hunt, S. L., Woodbury, B., Ray, C., Development and performance assessment of an integrated vermifiltration based treatment system for the treatment of feedlot runoff. *Journal of Cleaner Production,* 2021, 278, 123355.

Suslick, K. S., Mdleleni, M. M., Ries, J. T. Chemistry induced by hydrodynamic cavitation. *Journal of the American Chemical Society*, 1997, 119, 9303–9304.

Tomar, S., Suthar, S. Urban wastewater treatment using vermi-biofiltration system. *Desalination*, 2011, 282, 95–103.

Tsekours, G. J., Deligianni, P. M., Kanellos, F. D., Kontargyri, V. T., Kontaxis, P. A. Microbial fuel cell for wastewater treatment as power plant in smart grids: utopia or reality? *Frontiers in Energy Research,* 2022, 10. https://doi.org/10.3389/fenrg.2022.843768

Wang, B., Huijuan, S., Zhang, B. Hydrodynamic cavitation as a promising route for wastewater treatment – a review. *Chemical Engineering Journal*, 2021, 412, 128685.

Wang, L., Min, M., Li, Y., Chen, P., Chen, Y., Liu, Y., Wang, Y., Ruan, R. Cultivation of green algae Chlorella sp. in different wastewaters from municipal wastewater treatment plant. *Applied Biochemistry and Biotechnology*, 2010, 162(4), 1174–1186.

Wu, J., Huang, Y., Ye, W., Li, Y. CO_2 Reduction: from the electrochemical to photochemical approach. *Advanced Science,* 2017, 4, 1700194.

Xu, D., Li, K., Jia, B., Sun, W., Zhang, W., Liu, X., Ma, T. Electrocatalytic CO_2 reduction towards industrial applications. *Carbon Energy,* 2023, 5, e23.

Yakamercan, E., Bhatt, P., Aygun, A., Adesope, A. W., Simsek, H. Comprehensive understanding of electrochemical treatment systems combined with biological processes for wastewater remediation. *Environmental Pollution,* 2023, 1, 121680.

Zhang, F., Gao, S., Zhu, Y., Jin, J. Alkaline-induced superhydrophilic/underwater superoleophobic polyacrylonitrile membranes with ultralow oil-adhesion for high-efficient oil/water separation. *Journal of Membrane Science,* 2016, 513, 67–73.

Zhu, X., Lei, C., Qi, J., Zhen, G., Lu, X., Xu, S., Zhang, J., Liu, H., Zhang, X., Wu, Z. The role of microbiome in carbon sequestration and environment security during wastewater treatment. *Science of the Total Environment,* 2022, 837, 155793.

17 Circular Bioeconomy

An Environmentally Sustainable and Economically Viable Approach to Wastewater Treatment

Chetna Kumari, Sudesh Kumar, Nirmala Kumari Jangid, and Navjeet Kaur

17.1 INTRODUCTION

The basic component for people and all living things is water. Water pollution, on the other hand, is a problem that not only negatively affects aquatic species but also the larger biological community [1]. Every year, numerous industrial production or auxiliary processes, as well as other human activities, discharge heavy metals into the environment [2,3]. The term "heavy metals" refers to a group of more than 40 elements with masses of more than 5 g/cm^3 [4,5]. One of the biggest pollutants in water is heavy metal ions. Such contaminants, known as heavy metals, have gotten a lot of attention from the scientific, governmental, and community activist sectors due to their persistence and toxicity characteristics. All biological species, as well as the ecosystem as a whole, are extremely vulnerable to toxic metals, especially when their concentrations exceed the established tolerance limitations [6,7]. However, more than 10% of the dyes used in the chemical industry each year approximately 1 million tonnes are released as effluents, dangerously polluting the environment and harming both humans and aquatic life [8]. Due to the dyes' often complex and stable structures and the slow and difficult nature of their natural breakdown, they will inevitably accumulate [9–11]. In addition to heavy metals and dyes, radioactive waste; a waste byproduct made up of radioactive components often produced in a nuclear process like nuclear fission is one of the main contaminants substantially affecting water supplies. Even though such waste eventually loses its radioactivity, it does so very slowly, which means that it could have significant long-term effects on the environment and human health [12]. Additionally, pesticides have been widely employed in contemporary agriculture to protect crops and cattle, which have had many negative effects on human health. These pesticides, according to experts, have escaped into the environment through leaching, poor empty container disposal, and

DOI: 10.1201/9781003459286-17

direct agricultural runoff [13]. Additionally, the current discharge of nitrogen-containing substances (nitrate) into water sources is detrimental to both human and animal health. Since nitrates contribute to groundwater pollution and river eutrophication, they can spread throughout the water and result in several major issues [14]. Environmental protection requires the research and deployment of various techniques to eliminate harmful toxins from water.

Living in an environmentally friendly environment has become crucial for humans in the current climate of the era of eco-friendliness. Insoluble in water, hydrogels are three-dimensional networks of polymers that may absorb water to saturate the gap between macromolecules [15,16]. While the hydrophilic groups in the polymeric backbone of hydrogels allow for widespread water absorption, the cross-linked frameworks prevent hydrogels from dissolving in aqueous media [17]. The hydrogels maintain their precise forms and can absorb physiological fluids in biological systems. Due to their incredible potential in a wide range of applications, hydrogel development research has seen a great increase in interest [18]. In the 1950s, Wichterle and LI'm created the first hydrogel based on copolymers of 2-Hydroxyethyl Methacrylate (HEMA) and Ethylene Dimethyl Acrylate, which was used in medical implants (soft contact lenses) [19]. Numerous sectors, such as sanitary items, paints or resins, cosmetics, packaging material, agriculture water retention agents, separating membranes, and some medicinal uses, such as contact lenses or drug delivery systems, have found extensive use for hydrogels [20]. Some natural hydrogels have been replaced in recent years by artificial hydrogels, which have a longer lifespan, a better capacity for water absorption, and greater strength. It is possible to modify the chemical makeup of synthetic polymers to provide hydrogel functionality and degradability. Hydrogels can additionally be modified to maintain their stability in the presence of extreme and rapid temperature changes. The most priceless and necessary gift for the existence of human beings and other living things on this planet is water but unfortunately, water sources are being polluted through wastewater discharge from industries and other human activities. The threat that heavy metals and colors represent to both human health and the environment serves as a reminder of how crucial effective industrial wastewater treatment is. Since they are so good at removing contaminants from waste streams, hydrogels (HGs) have attracted a lot of attention [21]. Heavy metals, dyes, and other pollutants can be removed and recovered as a result of the high porosity and presence of hydrophilic functional groups, such as $-OH$, $-COOH$, $-NH_2$, $-SO_3H$, $-CONH_2$, etc., that allow them to absorb and retain a lot of water throughout the treatment process [22].

17.2 HYDROGELS IN WASTEWATER TREATMENT

The demand for water for daily usage is increasing daily as a result of industrial activity and technological advancement. Due to its superior capacity to dissolve more substances than any other solvent, water is quickly contaminated. Water quality is decreased when physical, chemical, or biological materials are added, which affects the organisms. The simple addition of dissolved and suspended particles to the process as well as the release of the sneakiest and most persistent harmful compounds are all included. Currently, generating pure water is essential for drinking, reusing, and agriculture. Numerous businesses, including those that produce harmful dyes, radioactive pollutants, deadly compounds, fluoride, and other toxins, are largely to blame for the thousands of organic, inorganic, and biological species that have been found in water. Several of the pollutants have significant negative impacts and toxicities, although only a small number of them are lethal and cancer-causing [23].

The presence of heavy metals, usually referred to as trace elements, poses a major threat to the water quality. These heavy metals Gd, Cr, Pb, Hg, Cu, Ni, Zn, and As are well-known [24]. Arsenic is a metalloid yet is nevertheless classified as a heavy metal. The body's soft and hard tissues, including the kidney, liver, and bones, as well as the food chain, are prone to accumulate these metal ions. These metals are dangerous to human health because they obstruct organism growth and development. They are harmful and can result in cancer in addition to these [25]. Due to the harm it does to all living creatures, water contamination has become a serious issue for the present and the future [26].

Due to a lack of available water resources, water treatment technologies are the only way to meet the demand for water in the next decades [27]. Therefore, developing wastewater treatment systems that are effective, rapid, and economical is essential for the 21st century. Making affordable adsorbents available to all social groups and economic sectors is the aim [22]. The danger that industrial wastewater contamination poses to both human health and the ecosystem has been recognized by many facets of society. Many habitats on earth have been seriously harmed by the poisons from industrial waste and human waste. It is essential to get rid of or degrade harmful compounds, chemicals, and elements that compromise the quality and safety of the environment. One of the most promising remediation techniques is the use of adsorbent materials to remove toxins from the aquatic environment [28]. Traditional water and wastewater treatment systems use a lot of energy and produce effluent water that is of low quality because of physical, chemical, and biological processes.

Because of its low cost, simple design, and efficiency against a range of pollutants, adsorption can be preferable to other procedures (such as coagulation, precipitation, filtering, reverse osmosis, and ion exchange) [29]. The adsorption technique primarily takes advantage of the adsorption activity of the adsorbent surface to remove the heavy metal ions. Heavy metal ions are transported from wastewater to the surface of the adsorbent through convection-diffusion, heavy metal ions are adsorbed on the adsorbent's outer surface, and heavy metal ions continue to permeate into the interior of the adsorbent [30]. It is extremely environmentally friendly, simple to use, and highly efficient in removing heavy metal ions at a low cost because the binding or chemical reaction takes place on the adsorbent, which ensures that it won't secondarily damage the sewage. Present-day adsorbent material preparation presents some difficulties, such as challenging preparation procedures and low recovery rates. Adopting a novel adsorbent with an easy preparation procedure, successful adsorption, and a high recovery rate is essential.

The ability of hydrogels to bind to pollutants through physical or chemical interactions is comparable to that of activated carbon, making them the most common adsorbent material used to remediate aqueous contaminants [12,31]. Contrary to other adsorbent materials, hydrogel's more adaptable structure enables the kind and quantity of functional groups to be altered to customize the characteristics for the intended application. For a range of water treatment procedures, hydrogels with qualities like low interfacial tension, high hydrophilicity, good expandability, super absorbency, high porosity, selective permeability, variable shape, simple handling, and programmable dimensions can be used. Additionally, the ease with which natural and synthetic polymers may be used to create hydrogels (mixed or not) is a huge benefit. Furthermore, the hydrogel matrix is issued to entrap various fillers, such as clays, metals, carbon-based materials, and organic fillers, resulting in composite materials with superior and increased thermal stability, chemical resistance, mechanical capabilities, and adsorption capacity [32]. Composite hydrogels may exhibit superior adsorption selectivity and capacity than traditional hydrogels due to the inherent filler features, such as many functional groups and a large specific area. Growing environmental concerns have led to the progressive replacement or blending of these synthetic monomers with natural polymers, primarily polysaccharides. In contrast to many synthetic monomers, polysaccharides are biodegradable, non-toxic, affordable, plentiful, and renewable [22].

The textile and fabric industries have utilized dye for a very long period. There are 0.1 million dyes that are readily available commercially. About 7.0×10^5 metric tonnes of dyes are created each year [33]. Dye contaminants are dangerous since it is difficult to get rid of them from water. While some synthetic colors can cause cancer, most of them are detrimental. The three colors that are known to cause human cancer are malachite green, rhodamine B, and Congo red [34]. The application of hydrogel biopolymers for wastewater treatment is currently receiving increased attention from scientists. The evaluation of the hydrogel's adsorption capacities on various dyes is shown in Table 17.1.

17.3 FACTOR AFFECTING ADSORPTION

Several factors affect the effectiveness of liquid-phase adsorption and, thus how well the water treatment process works. The kind of pollutant, its chemical makeup, and the characteristics of the utilized adsorbent all have an impact on how well sorption works. The interaction between the adsorbent and the adsorptive material, the adsorbent's surface chemistry and pore structure, particle size, the kind of adsorbent, the presence of additional ions in the aqueous solution, temperature, pH, pressure, and contact time are just a few examples of these physicochemical factors (Figure 17.1). It is also important to consider the adsorbate's characteristics, including its polarity, size, molecular weight, and molecular structure [8,58].

Pollutants found in wastewater include heavy metals, colloidal particles, herbicides, and pharmaceutical residues. All of these may interfere with the clearance of adsorption by competing for binding sites or in some other way. When the concentrations of competing pollutants rise, the target pollutant's removal by biosorption is frequently impeded [59]. It is crucial to take into account how each of these factors would affect the adsorption process when developing it. Enhancing these conditions will be particularly beneficial for the development of large-scale dye removal technology.

TABLE 17.1
The Evaluation of the Hydrogel's Adsorption Capacities on Various Dye

Sr No	Hydrogel	Dyes	Adsorption Capacity (mg/g)	References
1.	GO-hydroxypropyl cellulose (HPC) hydrogel	MB	118.483	[35]
2.	PAA-AgNPs hydrogel	CR, Rh B, and MB	15.17, 4.52, 14.08	[36]
3.	PHA hydrogel	Phenazine dyes like Safranin T, Janus Green, and Neutral Red	23	[37]
4.	PDDA-Graphene (GO) hydrogel	Anionic dyes (ponceau and trypan blue)	58.54	[38]
5.	Konjac Glucomannan (KGM)-GO hydrogel	MB and methyl orange (MO)	MO 51.6 and MB 92.3	[39]
6.	PVA/PAA/TiO_2-GO hydrogel	MB and CBB	—	[40]
7.	PDMDAAC/PAAm hydrogel	Anionic dyes such as CR	200–465	[41]
8.	CS-Cellulose-Rectorite (RT) interconnected hydrogel	Congo Red (CR)	166	[42]
9.	Laponite modified-PAAm hydrogel	Crystal Violet (CV)	40	[43]
10.	CS-g-poly (acrylic acid) (PAA)/halloysite (HT) hydrogel	Methylene Blue	1,336.05	[44]
11.	CS-Hexadecyl amine (HAD) (a cationic surfactant) hydrogel	Reactive Blue 4 (RB4)	454	[45]
12.	GO-TiO_2 hydrogel	MB dye	120	[46]
13.	Cellulose nano whiskers (CNWs) with starch (S) hydrogel	MB dye	2,050	[47]
14.	CS-starch (S) hydrogel	Direct Red 80 (DR 80)	312.77	[48]
15.	GO-Gum Tragacanth (GT)-PVA hydrogel	CV and CR	94.0 and 101.74 for CV and CR respectively	[49]
16.	CMC-g-PDMAEMA hydrogel	MO	1,825	[50]
17.	S-AAc-HEMA hydrogel	Acid-fast red dye	23	[51]
18.	PAAA-Gum Ghatti (Gg) (Gg-cl-PAAA) hydrogel	MG, MB, rhodamine	130	[52]
19.	SA-poly (AA-co-HEMA) hydrogel	CR and MV	89.83 and 92.07 for CR and MV dye respectively	[53]
20.	SA (Sodium Alginate)-g-poly (sodium acrylate-co-styrene)-p (AA-co-St)-(illite/smectite) organo-IS hydrogel	Methylene Blue (MB)	1,843.46	[54]
21.	Sodium Humate (SH)-laponite (LP) clay hydrogel	MB	21	[55]
22.	PVA-SA-CS-MMT hydrogel	MB dye	137.2	[56]
23.	S-g-AA/Na-Montmorillonite (MMA) hydrogel	Safranine-T	2,237	[57]

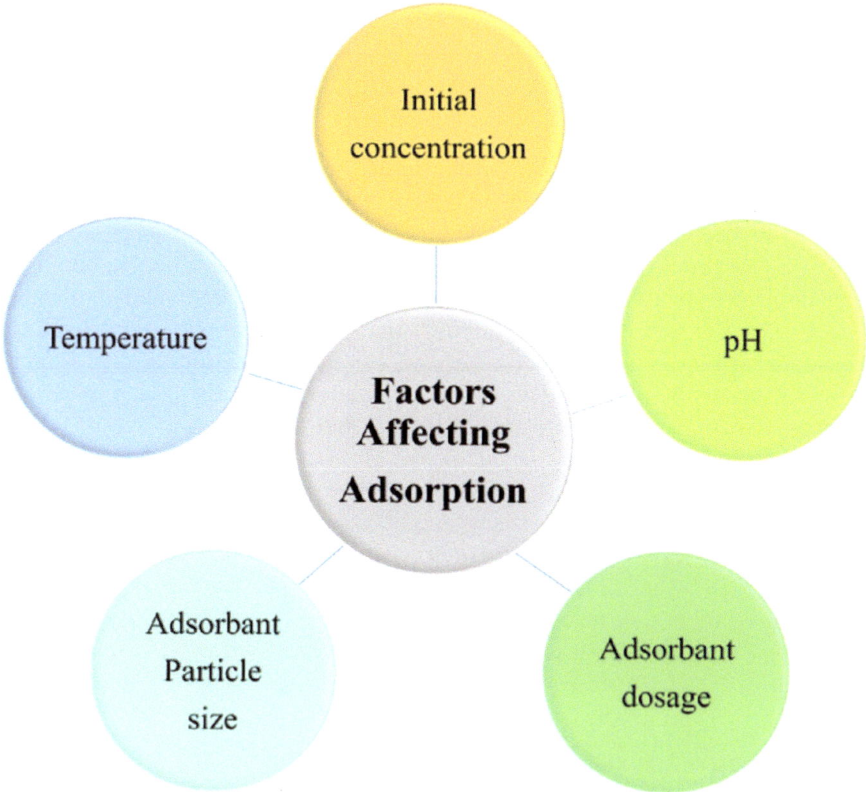

FIGURE 17.1 Factors affecting adsorption.

17.3.1 The Effect of Solution Temperature

The temperature has an impact on the healing process by converting exothermic reactions into endothermic ones or vice versa. As a result, temperature is a significant physicochemical element. It also affects adsorption significantly since it can change its magnitude [60]. Depending on the material that absorbs and the pollutant, the temperature can have a variety of effects on the sorption's effectiveness. By increasing kinetic energy and the adsorbate's surface activity, it generally enhances the biosorption of adsorption pollutants, but it can also cause structural damage to the bio-physical sorbent [61].

Exotherm: As the temperature rises, the adsorption process's effectiveness declines. It can be explained by the fact that dye removal effectiveness declines as the temperature rises and the adsorptive powers between the active sites of the adsorbent and the adsorbate weaken, which can be used to explain the phenomenon [62]. Exothermic adsorption is widely employed to regulate the mechanism of diffusion since adding heat to the solution increases the flexibility of the dye ions [63].

Endotherm: The adsorption process (efficiency) gets better as the temperature rises because there are more active sites available as a result of the adsorbent surface becoming activated at higher temperatures [64]. Raising the temperature may increase the large dye ions' mobility, which could be the reason why adsorption capacity values rise [65]. N-isopropyl acrylamide and some cellulose-based polymers are examples of hydrogels that are less soluble as the temperature rises while those with hydrocolloids, such as agar-agar and gelatin, are more soluble. Hydrogels with these properties include those that can form chain aggregation and hydrophobic associations. The values of entropy (S) and enthalpy (H) might increase or decrease when the temperature rises during the adsorption of dye molecules. Molecules can move in 3-dimensions before adsorption, but once they become adsorbed on a surface, their motion is constrained there and their disorder reduces, leading to a reduction in entropy that indicates an exothermic process. This could also be explained by the fact that at higher temperatures, dyes became more soluble while adsorbate-adsorbent interactions weakened, leading to less adsorption there, where their disorder decreases and an exothermic activity is indicated by a decrease in entropy [66]. This may also be accounted for by the fact that dyes became more soluble at higher temperatures, whereas adsorbate-adsorbent interactions weakened and less adsorption occurred [66]. The rise in entropy and enthalpy is indicative of an endothermic reaction [67]. Chiou and Li [68] cross-linked chitosan by epichlorohydrin (ECH) and obtained a high adsorption capacity (1600–1900 g/kg) of reactive dye (RR 189) on the cross-linked chitosan beads in acid-aqueous solutions at pH 3.0. It appears technically feasible to remove acid and reactive dyes from acid-aqueous solutions by cross-linked chitosan.

17.3.2 Size of Adsorbent Particles

Even though particle size is rarely examined in biosorption investigations, it can have a big impact on adsorption and heterogeneous chemical reactions. Specific surface area rises as a result of the small particle sizes. A key component of sorption processes is specific surface area, which is the entire surface area of a solid divided by its mass. SSA is impacted by particle size, material composition, and porosity [69]. Because there are greater active sites for binding, smaller particle sizes increase adsorption capacity and efficiency in static batch operations [61]. As the particle size reduces, the substance's BET surface rises. Depending on the kind of adsorbent, its adsorption capacity may be lower if the particle size is too tiny since the lighter particles float and can't make contact with the solution. After biosorption, it can be difficult to remove these tiny particles from water [70].

17.3.3 Effect of pH

Almost all adsorption techniques depend heavily on the solution's pH. This factor influences the process's effectiveness as well as the adsorbent's capacity. Since a medium's pH regulates the strength of the electrostatic charge provided by ionized dye molecules, the adsorption rate changes with its pH [63]. The pH of a solution has an impact on its surface charge, ionization level, and capacity to adsorb species. The cationic dye removal percentage usually decreases while the ionic dye adsorption removal percentage rises at low pH levels. While cationic dye adsorption is preferred for high pH solutions, anionic dye adsorption would have decreased efficiency. A rise in pH reduces the electrostatic repulsion between the adsorbent surface and the positively charged dye, which could increase the extent of adsorption [71]. In this case, hydrogels are ionic. As the pH value rises, anionic hydrogels having carboxyl groups, like poly-sodium alginate, become more capable of swelling. Chitosan and acrylate (dimethyl aminoethyl) are cationic hydrogels having amino groups that have less swelling potential. Up until a certain point, the amount of many metals adsorbed increases with solution pH before declining when pH is increased [58]. The formula can be used to determine pH.

$$\mathrm{pH} = \mathrm{pK_a} - \log[\mathrm{AH}]/[\mathrm{A^-}] \tag{17.1}$$

where [AH] and [A$^-$] represent the concentrations of protonated and deprotonated surface groups, respectively, and pK_a represents the equilibrium constants that resemble the carbonyl groups. Practical examples demonstrate that anionic dyes bind more tightly in acidic media while cationic dyes adhere to the adsorbent surface more strongly in basic conditions. To increase the pH of aqueous dye solutions, HCl and NaOH are routinely utilized. The outermost layer of the adsorbent becomes protonated after HCl is incorporated into the solution, making the anionic dye more electrostatically attracted to its surface. When NaOH is introduced to a basic solution, deprotonating the adsorbent's surface, the anionic dye interacts repellently with the adsorbent. As a result, cationic dyes exhibit the opposite behavior [58]. An inverse relationship exists between a solution's pH and its H^+ concentration. The capacity of metal ions to bind to surfaces declines with lower pH values because there are more metal ions and H^+ in the medium, which puts those ions in competition for sorption sites on the adsorbent. The anionic adsorbent surface interacts electrostatically with cationic metal ions more effectively when the pH of the medium increases due to the adsorbent's increased negative charge nature. Since metal hydroxide precipitates as the pH rises to a basic value, higher pH values have a lower adsorption capacity [72].

17.3.4 Initial Concentration

The initial dye concentration is likely one of the most significant parameters impacting the adsorption process because it influences the effectiveness of dye removal by lowering or increasing the availability of binding sites on the adsorbent surface. In these water treatment systems, the maximum amount of dye bound in equilibrium (*q*) and the efficacy of dye removal (*E*) are both directly related to the initial dye concentration [67,73].

$$E(\%) = \{(\mathrm{Ci} - \mathrm{Cf})/\mathrm{Ci}\} * 100 \tag{17.2}$$

$$q = \{(\mathrm{Ci} - \mathrm{Cf})/m\} * V \tag{17.3}$$

where: E (%) = efficiency; q (mg/g) = amount of dye bound in equilibrium; Ci (mg/L) = initial dye concentration; Cf (mg/L) = final dye concentration; m (g) = amount of adsorbent; and V (L) = volume of aqueous solution.

As concentration increases, the percentage of dye removal typically declines, which may be caused by saturation at the sites of adsorption on the surface of an adsorbent [71]. The concentration of the that are open on an adsorbent surface and the original dye concentration are directly correlated. There will be active patches on the adsorbent surface that are vacant at low concentrations. The loading capacity of an adsorbent rises with the initial dye concentration due to the strong driving force of mass at high-starting dye concentrations [61]. To put it another way, a higher initial dye concentration will lead to a higher final dye concentration. Lower concentrations result in low initial dye molecule-to-accessible adsorption site ratios, which make fractional adsorption independent of initial concentration. Zhang et al. [62] studied dye adsorption by the chitosan/alumina interface when the concentration of methyl orange was raised from 20 to 400 mg/L and found that, within the same Methylene Blue (MB) concentration range, the percentage of dye removal fell from 99.53% to 83.55%.

17.3.5 The Effect of Adsorbent Dosage

When assessing the adsorbent's efficiency for a specific amount of adsorbate under operational conditions, the adsorbent dose is essential. As the adsorbent dose is increased, more sorption sites are created at the adsorbent surface, increasing the percentage of dye removal [67]. As a result, dye removal generally increases with increasing adsorbent dosage. According to Farman et al.'s study on the adsorption of basic blue and acid blue dye on activated charcoal, adsorbent concentrations ranging from 2 to 12 mg at a constant dye concentration were investigated to determine how they affected the adsorption of colors. Since the dye absorption increased to that amount and then stabilized, it was determined that 10 mg would be the appropriate dosage for further research [74]. Kroeker's rule states that, at a fixed starting concentration, the specific adsorbed volume decreases with increasing adsorbent mass. As a result, increasing the adsorbent dose has an advantageous connection with color removal effectiveness and productivity. Increasing the adsorbent dosage results in more active adsorption sites and greater active surface areas for adsorption at constant contaminant concentrations [61,58]. The capacity may drop during the color removal procedure for two reasons [70]:

1. Even if the amount of adsorption sites increases, the sites remain vacant.
2. Agglomeration or particle aggregation of the adsorbent may form, decreasing the surface area that is accessible and extending the diffusion path.

17.4 MECHANISMS OF ACTION FOR HYDROGELS

Due to the negative effects of numerous organic dyes and heavy metals, much study has been done on their removal from various water reservoirs. Due to the porous, 3-D network structure of hydrogels, they have a huge specific surface area. Several polar

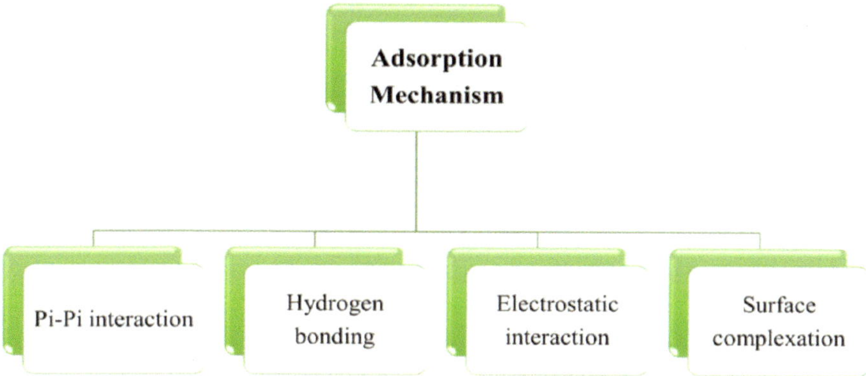

FIGURE 17.2 Adsorption mechanism of hydrogels.

functional groups are also present in the hydrogels' network, exposing more active sites for the adsorption of heavy metals. The ability of hydrogels to significantly expand in aqueous solutions, increasing their surface area in contact with contaminants and enhancing the efficiency of adsorption, is another advantage [75]. When removing metal ions and dyes using different hydrogel adsorbents, understanding the adsorption mechanism is crucial.

Hydrogen bonds, electrostatic interactions, ion exchange, *n*–*n* interactions, coordination/chelation, and surface complexation are only a few of the many mechanisms that contribute to the adsorption process (Figure 17.2). Understanding the adsorption process is essential when employing various hydrogel adsorbents to remove metal ions and dyes. The sections that followed provided in-depth introductions to each of these. The general adsorption process can involve a variety of processes, depending on the composition of the hydrogels (such as whether they contain chitosan, carboxymethyl cellulose, graphene, hectorite clay, etc.) and the particular adsorptive conditions (concentration, temperature, pH, etc.).

17.4.1 Hydrogen Bonding

The hydrogen bond, which is greater than van der Waals forces but lower than covalent or ionic bonds, is created when a hydrogen atom is wedged between two other atoms with a great attraction for electrons. Hydrogen bonds can be created between atoms in different molecules or parts of the same molecule. Covalent bonds between hydrogen atoms (–NH, –OH, or –FH) and one of the pair's (the donor's) typical fluorine, nitrogen, or oxygen atoms allow them to exchange electrons unevenly. Due to hydrogen's high attraction for electrons, it develops a little positive charge. The other element in a pair, which is frequently N, O, or F, has a slight negative charge when there is an unshared electron pair.

Together with electrostatic interactions, hydrogen bonds are one of the main interactions that remove dyes from water. RhB and methyl violet (MV) were removed using a dialdehyde carboxymethyl cellulose-gelatin hydrogel that was created by Sethi et al. [76]. Using a combination of hydrogen bonding, electrostatic interaction

(O-H—N), and van der Waals forces between dyes and hydrogels, the results demonstrated that the highest removal percentages of MV and RhB were 90% and 96.5% at pH 6, respectively. The author went on to say that a larger removal percentage of RhB than that of MV was caused by electrostatic interactions and stronger hydrogen bonds created by more carboxylic acid groups in RhB than MV.

17.4.2 Pi–Pi Interactions

The functional groups in hydrogels and organic dyes play a significant role in the non-covalent interactions that are created when acceptor and donor molecules connect via pi-electron interactions. Hydrogels treated with graphene are frequently more likely to adsorb aromatic dyes through such interactions, including methyl blue, methyl orange, and Congo red, among others [77]. This is because each carbon atom in the hexagonal lattice structure of graphene has a pi-electron. By tightly stacking graphene sheets and their derivatives, Jiao et al. were able to produce GO-based silver nanoparticles with composite high-gluten-content (HG) materials that had exceptionally high-mechanical strength [78]. The existence of many interconnected pores allowed for simple organic dye adsorption at the HG's vast water/solid interfaces. Hu et al. later uncovered the crucial function played by the graphene sheet's sp^2-conjugated carbon network in the separation of organic cations through pi–pi interactions [79]. To increase the mechanical strength and improve the adsorption capacities of aromatic organic pollutants, it would be helpful to find and characterize the pi–pi stacking interactions in various HGs.

Mittal et al. [80] produced chitosan/carboxymethyl cellulose hydrogels employing graphene oxide as a chemical crosslinker for the aim of cleaning wastewater contaminated with colors. The maximum adsorption capacities of MB and methyl orange (MO) were determined by the Langmuir models to be 655.98 and 404.52 mg/g, respectively. Due to the inclusion of CMC and CS, the composite hydrogel's capacity to adsorb cationic and anionic dyes via hydrogen bonding and electrostatic interactions, as well as the existence of graphene oxide and pi interactions might be utilized to explain the putative mechanism. In a different investigation, Tiwari et al. created 3D RGO-based HGs with remarkable adsorption capacities for MB (100%) and Rh B (97%), thanks to strong pi–pi stacking and anion-cation interactions [81]. The following are some examples of *n*-donating pollutants that are electron-rich and adsorb to *n*-accepting areas (electron-deficient) in various adsorbents: polyphenols, simple monoaromatic chemicals, complicated organic insecticides, and PAHs.

17.4.3 Surface Complexation

A crucial method for the removal of dyes and metal ions is surface complexation (outer and inner spheres). Complexation is the interaction of an electron pair source and an electron pair acceptor to produce a variety of complexes. All atoms and ions that may supply electron pairs, whether they are found in simple or complex compounds, qualify as donors, and metal ions and organic dyes are frequently acceptors. Li et al. employed guar gum and sodium lignosulfonate to make a composite hydrogel that removed Cu^{2+} and Co^{2+} and had maximal ion adsorption capacities of 709 and 601 mg/g, respectively [82]. The surface complexation between the lone pair electron of the oxygen atom from the $-COO^-$ group and the unoccupied electron orbitals of Cu^{2+} and Co^{2+} was the main mechanism by which these two metal ions were primarily adsorbed, according to their mechanistic study. The $-COO^-$ groups in the granular alginate-based hydrogel were shown to be complexed with Cu^{2+}, Ni^{2+}, Gd^{2+}, or Zn^{2+} ions by Wang et al. [83], who also discovered similar adsorption strategies.

For instance, complexation with the phenolic, carboxyl, and lactonic-functional groups in biochars generated at low-temperatures can efficiently bind heavy metals [84]. According to Wang et al., surface complexation, precipitation, and electrostatic interaction are key factors in regulating head sorption of both As (V) and Pb (II) onto biochar [85]. The divalent metal cations may be successfully adsorbed onto the hydrogel by complexing with functional groups like oxygen and nitrogen, which have more than one lone pair of electrons to interact with metal cations. Metal complexation removes toxic metal ions from the environment. Surface structures allow for the regulation of reduction and oxidation, which is not possible with transport processes. As a result, the surface structure has a big impact on surface reactivity, especially in the coordination environment, which alters the reaction's intensity and pace at the solid-water interface [86]. Given that inner-sphere complexation is not as much of advantageous entropically and only happens when there is a significant concentration of metals, it is reasonable to assume that the adsorption process for most transition metals will be outer-sphere complexation.

17.4.4 Electrostatic Interaction

Electrostatic Attractions are interactions between charges that are diametrically opposed. The attraction desires to bring the opposing charges together. The distance between charged particles and the amount of electric charge they carried were the only two parameters affecting the force between them; the closer the particles were to one another and the higher their surface charges, the stronger the attraction [86,87]. The amount of charge on a hydrogel's surface and the number of metal ions in a solution are both determined by pH-dependent processes called metal speciation and dissociation of the functional groups on the surface, which suggests that the pH of the solution is crucial when utilizing adsorbents [88]. In more detail, the surface of the hydrogels can have either positive charges or negative charges to remove metal ions or dyes with opposing charges when the pH of the solution varies from the pH of the hydrogels' point of zero charges. In contrast to the carboxylic group on chitosan, which has a pKa value of about 4.8, the amino group has a pK_a value of about 6.5. Based on this, Wang et al. created a carboxymethyl cellulose/chitosan hydrogel that responds to pH for the removal of anionic and cationic dyes. The composite hydrogels were able to adhere to the positive-charged MB, and at pH 11, when carboxylic groups were predominated and deprotonated as $-COO^-$, their adsorption capability peaked at 110 mg/g [89].

Alternately, the negatively charged acid orange II might be adsorbent to the extent of 100 mg/g in this composite hydrogel since the amino groups were in their protonated state (NH^{4+}) at pH 2. According to Chen et al., three-dimensional graphene foams constructed of few-layered, vertically oriented graphene sheets with a high-graphite structure successfully absorbed As (V) and Pb (II) ions [77]. The coordinated ion exchange and electrostatic attraction between hydroxyl groups and metal ions served to prove this.

Hu et al. recently published a paper on the development of sodium alginate-CMC cellulose gel beads for the removal of Pb^{2+} ions. The predominant adsorption process was complexation between the Pb^{2+} ions and the –OH and –COOH function groups of the HGs, as determined by the study, via sharing a solitary pair of electrons. Additionally, it was demonstrated that secondary interactions were brought about by electrostatic attraction between negatively charged gel beads and positively charged Pb^{2+} ions [90]. Similar to this, Wang and Li used cationic/free radical hybrid polymerization to remove Cu (II) to produce ion-imprinted HG (Cu(II)-IIH). The locations of Cu (II) interaction in HGs were the amido and ether groups [91].

17.5 COST EFFECTS

Many adsorption separation applications call for cost reduction, particularly those that entail significant treatment volumes, like sewage treatment. Since synthetic gel adsorbents are less cost-effective than traditional adsorbents like activated carbon and zeolites, it is anticipated that these materials would compete based on their ease of processing, superiority to traditional adsorbents, and a few other characteristics. Throughout an adsorbent's life cycle, green technology based on the optimal adsorption method and adsorbent design is required. Nevertheless, this project is still in its early phases. Therefore, gel adsorbents are expected to be affordable, easy to use, and highly effective in addition to being of academic interest. They also need to be justified as to why gels should be employed as the adsorbent. Hydrogel adsorbents can provide a functional adsorption process with minimum impact on the environment by incorporating diverse activities inside the gels. Gel adsorbents varied structural arrangements and applications would increase their utility. This is a period of carefully designed adsorbents and adsorption processes. Making use of underutilized biomass and natural resources could bring in money. Not many useful reports on this study exist. Accuracy, reproducibility, and endurance are also essential components for hydrogel adsorbents with stimuli-responsive functions [92].

17.6 ENVIRONMENTAL APPLICATION

As sorbents for heavy metals and ionic dyes, two common pollutants in effluent and industrial wastewater that must be removed before the effluent is released into the environment, hydrogels are primarily used in environmental applications [22]. Nanocomposite-reinforced hydrogels are frequently utilized to strengthen gel toughness and the potential for repeated sorption and desorption cycles since effective hydrogel reuse requires high-stability. The efficient handling of all sorption cycle phases required for the continuous removal of heavy metals and organic dyes by hydrogels is based on the combination of these properties [93]. The evaluation of the adsorbent's adsorption capacities is shown in Table 17.2.

TABLE 17.2
The Evaluation of the Adsorbent's Adsorption Capacities

Type	Pollutants	Adsorbents	Adsorption Mechanism	Adsorption Capacities (mg/g)	References
Antibiotics	Ciprofloxacin	Sodium alginate/K Carrageenan	Electrostatic interactions, H-bonds	229.00	[94]
	Ciprofloxacin	Graphene/ Alginate	Physical interactions, chemical interactions	344.83	[95]
Dyes	MB	Cu-MOFs/Fe_3O_4	Physical interactions	113.67	[96]
	MB	Polyvinyl alcohol/Alginate	Chemical interactions, H-bonds	1,437.48	[97]
Metal ions	Pb^{2+}	Sodium alginate/ Chitosan	Electrostatic interactions, coordination interactions	400.90	[98]
	Cd^{2+}	DN hydrogel	Coordination interactions, physical interactions	268.98	[99]

17.7 REUSABILITY

In addition to having a high adsorption capacity, hydrogels can renewed by releasing the heavy metals they have absorbed, allowing for reuse. After adding an adsorbent, hydrogels must have their active sites reinstated to continue adsorbing or desorbing the adsorbent molecules. The industrial use of hydrogels necessitates a high degree of reusability, which can lower treatment costs for wastewater containing adsorbent. Several strategies for improving the reusability of hydrogels employed for the adsorption of pollutants have been developed, despite the complexity of the factors controlling this property.

Several approaches have been used to investigate the efficient desorption of toxic substances from the hydrogel's 3-D network structure following each phase of elimination. Heavy metals will desorb in response to changes in the temperature, electric field, magnetic field, pH, and other factors [100]. Both internal and external factors have been looked into for improvements to hydrogels' reusability. The material that has absorbed the adsorbent is given a specific desorbing agent in the study's hydrogel regeneration process. After a predefined number of adsorption and desorption cycles have been finished, the factors that are evaluated to determine reusability include desorption and adsorption rates, adsorbent-removal rate, and equilibrium-adsorption capacity.

The amount of cross-linking was increased, and internal hydrogel factors were taken into account, improving mechanical properties and reducing mass loss brought on by consecutive washing procedures throughout the adsorption-desorption cycle. pH affects the desorption of heavy metals from a magnetized cellulose-chitosan hydrogel, according to Liu et al. [101]. At pH values ranging from 1.0 to 2.0, a desorption efficiency of 83%–86% was achievable. This demonstrates the benefits of utilizing hydrogels that are sensitive to pH for the adsorption of heavy metals during the elimination phase, such as arsenic and chromium for hydrogel recovery during desorption. The requirement for adjusting pH is a drawback [102].

Strong and mild acids (HCl, HNO_3, CH_3COOH, H_2SO_4, etc.) have allegedly been used in studies on the recovery of hydrogel adsorbents [103]. The desorption capability and durability of the hydrogel are significantly influenced by the type of acid used in the desorption process [104]. Mohammadi et al. developed a multi-functionalized hydrogel that mimics chelators and has exceptional reusability and great metal adsorption effectiveness (for lead, cadmium, and arsenic). A low-concentration of hydrochloric acid was used to elute the heavy metals that the hydrogel network had absorbed, and the hydrogel was then replenished for further use. After five adsorption/desorption reuse cycles, a clearance ratio of more than 60% was achieved [105]. A similar approach was used by Pourjavadi et al. to produce a novel hydrogel from chitosan, acrylic acid, and functionalized nano-silica. Using a 1 M hydrochloric acid solution, a hydrogel that was loaded with Pb^{2+} was recovered in this investigation. The hydrogel was filtered and washed with deionized water before being put to use for the second adsorption cycle. After three successive cycles, the regenerated hydrogel's efficacy was maintained between 687 and 715 mg/g [106]. Regarding extraneous factors, it has been proposed to dope titanium dioxide into a two-dimensional hydrogel made of montmorillonite, chitosan, and polyacrylic acid, where its photocatalytic properties enable the initial adsorption of MB to be broken down into inorganic acid and small-molecule inorganic substrates [107].

MB may afterward be entirely desorbed, greatly enhancing the hydrogel's potential for reuse. The MB removal efficiency was still reported to be 100% after four cycles. However, after four cycles, only 50% of the titanium dioxide was removed from the hydrogel. The findings showed good reusability, with little change in equilibrium-adsorption capacity after five cycles.

17.8 CONCLUSION

Due to population growth, greater urbanization, and industrialization, water pollution and other significant challenges associated with it are getting worse all over the world. How to provide pollutant-free water supplies to improve people's living conditions is one of the biggest challenges facing humanity in the 21st century. The ecology, growth, economic development, safety, and security are all significantly harmed by these problems. The issue of inadequate access to water and sanitary facilities must be taken more seriously because it contributes to cholera epidemics, malaria, and cholera. Many methods have been developed to eliminate these dangerous contaminants. Due to its many advantages, including simplicity, cost, and the avoidance of secondary waste creation, adsorption is a popular method for the removal of dyes or heavy metal ions. Cost is always the primary issue in the water treatment industry. The adsorption method may be efficient against a variety of contaminants, has an easy-to-understand structure, and is inexpensive and widely available. The effectiveness of several adsorbents in removing contaminants has been established. Hydrogels can be used for a variety of water treatment procedures due to their low interfacial tension, high hydrophilicity, good expandability, selective permeability, biocompatibility, low cost, degradability, super absorbency, high-water affinity, high porosity, diverse shape, and adaptable dimensions. It is evident that the use of hydrogel materials in water applications is a relatively recent trend, and research is being conducted to ascertain whether they are better than or worse than other real-world methods.

17.9 FUTURE OUTLOOK

Due to its exceptional qualities, including high adsorption capacity, quick kinetic, and reusability, chitosan hydrogels have attracted particular attention in the field of wastewater treatment. In addition to these drawbacks, chitosan hydrogels have poor mechanical and stability characteristics. To solve these issues, the hydrogel preparation process is crucial. This review highlights recent research on chitosan-based hydrogels for wastewater treatment and illustrates how they are categorized based on the technique of manufacture. The processes used to create and alter chitosan-based hydrogels have an impact on their mechanical characteristics and increase removal effectiveness. In the area of chitosan-based hydrogel for wastewater treatment, the studies examined in this study may offer novel suggestions. The following gaps should be taken into account in future works

- Chitosan and hydrogels based on chitosan should be modified to increase the adsorption capacity and rate of these adsorbents.
- Chitosan-based hydrogels' efficacy in treating industrial effluents, which contain a variety of contaminants, has to be studied.
- It is advised to design novel, multifunctional hydrogels based on chitosan that can remove different types of contaminants from wastewater.
- Researchers paid little attention to the existing complex relationships between adsorption's operational variables. In the future, it is advised to optimize the adsorption capacity of adsorbents using optimization techniques like response surface methodology (RSM).
- Chitosan-based hydrogels-pollutant adsorption molecular simulation is advised for better understanding and enhancing the adsorption process.

ABBREVIATIONS

CR	Congo red
CV	Crystal violet
GO	Graphene oxide
MB	Methylene blue
MO	Methyl orange
PAA	Polyacrylic acid
PDMAEMA	Poly (2-[dimethyl amino] ethyl methacrylate)
PDMDAAC	Polydimethyl diallyl ammonium chloride
PHA	Poly (hydroxamic acid)
PNIAm	Poly (N-isopropyl acrylamide)
PVP	Poly (N-vinyl pyrrolidone)
SA	Stearyl acrylate

CONFLICT OF INTEREST

No conflict of interest.

ACKNOWLEDGMENT

The authors are grateful to Banasthali Vidyapith for providing all the necessities to complete this work.

REFERENCES

1. Guo, Haiying, Tifeng Jiao, Qingrui Zhang, Wenfeng Guo, Qiuming Peng, and Xuehai Yan. "Preparation of graphene oxide-based hydrogels as efficient dye adsorbents for wastewater treatment." *Nanoscale Research Letters* 10 (2015): 1–10. https://doi.org/10.1186/s11671-015-0931-2
2. Chen, Suhong, Qinyan Yue, Baoyu Gao, and Xing Xu. "Equilibrium and kinetic adsorption study of the adsorptive removal of Cr (VI) using modified wheat residue." *Journal of Colloid and Interface Science* 349, no. 1 (2010): 256–264. https://doi.org/10.1016/j.jcis.2010.05.057
3. Patel, Ankit M., Ranjan G. Patel, and Manish P. Patel. "Nickel and copper removal study from aqueous solution using new cationic poly acrylamide/N,N-DAMB/N,N-DAPB] super absorbent hydrogel." *Journal of Applied Polymer Science* 119, no. 4 (2011): 2485–2493. https://doi.org/10.1002/app.32818
4. Srivastav Ratnesh Kumar, Sharad Kumar Gupta, Krishna Deo Prasad Nigam, and Padma Vasudevan. "Treatment of chromium and nickel in wastewater by using aquatic plants." *Water Research* 28, no. 7 (1994): 1631–1638. https://doi.org/10.1016/0043-1354(94)90231-3
5. Zenk, Meinhart H. "Heavy metal detoxification in higher plants-a review." *Gene* 179, no. 1 (1996): 21–30. https://doi.org/10.1016/S0378-1119(96)00422-2
6. Al-Abachi, Mouayad Q., Nagam S. Al-Awady, and Ahmed M. Al-Anbakey. "Evaluation of poly acrylic acid (PAA) hydrogel beads as adsorbent for the removal of lead (II) ion from water." *Al-Nahrain Journal of Science* 16, no. 3 (2013): 30–39.
7. Ali, Z. Ansar, Jayachandran Venkatesan, Se Kwon Kim, and P. N. Sudha. "Beneficial effect of chitosan-g-polyacrylamide copolymer in removal of heavy metals from industrial dye effluents." *International Journal of Environmental Sciences* 1, no. 5 (2011): 820–833.
8. Yagub, Mustafa T., Tushar Kanti Sen, Sharmeen Afroze, and Ha Ming Ang. "Dye and its removal from aqueous solution by adsorption: a review." *Advances in Colloid and Interface Science* 209 (2014): 172–184. https://doi.org/10.1016/j.cis.2014.04.002
9. Fu, Yuzhu, and T. Viraraghavan. "Fungal decolorization of dye wastewaters: a review." *Bioresource Technology* 79, no. 3 (2001): 251–262. https://doi.org/10.1016/S0960-8524(01)00028-1
10. Kadirvelu, Kavipriya, M. Kavipriya, C. Karthika, M. Radhika, N. Vennilamani, and Sangma Pattabhi. "Utilization of various agricultural wastes for activated carbon preparation and application for the removal of dyes and metal ions from aqueous solutions." *Bioresource Technology* 87, no. 1 (2003): 129–132. https://doi.org/10.1016/S0960-8524(02)00201-8
11. Zollinger, Heinrich. *Color Chemistry: Syntheses, Properties, and Applications of Organic Dyes and Pigments.* John Wiley & Sons (2003).
12. Jing, Guanghui, Li Wang, Haojie Yu, Wael A. Amer, and Lei Zhang. "Recent progress on study of hybrid hydrogels for water treatment." *Colloids and Surfaces A: Physicochemical and Engineering Aspects* 416 (2013): 86–94. https://doi.org/10.1016/j.colsurfa.2012.09.043

13. Dwivedi, Charu, Abhishek Gupta, Abhishek Chaudhary, and Chayan Kanti Nandi. "Gold nanoparticle chitosan composite hydrogel beads show efficient removal of methyl parathion from waste water." *RSC Advances* 4, no. 75 (2014): 39830–39838. https://doi.org/10.1039/C4RA03870C
14. Rossi, Federico, Oriana Motta, Simona Matrella, Antonio Proto, and Giovanni Vigliotta. "Nitrate removal from wastewater through biological denitrification with OGA 24 in a batch reactor." *Water* 7, no. 1 (2014): 51–62. https://doi.org/10.3390/w7010051
15. Chamkouri, Hossein, and Mahyodin Chamkouri. "A review of hydrogels, their properties and applications in medicine." *American Journal of Biomedical Science and Research* 11, no. 6 (2021): 485–493. https://doi.org/10.34297/AJBSR.2021.11.001682
16. Bahram, Morteza, Naimeh Mohseni, and Mehdi Moghtader. "An introduction to hydrogels and some recent applications." In Sutapa Biswas Majee (ed.), *Emerging Concepts in Analysis and Applications of Hydrogels*. IntechOpen (2017). https://doi.org/10.5772/64301.
17. Ahmed, Enas M. "Hydrogel: preparation, characterization, and applications: a review." *Journal of Advanced Research* 6, no. 2 (2015): 105–121. https://doi.org/10.1016/j.jare.2013.07.006
18. Li, Yuhui, Guoyou Huang, Xiaohui Zhang, Baoqiang Li, Yongmei Chen, Tingli Lu, Tian Jian Lu, and Feng Xu. "Magnetic hydrogels and their potential biomedical applications." *Advanced Functional Materials* 23, no. 6 (2013): 660–672. https://doi.org/10.1002/adfm.201201708
19. Akter, Maimuna, Maitry Bhattacharjee, Avik Kumar Dhar, Fahim Bin Abdur Rahman, Siddika Haque, Taslim Ur Rashid, and SM Fijul Kabir. "Cellulose-based hydrogels for wastewater treatment: a concise review." *Gels* 7, no. 1 (2021): 30. https://doi.org/10.3390/gels7010030
20. Khan, Musharib, and Irene M. C. Lo. "A holistic review of hydrogel applications in the adsorptive removal of aqueous pollutants: recent progress, challenges, and perspectives." *Water Research* 106 (2016): 259–271. https://doi.org/10.1016/j.watres.2017.10.008
21. Madduma-Bandarage, Ujith S. K., and Sundararajan V. Madihally. "Synthetic hydrogels: synthesis, novel trends, and applications." *Journal of Applied Polymer Science* 138, no. 19 (2021): 50376. https://doi.org/10.1002/app.50376
22. Peppas, Nikolas A., P. Bures, W. S. Leobandung, and H. Ichikawa. "Hydrogels in pharmaceutical formulations." *European Journal of Pharmaceutics and Biopharmaceutics* 50, no. 1 (2000): 27–46. https://doi.org/10.1016/S0939-6411(00)00090-4
23. Crini, Gregorio. "Non-conventional low-cost adsorbents for dye removal: a review." *Bioresource Technology* 97, no. 9 (2006): 1061–1085. https://doi.org/10.1016/j.biortech.2005.05.001
24. Rajasulochana, Perepi, and V. Preethy. "Comparison on efficiency of various techniques in treatment of waste and sewage water: a comprehensive review." *Resource-Efficient Technologies* 2, no. 4 (2016): 175–184. https://doi.org/10.1016/j.reffit.2016.09.004
25. Enriquez, Carlos, Nena Nwachuku, and Charles P. Gerba. "Direct exposure to animal enteric pathogens." *Reviews on Environmental Health* 16, no. 2 (2001): 117–132. https://doi.org/10.1515/REVEH.2001.16.2.117
26. Tchounwou, Paul B., Clement G. Yedjou, Anita K. Patlolla, and Dwayne J. Sutton. "Heavy metal toxicity and the environment." In Andreas Luch (ed.), *Molecular, Clinical and Environmental Toxicology: Volume 3: Environmental Toxicology* (2012), pp. 133–164. Springer. https://doi.org/10.1007/978-3-7643-8340-4_6
27. Cosgrove, William J., and Daniel P. Loucks. "Water management: current and future challenges and research directions." *Water Resources Research* 51, no. 6 (2015): 4823–4839. https://doi.org/10.1002/2014WR016869

28. Mu, Ruihua, Bin Liu, Xi Chen, Ning Wang, and Jing Yang. "Hydrogel adsorbent in industrial wastewater treatment and ecological environment protection." *Environmental Technology & Innovation* 20 (2020): 101107. https://doi.org/10.1016/j.eti.2020.101107
29. Amin, Muhammad Tahir, Abdulrehman Ali Alazba, and Umair Manzoor. "A review of removal of pollutants from water/wastewater using different types of nanomaterials." *Advances in Materials Science and Engineering* 2014 (2014): 1–24. https://doi.org/10.1155/2014/825910
30. Sharshir, Swellam W., Almoataz M. Algazzar, K. A. Elmaadawy, A. W. Kandeal, M. R. Elkadeem, T. Arunkumar, Jianfeng Zang, and Nuo Yang. "New hydrogel materials for improving solar water evaporation, desalination and wastewater treatment: a review." *Desalination* 491 (2020): 114564. https://doi.org/10.1016/j.desal.2020.114564
31. Chang, Xiaojing, Zhenxing Wang, Shuai Quan, Yanchao Xu, Zaixing Jiang, and Lu Shao. "Exploring the synergetic effects of graphene oxide (GO) and polyvinylpyrrodione (PVP) on poly (vinylylidenefluoride)(PVDF) ultrafiltration membrane performance." *Applied Surface Science* 316 (2014): 537–548. https://doi.org/10.1016/j.apsusc.2014.07.202
32. Muya, Francis Ntumba, Christopher Edoze Sunday, Priscilla Baker, and Emmanuel Iwuoha. "Environmental remediation of heavy metal ions from aqueous solution through hydrogel adsorption: a critical review." *Water Science and Technology* 73, no. 5 (2016): 983–992. https://doi.org/10.2166/wst.2015.567
33. Ardila-Leal, Leidy D., Raúl A. Poutou-Piñales, Aura M. Pedroza-Rodríguez, and Balkys E. Quevedo-Hidalgo. "A brief history of colour, the environmental impact of synthetic dyes and removal by using laccases." *Molecules* 26, no. 13 (2021): 3813. https://doi.org/10.3390/molecules26133813
34. Jaishankar, Monisha, Tenzin Tseten, Naresh Anbalagan, Blessy B. Mathew, and Krishnamurthy N. Beeregowda. "Toxicity, mechanism and health effects of some heavy metals." *Interdisciplinary Toxicology* 7, no. 2 (2014): 60. https://doi.org/10.2478/intox-2014-0009
35. Liu, Xiaoyan, Yifeng Zhou, Wangyan Nie, Linyong Song, and Pengpeng Chen. "Fabrication of hydrogel of hydroxypropyl cellulose (HPC) composited with graphene oxide and its application for methylene blue removal." *Journal of Materials Science* 50 (2015): 6113–6123. https://doi.org/10.1007/s10853-015-9166-y
36. Hou, Caili, Kai Ma, Tifeng Jiao, Ruirui Xing, Kaikai Li, Jingxin Zhou, and Lexin Zhang. "Preparation and dye removal capacities of porous silver nanoparticle-containing composite hydrogels via poly (acrylic acid) and silver ions." *RSC Advances* 6, no. 112 (2016): 110799–110807. https://doi.org/10.1039/C6RA23371F
37. Saraydın, Dursun, Yasemin Işıkver, and Erdener Karadağ. "Adsorption of phenazine dyes using poly (hydroxamic acid) hydrogels from aqueous solutions." *Polymer Engineering & Science* 58, no. 3 (2018): 310–318. https://doi.org/10.1002/pen.24574
38. Wang, Xinpeng, Zhiming Liu, Xiangping Ye, Kun Hu, Huiqing Zhong, Jianfeng Yu, Mei Jin, and Zhouyi Guo. "A facile one-step approach to functionalized graphene oxide-based hydrogels used as effective adsorbents toward anionic dyes." *Applied Surface Science* 308 (2014): 82–90. https://doi.org/10.1016/j.apsusc.2014.04.103
39. Gan, Lu, Songmin Shang, Enling Hu, Chun Wah Marcus Yuen, and Shou-Xiang Jiang. "Konjac glucomannan/graphene oxide hydrogel with enhanced dyes adsorption capability for methyl blue and methyl orange." *Applied Surface Science* 357 (2015): 866–872. https://doi.org/10.1016/j.apsusc.2015.09.106
40. Moon, Y.-E., Gowun Jung, Jumi Yun, and Hyung-Il Kim. "Poly (vinyl alcohol)/poly (acrylic acid)/TiO_2/graphene oxide nanocomposite hydrogels for pH-sensitive photocatalytic degradation of organic pollutants." *Materials Science and Engineering:* B 178, no. 17 (2013): 1097–1103. https://doi.org/10.1016/j.mseb.2013.07.002

41. Liu, Yi, Yian Zheng, Dajian Huang, and Aiqin Wang. "Adsorption of Congo red by poly (dimethyl diallyl ammonium chloride)/polyacrylamide hydrogels with excellent acid and alkali resistance." *Separation Science and Technology* 47, no. 12 (2012): 1828–1836. https://doi.org/10.1080/01496395.2012.662259
42. Tu, Hu, Yi Yu, Jiajia Chen, Xiaowen Shi, Jialin Zhou, Hongbing Deng, and Yumin Du. "Highly cost-effective and high-strength hydrogels as dye adsorbents from natural polymers: chitosan and cellulose." *Polymer Chemistry* 8, no. 19 (2017): 2913–2921. https://doi.org/10.1039/C7PY00223H
43. Li, Peng, Nam Hoon Kim, Seok-Bong Heo, and Joong-Hee Lee. "Novel PAAm/Laponite clay nanocomposite hydrogels with improved cationic dye adsorption behavior." *Composites Part B: Engineering* 39, no. 5 (2008): 756–763. https://doi.org/10.1016/j.compositesb.2007.11.003
44. Liu, Yi, Yian Zheng, and Aiqin Wang. "Response surface methodology for optimizing adsorption process parameters for methylene blue removal by a hydrogel composite." *Adsorption Science & Technology* 28, no. 10 (2010): 913–922. https://doi.org/10.1260/0263-6174.28.10.913
45. Vakili, Mohammadtaghi, Mohd Rafatullah, Mahamad Hakimi Ibrahim, Ahmad Zuhairi Abdullah, Babak Salamatinia, and Zahra Gholami. "Chitosan hydrogel beads impregnated with hexadecylamine for improved reactive blue 4 adsorption." *Carbohydrate Polymers* 137 (2016): 139–146. https://doi.org/10.1016/j.carbpol.2015.09.017
46. Zhang, Zheye, Fei Xiao, Yunlong Guo, Shuai Wang, and Yunqi Liu. "One-pot self-assembled three-dimensional TiO_2-graphene hydrogel with improved adsorption capacities and photocatalytic and electrochemical activities." *ACS Applied Materials &* Interfaces 5, no. 6 (2013): 2227–2233. https://doi.org/10.1021/am303299r
47. Gomes, Raelle F., Antonio C. Neto de Azevedo, Antonio G. B. Pereira, Edvani C. Muniz, André R. Fajardo, and Francisco H. A. Rodrigues. "Fast dye removal from water by starch-based nanocomposites." *Journal of Colloid and Interface Science* 454 (2015): 200–209. https://doi.org/10.1016/j.jcis.2015.05.026
48. Ngwabebhoh, FahanwiAsabuwa, Mustafa Gazi, and Akeem Adeyemi Oladipo. "Adsorptive removal of multi-azo dye from aqueous phase using a semi-IPN superabsorbent chitosan-starch hydrogel." *Chemical Engineering Research and Design* 112 (2016): 274–288. https://doi.org/10.1016/j.cherd.2017.06.023
49. Sahraei, Razieh, Zahra Sekhavat Pour, and Mousa Ghaemy. "Novel magnetic bio-sorbent hydrogel beads based on modified gum tragacanth/graphene oxide: removal of heavy metals and dyes from water." *Journal of Cleaner Production* 142 (2017): 2973–2984. https://doi.org/10.1016/j.jclepro.2017.10.170
50. Salama, Ahmed, Nadia Shukry, and Mohamed El-Sakhawy. "Carboxymethyl cellulose-g-poly (2-(dimethylamino) ethyl methacrylate) hydrogel as adsorbent for dye removal." *International Journal of Biological Macromolecules* 73 (2015): 72–75. https://doi.org/10.1016/j.ijbiomac.2014.11.002
51. Mahmoud, Ghada A., Samia E. Abdel-Aal, Nagwa A. Badway, Samia A. Abo Farha, and Esraa A. Alshafei. "Radiation synthesis and characterization of starch-based hydrogels for removal of acid dye." *Starch-Stärke* 66, no. 3–4 (2014): 400–408. https://doi.org/10.1002/star.201300117
52. Sharma, Kashma, B. S. Kaith, Vijay Kumar, Susheel Kalia, Vinod Kumar, and H. C. Swart. "Water retention and dye adsorption behavior of Gg-cl-poly (acrylic acid-aniline) based conductive hydrogels." *Geoderma* 232 (2014): 45–55. https://doi.org/10.1016/j.geoderma.2014.04.035
53. Mandal, Bidyadhar, and Samit Kumar Ray. "Synthesis of interpenetrating network hydrogel from poly (acrylic acid-co-hydroxyethyl methacrylate) and sodium alginate: modeling and kinetics study for removal of synthetic dyes from water." *Carbohydrate Polymers* 98, no. 1 (2013): 257–269. https://doi.org/10.1016/j.carbpol.2013.05.093

54. Wang, Yizhe, Wenbo Wang, and Aiqin Wang. "Efficient adsorption of methylene blue on an alginate-based nanocomposite hydrogel enhanced by organo-illite/smectite clay." *Chemical Engineering Journal* 228 (2013): 132–139. https://doi.org/10.1016/j.cej.2013.04.090
55. Yi, Ju-Zhen, and Li-Ming Zhang. "Removal of methylene blue dye from aqueous solution by adsorption onto sodium humate/polyacrylamide/clay hybrid hydrogels." *Bioresource Technology* 99, no. 7 (2008): 2182–2186. https://doi.org/10.1016/j.biortech.2007.05.028
56. Wang, Wei, Yunliang Zhao, Haoyu Bai, Tingting Zhang, Valentin Ibarra-Galvan, and Shaoxian Song. "Methylene blue removal from water using the hydrogel beads of poly (vinyl alcohol)-sodium alginate-chitosan-montmorillonite." *Carbohydrate Polymers* 198 (2018): 518–528. https://doi.org/10.1016/j.carbpol.2018.06.124
57. Al, Ebru, Gamze Güçlü, Tülin Banu İyim, Serkan Emik, and Saadet Özgümüş. "Synthesis and properties of starch-graft-acrylic acid/Na-montmorillonite superabsorbent nanocomposite hydrogels." *Journal of Applied Polymer Science* 109, no. 1 (2008): 16–22. https://doi.org/10.1002/app.27968
58. Razi, Mohd Adib Mohammad, Mimi Nur Attahirah Mohd Hishammudin, and Rafidah Hamdan. "Factor affecting textile dye removal using adsorbent from activated carbon: a review." In *MATEC Web of Conferences* (2017), vol. 103, p. 06015. EDP Sciences. https://doi.org/10.1051/matecconf/201710306015
59. Legorreta-Castañeda, Adriana Jazmín, Carlos Alexander Lucho-Constantino, Rosa Icela Beltrán-Hernández, Claudia Coronel-Olivares, and Gabriela A. Vázquez-Rodríguez. "Biosorption of water pollutants by fungal pellets." *Water* 12, no. 4 (2020): 1155. https://doi.org/10.3390/w12041155
60. Badawy, Abdelrahman A., Shaimaa M. Ibrahim, and Hisham A. Essawy. "Enhancing the textile dye removal from aqueous solution using cobalt ferrite nanoparticles prepared in presence of fulvic acid." *Journal of Inorganic and Organometallic Polymers and Materials* 30 (2020): 1798–1813. https://doi.org/10.1007/s10904-019-01355-1
61. Rápó, Eszter, and Szende Tonk. "Factors affecting synthetic dye adsorption; desorption studies: a review of results from the last five years (2017–2021)." *Molecules* 26, no. 17 (2021): 5419. https://doi.org/10.3390/molecules26175419
62. Zhang, Jixiang, Qiuxiang Zhou, and Lailiang Ou. "Kinetic, isotherm, and thermodynamic studies of the adsorption of methyl orange from aqueous solution by chitosan/alumina composite." *Journal of Chemical & Engineering Data* 57, no. 2 (2012): 412–419. https://doi.org/10.1021/je2009945
63. Etim, Ubong Jerome., Saviour Alphonsus Umoren, and Eduok Ubong M. "Coconut coir dust as a low-cost adsorbent for the removal of cationic dye from aqueous solution." *Journal of Saudi Chemical Society* 20 (2016): S67–S76. https://doi.org/10.1016/j.jscs.2012.09.014
64. Abualnaja, Khamael M., Ahmed E. Alprol, M. A. Abu-Saied, Mohamed Ashour, and Abdallah Tageldein Mansour. "Removing of anionic dye from aqueous solutions by adsorption using of multiwalled carbon nanotubes and poly (Acrylonitrile-styrene) impregnated with activated carbon." *Sustainability* 13, no. 13 (2021): 7077. https://doi.org/10.3390/su13137077
65. Sharma, Kashma, Shreya Sharma, Vipasha Sharma, Pawan Kumar Mishra, Adam Ekielski, Vishal Sharma, and Vijay Kumar. "Methylene blue dye adsorption from wastewater using hydroxyapatite/gold nanocomposite: kinetic and thermodynamics studies." *Nanomaterials* 11, no. 6 (2021): 1403. https://doi.org/10.3390/nano11061403
66. El-Harby, Nouf F., Shaimaa M. A. Ibrahim, and Nadia A. Mohamed. "Adsorption of Congo red dye onto antimicrobial terephthaloyl thiourea cross-linked chitosan hydrogels." *Water Science and Technology* 76, no. 10 (2017): 2719–2732. https://doi.org/10.2166/wst.2017.442

67. Ali, Farman, Nisar Ali, Iram Bibi, Amir Said, Shahid Nawaz, Zarshad Ali, Syed Muhammad Salman, Hafiz M. N. Iqbal, and Muhammad Bilal. "Adsorption isotherm, kinetics and thermodynamic of acid blue and basic blue dyes onto activated charcoal." *Case Studies in Chemical and Environmental Engineering* 2 (2020): 100040. https://doi.org/10.1016/j.cscee.2020.100040
68. Chiou, M. S., & Li, H. Y. (2003). Adsorption behavior of reactive dye in aqueous solution on chemical cross-linked chitosan beads. *Chemosphere*, 50(8), 1095–1105. https://doi.org/10.1016/S0045-6535(02)00636-7
69. Šljivić-Ivanović, Marija, and Ivana Smičiklas. "Utilization of C&D waste in radioactive waste treatment—current knowledge and perspectives." In Fernando Pacheco Torgal, Yining Ding, Francesco Colangelo, Rabin Tuladhar, and Alexander Koutamanis (eds.), *Advances in Construction and Demolition Waste Recycling* (2020), pp. 475–500. Woodhead Publishing. https://doi.org/10.1016/B978-0-12-819055-5.00023-1
70. Stjepanović, Marija, Natalija Velić, Antonela Galić, Indira Kosović, Tamara Jakovljević, and Mirna Habuda-Stanić. "From waste to biosorbent: removal of Congo red from water by waste wood biomass." *Water* 13, no. 3 (2021): 279. https://doi.org/10.3390/w13030279
71. Ariffin, Nurliyana, Mohd Mustafa Al Bakri Abdullah, Mohd Remy Rozainy Mohd Arif Zainol, Mohd Fared Murshed, Meor Ahmad Faris, and Ridho Bayuaji. "Review on adsorption of heavy metal in wastewater by using geopolymer." In *MATEC Web of Conferences* (2017), vol. 97, p. 01023. EDP Sciences. https://doi.org/10.1051/matecconf/20179701023
72. Dharmapriya, Thakshila Nadeeshani, Ding Yang Li, Yi-Chen Chung, and Po-Jung Huang. "Green synthesis of reusable adsorbents for the removal of heavy metal ions." *ACS Omega* 6, no. 45 (2021): 30478–30487. https://doi.org/10.1021/acsomega.1c03879
73. Terangpi, Praisy, and Saswati Chakraborty. "Adsorption kinetics and equilibrium studies for removal of acid azo dyes by aniline formaldehyde condensate." *Applied Water Science* 7 (2017): 3661–3671. https://doi.org/10.1007/s13201-016-0510-4
74. Popa, Simona, Maria Elena Radulescu-Grad, Alina Perdivara, and GianninMosoarca. "Aspects regarding colour fastness and adsorption studies of a new azo-stilbene dye for acrylic resins." *Scientific Reports* 11, no. 1 (2021): 5889. https://doi.org/10.1038/s41598-021-85452-7
75. Eltaweil, Abdelazeem S., Gehad S. Elgarhy, Gehan M. El-Subruiti, and Ahmed M. Omer. "Carboxymethyl cellulose/carboxylated graphene oxide composite microbeads for efficient adsorption of cationic methylene blue dye." *International Journal of Biological Macromolecules* 154 (2020): 307–318. https://doi.org/10.1016/j.ijbiomac.2020.03.122
76. Sethi, Sapna, Balbir Singh Kaith, Mandeep Kaur, Neeraj Sharma, and Sadhika Khullar. "A hydrogel based on dialdehyde carboxymethyl cellulose–gelatin and its utilization as a bio adsorbent." *Journal of Chemical Sciences* 132 (2020): 1–17. https://doi.org/10.1007/s12039-019-1700-z
77. Chen, Guiqiang, Yanxia Liu, Fei Liu, and Xiao Zhang. "Fabrication of three-dimensional graphene foam with high electrical conductivity and large adsorption capability." *Applied Surface Science* 311 (2014): 808–815. https://doi.org/10.1016/j.apsusc.2014.05.171
78. Jiao, Tifeng, Haiying Guo, Qingrui Zhang, Qiuming Peng, Yongfu Tang, Xuehai Yan, and Bingbing Li. "Reduced graphene oxide-based silver nanoparticle-containing composite hydrogel as highly efficient dye catalysts for wastewater treatment." *Scientific Reports* 5, no. 1 (2015): 11873. https://doi.org/10.1038/srep11873
79. Hu, Huawen, Menglei Chang, Min Zhang, Xiaowen Wang, and Dongchu Chen. "A new insight into PAM/graphene-based adsorption of water-soluble aromatic pollutants." *Journal of Materials Science* 52, no. 14 (2017): 8650–8664. https://doi.org/10.1007/s10853-017-1090-x

80. Mittal, Hemant, Ali Al Alili, Pranay P. Morajkar, and Saeed M. Alhassan. "GO crosslinked hydrogel nanocomposites of chitosan/carboxymethyl cellulose–a versatile adsorbent for the treatment of dyes contaminated wastewater." *International Journal of Biological M*acromolecules 167 (2021): 1248–1261. https://doi.org/10.1016/j.ijbiomac.2020.11.079
81. Tiwari, Jitendra N., Kandula Mahesh, Nhien H. Le, K. Christian Kemp, Rupak Timilsina, Rajanish N. Tiwari, and Kwang S. Kim. "Reduced graphene oxide-based hydrogels for the efficient capture of dye pollutants from aqueous solutions." *Carbon* 56 (2013): 173–182. https://doi.org/10.1016/j.carbon.2013.01.001
82. Li, Xin, Xiaohong Wang, Tiantian Han, Chen Hao, Shiqi Han, and Xiangbo Fan. "Synthesis of sodium lignosulfonate-guar gum composite hydrogel for the removal of Cu^{2+} and Co^{2+}." *International Journal of Biological Macromolecules* 175 (2021): 459–472. https://doi.org/10.1016/j.ijbiomac.2021.02.018
83. Wang, Wenbo, Yuru Kang, and Aiqin Wang. "One-step fabrication in aqueous solution of a granular alginate-based hydrogel for fast and efficient removal of heavy metal ions." *Journal of Polymer Research* 20 (2013): 1–10. https://doi.org/10.1007/s10965-013-0101-0
84. Mohan, Dinesh, Charles U. Pittman Jr, Mark Bricka, Fran Smith, Ben Yancey, Javeed Mohammad, Philip H. Steele, Maria F. Alexandre-Franco, Vicente Gómez-Serrano, and Henry Gong. "Sorption of arsenic, cadmium, and lead by chars produced from fast pyrolysis of wood and bark during bio-oil production." *Journal of Colloid and Interface Science* 310, no. 1 (2007): 57–73. https://doi.org/10.1016/j.jcis.2007.01.020
85. Wang, Shengsen, Bin Gao, Andrew R. Zimmerman, Yuncong Li, Lena Ma, Willie G. Harris, and Kati W. Migliaccio. "Physicochemical and sorptive properties of biochars derived from woody and herbaceous biomass." *Chemosphere* 134 (2015): 257–262. https://doi.org/10.1016/j.chemosphere.2015.04.062
86. Badsha, Mohammad A. H., Musharib Khan, Baile Wu, Ashutosh Kumar, and Irene M. C. Lo. Role of surface functional groups of hydrogels in metal adsorption: from performance to mechanism. *Journal of Hazardous Materials* 408 (2021): 124463. https://doi.org/10.1016/j.jhazmat.2020.124463
87. Yang, Xiaodong, Yongshan Wan, Yulin Zheng, Feng He, Zebin Yu, Jun Huang, Hailong Wang, Yong Sik Ok, Yinshan Jiang, and Bin Gao. "Surface functional groups of carbon-based adsorbents and their roles in the removal of heavy metals from aqueous solutions: a critical review." *Chemical Engineering Journal* 366 (2019): 608–621. https://doi.org/10.1016/j.cej.2019.02.119
88. Sivakumar, Rajamanickam, and Nae Yoon Lee. "Adsorptive removal of organic pollutant methylene blue using polysaccharide-based composite hydrogels." *Chemosphere* 286 (2022): 131890. https://doi.org/10.1016/j.chemosphere.2021.131890
89. Wang, Wei, Jinjing Hu, Rundong Zhang, Chao Yan, Li Cui, and Junjiang Zhu. "A pH-responsive carboxymethyl cellulose/chitosan hydrogel for adsorption and desorption of anionic and cationic dyes." *Cellulose* 28 (2021): 897–909. https://doi.org/10.1007/s10570-020-03561-4
90. Hu, Zhao-Hong, Ahmed Mohamed Omer, Xiao–kun Ouyang, and Di Yu. "Fabrication of carboxylated cellulose nanocrystal/sodium alginate hydrogel beads for adsorption of Pb (II) from aqueous solution." *International Journal of Biological Macromolecules* 108 (2018): 149–157. https://doi.org/10.1016/j.ijbiomac.2017.11.171
91. Wang, Jingjing, and Zhengkui Li. "Enhanced selective removal of Cu (II) from aqueous solution by novel polyethylenimine-functionalized ion imprinted hydrogel: behaviors and mechanisms." *Journal of Hazardous Materials* 300 (2015): 18–28. https://doi.org/10.1016/j.jhazmat.2015.06.043

92. Seida, Yoshimi, and Hideaki Tokuyama. "Hydrogel adsorbents for the removal of hazardous pollutants—requirements and available functions as adsorbent." *Gels* 8, no. 4 (2022): 220. https://doi.org/10.3390/gels8040220
93. Peng, Na, Danning Hu, Jian Zeng, Yu Li, Lei Liang, and Chunyu Chang. "Superabsorbent cellulose–clay nanocomposite hydrogels for highly efficient removal of dye in water." *ACS Sustainable Chemistry & Engineering* 4, no. 12 (2016): 7217–7224. https://doi.org/10.1021/acssuschemeng.6b02178
94. Yu, Fei, Tianran Cui, Changfu Yang, Xiaohu Dai, and Jie Ma. "κ-Carrageenan/sodium alginate double-network hydrogel with enhanced mechanical properties, anti-swelling, and adsorption capacity." *Chemosphere* 237 (2019): 124417. https://doi.org/10.1016/j.chemosphere.2019.124417
95. Zhuang, Yuan, Fei Yu, Jie Ma, and Junhong Chen. "Enhanced adsorption removal of antibiotics from aqueous solutions by modified alginate/graphene double network porous hydrogel." *Journal of Colloid and Interface Science* 507 (2017): 250–259. https://doi.org/10.1016/j.jcis.2017.07.033
96. Shi, Zhennan, Chen Xu, Han Guan, Ling Li, Lu Fan, Yingxi Wang, Li Liu, Qingtao Meng, and Run Zhang. "Magnetic metal organic frameworks (MOFs) composite for removal of lead and malachite green in wastewater." *Colloids and Surfaces A: Physicochemical and Engineering Aspects* 539 (2018): 382–390. https://doi.org/10.1016/j.colsurfa.2017.12.043
97. Kong, Yan, Yuan Zhuang, Zhiyong Han, Jianwei Yu, Baoyou Shi, Kun Han, and Haotian Hao. "Dye removal by eco-friendly physically cross-linked double network polymer hydrogel beads and their functionalized composites." *Journal of Environmental Sciences* 78 (2019): 81–91. https://doi.org/10.1016/j.jes.2018.07.006
98. Tang, Shuxian, Jueying Yang, Lizhi Lin, Kelin Peng, Yu Chen, Shaohua Jin, and Weishang Yao. "Construction of physically crosslinked chitosan/sodium alginate/calcium ion double-network hydrogel and its application to heavy metal ions removal." *Chemical Engineering Journal* 393 (2020): 124728. https://doi.org/10.1016/j.cej.2020.124728
99. Ma, Jianhong, Jinming Luo, Yutang Liu, Yuanfeng Wei, Tao Cai, Xingwen Yu, Hui Liu, Chengbin Liu, and John C. Crittenden. "Pb (II), Cu (II) and Cd (II) removal using a humic substance-based double network hydrogel in individual and multicomponent systems." *Journal of Materials Chemistry A* 6, no. 41 (2018): 20110–20120. https://doi.org/10.1039/C8TA07250G
100. Darban, Zenab, Syed Shahabuddin, Rama Gaur, Irfan Ahmad, and Nanthini Sridewi. "Hydrogel-based adsorbent material for the effective removal of heavy metals from wastewater: a comprehensive review." *Gels* 8, no. 5 (2022): 263. https://doi.org/10.3390/gels8050263
101. Liu, Zhen, Haisong Wang, Chao Liu, Yijun Jiang, Guang Yu, Xindong Mu, and Xiaoyan Wang. "Magnetic cellulose–chitosan hydrogels prepared from ionic liquids as reusable adsorbent for removal of heavy metal ions." *Chemical Communications* 48, no. 59 (2012): 7350–7352. https://doi.org/10.1039/C2CC17795A
102. Barakat, M. A., and Nurettin Sahiner. "Cationic hydrogels for toxic arsenate removal from aqueous environment." *Journal of Environmental Management* 88, no. 4 (2008): 955–961. https://doi.org/10.1016/j.jenvman.2007.05.003
103. Ozay, Ozgur, Sema Ekici, Yakup Baran, Senol Kubilay, NahitAktas, and Nurettin Sahiner. "Utilization of magnetic hydrogels in the separation of toxic metal ions from aqueous environments." *Desalination* 260, no. 1–3 (2010): 57–64. https://doi.org/10.1016/j.desal.2010.04.067

104. Weerasundara, Lakshika, Bartolo Gabriele, Alberto Figoli, Yong-Sik Ok, and Jochen Bundschuh. "Hydrogels: novel materials for contaminant removal in water: a review." *Critical Reviews in Environmental Science and Technology* 51, no. 17 (2021): 1970–2014. https://doi.org/10.1080/10643389.2020.1776055
105. Mohammadi, Zahra, Sun Shangbin, Cory Berkland, and Jenn-Tai Liang. "Chelator-mimetic multi-functionalized hydrogel: highly efficient and reusable sorbent for Cd, Pb, and As removal from waste water." *Chemical Engineering Journal* 307 (2017): 496–502. https://doi.org/10.1016/j.cej.2017.08.121
106. Pourjavadi, Ali, Zahra Mazaheri Tehrani, Hamid Salimi, Alireza Banazadeh, and Niloufar Abedini. "Hydrogel nanocomposite based on chitosan-g-acrylic acid and modified nanosilica with high adsorption capacity for heavy metal ion removal." *Iranian Polymer Journal* 24 (2015): 725–734. https://doi.org/10.1007/s13726-015-0360-1
107. Mittal, Hemant, and Suprakas Sinha Ray. "A study on the adsorption of methylene blue onto gum ghatti/TiO_2 nanoparticles-based hydrogel nanocomposite." *International Journal of Biological Macromolecules* 88 (2016): 66–80. https://doi.org/10.1016/j.ijbiomac.2017.03.032

Index

Note: **Bold** page numbers refer to tables and *italic* page numbers refer to figures.